LINEAR ALGEBRA AND ITS APPLICATIONS

THIRD EDITION

LINEAR ALGEBRA
AND ITS
APPLICATIONS
THIRD EDITION

GILBERT STRANG
Massachusetts Institute of Technology

HARCOURT BRACE JOVANOVICH, PUBLISHERS
San Diego

The lithographs on the cover were contributed by the distinguished artist Susumu Endo.
Back cover: Space and Space/Book and Pencil
Front cover: Space and Space/Pencil VI
Susumu Endo: 3-13-3 Jingumae, Shibuyaku Tokyo 150, Japan

ISBN: 0-15-551005-3
Library of Congress Catalog Card Number: 87-083019

Printed in the United States of America

Portions of this material are adapted from
Introduction to Applied Mathematics, © 1986 Gilbert Strang

CONTENTS

Preface   ix

1 MATRICES AND GAUSSIAN ELIMINATION

1.1   Introduction   1
1.2   The Geometry of Linear Equations   3
1.3   An Example of Gaussian Elimination   12
1.4   Matrix Notation and Matrix Multiplication   19
1.5   Triangular Factors and Row Exchanges   31
1.6   Inverses and Transposes   42
1.7   Special Matrices and Applications   52
      Review Exercises   60

2 VECTOR SPACES AND LINEAR EQUATIONS

2.1   Vector Spaces and Subspaces   63
2.2   The Solution of m Equations in n Unknowns   71
2.3   Linear Independence, Basis, and Dimension   80
2.4   The Four Fundamental Subspaces   90
2.5   Networks and Incidence Matrices   102
2.6   Linear Transformations   116
      Review Exercises   128

3   ORTHOGONALITY

3.1   Perpendicular Vectors and Orthogonal Subspaces   132
3.2   Inner Products and Projections Onto Lines   144
3.3   Projections and Least Squares Approximations   153
3.4   Orthogonal Bases, Orthogonal Matrices, and Gram-Schmidt
      Orthogonalization   166
3.5   The Fast Fourier Transform   183
3.6   Review and Preview   195
      Review Exercises   208

4   DETERMINANTS

4.1   Introduction   211
4.2   The Properties of the Determinant   214
4.3   Formulas for the Determinant   222
4.4   Applications of Determinants   231
      Review Exercises   241

5   EIGENVALUES AND EIGENVECTORS

5.1   Introduction   243
5.2   The Diagonal Form of a Matrix   254
5.3   Difference Equations and the Powers A^k 262
5.4   Differential Equations and the Exponential e^{At} 275
5.5   Complex Matrices: Symmetric vs. Hermitian and
      Orthogonal vs. Unitary   290
5.6   Similarity Transformations   304
      Review Exercises   319

6   POSITIVE DEFINITE MATRICES

6.1   Minima, Maxima, and Saddle Points   322
6.2   Tests for Positive Definiteness   330
6.3   Semidefinite and Indefinite Matrices; $Ax = \lambda Mx$ 339
6.4   Minimum Principles and the Rayleigh Quotient   347
6.5   The Finite Element Method   355

7   COMPUTATIONS WITH MATRICES

7.1   Introduction   361
7.2   The Norm and Condition Number of a Matrix   363
7.3   The Computation of Eigenvalues   370
7.4   Iterative Methods for $Ax = b$ 380

8 LINEAR PROGRAMMING AND GAME THEORY

8.1	Linear Inequalities	388
8.2	The Simplex Method and Karmarkar's Method	395
8.3	The Theory of Duality	412
8.4	Network Models	424
8.5	Game Theory and the Minimax Theorem	433

APPENDIX A   THE SINGULAR VALUE DECOMPOSITION   442
AND THE PSEUDOINVERSE

APPENDIX B   THE JORDAN FORM   453

APPENDIX C   COMPUTER CODES FOR LINEAR ALGEBRA   461

References   472

Solutions to Selected Exercises   473

Index   499

PREFACE

Linear algebra is a fantastic subject. On the one hand it is clean and beautiful. If you have three vectors in 12-dimensional space, you can almost see them. A combination like the first plus the second minus twice the third is harder to visualize, but may still be possible. I don't know if anyone can see *all* such combinations, but somehow (in this course) you begin to do it. Certainly the combinations of those three vectors will not fill the whole 12-dimensional space. (I'm afraid the course has already begun; usually you get to read the preface for free!) What those combinations do fill out is something important—and not just to pure mathematicians.

That is the other side of linear algebra. It is *needed* and *used*. Ten years ago it was taught too abstractly, and the crucial importance of the subject was missed. Such a situation could not continue. Linear algebra has become as basic and as applicable as calculus, and fortunately it is easier. The curriculum had to change, and this is now widely accepted as an essential sophomore or junior course—a requirement for engineering and science, and a central part of mathematics.

The goal of this book is to show both sides—the beauty of linear algebra, and its value. The effort is not all concentrated on theorems and proofs, although the mathematics is there. The emphasis is less on rigor, and much more on understanding. *I try to explain rather than to deduce.* In the book, and also in class, ideas come with examples. Once you work with subspaces, you understand them. The ability to reason mathematically will develop, if it is given enough to do. And the essential ideas of linear algebra are *not too hard*.

I would like to say clearly that this is a book about mathematics. It is not so totally watered down that all the purpose is drained out. I do not believe that students or instructors want an empty course; three hours a week can achieve something worthwhile, provided the textbook helps. I hope and believe that you will see, behind the informal and personal style of this book, that it is written to teach real mathematics. There will be sections you omit, and there might be explanations you don't need,† but you cannot miss the underlying force of this subject. It moves simply and naturally from a line or a plane to the n-dimensional space $\mathbf{R}^n$. That step is mathematics at its best, and every student can take it.

One question is hard to postpone: How should the course start? Most students come to the first class already knowing something about linear equations. Still I believe that we must begin with n equations in n unknowns, $Ax = b$, and with the simplest and most useful method of solution—*Gaussian elimination* (not determinants!). It is a perfect introduction to matrix multiplication. And fortunately, even though the method is so straightforward, there are insights that are central to its understanding and new to almost every student. One is to recognize, as elimination goes from the original matrix A to an upper triangular U, that A is being factored into two triangular matrices: $A = LU$. That observation is not deep, and it is easy to verify, but it is tremendously important in practice. For me this is one indicator of a serious course, a dividing line from a presentation that deals only with row operations or A^{-1}.

Another question is to find the right speed. If matrix calculations are familiar, then *Chapter 1 must not be too slow*. It is Chapter 2 that demands more work, and that means work of a different kind—not the crunching of numbers which a computer can do, but an understanding of $Ax = b$ which starts with elimination and goes deeper. The class has to know that the gears have changed; ideas are coming. Instead of individual vectors, we need vector spaces. I am convinced that the four fundamental subspaces—the column space of A, its row space, and the nullspaces of A and A^{T}—are the most effective way to illustrate linear dependence and independence, and to understand "basis" and "dimension" and "rank." Those are developed gradually but steadily, and they generate examples in a completely natural way. They are also the key to $Ax = b$.

May I take one example, to show how an idea can be seen in different ways? It is the fundamental step of multiplying A times x, a matrix times a vector. At one level Ax is just numbers. At the next level it is a combination of the columns of A. At a third level it is a vector in the column space. (We are seeing a space of vectors, containing all combinations and not only this one.) To an algebraist, A represents a linear transformation and Ax is the result of applying it to x. All four are important, and the book must make the connections.

Chapters 1–5 are really the heart of a course in linear algebra. They contain a large number of applications to physics, engineering, probability and statistics, economics, and biology. Those are not tacked on at the end; they are part of the mathematics. Networks and graphs are a terrific source of rectangular matrices, essential in engineering and computer science and also perfect examples for teaching. What mathematics can do, and what linear algebra does so well, is to see patterns that are partly hidden in the applications. *This is a book that allows*

---

† My favorite proof comes in a book by Ring Lardner: " *'Shut up,' he explained."*

pure mathematicians to teach applied mathematics. I believe the faculty can do the moving, and teach what students need. The effort is absolutely rewarding.

If you know earlier editions of the book, you will see changes. Section 1.1 is familiar but not 1.2. Certainly the spirit has not changed; this course is alive because its subject is. By teaching it constantly, I found a whole series of improvements—in the organization and the exercises (hundreds are new, over a very wide range), and also in the content. Most of these improvements will be visible only in the day-by-day effort of teaching and learning this subject—when the right explanation or the right exercise makes the difference. I mention two changes that are visible in the table of contents: *Linear transformations* are integrated into the text, and there is a new (and optional) section on the *Fast Fourier Transform.* That is perhaps the outstanding algorithm in modern mathematics, and it has revolutionized digital processing. It is nothing more than a fast way of multiplying by a certain matrix! You may only know (as I did) that the idea exists and is important. It is a pleasure to discover how it fits into linear algebra (and introduces complex numbers).

This is a first course in linear algebra. The theory is motivated, and reinforced, by genuine applications. At the same time, the goal is understanding—and the subject is well established. After Chapter 2 reaches beyond elimination and A^{-1} to the idea of a vector space, Chapter 3 concentrates on *orthogonality.* Geometrically, that is understood before the first lecture. Algebraically, the steps are familiar but crucial—to know when vectors are perpendicular, or which subspaces are orthogonal, or how to project onto a subspace, or how to construct an orthonormal basis. Do not underestimate that chapter. Then Chapter 4 presents *determinants,* the key link between $Ax = b$ and $Ax = \lambda x$. They give a test for invertibility which picks out the eigenvalues. That introduces the last big step in the course.

Chapter 5 puts diagonalization ahead of the Jordan form. The *eigenvalues* and *eigenvectors* take us directly from a matrix A to its powers A^k. They solve equations that evolve in time—dynamic problems, in contrast to the static problem $Ax = b$. They also carry information which is not obvious from the matrix itself—a Markov matrix has $\lambda_{max} = 1$, an orthogonal matrix has all $|\lambda| = 1$, and a symmetric matrix has real eigenvalues. If your course reaches the beginning of Chapter 6, the connections between eigenvalues and pivots and determinants of symmetric matrices tie the whole subject together. (The last section of each chapter is optional.) Then Chapter 7 gives more concentrated attention to numerical linear algebra, which has become the foundation of scientific computing. And I believe that even a brief look at Chapter 8 allows a worthwhile but relaxed introduction to linear programming—my class is happy because it comes at the end, without examination.

I would like to mention the **Manual for Instructors**, and another book. The manual contains solutions to all exercises (including the Review Exercises at the ends of Chapters 1–5), and also a collection of ideas and suggestions about applied linear algebra. I hope instructors will request a copy from the publisher (HBJ College Department, 7555 Caldwell Avenue, Chicago IL 60648). I also hope that readers of this book will go forward to the next one. That is called **Introduction to Applied Mathematics**, and it combines linear algebra with differential equations into a text for modern applied mathematics and engineering mathematics. It includes Fourier analysis, complex variables, partial differential equations, numerical methods, and optimization—but the starting point is linear algebra. It is published by Wellesley-Cambridge Press (Box 157, Wellesley MA 02181) and the response has been tremendous—many departments have wanted a renewal of that course, to teach what is most needed.

This book, like that next one, aims to recognize what the computer can do (without being dominated by it). Solving a problem no longer means writing down an infinite series, or finding a formula like Cramer's rule, but constructing an effective algorithm. That needs good ideas: mathematics survives! The algebra stays clear and simple and stable. For elimination, the operation count in Chapter 1 has also a second purpose—to reinforce a detailed grasp of the n by n case, by actually counting the steps. But I do not do everything in class. The text should supplement as well as summarize the lectures.

In short, a book is needed that will permit the applications to be taught successfully, in combination with the underlying mathematics. That is the book I have tried to write.

In closing, this is a special opportunity for me to say thank you. I am extremely grateful to readers who have liked the book, and have seen what it stands for. Many have written with ideas and encouragement, and I mention only five names: Dan Drucker, Vince Giambalvo, Steve Kleiman, Beresford Parlett, and Jim Simmonds. Behind them is an army of friends and critics that I am proud to have. This third edition has been made better by what they have taught—to students and to the author. It was a very great pleasure to work with Sophia Koulouras, who typed the manuscript, and Michael Michaud, who designed the book and the cover. And above all, my gratitude goes to my wife and children and parents. The book is theirs too, and so is the spirit behind it—which in the end is everything. May I rededicate this book to my mother and father, who gave so much to it: Thank you both.

GILBERT STRANG

LINEAR ALGEBRA AND ITS APPLICATIONS

THIRD EDITION

1

MATRICES AND GAUSSIAN ELIMINATION

The central problem of linear algebra is the solution of linear equations. The most important case, and the simplest, is when the number of unknowns equals the number of equations. Therefore we begin with this basic problem: *n **equations in** n **unknowns**.*

There are two well-established ways to solve linear equations. One is the method of ***elimination***, in which multiples of the first equation are subtracted from the other equations—so as to remove the first unknown from those equations. This leaves a smaller system, of $n - 1$ equations in $n - 1$ unknowns. The process is repeated until there is only one equation and one unknown, which can be solved immediately. Then it is not hard to go backward, and find the other unknowns in reverse order; we shall work out an example in a moment. A second and more sophisticated way introduces the idea of ***determinants***. There is an exact formula called Cramer's rule, which gives the solution (the correct values of the unknowns) as a ratio of two n by n determinants. From the examples in a textbook it is not obvious which way is better ($n = 3$ or $n = 4$ is about the upper limit on the patience of a reasonable human being).

In fact, the more sophisticated formula involving determinants is a disaster in practice, and elimination is the algorithm that is constantly used to solve large systems of equations. Our first goal is to understand this algorithm. It is generally called ***Gaussian elimination***.

The idea is deceptively simple, and in some form it may already be familiar to the reader. But there are four aspects that lie deeper than the simple mechanics of elimination. Together with the algorithm itself, we want to explain them in this chapter. They are:

(1)   The **geometry** of linear equations. It is not easy to visualize a 10-dimensional plane in 11-dimensional space. It is harder to see eleven of those planes intersecting at a single point in that space—but somehow it is almost possible. With three planes in three dimensions it can certainly be done. Then linear algebra moves the problem into four dimensions, or eleven dimensions, where the intuition has to imagine the geometry (and gets it right).

(2)   The interpretation of elimination as a **factorization** of the coefficient matrix. We shall introduce *matrix notation* for the system of n equations, writing the unknowns as a vector x and the equations in the matrix shorthand $Ax = b$. Then *elimination amounts to factoring A into a product LU, of a lower triangular matrix L and an upper triangular matrix U.*

First we have to introduce matrices and vectors in a systematic way, as well as the rules for their multiplication. We also define the **transpose** A^{T} and the **inverse** A^{-1} of a matrix A.

(3)   In most cases elimination goes forward without difficulties. In some exceptional cases it will **break down**—either because the equations were written in the wrong order, which is easily fixed by exchanging them, or else because the equations fail to have a unique solution. In the latter case there may be **no solution**, **or infinitely many**. We want to understand how, at the time of breakdown, the elimination process identifies each of these possibilities.

(4)   It is essential to have a rough count of the **number of operations** required to solve a system by elimination. The expense in computing often determines the accuracy in the model. The computer can do millions of operations, but not very many trillions. And already after a million steps, roundoff error can be significant. (Some problems are sensitive; others are not.) Without trying for full detail, we want to see what systems arise in practice and how they are actually solved.

The final result of this chapter will be an elimination algorithm which is about as efficient as possible. It is essentially the algorithm that is in constant use in a tremendous variety of applications. And at the same time, understanding it in terms of matrices—the coefficient matrix, the matrices that carry out an elimination step or an exchange of rows, and the final triangular factors L and U—is an essential foundation for the theory.

THE GEOMETRY OF LINEAR EQUATIONS ■ 1.2

The way to understand this subject is by example. We begin with two extremely humble equations, recognizing that you could solve them without a course in linear algebra. Nevertheless I hope you will give Gauss a chance:

$$2x - y = 1$$
$$x + y = 5.$$

There are two ways to look at that system, and our main point is to see them both.

The first approach concentrates on the separate equations, in other words on the *rows*. That is the most familiar, and in two dimensions we can do it quickly. The equation $2x - y = 1$ is represented by a *straight line* in the x-y plane. The line goes through the points $x = 1$, $y = 1$ and $x = \frac{1}{2}$, $y = 0$ (and also through $(0, -1)$ and $(2, 3)$ and all intermediate points). The second equation $x + y = 5$ produces a second line (Fig. 1.1a). Its slope is $dy/dx = -1$ and it crosses the first line at the solution. The point of intersection is the only point on both lines, and therefore it is the only solution to both equations. It has the coordinates $x = 2$ and $y = 3$—which will soon be found by a systematic elimination.

The second approach is not so familiar. It looks at the **columns** of the linear system. The two separate equations are really one **vector equation**

$$x \begin{bmatrix} 2 \\ 1 \end{bmatrix} + y \begin{bmatrix} -1 \\ 1 \end{bmatrix} = \begin{bmatrix} 1 \\ 5 \end{bmatrix}.$$

The problem is *to find the combination of the column vectors on the left side which produces the vector on the right side*. Those two-dimensional vectors are represented by the bold lines in Fig. 1.1b. The unknowns are the numbers x and y which

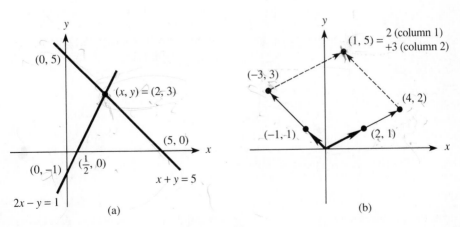

Fig. 1.1. The geometry by rows and by columns.

multiply the column vectors. The whole idea can be seen in that figure, where 2 times column 1 is added to 3 times column 2. Geometrically this produces a famous parallelogram. Algebraically it produces the correct vector $\begin{bmatrix} 1 \\ 5 \end{bmatrix}$, on the right side of our equations. The column picture confirms that the solution is $x = 2$, $y = 3$.

More time could be spent on that example, but I would rather move forward to $n = 3$. Three equations are still manageable, and they have much more variety. As a specific example, consider

$$
\begin{aligned}
2u + v + w &= 5 \\
4u - 6v &= -2 \\
-2u + 7v + 2w &= 9.
\end{aligned}
\tag{1}
$$

Again we can study the rows or the columns, and we start with the rows. Each equation describes a *plane* in three dimensions. The first plane is $2u + v + w = 5$, and it is sketched in Fig. 1.2. It contains the points $(\frac{5}{2}, 0, 0)$ and $(0, 5, 0)$ and $(0, 0, 5)$. It is determined by those three points, or by any three of its points—provided they do not lie on a line. We mention in passing that *the plane $2u + v + w = 10$ is parallel to this one*. The corresponding points are $(5, 0, 0)$ and $(0, 10, 0)$ and $(0, 0, 10)$, twice as far away from the origin—which is the center point $u = 0$, $v = 0$, $w = 0$. Changing the right hand side moves the plane parallel to itself, and the plane $2u + v + w = 0$ goes through the origin.†

The second plane is $4u - 6v = -2$. It is drawn vertically, because w can take any value. The coefficient of w happens to be zero, but this remains a plane in

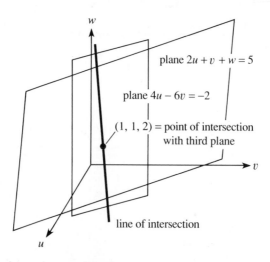

Fig. 1.2. The row picture: intersecting planes.

---

† If the first two equations were $2u + v + w = 5$ and $2u + v + w = 10$, the planes would not intersect and there would be no solution.

3-space. (If the equation were $4u = 3$, or even the extreme case $u = 0$, it would still describe a plane.) The figure shows the intersection of the second plane with the first. That intersection is a line. *In three dimensions a line requires two equations*; in n dimensions it will require $n - 1$.

Finally the third plane intersects this line in a point. The plane (not drawn) represents the third equation $-2u + 7v + 2w = 9$, and it crosses the line at $u = 1$, $v = 1$, $w = 2$. That point solves the linear system.

How does this picture extend into n dimensions? We will have n equations, and they contain n unknowns. The first equation still determines a "*plane*." It is no longer an ordinary two-dimensional plane in 3-space; somehow it has "*dimension* $n - 1$." It must be extremely thin and flat within n-dimensional space, although it would look solid to us. If time is the fourth dimension, then the plane $t = 0$ cuts through 4-dimensional space and produces the 3-dimensional universe we live in (or rather, the universe as it was at $t = 0$). Another plane is $z = 0$, which is also 3-dimensional; it is the ordinary x-y plane taken over all time. Those three-dimensional planes will intersect! What they have in common is the ordinary x-y plane at $t = 0$. We are down to two dimensions, and the next plane leaves a line. Finally a fourth plane leaves a single point. It is the point at the intersection of 4 planes in 4 dimensions, and it solves the 4 underlying equations.

I will be in trouble if that example from relativity goes any further. The point is that linear algebra can operate with any number of equations. The first one produces an $n - 1$-dimensional plane in n dimensions. The second equation determines another plane, and they intersect (we hope) in a smaller set of "dimension $n - 2$." Assuming all goes well, every new plane (every new equation) reduces the dimension by one. At the end, when all n planes are accounted for, the intersection has dimension zero. It is a *point*, it lies on all the planes, and its coordinates satisfy all n equations. It is the solution! That picture is intuitive—the geometry will need support from the algebra—but it is basically correct.

Column Vectors

We turn to the columns. This time the vector equation (the same equation as (1)) is

$$u \begin{bmatrix} 2 \\ 4 \\ -2 \end{bmatrix} + v \begin{bmatrix} 1 \\ -6 \\ 7 \end{bmatrix} + w \begin{bmatrix} 1 \\ 0 \\ 2 \end{bmatrix} = \begin{bmatrix} 5 \\ -2 \\ 9 \end{bmatrix}. \tag{2}$$

Those are *three-dimensional column vectors*. The vector b on the right side has components $5, -2, 9$, and these components allow us to draw the vector. *The vector b is identified with the point whose coordinates are $5, -2, 9$.* Every point in three-dimensional space is matched to a vector, and vice versa. That was the idea of Descartes, who turned geometry into algebra by working with the coordinates of the point. We can write the vector in a column, or we can list its components

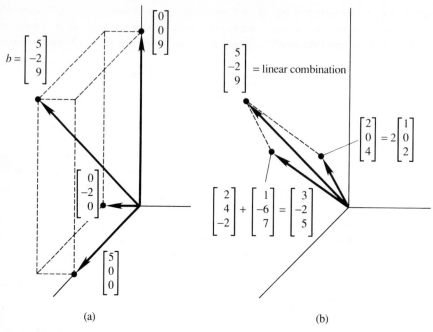

Fig. 1.3. The column picture: linear combination of columns equals b.

as $b = (5, -2, 9)$, or we can represent it geometrically by an arrow from the origin.†
Throughout the book we use parentheses and commas when the components are
listed horizontally, and square brackets (with no commas) when a column vector
is printed vertically.

What really matters is ***addition of vectors*** and ***multiplication by a scalar*** (a number).
In Fig. 1.3a you see a vector addition, which is carried out component by
component:

$$
\begin{bmatrix} 5 \\ 0 \\ 0 \end{bmatrix} + \begin{bmatrix} 0 \\ -2 \\ 0 \end{bmatrix} + \begin{bmatrix} 0 \\ 0 \\ 9 \end{bmatrix} = \begin{bmatrix} 5 \\ -2 \\ 9 \end{bmatrix}.
$$

In the right figure there is a multiplication by 2 (and if it had been -2 the vector
would have gone in the reverse direction):

$$
2\begin{bmatrix} 1 \\ 0 \\ 2 \end{bmatrix} = \begin{bmatrix} 2 \\ 0 \\ 4 \end{bmatrix}, \qquad -2\begin{bmatrix} 1 \\ 0 \\ 2 \end{bmatrix} = \begin{bmatrix} -2 \\ 0 \\ -4 \end{bmatrix}.
$$

† Some authors prefer to say that the arrow is really the vector, but I think it doesn't
matter; you can choose the arrow, or the point, or the three numbers. (They all start with
the same origin $(0, 0, 0)$.) In six dimensions it is probably easiest to choose the six numbers.

Also in the right figure is one of the central ideas of linear algebra. It uses *both* of the basic operations; vectors are *multiplied by numbers and then added*. The result is called a ***linear combination***, and in this case the linear combination is

$$\begin{bmatrix} 2 \\ 4 \\ -2 \end{bmatrix} + \begin{bmatrix} 1 \\ -6 \\ 7 \end{bmatrix} + 2\begin{bmatrix} 1 \\ 0 \\ 2 \end{bmatrix} = \begin{bmatrix} 5 \\ -2 \\ 9 \end{bmatrix}.$$

You recognize the significance of that special combination; it solves equation (2). The equation asked for multipliers u, v, w which produce the right side b. Those numbers are $u = 1$, $v = 1$, $w = 2$. They give the correct linear combination of the column vectors, and they also gave the point (1, 1, 2) in the row picture (where the three planes intersect).

Multiplication and addition are carried out on each component separately. Therefore linear combinations are possible provided the vectors have the same number of components. Note that all vectors in the figure were three-dimensional, even though some of their components were zero.

Do not forget the goal. It is to look beyond two or three dimensions into n dimensions. With n equations in n unknowns, there were n planes in the row picture. There are n vectors in the column picture, plus a vector b on the right side. The equations ask for a ***linear combination of the n vectors that equals b***. In this example we found one such combination (there are no others) but for certain equations that will be impossible. Paradoxically, the way to understand the good case is to study the bad one. Therefore we look at the geometry exactly when it breaks down, in what is called the *singular case*.

First we summarize:

Row picture: Intersection of n planes
Column picture: The right side b is a combination of the column vectors
Solution to equations: Intersection point of planes = coefficients in the combination of columns

The Singular Case

Suppose we are again in three dimensions, and the three planes in the row picture *do not intersect*. What can go wrong? One possibility, already noted, is that two planes may be parallel. Two of the equations, for example $2u + v + w = 5$ and $4u + 2v + 2w = 11$, may be inconsistent—and there is no solution (Fig. 1.4a shows an end view). In the two-dimensional problem, with lines instead of planes, this is the only possibility for breakdown. That problem is singular if the lines are parallel, and it is nonsingular if they meet. But three planes in three dimensions can be in trouble without being parallel.

The new difficulty is shown in Fig. 1.4b. All three planes are perpendicular to the page; from the end view they form a triangle. Every pair of planes intersects,

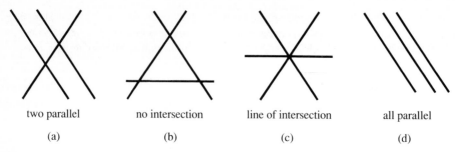

two parallel	no intersection	line of intersection	all parallel
(a)	(b)	(c)	(d)

Fig. 1.4. Singular cases: no solution or an infinity of solutions.

but no point is common to all three.† This is more typical than parallel planes, and it corresponds to a singular system like

$$
\begin{aligned}
u + v + \ w &= 2 \\
2u \quad\ \ + 3w &= 5 \\
3u + v + 4w &= 6
\end{aligned}
\tag{3}
$$

The first two left hand sides add up to the third. On the right side that fails. Equation 1 plus equation 2 minus equation 3 is the impossible statement $0 = 1$. Thus the equations are **inconsistent**, as Gaussian elimination will systematically discover.

There is another singular system, close to this one, with an *infinity of solutions* instead of none. If the 6 in the last equation becomes 7, then the three equations combine to give $0 = 0$. It looks OK, because the third equation is the sum of the first two. In that case the three planes have a whole *line in common* (Fig. 1.4c). The effect of changing the right sides is to move the planes parallel to themselves, and for the right hand side $b = (2, 5, 7)$ the figure is suddenly different. The lowest plane moves up to meet the others, and there is a line of solutions. The problem is still singular, but now it suffers from **too many solutions** instead of too few.

Of course there is the extreme case of three parallel planes. For most right sides there is no solution (Fig. 1.4d). For special right sides (like $b = (0, 0, 0)$!) there is a whole plane of solutions—because all three planes become the same.

What happens to the *column picture* when the system is singular? It has to go wrong; the question is how. There are still three columns on the left side of the equations, and we try to combine them to produce b:

$$
u \begin{bmatrix} 1 \\ 2 \\ 3 \end{bmatrix} + v \begin{bmatrix} 1 \\ 0 \\ 1 \end{bmatrix} + w \begin{bmatrix} 1 \\ 3 \\ 4 \end{bmatrix} = b.
\tag{4}
$$

For $b = (2, 5, 7)$ this was possible; for $b = (2, 5, 6)$ it was not. The reason is that *those three columns lie in a plane*. Therefore all their linear combinations are also in the plane (which goes through the origin). If the vector b is not in that plane,

---

† The third plane is not parallel to the other planes, but it is parallel to their line of intersection.

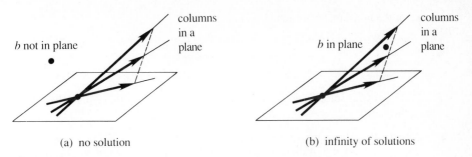

(a) no solution (b) infinity of solutions

Fig. 1.5. Singular cases: b outside or inside the plane of the columns.

then no solution is possible. That is by far the most likely event; a singular system generally has no solution. But there is also the chance that b does lie in the plane of the columns, in which case there are too many solutions. In that case the three columns can be combined in *infinitely many* ways to produce b. That gives the column picture 1.5b which corresponds to the row picture 1.4c.

Those are true statements, but they have not been justified. How do we know that the three columns lie in the same plane? One answer is to find a combination of the columns that adds to zero. After some calculation, it is $u = 3$, $v = -1$, $w = -2$. Three times column 1 equals column 2 plus twice column 3. Column 1 is in the plane of the others, and only two columns are independent. When a vector like $b = (2, 5, 7)$ is also in that plane—it is column 1 plus column 3—the system can be solved. However many other combinations are also possible. The same vector b is also 4 (column 1) $-$ (column 2) $-$ (column 3), by adding to the first solution the combination $(3, -1, -2)$ that gave zero. Because we can add any multiple of $(3, -1, -2)$, there is a whole line of solutions—as we know from the row picture.

That is numerically convincing, but it is not the real reason we expected the columns to lie in a plane. The truth is that we *knew* the columns would combine to give zero, because the rows did. That is a fact of mathematics, not of computation—and it remains true in dimension n. **If the n planes have no point in common, then the n columns lie in the same plane.** If the row picture breaks down, so does the column picture. That is a fundamental conclusion of linear algebra, and it brings out the difference between Chapter 1 and Chapter 2. This chapter studies the most important problem—the *nonsingular* case—where there is one solution and it has to be found. The next chapter studies the general case, where there may be many solutions or none. In both cases we cannot continue without a decent notation (*matrix notation*) and a decent algorithm (*elimination*). After the exercises, we start with the algorithm.

EXERCISES

1.2.1 For the equations $x + y = 4$, $2x - 2y = 4$, draw the row picture (two intersecting lines) and the column picture (combination of two columns equal to the column vector $(4, 4)$ on the right side).

1.2.2 Solve the nonsingular triangular system

$$u + v + w = b_1$$
$$v + w = b_2$$
$$w = b_3.$$

Show that your solution gives a combination of the columns that equals the column on the right.

1.2.3 Describe the intersection of the three planes $u + v + w + z = 6$ and $u + w + z = 4$ and $u + w = 2$ (all in 4-dimensional space). Is it a line or a point or an empty set? What is the intersection if the fourth plane $u = -1$ is included?

1.2.4 Sketch the three lines

$$x + 2y = 2$$
$$x - y = 2$$
$$y = 1.$$

Can the three equations be solved simultaneously? What happens to the figure if all right hand sides are zero? Is there any nonzero choice of right hand sides which allows the three lines to intersect at the same point and the three equations to have a solution?

1.2.5 Find two points on the line of intersection of the three planes $t = 0$ and $z = 0$ and $x + y + z + t = 1$ in four-dimensional space.

1.2.6 When $b = (2, 5, 7)$, find a solution (u, v, w) to equation (4) other than the solutions $(1, 0, 1)$ and $(4, -1, -1)$ mentioned in the text.

1.2.7 Give two more right hand sides in addition to $b = (2, 5, 7)$ for which equation (4) can be solved. Give two more right hand sides in addition to $b = (2, 5, 6)$ for which it cannot be solved.

1.2.8 Explain why the system

$$u + v + w = 2$$
$$u + 2v + 3w = 1$$
$$v + 2w = 0$$

is singular, by finding a combination of the three equations that adds up to $0 = 1$. What value should replace the last zero on the right side, to allow the equations to have solutions—and what is one of the solutions?

1.2.9 The column picture for the previous exercise is

$$u \begin{bmatrix} 1 \\ 1 \\ 0 \end{bmatrix} + v \begin{bmatrix} 1 \\ 2 \\ 1 \end{bmatrix} + w \begin{bmatrix} 1 \\ 3 \\ 2 \end{bmatrix} = b.$$

Show that the three columns on the left lie in the same plane, by expressing the third column as a combination of the first two. What are all the solutions (u, v, w) if b is the zero vector $(0, 0, 0)$?

1.2.10 Under what condition on y_1, y_2, y_3 do the points $(0, y_1)$, $(1, y_2)$, $(2, y_3)$ lie on a straight line?

1.2.11 The equations

$$ax + 2y = 0$$
$$2x + ay = 0$$

are certain to have the solution $x = y = 0$. For which values of a is there a whole line of solutions?

1.2.12 Sketch the plane $x + y + z = 1$, or the part of the plane that is in the positive octant where $x \geq 0$, $y \geq 0$, $z \geq 0$. Do the same for $x + y + z = 2$ in the same figure. What vector is *perpendicular* to those planes?

1.2.13 Starting with the line $x + 4y = 7$, find the equation for the parallel line through $x = 0$, $y = 0$. Find the equation of another line that meets the first at $x = 3$, $y = 1$.

1.3 ■ AN EXAMPLE OF GAUSSIAN ELIMINATION

The way to understand this subject is by example. We begin in three dimensions with the system

$$\begin{aligned} 2u + v + w &= 5 \\ 4u - 6v \quad\quad &= -2 \\ -2u + 7v + 2w &= 9. \end{aligned} \qquad (1)$$

The problem is to find the unknown values of u, v, and w, and we shall apply Gaussian elimination. (Gauss is recognized as the greatest of all mathematicians, but certainly not because of this invention, which probably took him ten minutes. Ironically, however, it is the most frequently used of all the ideas that bear his name.) The method starts by *subtracting multiples of the first equation from the others, so as to eliminate u from the last two equations.* This requires that we

(a)   subtract 2 times the first equation from the second;
(b)   subtract −1 times the first equation from the third.

The result is an equivalent system of equations

$$\begin{aligned} 2u + v + w &= 5 \\ -8v - 2w &= -12 \\ 8v + 3w &= 14. \end{aligned} \qquad (2)$$

The coefficient 2, which multiplied the first unknown u in the first equation, is known as the ***first pivot***. Elimination is constantly dividing the pivot into the numbers underneath it, to find out the right multipliers.

At the second stage of elimination, we ignore the first equation. The other two equations involve only the two unknowns v and w, and elimination can be applied to them. *The pivot for this stage is −8*, and a multiple of this second equation will be subtracted from the remaining equations (in this case there is only the third one remaining) so as to eliminate v. We add the second equation to the third or, in other words, we

(c)   subtract −1 times the second equation from the third.

The elimination process is now complete, at least in the "forward" direction. It leaves the simplified system

$$\begin{aligned} 2u + v + w &= 5 \\ -8v - 2w &= -12 \\ w &= 2. \end{aligned} \qquad (3)$$

There is an obvious order in which to solve this system. The last equation gives $w = 2$. Substituting into the second equation, we find $v = 1$. Then the first equation gives $u = 1$. This process is called ***back-substitution***.

To repeat: Forward elimination produced the pivots 2, -8, 1. It subtracted multiples of each row from the rows beneath. It reached the "*triangular*" system (3). Then this system was solved in reverse order, from bottom to top, by substituting each newly computed value into the equation above.

Remark   One good way to write down the forward elimination steps is to include the right hand side as an extra column. There is no need to copy u and v and w and $=$ at every step, so we are left with the bare minimum:

$$\begin{bmatrix} 2 & 1 & 1 & 5 \\ 4 & -6 & 0 & -2 \\ -2 & 7 & 2 & 9 \end{bmatrix} \rightarrow \begin{bmatrix} 2 & 1 & 1 & 5 \\ 0 & -8 & -2 & -12 \\ 0 & 8 & 3 & 14 \end{bmatrix} \rightarrow \begin{bmatrix} 2 & 1 & 1 & 5 \\ 0 & -8 & -2 & -12 \\ 0 & 0 & 1 & 2 \end{bmatrix}.$$

At the end is the triangular system, ready for back-substitution. You may prefer this arrangement, which guarantees that operations on the left side of the equations are also done on the right side—because *both sides are there together*.

In a larger problem, forward elimination takes most of the effort. It is governed by the left sides of the equations, where back-substitution depends also on the right sides. At the first stage, we use multiples of the first equation to produce zeros below the first pivot. Then the second column is cleared out below the second pivot. Finally, equation n contains only the last unknown x_n, multiplied by the last pivot. The forward step is finished when the system is triangular. Back-substitution yields the complete solution in the opposite order—beginning with the last unknown, then solving for the next to last, and eventually for the first.
By definition, ***pivots cannot be zero***. We need to divide by them.

The Breakdown of Elimination

We want to ask two questions. They may seem a little premature—after all, we have barely got the algorithm working—but their answers shed light on the method itself. The first question is: ***Under what circumstances could the process break down***? Something *must* go wrong in the singular case, and something *might* go wrong in the nonsingular case. The question is not geometric but algebraic.
The answer is: If the algorithm produces n pivots, then there is only one solution to the equations. The system is nonsingular, and it is solved by forward elimination and back-substitution. But *if a zero appears* in a pivot position, elimination has to stop—either temporarily or permanently. The system might or might not be singular.
If the first coefficient is zero, in the upper left corner, the elimination of u from the other equations will be impossible. The same is true at every intermediate stage. Notice that a zero can appear in a pivot position, even if the original coefficient in that place was not zero. Roughly speaking, ***we do not know whether a zero will appear until we try***, by actually going through the elimination process.

In many cases this problem can be cured, and elimination can proceed. Such a system still counts as nonsingular; it is only the algorithm that needs repair. In other cases a breakdown is unavoidable. Those incurable systems are singular, they have no solution or else infinitely many, and a full set of pivots cannot be found.

NONSINGULAR EXAMPLE   (cured by exchanging equations 2 and 3)

$$
\begin{array}{ccc}
\begin{aligned}
u + v + w &= \underline{\ \ } \\
2u + 2v + 5w &= \underline{\ \ } \\
4u + 6v + 8w &= \underline{\ \ }
\end{aligned}
\quad\longrightarrow\quad
\begin{aligned}
u + v + w &= \underline{\ \ } \\
3w &= \underline{\ \ } \\
2v + 4w &= \underline{\ \ }
\end{aligned}
\quad\longrightarrow\quad
\begin{aligned}
u + v + w &= \underline{\ \ } \\
2v + 4w &= \underline{\ \ } \\
3w &= \underline{\ \ }
\end{aligned}
\end{array}
$$

The system is now triangular, and back-substitution will solve it.

SINGULAR EXAMPLE   (incurable)

$$
\begin{aligned}
u + v + w &= \underline{\ \ } \\
2u + 2v + 5w &= \underline{\ \ } \\
4u + 4v + 8w &= \underline{\ \ }
\end{aligned}
\quad\longrightarrow\quad
\begin{aligned}
u + v + w &= \underline{\ \ } \\
3w &= \underline{\ \ } \\
4w &= \underline{\ \ }
\end{aligned}
$$

Now there is no exchange of equations that can avoid zero in the second pivot position. The equations themselves may be solvable or unsolvable. If the last two equations are $3w = 6$ and $4w = 7$, there is no solution. If those two equations happen to be consistent—as in $3w = 6$ and $4w = 8$—then this singular case has an infinity of solutions. We know that $w = 2$, but the first equation cannot decide both u and v.

Section 1.5 will discuss row exchanges when the system is not singular. Then the exchanges produce a full set of pivots. Chapter 2 admits the singular case, and limps forward with elimination. The $3w$ can still eliminate the $4w$, and we will call 3 the second pivot. (There won't be a third pivot.) For the present we trust all n pivot entries to be nonzero, without changing the order of the equations. That is the best case, with which we continue.

The Cost of Elimination

Our other question is very practical. *How many separate arithmetical operations does elimination require, for n equations in n unknowns?* If n is large, a computer is going to take our place in carrying out the elimination (you may have a program available, or you could use the codes in Appendix C). Since all the steps are known, we should be able to predict the number of operations a computer will take. For the moment, ignore the right-hand sides of the equations, and count only the operations on the left. These operations are of two kinds. One is a division by the pivot, to find out what multiple (say *l*) of the pivot equation is to be subtracted. Then when we actually do this subtraction, we continually meet a "multiply-subtract" combination; the terms in the pivot equation are multiplied by l, and then subtracted from the equation beneath it.

Suppose we call each division, and each multiplication-subtraction, a single operation. At the beginning, when the first equation has length n, *it takes n operations for every zero we achieve in the first column*—one to find the multiple l, and the others to find the new entries along the row. There are $n - 1$ rows underneath the first one, so the first stage of elimination needs $n(n - 1) = n^2 - n$ operations. (Another approach to $n^2 - n$ is this: *All n^2 entries need to be changed, except the n in the first row.*) Now notice that later stages are faster because the equations are becoming progressively shorter. When the elimination is down to k equations, only $k^2 - k$ operations are needed to clear out the column below the pivot—by the same reasoning that applied to the first stage, when k equaled n. Altogether, the total number of operations on the left side of the equations is the sum of $k^2 - k$ over all values of k from 1 to n:

$$(1^2 + \cdots + n^2) - (1 + \cdots + n) = \frac{n(n + 1)(2n + 1)}{6} - \frac{n(n + 1)}{2} = \frac{n^3 - n}{3}.$$

Those are standard formulas for the sum of the first n numbers and the sum of the first n squares. Substituting $n = 1$ and $n = 2$ and $n = 100$ into the formula $\frac{1}{3}(n^3 - n)$, forward elimination can take no steps or two steps or about a third of a million steps (which means 41 seconds with a good code on a PC). What is important is the conclusion:

If n is at all large, ***a good estimate for the number of operations is $\frac{1}{3}n^3$.***

If the size of a system is doubled, and few of the coefficients are zero, then the cost goes up by a factor of eight.

Back-substitution is considerably faster. The last unknown is found in only one operation (a division by the last pivot). The second to last unknown requires two operations, and so on. Then the total for back-substitution is

$$1 + 2 + \cdots + n = \frac{n(n + 1)}{2} \approx \frac{n^2}{2}.$$

We will see that another $n^2/2$ steps prepare the right side for back-substitution, so ***the right side is responsible for n^2 operations***—much less than the $n^3/3$ on the left.

A few years ago, almost every mathematician would have guessed that these numbers were essentially optimal, in other words that a general system of order n could not be solved with much fewer than $n^3/3$ multiplications. (There were even theorems to demonstrate it, but they did not allow for all possible methods.) Astonishingly, that guess has been proved wrong, and *there now exists a method that requires only $Cn^{\log_2 7}$ multiplications*! It depends on a simple fact: Two combinations of two vectors in two-dimensional space would seem to take 8 multiplications, but they can be done in 7. That lowered the exponent from $\log_2 8$, which is 3, to $\log_2 7 \approx 2.8$. This discovery produced tremendous activity to find the small-est possible power of n. The exponent finally fell (at the IBM Research Center) below 2.5, where it remains as I write.† Fortunately for elimination, the constant C is so large and the coding is so awkward that the new method is largely (or en-tirely) of theoretical interest. The newest problem is the cost with *many processors in parallel*, and that is not yet known.

EXERCISES

1.3.1   Apply elimination and back-substitution to solve

$$2u - 3v \quad\quad = 3$$
$$4u - 5v + \ w = 7$$
$$2u - \ v - 3w = 5.$$

What are the pivots? List the three operations in which a multiple of one row is subtracted from another.

1.3.2   For the system

$$u + \ v + \ w = 2$$
$$u + 3v + 3w = 0$$
$$u + 3v + 5w = 2,$$

what is the triangular system after forward elimination, and what is the solution?

1.3.3   Solve the system and find the pivots when

$$2u - \ v \quad\quad\quad = 0$$
$$-u + 2v - \ w \quad\quad = 0$$
$$- \ v + 2w - \ z = 0$$
$$- \ w + 2z = 5.$$

You may carry the right side as a fifth column (and omit writing u, v, w, z until the solution at the end).

1.3.4   Apply elimination to the system

$$u + \ v + w = -2$$
$$3u + 3v - w = \quad 6$$
$$u - \ v + w = -1.$$

---

† With help from Zurich it just went below 2.376. The lower bound seems most likely to be 2, since no number in between looks special. That is just a personal opinion.

When a zero arises in the pivot position, exchange that equation for the one below it and proceed. What coefficient of v in the third equation, in place of the present -1, would make it impossible to proceed—and force elimination to break down?

1.3.5 Solve by elimination the system of two equations

$$x - \ y = \ 0$$
$$3x + 6y = 18.$$

Draw a graph representing each equation as a straight line in the x-y plane; the lines intersect at the solution. Also, add one more line—the graph of the new second equation which arises after elimination.

1.3.6 Find three values of a for which elimination breaks down, temporarily or permanently, in

$$au + \ v = 1$$
$$4u + av = 2.$$

Breakdown at the first step can be fixed by exchanging rows—but not breakdown at the last step.

1.3.7 (a) If the first two *rows* of A are the same, when will elimination discover that A is singular? Do a 3 by 3 example, allowing row exchanges.
(b) If the first two *columns* of A are the same, when will elimination discover that A is singular?

1.3.8 How many multiplication-subtractions would it take to solve a system of order $n = 600$? How many seconds, on a PC that can do 8000 a second or a VAX that can do 80,000 or a CRAY X-MP/2 that can do 12 million? (Those are in double precision—I think the CRAY is cheapest, if you can afford it.)

1.3.9 True or false: (a) If the third equation starts with a zero coefficient (it begins with $0u$) then no multiple of equation 1 will be subtracted from equation 3.
(b) If the third equation has zero as its second coefficient (it contains $0v$) then no multiple of equation 2 will be subtracted from equation 3.
(c) If the third equation contains $0u$ *and* $0v$, then no multiple of equation 1 or equation 2 will be subtracted from equation 3.

1.3.10 (very optional)   Normally the multiplication of two complex numbers

$$(a + ib)(c + id) = (ac - bd) + i(bc + ad)$$

involves the four separate multiplications ac, bd, bc, ad. Ignoring i, can you compute the quantities $ac - bd$ and $bc + ad$ with only three multiplications? (You may do additions, such as forming $a + b$ before multiplying, without any penalty. We note however that addition takes six clock cycles on a CRAY X-MP/48, and multiplication is only one more.)

1.3.11 Use elimination to solve

$$
\begin{array}{ll}
\begin{aligned}
u + \ v + \ w &= \ 6 \\
u + 2v + 2w &= 11 \\
2u + 3v - 4w &= \ 3
\end{aligned}
& \text{and}
\begin{aligned}
u + \ v + \ w &= \ 7 \\
u + 2v + 2w &= 10 \\
2u + 3v - 4w &= \ 3
\end{aligned}
\end{array}
$$

The final exercises give experience in setting up linear equations. Suppose that
(a)   of those who start a year in California, 80 percent stay in and 20 percent move out;
(b)   of those who start a year outside California, 90 percent stay out and 10 percent move in.
If we know the situation at the beginning, say 200 million outside and 30 million in, then it is easy to find the numbers u and v that are outside and inside at the end:

$$.9(200,000,000) + .2(30,000,000) = u$$
$$.1(200,000,000) + .8(30,000,000) = v$$

The real problem is to go backwards, and compute the start from the finish.

1.3.12   If $u = 200$ million and $v = 30$ million at the end, set up (without solving) the equations to find u and v at the beginning.

1.3.13   If u and v at the end are the same as u and v at the beginning, what equations do you get? What is the ratio of u to v in this "steady state"?

MATRIX NOTATION AND MATRIX MULTIPLICATION ■ 1.4

So far, with our 3 by 3 example, we have been able to write out all the equations in full. We could even list the elimination steps, which subtract a multiple of one equation from another and reach a triangular form. For a large system, this way of keeping track of elimination would be hopeless; a much more concise record is needed. We shall now introduce matrix notation to describe the original system, and matrix multiplication to describe the operations that make it simpler.

Notice that in our example

$$
\begin{aligned}
2u + v + w &= 5 \\
4u - 6v &= -2 \\
-2u + 7v + 2w &= 9
\end{aligned}
\tag{1}
$$

three different types of quantities appear. On the right side is the column vector b. On the left side are the unknowns u, v, w. And also on the left side there are nine coefficients (one of which happens to be zero). It is natural to represent the three unknowns by a vector:

$$
\text{the unknown is } x = \begin{bmatrix} u \\ v \\ w \end{bmatrix}; \quad \text{the solution is } x = \begin{bmatrix} 1 \\ 1 \\ 2 \end{bmatrix}.
$$

As for the nine coefficients, which fall into three rows and three columns, the right format is a ***three by three matrix***. It is called the *coefficient matrix*:

$$
A = \begin{bmatrix} 2 & 1 & 1 \\ 4 & -6 & 0 \\ -2 & 7 & 2 \end{bmatrix}.
$$

A is a *square* matrix, because the number of equations agrees with the number of unknowns. More generally, if there are n equations in n unknowns, we have a square coefficient matrix *of order n*. Still more generally, we might have m equations and n unknowns. In this case the matrix is *rectangular*, with m rows and n columns. In other words, it will be an "m by n matrix."

Matrices are added to each other, or multiplied by numerical constants, exactly as vectors are—one component at a time. In fact we may regard vectors as special cases of matrices; *they are matrices with only one column*. As with vectors, two matrices can be added only if they have the same shape:

$$
\begin{bmatrix} 2 & 1 \\ 3 & 0 \\ 0 & 4 \end{bmatrix} + \begin{bmatrix} 1 & 2 \\ -3 & 1 \\ 1 & 2 \end{bmatrix} = \begin{bmatrix} 3 & 3 \\ 0 & 1 \\ 1 & 6 \end{bmatrix}, \quad 2 \begin{bmatrix} 2 & 1 \\ 3 & 0 \\ 0 & 4 \end{bmatrix} = \begin{bmatrix} 4 & 2 \\ 6 & 0 \\ 0 & 8 \end{bmatrix}.
$$

Multiplication of a Matrix and a Vector

Now we put this notation to use. We propose to rewrite the system (1) of three equations in three unknowns in the simplified matrix form $Ax = b$. Written out in full, this form is

$$\begin{bmatrix} 2 & 1 & 1 \\ 4 & -6 & 0 \\ -2 & 7 & 2 \end{bmatrix} \begin{bmatrix} u \\ v \\ w \end{bmatrix} = \begin{bmatrix} 5 \\ -2 \\ 9 \end{bmatrix}.$$

The right side b is clear enough; it is the column vector of "inhomogeneous terms." The left side consists of the vector x, premultiplied by the matrix A. This multiplication will be defined *exactly so as to reproduce the original system*. Therefore the first component of the product Ax must come from "multiplying" the first row of A into the column vector x:

$$[2 \quad 1 \quad 1] \begin{bmatrix} u \\ v \\ w \end{bmatrix} = [2u + v + w]. \tag{2}$$

This equals the first component of b; $2u + v + w = 5$ is the first equation in our system. The second component of the product Ax is determined by the second row of A—it is $4u - 6v$—and the third component comes from the third row. Thus the matrix equation $Ax = b$ is precisely equivalent to the three simultaneous equations with which we started.

The operation in Eq. (2) is fundamental to all matrix multiplications. It starts with a row vector and a column vector of matching lengths, and it produces a single number. This single quantity is called the ***inner product*** of the two vectors. In other words, the product of a 1 by n matrix, which is a *row vector*, and an n by 1 matrix, alias a *column vector*, is a 1 by 1 matrix:

$$[2 \quad 1 \quad 1] \begin{bmatrix} 1 \\ 1 \\ 2 \end{bmatrix} = [2 \cdot 1 + 1 \cdot 1 + 1 \cdot 2] = [5].$$

That confirms that the proposed solution $x = (1, 1, 2)$ does satisfy the first equation.

If we look at the whole computation, multiplying a matrix by a vector, there are two ways to do it. One is to continue *a row at a time*. Each row of the matrix combines with the vector to give a component of the product. There are three inner products when there are three rows: for example

$$\textbf{\textit{by rows}} \qquad Ax = \begin{bmatrix} 1 & 1 & 6 \\ 3 & 0 & 3 \\ 1 & 1 & 4 \end{bmatrix} \begin{bmatrix} 2 \\ 5 \\ 0 \end{bmatrix} = \begin{bmatrix} 1 \cdot 2 + 1 \cdot 5 + 6 \cdot 0 \\ 3 \cdot 2 + 0 \cdot 5 + 3 \cdot 0 \\ 1 \cdot 2 + 1 \cdot 5 + 4 \cdot 0 \end{bmatrix} = \begin{bmatrix} 7 \\ 6 \\ 7 \end{bmatrix}.$$

That is how it is usually explained, but the second way is equally important. In fact it is more important. It does the multiplication *a column at a time*. The product

Ax is found all at once, and it is ***a combination of the three columns of*** A:

$$\textbf{\textit{by columns}} \qquad Ax = 2\begin{bmatrix} 1 \\ 3 \\ 1 \end{bmatrix} + 5\begin{bmatrix} 1 \\ 0 \\ 1 \end{bmatrix} + 0\begin{bmatrix} 6 \\ 3 \\ 4 \end{bmatrix} = \begin{bmatrix} 7 \\ 6 \\ 7 \end{bmatrix}. \qquad (3)$$

The answer is twice column 1 plus 5 times column 2. It corresponds to the "column picture" of a linear system $Ax = b$. If the right side b has components 7, 6, 7, then the solution x has components 2, 5, 0. Of course the row picture agrees with that (and we eventually have to do the same multiplications).

The column rule will be used over and over throughout the book, and we repeat it for emphasis:

1A   The product Ax can be found by using whole columns as in (3). Therefore Ax is ***a combination of the columns of*** A. The coefficients which multiply the columns are the components of x.

If we try to write down the general rule in n dimensions, we need a notation for the individual entries in A. It is easy to learn. The entry in the ith row and jth column is always denoted by a_{ij}. The first subscript gives the row number, and the second subscript indicates the column. (In the matrix above, a_{21} was 3 and a_{13} was 6.) If A is an m by n matrix, then the index i goes from 1 to m—there are m rows—and the index j goes from 1 to n. Altogether the matrix has mn entries, forming a rectangular array, and a_{mn} is in the lower right corner.

One subscript is enough for a vector. The jth component of x is denoted by x_j. (The multiplication above had $x_1 = 2$, $x_2 = 5$, $x_3 = 0$.) Normally x is written as a column vector—like an n by 1 matrix—but sometimes it is printed on a line, as in $x = (2, 5, 0)$. The parentheses and commas emphasize that it is not a 1 by 3 matrix. It is a column vector, and it is just temporarily lying down.

To describe the product Ax, we use the summation symbol "*sigma:*"

$$\sum_{j=1}^{n} a_{ij}x_j \text{ is the } i\text{th component of } Ax.$$

This sum takes us along the ith row of A, forming its inner product with x. It gives the index j each value from 1 to n and adds up the results—the sum is $a_{i1}x_1 + a_{i2}x_2 + \cdots + a_{in}x_n$. We see again that the length of the rows (the number of columns in A) must match the length of x. An m by n matrix multiplies an n-dimensional vector (and produces an m-dimensional vector). Summations are simpler to work with than writing everything out in full, but they are not as good as matrix notation itself.†

---

  † Einstein introduced "tensor notation," in which a repeated index automatically means summation. He wrote $a_{ij}x_j$, or even $a_i^j x_j$, without the Σ.

The Matrix Form of One Elimination Step

So far we have a convenient shorthand $Ax = b$ for the original system of equations. What about the operations that are carried out during elimination? In our example, the first step subtracted 2 times the first equation from the second. On the right side of the equation, this means that 2 times the first component of b was subtracted from the second component. We claim that *this same result is achieved if we multiply b by the following matrix:*

$$E = \begin{bmatrix} 1 & 0 & 0 \\ -2 & 1 & 0 \\ 0 & 0 & 1 \end{bmatrix}.$$

This is verified just by obeying the rule for multiplying a matrix and a vector:

$$Eb = \begin{bmatrix} 1 & 0 & 0 \\ -2 & 1 & 0 \\ 0 & 0 & 1 \end{bmatrix} \begin{bmatrix} 5 \\ -2 \\ 9 \end{bmatrix} = \begin{bmatrix} 5 \\ -12 \\ 9 \end{bmatrix}.$$

The first and third components, 5 and 9, stayed the same (because of the special form for the first and third rows of E). The new second component is the correct value -12; it appeared after the first elimination step.

It is easy to describe the matrices like E, which carry out the separate elimination steps. We also notice the "identity matrix," which does nothing at all.

1B   The matrix that leaves every vector unchanged is the *identity matrix I*, with 1's on the diagonal and 0's everywhere else. The matrix that subtracts a multiple l of row j from row i is the *elementary matrix E_{ij}*, with 1's on the diagonal and the number $-l$ in row i, column j.

EXAMPLE $I = \begin{bmatrix} 1 & 0 & 0 \\ 0 & 1 & 0 \\ 0 & 0 & 1 \end{bmatrix}$ and $E_{31} = \begin{bmatrix} 1 & 0 & 0 \\ 0 & 1 & 0 \\ -l & 0 & 1 \end{bmatrix}.$

If you multiply any vector b by the identity matrix you get b again. It is the matrix analogue of multiplying a number by 1: $Ib = b$. If you multiply instead by E_{31}, you get $E_{31}b = (b_1, b_2, b_3 - lb_1)$. That is a typical operation on the right side of the equations, and the important thing is what happens on the left side.

To maintain equality, we must apply the same operation to both sides of $Ax = b$. In other words, we must also multiply the vector Ax by the matrix E. Our original matrix E subtracts 2 times the first component from the second, leaving the first and third components unchanged. After this step the new and simpler system (equivalent to the old) is just $E(Ax) = Eb$. It is simpler because of the zero that was created below the first pivot. It is equivalent because we can recover the original system (by adding 2 times the first equation back to the second). So the two systems have exactly the same solution x.

Now we come to the most important question: *How do we multiply two matrices?* There is already a partial clue from Gaussian elimination: We know the original coefficient matrix A, we know what it becomes after an elimination step, and now we know the matrix E which carries out that step. Therefore, we hope that

$$E = \begin{bmatrix} 1 & 0 & 0 \\ -2 & 1 & 0 \\ 0 & 0 & 1 \end{bmatrix} \quad \text{times} \quad A = \begin{bmatrix} 2 & 1 & 1 \\ 4 & -6 & 0 \\ -2 & 7 & 2 \end{bmatrix} \quad \text{gives} \quad EA = \begin{bmatrix} 2 & 1 & 1 \\ 0 & -8 & -2 \\ -2 & 7 & 2 \end{bmatrix}.$$

The first and third rows of A appear unchanged in EA, while *twice the first row has been subtracted from the second*. Thus the matrix multiplication is consistent with the row operations of elimination. We can write the result either as $E(Ax) = Eb$, applying E to both sides of our equation, or as $(EA)x = Eb$. The matrix EA is constructed exactly so that these equations agree. In other words, the parentheses are superfluous, and we can just write $EAx = Eb$.†

There is also another requirement on matrix multiplication. We know how to multiply Ax, a matrix and a vector, and the new definition should remain consistent with that one. When a matrix B contains only a single column x, the matrix–matrix product AB should be identical with the matrix–vector product Ax. It is even better if this goes further: When the matrix B contains several columns, say x_1, x_2, x_3, we hope that the columns of AB are just Ax_1, Ax_2, Ax_3. Then matrix multiplication is completely straightforward; B contains several columns side by side, and we can take each one separately. This rule works for the matrices multiplied above. The first column of EA equals E times the first column of A,

$$\begin{bmatrix} 1 & 0 & 0 \\ -2 & 1 & 0 \\ 0 & 0 & 1 \end{bmatrix} \begin{bmatrix} 2 \\ 4 \\ -2 \end{bmatrix} = \begin{bmatrix} 2 \\ 0 \\ -2 \end{bmatrix},$$

and the same is true for the other columns.

Notice that our first requirement had to do with rows, whereas the second was concerned with columns. A third approach is to describe each individual entry in AB and hope for the best. In fact, there is only one possible rule, and I am not sure who discovered it. It makes everything work. It does not allow us to multiply every pair of matrices; if they are square, as in our example, then they must be of

---

† This is the whole point of an "associative law" like $2 \times (3 \times 4) = (2 \times 3) \times 4$, which seems so obvious that it is hard to imagine it could be false. But the same could be said of the "commutative law" $2 \times 3 = 3 \times 2$—and for matrices that law really is false.

the same size. If they are rectangular, then they must *not* have the same shape; *the number of columns in A has to equal the number of rows in B*. Then *A* can be multiplied into each column of *B*. In other words, if *A* is *m* by *n*, and *B* is *n* by *p*, then multiplication is possible, and *the product AB will be m by p*.

We now describe how to find the entry in row *i* and column *j* of *AB*.

1C   The i, j entry of AB is the inner product of the ith row of A and the jth column of B. For the example in Fig. 1.6,

$$(AB)_{32} = a_{31}b_{12} + a_{32}b_{22} + a_{33}b_{32} + a_{34}b_{42}. \tag{4}$$

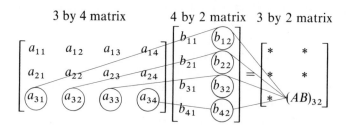

Fig. 1.6. An illustration of matrix multiplication.

Note.   We write *AB* when the matrices have nothing special to do with Gaussian elimination. Our earlier example was *EA*, because of the "elementary" matrix *E*; later it will be *PA*, or *LU*, or even *LDU*. In every case we use the same general rule for matrix multiplication.

EXAMPLE 1

$$\begin{bmatrix} 2 & 3 \\ 4 & 0 \end{bmatrix} \begin{bmatrix} 1 & 2 & 0 \\ 5 & -1 & 0 \end{bmatrix} = \begin{bmatrix} 17 & 1 & 0 \\ 4 & 8 & 0 \end{bmatrix}.$$

The entry 17 is $(2)(1) + (3)(5)$, the inner product of the first row and first column. The entry 8 is $(4)(2) + (0)(-1)$, from the second row and second column. The third column is zero in *B*, so it is zero in *AB*.

EXAMPLE 2

$$\begin{bmatrix} 0 & 1 \\ 1 & 0 \end{bmatrix} \begin{bmatrix} 2 & 3 \\ 7 & 8 \end{bmatrix} = \begin{bmatrix} 7 & 8 \\ 2 & 3 \end{bmatrix}.$$

The matrix on the left produced a *row exchange* in the matrix on the right.

EXAMPLE 3 The identity matrix leaves every vector unchanged, and it also leaves every matrix unchanged:

$$IA = A \quad \text{and} \quad BI = B.$$

EXAMPLE 4

$$AB = \begin{bmatrix} 1 & 0 \\ 2 & 3 \end{bmatrix} \begin{bmatrix} a & b \\ c & d \end{bmatrix} = \begin{bmatrix} a & b \\ 2a + 3c & 2b + 3d \end{bmatrix}.$$

By columns, this illustrates the property we hoped for. B consists of two columns side by side, and A multiplies each of them separately. Therefore *each column of AB is a combination of the columns of A*. Just as in a matrix-vector multiplication, the columns of A are multiplied by the individual numbers (or letters) in B. The first column of AB is "a" times column 1 plus "c" times column 2.

Now we ask about the *rows* of AB. The multiplication can also be done *a row at a time*. The second row of the answer uses the numbers 2 and 3 from the second row of A. Those numbers multiply the rows of B, to give $2\begin{bmatrix} a & b \end{bmatrix} + 3\begin{bmatrix} c & d \end{bmatrix}$. Similarly the first row of the answer uses the numbers 1 and 0, to give $1\begin{bmatrix} a & b \end{bmatrix} + 0\begin{bmatrix} c & d \end{bmatrix}$. Exactly as in elimination, where all this started, each row of AB is a *combination of the rows of B*.

We summarize these three different ways to look at matrix multiplication.

1D (i) Each entry of AB is the product of a *row* and a *column*:

$$(AB)_{ij} = \text{row } i \text{ of } A \text{ times column } j \text{ of } B$$

(ii) Each column of AB is the product of a *matrix* and a *column*:

$$\text{column } j \text{ of } AB = A \text{ times column } j \text{ of } B$$

(iii) Each row of AB is the product of a *row* and a *matrix*:

$$\text{row } i \text{ of } AB = \text{row } i \text{ of } A \text{ times } B.$$

This observation is useful in verifying one of the key properties of matrix multiplication. Suppose we are given three matrices A, B, and C, possibly rectangular, and suppose their shapes permit them to be multiplied in that order: The number of columns in A and B match, respectively, the number of rows in B and C. Then the property is this:

1E Matrix multiplication is associative: $(AB)C = A(BC)$.

If C happens to be just a vector (a matrix with only one column) this is the requirement $(EA)x = E(Ax)$ mentioned earlier. It is the whole basis for the laws of matrix multiplication. And if C has several columns, we have only to think of them placed side by side, and apply the same rule several times. Thus parentheses are not needed when we have a product of several matrices. This can also be verified by comparing each entry of $(AB)C$ and $A(BC)$ (Ex. 1.4.20); but you will see why matrix notation is preferred.

We want to get on with the connection between matrix multiplication and Gaussian elimination, but there are two more properties to mention first—one property that matrix multiplication has, and another which it *does not have*. The property that it does possess is:

1F   Matrix operations are distributive:

$$A(B + C) = AB + AC \qquad \text{and} \qquad (B + C)D = BD + CD.$$

Of course the shapes of these matrices must be property matched—B and C have the same shape, so they can be added, and A and D are the right size for premultiplication and postmultiplication. The proof of this law is too boring for words.

The property that fails to hold is a little more interesting:

1G   Matrix multiplication is not commutative: Usually $FE \neq EF$.

EXAMPLE 5   Suppose E is the matrix introduced earlier, to subtract twice the first equation from the second—and suppose F is the matrix for the next step, *to add row 1 to row 3*:

$$E = \begin{bmatrix} 1 & 0 & 0 \\ -2 & 1 & 0 \\ 0 & 0 & 1 \end{bmatrix} \qquad \text{and} \qquad F = \begin{bmatrix} 1 & 0 & 0 \\ 0 & 1 & 0 \\ 1 & 0 & 1 \end{bmatrix}.$$

These two matrices do commute:

$$EF = \begin{bmatrix} 1 & 0 & 0 \\ -2 & 1 & 0 \\ 1 & 0 & 1 \end{bmatrix} = FE.$$

That product does both steps—in either order, or both at once, because in this special case the order doesn't matter.

EXAMPLE 6   Suppose E is the same but G is the matrix for the final step—*it adds row 2 to row 3*. Now the order makes a difference. In one case, where we apply E and then G, the second equation is altered *before* it affects the third. That is the order actually met in elimination. The first equation affects the second which affects the third. If E comes *after* G, then the third equation feels no effect from the first. You will see a zero in the $(3, 1)$ entry of EG, where there is a -2 in GE:

$$GE = \begin{bmatrix} 1 & 0 & 0 \\ 0 & 1 & 0 \\ 0 & 1 & 1 \end{bmatrix} \begin{bmatrix} 1 & 0 & 0 \\ -2 & 1 & 0 \\ 0 & 0 & 1 \end{bmatrix} = \begin{bmatrix} 1 & 0 & 0 \\ -2 & 1 & 0 \\ -2 & 1 & 1 \end{bmatrix} \quad \text{but} \quad EG = \begin{bmatrix} 1 & 0 & 0 \\ -2 & 1 & 0 \\ 0 & 1 & 1 \end{bmatrix}.$$

Thus $EG \neq GE$. A random example would almost certainly have shown the same thing—most matrices don't commute—but here the matrices have meaning. There was a reason for $EF = FE$, and a reason for $EG \neq GE$. It is worth taking one more step, to see what happens with *all three elimination matrices at once*:

$$GFE = \begin{bmatrix} 1 & 0 & 0 \\ -2 & 1 & 0 \\ -1 & 1 & 1 \end{bmatrix} \quad \text{and} \quad EFG = \begin{bmatrix} 1 & 0 & 0 \\ -2 & 1 & 0 \\ 1 & 1 & 1 \end{bmatrix}.$$

The product GFE is in the true order of elimination. *It is the matrix that takes the original A to the upper triangular U.* We will see it again in the next section.

The other matrix EFG is nicer, because in that order the numbers -2 from E, 1 from F, and 1 from G were not disturbed. They went straight into the product. Unfortunately it is the wrong order for elimination. But fortunately *it is the right order for reversing the elimination steps*—which also comes in the next section.

Notice that the product of lower triangular matrices is again lower triangular.

EXERCISES

1.4.1   Compute the products

$$\begin{bmatrix} 4 & 0 & 1 \\ 0 & 1 & 0 \\ 4 & 0 & 1 \end{bmatrix} \begin{bmatrix} 3 \\ 4 \\ 5 \end{bmatrix} \quad \text{and} \quad \begin{bmatrix} 1 & 0 & 0 \\ 0 & 1 & 0 \\ 0 & 0 & 1 \end{bmatrix} \begin{bmatrix} 5 \\ -2 \\ 3 \end{bmatrix} \quad \text{and} \quad \begin{bmatrix} 2 & 0 \\ 1 & 3 \end{bmatrix} \begin{bmatrix} 1 \\ 1 \end{bmatrix}.$$

Draw a pair of perpendicular axes and mark off the vectors to the points $x = 2$, $y = 1$ and $x = 0$, $y = 3$. Add the two vectors by completing the parallelogram.

1.4.2   Working a column at a time, compute the products

$$\begin{bmatrix} 4 & 1 \\ 5 & 1 \\ 6 & 1 \end{bmatrix} \begin{bmatrix} 1 \\ 3 \end{bmatrix} \quad \text{and} \quad \begin{bmatrix} 1 & 2 & 3 \\ 4 & 5 & 6 \\ 7 & 8 & 9 \end{bmatrix} \begin{bmatrix} 0 \\ 1 \\ 0 \end{bmatrix} \quad \text{and} \quad \begin{bmatrix} 4 & 3 \\ 6 & 6 \\ 8 & 9 \end{bmatrix} \begin{bmatrix} \frac{1}{2} \\ \frac{1}{3} \end{bmatrix}.$$

1.4.3   Find two inner products and a matrix product:

$$[1 \quad -2 \quad 7] \begin{bmatrix} 1 \\ -2 \\ 7 \end{bmatrix} \quad \text{and} \quad [1 \quad -2 \quad 7] \begin{bmatrix} 3 \\ 5 \\ 1 \end{bmatrix} \quad \text{and} \quad \begin{bmatrix} 1 \\ -2 \\ 7 \end{bmatrix} [3 \quad 5 \quad 1].$$

The first gives the length of the vector (squared).

1.4.4   If an m by n matrix A multiplies an n-dimensional vector x, how many separate multiplications are involved? What if A multiplies an n by p matrix B?

1.4.5   Compute the product

$$Ax = \begin{bmatrix} 3 & -6 & 0 \\ 0 & 2 & -2 \\ 1 & -1 & -1 \end{bmatrix} \begin{bmatrix} 2 \\ 1 \\ 1 \end{bmatrix}.$$

For this matrix A, find a solution vector x to the system $Ax = 0$, with zeros on the right side of all three equations. Can you find more than one solution?

1.4.6   Write down the 3 by 2 matrices A and B which have entries $a_{ij} = i + j$ and $b_{ij} = (-1)^{i+j}$.

1.4.7   Express the inner product of the row vector $y = [y_1 \ y_2 \ \cdots \ y_n]$ and the column vector x in summation notation.

1.4.8   Give 3 by 3 examples (not just $A = 0$) of
(a)   a diagonal matrix: $a_{ij} = 0$ if $i \neq j$;
(b)   a symmetric matrix: $a_{ij} = a_{ji}$ for all i and j;
(c)   an upper triangular matrix: $a_{ij} = 0$ if $i > j$;
(d)   a skew-symmetric matrix: $a_{ij} = -a_{ji}$ for all i and j.

1.4.9   Do the following subroutines multiply Ax by rows or columns?

```
DO 10 I = 1,N              DO 10 J = 1,N
DO 10 J = 1,N              DO 10 I = 1,N
10 B(I) = B(I) + A(I,J) * X(J)   10 B(I) = B(I) + A(I,J) * X(J)
```

The results are mathematically equivalent, assuming that initially all $B(I) = 0$, but the structure of FORTRAN makes the second code slightly more efficient (Appendix C). It is much more efficient on a vector machine like the CRAY, since the inner loop can change many $B(I)$ at once—where the first code makes many changes to a single $B(I)$ and does not vectorize.

1.4.10   If the entries of A are a_{ij}, use subscript notation to write down
(1)   the first pivot
(2)   the multiplier l_{i1} of row 1 to be subtracted from row i
(3)   the new entry that replaces a_{ij} after that subtraction
(4)   the second pivot.

1.4.11   True or false; give a specific counterexample when false.
(a)   If the first and third columns of B are the same, so are the first and third columns of AB.
(b)   If the first and third rows of B are the same, so are the first and third rows of AB.
(c)   If the first and third rows of A are the same, so are the first and third rows of AB.
(d) $(AB)^2 = A^2 B^2$.

1.4.12 The first row of AB is a linear combination of all the rows of B. What are the coefficients in this combination, and what is the first row of AB, if

$$A = \begin{bmatrix} 2 & 1 & 4 \\ 0 & -1 & 1 \end{bmatrix}, \qquad B = \begin{bmatrix} 1 & 1 \\ 0 & 1 \\ 1 & 0 \end{bmatrix}?$$

1.4.13 The product of two lower triangular matrices is again lower triangular (all its entries above the main diagonal are zero). Confirm this with a 3 by 3 example, and then explain how it follows from the laws of matrix multiplication.

1.4.14 By trial and error find examples of 2 by 2 matrices such that
   (a) $A^2 = -I$, A having only real entries;
   (b) $B^2 = 0$, although $B \neq 0$;
   (c) $CD = -DC$, not allowing the case $CD = 0$;
   (d) $EF = 0$, although no entries of E or F are zero.

1.4.15 Describe the rows of EA and the *columns* of AE if $E = \begin{bmatrix} 1 & 7 \\ 0 & 1 \end{bmatrix}$.

1.4.16 Check the associative law $(EF)G = E(FG)$ for the matrices in the text.

1.4.17 Suppose A commutes with every 2 by 2 matrix $(AB = BA)$, and in particular

$$A = \begin{bmatrix} a & b \\ c & d \end{bmatrix} \quad \text{commutes with} \quad B_1 = \begin{bmatrix} 1 & 0 \\ 0 & 0 \end{bmatrix} \quad \text{and} \quad B_2 = \begin{bmatrix} 0 & 1 \\ 0 & 0 \end{bmatrix}.$$

Show that $a = d$ and $b = c = 0$. If $AB = BA$ for all matrices B, then A is a multiple of the identity.

1.4.18 Let x be the column vector with components $1, 0, \ldots, 0$. Show that the rule $(AB)x = A(Bx)$ forces the first column of AB to equal A times the first column of B.

1.4.19 Which of the following matrices are guaranteed to equal $(A + B)^2$?

$$(B+A)^2, \quad A^2 + 2AB + B^2, \quad A(A+B) + B(A+B), \quad (A+B)(B+A), \quad A^2 + AB + BA + B^2.$$

1.4.20 In summation notation, the i, j entry of AB is

$$(AB)_{ij} = \sum_k a_{ik} b_{kj}.$$

If A and B are n by n matrices with all entries equal to 1, find $(AB)_{ij}$.
   The same notation turns the associative law $(AB)C = A(BC)$ into

$$\sum_j \left(\sum_k a_{ik} b_{kj} \right) c_{jl} = \sum_k a_{ik} \left(\sum_j b_{kj} c_{jl} \right).$$

Compute both sides if C is also n by n, with every $c_{jl} = 2$.

1.4.21 There is a fourth way of looking at matrix multiplication, as *columns* times *rows*. If the columns of A are $c_1, \ldots, c_n$ and the rows of B are the row vectors $r_1, \ldots, r_n$, then $c_1 r_1$ is a matrix and

$$AB = c_1 r_1 + c_2 r_2 + \cdots + c_n r_n.$$

(a) Give a 2 by 2 example of this rule for matrix multiplication.

(b) Explain why the right side gives the correct value $\sum_{k=1}^{n} a_{ik}b_{kj}$ for the entry $(AB)_{ij}$.

1.4.22 The matrices that "rotate" the x-y plane are

$$A(\theta) = \begin{bmatrix} \cos\theta & -\sin\theta \\ \sin\theta & \cos\theta \end{bmatrix}.$$

(a) Verify $A(\theta_1)A(\theta_2) = A(\theta_1 + \theta_2)$ from the identities for $\cos(\theta_1 + \theta_2)$ and $\sin(\theta_1 + \theta_2)$.
(b) What is $A(\theta)$ times $A(-\theta)$?

1.4.23 For the matrices

$$A = \begin{bmatrix} \frac{1}{2} & \frac{1}{2} \\ \frac{1}{2} & \frac{1}{2} \end{bmatrix} \quad \text{and} \quad B = \begin{bmatrix} 1 & 0 \\ 0 & -1 \end{bmatrix} \quad \text{and} \quad C = AB = \begin{bmatrix} \frac{1}{2} & -\frac{1}{2} \\ \frac{1}{2} & -\frac{1}{2} \end{bmatrix}$$

find all the powers A^2, A^3 (which is A^2 times A), ... and B^2, B^3, ... and C^2, C^3, ...

1.4.24 More general than multiplication by columns is *block multiplication*. If matrices are separated into blocks (submatrices) and their shapes make block multiplication possible, then it is allowed:

$$\begin{bmatrix} x & x & | & x \\ x & x & | & x \\ \hline x & x & | & x \end{bmatrix}\begin{bmatrix} x & x & | & x \\ x & x & | & x \\ \hline x & x & | & x \end{bmatrix} \quad \text{or} \quad \begin{bmatrix} x & x & | & x & x \\ x & x & | & x & x \end{bmatrix}\begin{bmatrix} x & x \\ x & x \\ \hline x & x \\ x & x \end{bmatrix} \quad \text{or} \quad \cdots$$

(a) Replace those x's by numbers and confirm that block multiplication succeeds.
(b) Give two more examples (with x's) if A is 3 by 4 and B is 4 by 2. Vertical cuts in A must be matched by horizontal cuts in B.

TRIANGULAR FACTORS AND ROW EXCHANGES ■ 1.5

We want to look again at elimination, to see what it means in terms of matrices. The starting point was the system $Ax = b$:

$$Ax = \begin{bmatrix} 2 & 1 & 1 \\ 4 & -6 & 0 \\ -2 & 7 & 2 \end{bmatrix} \begin{bmatrix} u \\ v \\ w \end{bmatrix} = \begin{bmatrix} 5 \\ -2 \\ 9 \end{bmatrix} = b. \tag{1}$$

Then there were three elimination steps:

  (i)   Subtract   2 times the first equation from the second;
  (ii) Subtract -1 times the first equation from the third;
  (iii) Subtract -1 times the second equation from the third.

The result was an equivalent but simpler system, with a new coefficient matrix U:

$$Ux = \begin{bmatrix} 2 & 1 & 1 \\ 0 & -8 & -2 \\ 0 & 0 & 1 \end{bmatrix} \begin{bmatrix} u \\ v \\ w \end{bmatrix} = \begin{bmatrix} 5 \\ -12 \\ 2 \end{bmatrix} = c. \tag{2}$$

This matrix U is **upper triangular**—all entries below the diagonal are zero.

The right side, which is a new vector c, was derived from the original vector b by the same steps that took A into U. Thus, forward elimination amounted to:

> Start with A and b
> Apply steps (i), (ii), (iii) in that order
> End with U and c.

The last stage is the solution of $Ux = c$ by back-substitution, but for the moment we are not concerned with that. We concentrate on the relation of A to U.

The matrices E for step (i), F for step (ii), and G for step (iii) were introduced in the previous section. They are called **elementary matrices**, and it is easy to see how they work. To subtract a multiple l of equation j from equation i, put the number $-l$ into the (i, j) position. Otherwise keep the identity matrix, with 1's on the diagonal and 0's elsewhere. Then matrix multiplication executes the step.

The result of all three steps is $GFEA = U$. Note that E is the first to multiply A, then F, then G. We could multiply GFE together to find the single matrix that takes A to U (and also takes b to c). Omitting the zeros it is

$$GFE = \begin{bmatrix} 1 & & \\ & 1 & \\ & 1 & 1 \end{bmatrix} \begin{bmatrix} 1 & & \\ & 1 & \\ 1 & & 1 \end{bmatrix} \begin{bmatrix} 1 & & \\ -2 & 1 & \\ & & 1 \end{bmatrix} = \begin{bmatrix} 1 & & \\ -2 & 1 & \\ -1 & 1 & 1 \end{bmatrix}. \tag{3}$$

This is good, but the most important question is exactly the opposite: How would we get from U back to A? *How can we undo the steps of Gaussian elimination?*

A single step, say step (i), is not hard to undo. Instead of subtracting, we *add* twice the first row to the second. (Not twice the second row to the first!) The result of doing both the subtraction and the addition is to bring back the identity matrix:

$$\begin{bmatrix} 1 & 0 & 0 \\ 2 & 1 & 0 \\ 0 & 0 & 1 \end{bmatrix} \begin{bmatrix} 1 & 0 & 0 \\ -2 & 1 & 0 \\ 0 & 0 & 1 \end{bmatrix} = \begin{bmatrix} 1 & 0 & 0 \\ 0 & 1 & 0 \\ 0 & 0 & 1 \end{bmatrix}. \tag{4}$$

One operation cancels the other. In matrix terms, one matrix is the *inverse* of the other. If the elementary matrix E has the number $-l$ in the (i, j) position, then its inverse has $+l$ in that position.† That matrix is denoted by E^{-1}. Thus E^{-1} times E is the identity matrix; that is equation (4).

We can invert each step of elimination, by using E^{-1} and F^{-1} and G^{-1}. The final problem is to undo the whole process at once, and see what matrix takes U back to A. Notice that *since step* (iii) *was last in going from A to U, its matrix G must be the first to be inverted in the reverse direction*. Inverses come in the opposite order, so the second reverse step is F^{-1} and the last is E^{-1}:

$$E^{-1}F^{-1}G^{-1}U = A. \tag{5}$$

You can mentally substitute $GFEA$ for U, to see how the inverses knock out the original steps.

Now we recognize the matrix that takes U back to A. *It has to be $E^{-1}F^{-1}G^{-1}$*, and it is the key to elimination. It is the link between the A we start with and the U we reach. It is called L, because it is *lower triangular*. But it also has a special property that can be seen only by multiplying it out:

$$\begin{bmatrix} 1 & & \\ 2 & 1 & \\ & & 1 \end{bmatrix} \begin{bmatrix} 1 & & \\ & 1 & \\ -1 & & 1 \end{bmatrix} \begin{bmatrix} 1 & & \\ & 1 & \\ & -1 & 1 \end{bmatrix} = \begin{bmatrix} 1 & & \\ 2 & 1 & \\ -1 & -1 & 1 \end{bmatrix} = L. \tag{6}$$

Certainly L is lower triangular, with 1's on the main diagonal and 0's above. The special thing is that *the entries below the diagonal are exactly the multipliers $l = 2$*, -1, and -1. Normally we expect, when matrices are multiplied, that there is no direct way to read off the answer. Here the matrices come in just the right order so that their product can be written down immediately. If the computer stores each multiplier l_{ij}—the number that multiplies the pivot row when it is subtracted from row i, and produces a zero in the i, j position—then these multipliers give a complete record of elimination. *They fit right into the matrix L that takes U back to A.*

1H (***Triangular factorization*** $A = LU$): If no row exchanges are required, the original matrix A can be written as a product $A = LU$. The matrix L is lower triangular, with 1's on the diagonal and the multipliers l_{ij} (taken from elimination) below the diagonal. U is the upper triangular matrix which appears after forward elimination and before back-substitution; its diagonal entries are the pivots.

---

† Most matrices are not so easy to invert! We present inverses more systematically in the next section, but I think it's not bad to see the simplest case here.

EXAMPLE 1

$$A = \begin{bmatrix} 1 & 2 \\ 3 & 4 \end{bmatrix} \quad \text{goes to} \quad U = \begin{bmatrix} 1 & 2 \\ 0 & -2 \end{bmatrix} \quad \text{with} \quad L = \begin{bmatrix} 1 & 0 \\ 3 & 1 \end{bmatrix}.$$

EXAMPLE 2   (which needs a row exchange)

$$A = \begin{bmatrix} 0 & 2 \\ 3 & 4 \end{bmatrix} \quad \text{cannot be factored into} \quad A = LU$$

EXAMPLE 3   (with all pivots and multipliers equal to 1)

$$A = \begin{bmatrix} 1 & 1 & 1 \\ 1 & 2 & 2 \\ 1 & 2 & 3 \end{bmatrix} = \begin{bmatrix} 1 & 0 & 0 \\ 1 & 1 & 0 \\ 1 & 1 & 1 \end{bmatrix} \begin{bmatrix} 1 & 1 & 1 \\ 0 & 1 & 1 \\ 0 & 0 & 1 \end{bmatrix} = LU$$

There are subtractions from A to U, and additions from U to A.

EXAMPLE 4   (when U is the identity and L is the same as A)

$$A = \begin{bmatrix} 1 & 0 & 0 \\ l_{21} & 1 & 0 \\ l_{31} & l_{32} & 1 \end{bmatrix}.$$

The elimination steps are (i) E subtracts l_{21} times row 1 from row 2 (ii) F subtracts l_{31} times row 1 from row 3 (iii) G subtracts l_{32} times row 2 from row 3. The result is the identity matrix: in this example $U = I$. In the reverse direction, if our rule is correct, the inverses should bring back A:

$$E^{-1} \text{ applied to } (F^{-1} \text{ applied to } (G^{-1} \text{ applied to } I)) = A.$$

$$\begin{bmatrix} 1 & & \\ l_{21} & 1 & \\ & & 1 \end{bmatrix} \text{ times } \begin{bmatrix} 1 & & \\ & 1 & \\ l_{31} & & 1 \end{bmatrix} \text{ times } \begin{bmatrix} 1 & & \\ & 1 & \\ & l_{32} & 1 \end{bmatrix} \text{ times } I = A.$$

The order is right to avoid interactions between the matrices. Note that parentheses were not necessary because of the associative law.

$A = LU$: The n by n case

That factorization $A = LU$ is so important that we ought to say more. It used to be missing in linear algebra courses, especially if they concentrated on the abstract side. Or maybe it was thought to be too hard—but you have got it. If the last Example 4 is pushed one step further, to allow any U instead of the particular case $U = I$, it becomes possible to see how the rule works in general. Then we need to say how this LU factorization is used in practice.

The rule is that *the matrix L, applied to U, brings back A*:

$$
\begin{bmatrix} 1 & 0 & 0 \\ l_{21} & 1 & 0 \\ l_{31} & l_{32} & 1 \end{bmatrix}
\begin{bmatrix} \text{row 1 of } U \\ \text{row 2 of } U \\ \text{row 3 of } U \end{bmatrix} = A.
\tag{7}
$$

The proof is to *apply the steps of elimination*. On the right side they will take A to U. We show that they do the same to the left side; they wipe out L. Since both sides of (7) lead to the same U, and the steps to get there are all reversible, those two sides are equal and (7) is correct.

> The first step subtracts l_{21} times $(1, 0, 0)$ from the second row, which removes l_{21}. The second operation subtracts l_{31} times row 1 from row 3. The third subtracts l_{32} times the *new* row 2 (which is 0, 1, 0) from row 3. The order is right, and we now have the identity matrix. The associative law allows us to carry along the matrix U in all those multiplications. At the end, when L has been changed to I, only U is left. This confirms that both sides of (7) end up equal to U—so they must be equal now.

This factorization $A = LU$ is so crucial, and so beautiful, that the exercises suggest a second approach. We are writing down 3 by 3 matrices, but you can see how the arguments apply to larger matrices. A third proof reaches $A = LU$ by induction, reducing every n by n problem to the next lower order $n - 1$. My textbook *Introduction to Applied Mathematics* shows how elimination peels off a simple matrix at every step, and the sum of those matrices is exactly $A = LU$. Here we give one more example and then put $A = LU$ to use.

EXAMPLE   ($A = LU$, with zeros in the empty spaces)

$$
A = \begin{bmatrix} 1 & -1 & & \\ -1 & 2 & -1 & \\ & -1 & 2 & -1 \\ & & -1 & 2 \end{bmatrix} = \begin{bmatrix} 1 & & & \\ -1 & 1 & & \\ & -1 & 1 & \\ & & -1 & 1 \end{bmatrix}\begin{bmatrix} 1 & -1 & & \\ & 1 & -1 & \\ & & 1 & -1 \\ & & & 1 \end{bmatrix}.
$$

That shows how a matrix with three nonzero diagonals has factors with two nonzero diagonals. This example comes from an important problem in differential equations (Section 1.7).

One Linear System = Two Triangular Systems

There is a serious practical point about $A = LU$. It is more than just a record of elimination steps; it also gives the right matrices to complete the solution of $Ax = b$. In fact A could be thrown away, when L and U are known. We go from b to c by forward elimination (that uses L) and we go from c to x by back-substitution (that uses U). We can and should do without A, when its factors have been found.

In matrix terms, elimination splits $Ax = b$ into *two triangular systems*:

$$first \quad Lc = b \quad and \ then \quad Ux = c. \tag{8}$$

This is identical to $Ax = b$. *Multiply the second equation by* L *to give* $LUx = Lc$, *which is* $Ax = b$. Each triangular system is quickly solved. That is exactly what elimination does, and a good code does it in that order:

1. ***Factor*** (from A find L and U)
2. ***Solve*** (from L and U and b find x)

The separation of these steps means that a whole series of right hand sides can be processed. The subroutine ***solve*** obeys equation (8). It solves the two triangular systems in $n^2/2$ steps each. ***The solution for any new right side*** b' ***can be found in only*** n^2 ***operations***. That is far below the $n^3/3$ steps needed to factor A on the left hand side.

EXAMPLE   (the previous example with a right side b)

$$Ax = b \qquad \begin{matrix} x_1 - & x_2 & & = 1 \\ -x_1 + & 2x_2 - & x_3 & = 1 \\ & - x_2 + & 2x_3 - & x_4 = 1 \\ & & - x_3 + & 2x_4 = 1 \end{matrix} \qquad \text{splits into two systems}$$

$$Lc = b \qquad \begin{matrix} c_1 & & & = 1 \\ -c_1 + & c_2 & & = 1 \\ & - c_2 + & c_3 & = 1 \\ & & - c_3 + & c_4 = 1 \end{matrix} \qquad \text{which gives} \quad c = \begin{bmatrix} 1 \\ 2 \\ 3 \\ 4 \end{bmatrix}$$

$$Ux = c \qquad \begin{matrix} x_1 - x_2 & & = 1 \\ x_2 - x_3 & & = 2 \\ x_3 - x_4 & = 3 \\ x_4 & = 4 \end{matrix} \qquad \text{which gives} \quad x = \begin{bmatrix} 10 \\ 9 \\ 7 \\ 4 \end{bmatrix}$$

For these special "band matrices" the operation count drops down to $2n$. You see how $Lc = b$ is solved *forward*; it is precisely what happens during elimination. Then $Ux = c$ is solved by back-substitution—as always.

Remark 1   The LU form is "unsymmetric" in one respect: U has the pivots along its main diagonal, where L always has 1's. This is easy to correct. We divide out of U a *diagonal matrix* D made up entirely of the pivots $d_1, d_2, \ldots, d_n$.

$$U = \begin{bmatrix} d_1 & & & \\ & d_2 & & \\ & & \cdot & \\ & & & d_n \end{bmatrix} \begin{bmatrix} 1 & u_{12}/d_1 & u_{13}/d_1 & \cdot \\ & 1 & u_{23}/d_2 & \cdot \\ & & \cdot & \cdot \\ & & & 1 \end{bmatrix}. \tag{9}$$

In the last example all pivots were $d_i = 1$. In that case D is the identity matrix and this present remark has no importance. But that was very exceptional, and normally LDU is different from LU.

> *The triangular factorization is often written $A = LDU$, where L and U have* 1's *on the diagonal and D is the diagonal matrix of pivots.*

It is conventional, although completely confusing, to go on denoting this new upper triangular matrix by the same letter U. Whenever you see LDU, it is understood that U has 1's on the diagonal—in other words that each row was divided by the pivot. Then L and U are treated evenly. An example is

$$A = \begin{bmatrix} 1 & 2 \\ 3 & 4 \end{bmatrix} = \begin{bmatrix} 1 & \\ 3 & 1 \end{bmatrix} \begin{bmatrix} 1 & 2 \\ & -2 \end{bmatrix} = \begin{bmatrix} 1 & \\ 3 & 1 \end{bmatrix} \begin{bmatrix} 1 & \\ & -2 \end{bmatrix} \begin{bmatrix} 1 & 2 \\ & 1 \end{bmatrix} = LDU.$$

That has the 1's on the diagonals of L and U, and the pivots 1 and -2 in D.

Remark 2   We may have given the impression, in describing each step of the elimination process, that there was no freedom to do the calculations in a different order. That is wrong. There is *some* freedom, and there is a "Crout algorithm" which arranges the calculations in a slightly different way. But there is certainly not complete freedom since row operations in a random order could easily destroy at one step the zeros that were created at a previous step. And also, *there is no freedom in the final L, D, and U*. That is our main point:

1I   If $A = L_1 D_1 U_1$ and $A = L_2 D_2 U_2$, where the L's are lower triangular with unit diagonal, the U's are upper triangular with unit diagonal, and the D's are diagonal matrices with no zeros on the diagonal, then $L_1 = L_2$, $D_1 = D_2$, $U_1 = U_2$. The LDU factorization and the LU factorization are uniquely determined by A.

The proof is a good exercise with inverse matrices, in the next section.

Row Exchanges and Permutation Matrices

We now have to face a problem that has so far been avoided: the number we expect to use as a pivot might be zero. This could occur in the middle of a calculation, or it can happen at the very beginning (in case $a_{11} = 0$). A simple example is

$$\begin{bmatrix} 0 & 2 \\ 3 & 4 \end{bmatrix} \begin{bmatrix} u \\ v \end{bmatrix} = \begin{bmatrix} b_1 \\ b_2 \end{bmatrix}.$$

The difficulty is clear; no multiple of the first equation will remove the coefficient 3.

The remedy is equally clear. ***Exchange the two equations***, moving the entry 3 up into the pivot. In this simple case the matrix would then be upper triangular already, and the system

$$3u + 4v = b_2$$
$$2v = b_1$$

can be solved immediately by back-substitution.

To express this in matrix terms, we need to find the **permutation matrix** that produces the row exchange. It is

$$P = \begin{bmatrix} 0 & 1 \\ 1 & 0 \end{bmatrix},$$

and multiplying by P does exchange the rows:

$$PA = \begin{bmatrix} 0 & 1 \\ 1 & 0 \end{bmatrix} \begin{bmatrix} 0 & 2 \\ 3 & 4 \end{bmatrix} = \begin{bmatrix} 3 & 4 \\ 0 & 2 \end{bmatrix}.$$

P has the same effect on b, exchanging b_1 and b_2; the new system is $PAx = Pb$. The unknowns u and v are *not* reversed in a row exchange.

Now we go to a more difficult case. Suppose A is a 3 by 3 matrix of the form

$$A = \begin{bmatrix} 0 & a & b \\ 0 & 0 & c \\ d & e & f \end{bmatrix}.$$

A zero in the pivot location raises two possibilities: *The trouble may be easy to fix, or it may be serious.* This is decided by looking *below the zero*. If there is a nonzero entry lower down in the same column, then a row exchange is carried out; the nonzero entry becomes the needed pivot, and elimination can get going again.† In our case everything depends on the number d. If $d = 0$, the problem is incurable and the matrix is **singular**. There is no hope for a unique solution. If d is *not* zero, an exchange of rows 1 and 3 will move d into the pivot, and stage 1 is complete. However the next pivot position also contains a zero. The number a is now below it (the e above it is useless), and if a is not zero then another row exchange is called for. The first permutation P_{13} and this second permutation P_{23} are produced by the matrices

$$P_{13} = \begin{bmatrix} 0 & 0 & 1 \\ 0 & 1 & 0 \\ 1 & 0 & 0 \end{bmatrix} \quad \text{and} \quad P_{23} = \begin{bmatrix} 1 & 0 & 0 \\ 0 & 0 & 1 \\ 0 & 1 & 0 \end{bmatrix}.$$

They can be found by a neat trick: apply the permutation to the identity matrix. For example $P_{13}I = P_{13}$ must be the identity matrix with rows 1 and 3 reversed.

---

† In practice, we also consider a row exchange when the original pivot is *near* zero—even if it is not exactly zero. Choosing a larger pivot reduces the roundoff error.

One more point: There is a permutation matrix that will do both of the row exchanges at once. It is the product of the two separate permutations (with P_{13} acting first!):

$$P_{23}P_{13} = \begin{bmatrix} 1 & 0 & 0 \\ 0 & 0 & 1 \\ 0 & 1 & 0 \end{bmatrix} \begin{bmatrix} 0 & 0 & 1 \\ 0 & 1 & 0 \\ 1 & 0 & 0 \end{bmatrix} = \begin{bmatrix} 0 & 0 & 1 \\ 1 & 0 & 0 \\ 0 & 1 & 0 \end{bmatrix} = P.$$

If we had known what to do, we could have multiplied our matrix by P in the first place. Then elimination would have no difficulty with PA; it was only that the original order was unfortunate. With the rows in the right order,

$$PA = \begin{bmatrix} d & e & f \\ 0 & a & b \\ 0 & 0 & c \end{bmatrix} \quad \text{is already triangular.}$$

The fact that it is triangular is an accident; usually there is elimination left to do. But the main point is this: If elimination can be completed with the help of row exchanges, then we can imagine that those exchanges are done ahead of time. That produces PA, and *this matrix will not need row exchanges*. Elimination works with whole rows, and now they are in an order that produces nonzeros in the pivot positions. In other words, PA allows the standard factorization into L times U. The theory of Gaussian elimination can be summarized as follows:

1J   In the *nonsingular* case, there is a permutation matrix P that reorders the rows of A to avoid zeros in the pivot positions. In this case

(1) $Ax = b$ has a unique solution
(2)   it is found by elimination with row exchanges
(3)   with the rows reordered in advance, PA can be factored into LU.

In the *singular* case, no reordering can produce a full set of pivots.

Note that *a permutation matrix P has the same rows as the identity*, in some order. In fact $P = I$ is the simplest and most common permutation matrix (it exchanges nothing). The product of two permutation matrices is another permutation. And they need not commute: P_{13} times P_{23}, multiplied in the opposite order from the example above, would have given a different P.

Remark   You have to be careful with L. Suppose elimination subtracts row 1 from row 2, creating $l_{21} = 1$. Then suppose it exchanges rows 2 and 3. If that exchange is done in advance, the multiplier will change to $l_{31} = 1$ in $PA = LU$.†

---

† Algebraists tell me that P should go between L and U, keeping $l_{21} = 1$, but it's too late.

EXAMPLE $A = \begin{bmatrix} 1 & 1 & 1 \\ 1 & 1 & 3 \\ 2 & 5 & 8 \end{bmatrix} \rightarrow \begin{bmatrix} 1 & 1 & 1 \\ 0 & 0 & 2 \\ 0 & 3 & 6 \end{bmatrix} \rightarrow \begin{bmatrix} 1 & 1 & 1 \\ 0 & 3 & 6 \\ 0 & 0 & 2 \end{bmatrix} = U.$

With the rows exchanged, we recover LU—but now $l_{31} = 1$ and $l_{21} = 2$:

$$P = \begin{bmatrix} 1 & 0 & 0 \\ 0 & 0 & 1 \\ 0 & 1 & 0 \end{bmatrix} \quad \text{and} \quad L = \begin{bmatrix} 1 & 0 & 0 \\ 2 & 1 & 0 \\ 1 & 0 & 1 \end{bmatrix} \quad \text{and} \quad PA = LU.$$

To summarize: A good code for Gaussian elimination keeps a record of L and U and P. Those matrices carry the information that originally came in A—and they carry it in a more usable form. They allow the solution of $Ax = b$ from two triangular systems. They are the practical equivalent of the calculation we do next—*to find the inverse matrix and the solution* $x = A^{-1}b$.

EXERCISES

1.5.1   When is an upper triangular matrix nonsingular?

1.5.2   What multiple of row 2 of A will elimination subtract from row 3?

$$A = \begin{bmatrix} 1 & 0 & 0 \\ 2 & 1 & 0 \\ 1 & 4 & 1 \end{bmatrix} \begin{bmatrix} 5 & 7 & 8 \\ 0 & 2 & 3 \\ 0 & 0 & 6 \end{bmatrix}.$$

What will be the pivots? Will a row exchange be required?

1.5.3   Multiply the matrix $L = E^{-1}F^{-1}G^{-1}$ in equation (6) by GFE in equation (3):

$$\begin{bmatrix} 1 & 0 & 0 \\ 2 & 1 & 0 \\ -1 & -1 & 1 \end{bmatrix} \quad \text{times} \quad \begin{bmatrix} 1 & 0 & 0 \\ -2 & 1 & 0 \\ -1 & 1 & 1 \end{bmatrix}.$$

Multiply also in the opposite order. *Why are the answers what they are?*

1.5.4   Apply elimination to produce the factors L and U for

$$A = \begin{bmatrix} 2 & 1 \\ 8 & 7 \end{bmatrix} \quad \text{and} \quad A = \begin{bmatrix} 3 & 1 & 1 \\ 1 & 3 & 1 \\ 1 & 1 & 3 \end{bmatrix} \quad \text{and} \quad A = \begin{bmatrix} 1 & 1 & 1 \\ 1 & 4 & 4 \\ 1 & 4 & 8 \end{bmatrix}.$$

1.5.5   Factor A into LU, and write down the upper triangular system $Ux = c$ which appears after elimination, for

$$Ax = \begin{bmatrix} 2 & 3 & 3 \\ 0 & 5 & 7 \\ 6 & 9 & 8 \end{bmatrix} \begin{bmatrix} u \\ v \\ w \end{bmatrix} = \begin{bmatrix} 2 \\ 2 \\ 5 \end{bmatrix}.$$

1.5.6   Find E^2 and E^8 and E^{-1} if

$$E = \begin{bmatrix} 1 & 0 \\ 6 & 1 \end{bmatrix}.$$

1.5.7   Find the products FGH and HGF if (with upper triangular zeros omitted)

$$F = \begin{bmatrix} 1 & & & \\ 2 & 1 & & \\ 0 & 0 & 1 & \\ 0 & 0 & 0 & 1 \end{bmatrix}, \quad G = \begin{bmatrix} 1 & & & \\ 0 & 1 & & \\ 0 & 2 & 1 & \\ 0 & 0 & 0 & 1 \end{bmatrix}, \quad H = \begin{bmatrix} 1 & & & \\ 0 & 1 & & \\ 0 & 0 & 1 & \\ 0 & 0 & 2 & 1 \end{bmatrix}.$$

1.5.8   (*Second proof of $A = LU$*) The third row of U comes from the third row of A by subtracting multiples of rows 1 and 2 (*of U!*):

   row 3 of U = row 3 of $A - l_{31}$(row 1 of U) $- l_{32}$(row 2 of U).

(a)   Why are rows of U subtracted off and not rows of A? Answer: Because by the time a pivot row is used, ...
(b)   The equation above is the same as

   row 3 of $A = l_{31}$(row 1 of U) $+ l_{32}$(row 2 of U) $+ 1$(row 3 of U).

Which rule for matrix multiplication (shaded in 1D) makes this exactly row 3 of L times U?
   The other rows of LU agree similarly with the rows of A.

1.5.9   (a)   Under what conditions is A nonsingular, if A is the product

$$A = \begin{bmatrix} 1 & 0 & 0 \\ -1 & 1 & 0 \\ 0 & -1 & 1 \end{bmatrix} \begin{bmatrix} d_1 & & \\ & d_2 & \\ & & d_3 \end{bmatrix} \begin{bmatrix} 1 & -1 & 0 \\ 0 & 1 & -1 \\ 0 & 0 & 1 \end{bmatrix}?$$

(b)   Solve the system $Ax = b$ starting with $Lc = b$:

$$\begin{bmatrix} 1 & 0 & 0 \\ -1 & 1 & 0 \\ 0 & -1 & 1 \end{bmatrix} \begin{bmatrix} c_1 \\ c_2 \\ c_3 \end{bmatrix} = \begin{bmatrix} 0 \\ 0 \\ 1 \end{bmatrix} = b.$$

1.5.10   (a)   Why does it take approximately $n^2/2$ multiplication-subtraction steps to solve each of $Lc = b$ and $Ux = c$?
(b)   How many steps does elimination use in solving 10 systems with the same 60 by 60 coefficient matrix A?

1.5.11   Solve $LUx = \begin{bmatrix} 1 & 0 & 0 \\ 1 & 1 & 0 \\ 1 & 0 & 1 \end{bmatrix} \begin{bmatrix} 2 & 4 & 4 \\ 0 & 1 & 2 \\ 0 & 0 & 1 \end{bmatrix} \begin{bmatrix} u \\ v \\ w \end{bmatrix} = \begin{bmatrix} 2 \\ 0 \\ 2 \end{bmatrix}$

without multiplying LU to find A.

1.5.12 How could you factor A into a product UL, upper triangular times lower triangular? Would they be the same factors as in $A = LU$?

1.5.13 Solve by elimination, exchanging rows when necessary:

$$
\begin{aligned}
u + 4v + 2w &= -2 \\
-2u - 8v + 3w &= 32 \\
v + w &= 1
\end{aligned}
\qquad \text{and} \qquad
\begin{aligned}
v + w &= 0 \\
u + v \quad &= 0 \\
u + v + w &= 1
\end{aligned}
$$

Which permutation matrices are required?

1.5.14 Write down all six of the 3 by 3 permutation matrices, including $P = I$. Identify their inverses, which are also permutation matrices—they satisfy $PP^{-1} = I$ and are on the same list.

1.5.15 Find the $PA = LDU$ factorizations (and check them) for

$$
A = \begin{bmatrix} 0 & 1 & 1 \\ 1 & 0 & 1 \\ 2 & 3 & 4 \end{bmatrix}
\qquad \text{and} \qquad
A = \begin{bmatrix} 1 & 2 & 1 \\ 2 & 4 & 2 \\ 1 & 1 & 1 \end{bmatrix}.
$$

1.5.16 Find a nonsingular 4 by 4 matrix that requires three row exchanges to reach the end of elimination. If possible, let the example be a permutation matrix.

1.5.17 In the LPU order that algebraists prefer, elimination exchanges rows only at the end:

$$
A = \begin{bmatrix} 1 & 1 & 1 \\ 1 & 1 & 3 \\ 2 & 5 & 8 \end{bmatrix} \rightarrow \begin{bmatrix} 1 & 1 & 1 \\ 0 & 0 & 2 \\ 0 & 3 & 6 \end{bmatrix} = PU = \begin{bmatrix} 1 & 0 & 0 \\ 0 & 0 & 1 \\ 0 & 1 & 0 \end{bmatrix} \begin{bmatrix} 1 & 1 & 1 \\ 0 & 3 & 6 \\ 0 & 0 & 2 \end{bmatrix}.
$$

What is L is this case? Unlike $PA = LU$ and the example after 1J, the multipliers stay in place (l_{21} is 1 and l_{31} is 2).

1.5.18 Decide whether the following systems are singular or nonsingular, and whether they have no solution, one solution, or infinitely many solutions:

$$
\begin{aligned}
v - w &= 2 \\
u - v \quad &= 2 \\
u \quad - w &= 2
\end{aligned}
\qquad \text{and} \qquad
\begin{aligned}
v - w &= 0 \\
u - v \quad &= 0 \\
u \quad - w &= 0
\end{aligned}
\qquad \text{and} \qquad
\begin{aligned}
v + w &= 1 \\
u + v \quad &= 1 \\
u \quad + w &= 1
\end{aligned}
$$

1.5.19 Which values of a, b, c lead to row exchanges, and which make the matrices singular?

$$
A = \begin{bmatrix} 1 & 2 & 0 \\ a & 8 & 3 \\ 0 & b & 5 \end{bmatrix}
\qquad \text{and} \qquad
A = \begin{bmatrix} c & 2 \\ 6 & 4 \end{bmatrix}.
$$

1.6 ■ INVERSES AND TRANSPOSES

The inverse of an n by n matrix is another n by n matrix. If the first matrix is A, its inverse is written A^{-1} (and pronounced "A inverse"). The fundamental property is simple: *If you multiply by A and then multiply by A^{-1}, you are back where you started*:

$$\text{if}\quad Ax = b \quad \text{then}\quad A^{-1}b = x.$$

Thus $A^{-1}Ax = x$. The matrix A times the matrix A^{-1} is the identity matrix. But *not all matrices have inverses*. The problem comes when Ax is zero, but x is non-zero. The inverse would have to get back from Ax to x. No matrix can multiply the zero vector and produce a nonzero vector—in this case A^{-1} will not exist. Our goals are to define the inverse and compute it and use it, in the cases when it exists—and then to understand which matrices have inverses.

1K   The matrix A is ***invertible*** if there exists a matrix B such that $BA = I$ and $AB = I$. There is at most one such B, called the inverse of A and denoted by A^{-1}:

$$A^{-1}A = I \quad \text{and} \quad AA^{-1} = I. \tag{1}$$

Note 1   There could not be two different inverses, because if $BA = I$ and $AC = I$ then we have

$$B = B(AC) = (BA)C = C.$$

If there is a left-inverse and a right-inverse (B multiplies from the left and C from the right) then they are the same.

Note 2   The inverse of A^{-1} is A itself. It satisfies equation (1)!

Note 3   A 1 by 1 matrix is invertible when it is not zero: if $A = [a]$ then $A^{-1} = [1/a]$. The inverse of a 2 by 2 matrix can be written down once and for all (*provided $ad - bc$ is not zero*):

$$\begin{bmatrix} a & b \\ c & d \end{bmatrix}^{-1} = \frac{1}{ad - bc} \begin{bmatrix} d & -b \\ -c & a \end{bmatrix}.$$

Note 4   A diagonal matrix is invertible when none of its diagonal entries are zero:

$$\text{if}\quad A = \begin{bmatrix} d_1 & & \\ & \ddots & \\ & & d_n \end{bmatrix} \quad \text{then}\quad A^{-1} = \begin{bmatrix} 1/d_1 & & \\ & \ddots & \\ & & 1/d_n \end{bmatrix}.$$

Note 5   The 2 by 2 matrix $A = \begin{bmatrix} 1 & 1 \\ 1 & 1 \end{bmatrix}$ is not invertible. The columns of B times A are certain to be the same, and they cannot be the two columns of I—which are different. (You will see other and better reasons why this A is not invertible).

When two matrices are involved, not much can be done about the inverse of $A + B$. The sum might or might not be invertible, independent of the separate invertibility of A and B. Instead, it is the inverse of their *product AB* which is the key formula in matrix computations. For ordinary numbers the situation is the same: $(a + b)^{-1}$ is hard to simplify, while $1/ab$ splits into $1/a$ times $1/b$. But for matrices *the order of multiplication must be correct*—if $ABx = y$ then $Bx = A^{-1}y$ and $x = B^{-1}A^{-1}y$, which shows how the inverses come in reverse order.

1L   A product AB of invertible matrices has an inverse. It is found by multiplying the individual inverses in reverse order:

$$(AB)^{-1} = B^{-1}A^{-1}. \tag{2}$$

Proof   To show that $B^{-1}A^{-1}$ is the inverse of AB, we multiply them and use the associative law to remove parentheses:

$$(AB)(B^{-1}A^{-1}) = ABB^{-1}A^{-1} = AIA^{-1} = AA^{-1} = I$$
$$(B^{-1}A^{-1})(AB) = B^{-1}A^{-1}AB = B^{-1}IB = B^{-1}B = I.$$

A similar rule holds with three or more matrices:

$$(ABC)^{-1} = C^{-1}B^{-1}A^{-1}.$$

We saw the change of order earlier, when the elementary matrices E, F, G in elimination were inverted to come back from U to A. In the forward direction, $GFEA$ was U. In the backward direction, $L = E^{-1}F^{-1}G^{-1}$ was the product of the inverses. Since G was last to be applied to A, G^{-1} comes first.

A matrix equation like $GFEA = U$ can be multiplied through by G^{-1} (on the left) or A^{-1} (on the right) or U^{-1} (on either side). If we needed A^{-1}, it would be $U^{-1}GFE$. Please check that, it's easy to get wrong.

The Calculation of A^{-1}

Consider the equation $AA^{-1} = I$. If it is taken *a column at a time*, that equation determines the columns of A^{-1}. The first column of A^{-1} is multiplied by A, to yield the first column of the identity: $Ax_1 = e_1$. Similarly $Ax_2 = e_2$ and $Ax_3 = e_3$; the e's are the columns of I. In a 3 by 3 example, A times A^{-1} is

$$\begin{bmatrix} 2 & 1 & 1 \\ 4 & -6 & 0 \\ -2 & 7 & 2 \end{bmatrix} \begin{bmatrix} x_1 & x_2 & x_3 \end{bmatrix} = \begin{bmatrix} e_1 & e_2 & e_3 \end{bmatrix} = \begin{bmatrix} 1 & 0 & 0 \\ 0 & 1 & 0 \\ 0 & 0 & 1 \end{bmatrix}. \tag{3}$$

Thus we have three systems of equations (or n systems) and they all have the same coefficient matrix A. The right sides are different, but it is possible to carry out elimination *on all systems simultaneously*. This is called the ***Gauss-Jordan method***. Instead of stopping at U and switching to back-substitution, it continues by subtracting multiples of a row *from the rows above*. It produces zeros above the diagonal as well as below, and when it reaches the identity matrix we have found A^{-1}.

The example keeps all three columns e_1, e_2, e_3, and operates on rows of length six:

EXAMPLE OF THE GAUSS-JORDAN METHOD TO FIND A^{-1}

$$[A \quad e_1 \quad e_2 \quad e_3] = \begin{bmatrix} 2 & 1 & 1 & 1 & 0 & 0 \\ 4 & -6 & 0 & 0 & 1 & 0 \\ -2 & 7 & 2 & 0 & 0 & 1 \end{bmatrix}$$

$$\rightarrow \begin{bmatrix} 2 & 1 & 1 & 1 & 0 & 0 \\ 0 & -8 & -2 & -2 & 1 & 0 \\ 0 & 8 & 3 & 1 & 0 & 1 \end{bmatrix}$$

$$\rightarrow \begin{bmatrix} 2 & 1 & 1 & 1 & 0 & 0 \\ 0 & -8 & -2 & -2 & 1 & 0 \\ 0 & 0 & 1 & -1 & 1 & 1 \end{bmatrix} = [U \quad L^{-1}].$$

This completes the forward elimination. In the last matrix, the first three columns give the familiar upper triangular U. The other three columns, which are the three right-hand sides after they have been prepared for back-substitution, are the same as L^{-1}. (This is the effect of applying the elementary operations to the identity matrix: $L^{-1} = GFE$.) The first half of elimination has gone from A to U, and now the second half will go from U to I. Creating zeros *above* the pivots in the last matrix, we reach A^{-1}:

$$[U \quad L^{-1}] \rightarrow \begin{bmatrix} 2 & 1 & 0 & 2 & -1 & -1 \\ 0 & -8 & 0 & -4 & 3 & 2 \\ 0 & 0 & 1 & -1 & 1 & 1 \end{bmatrix}$$

$$\rightarrow \begin{bmatrix} 2 & 0 & 0 & \frac{12}{8} & -\frac{5}{8} & -\frac{6}{8} \\ 0 & -8 & 0 & -4 & 3 & 2 \\ 0 & 0 & 1 & -1 & 1 & 1 \end{bmatrix}$$

$$\rightarrow \begin{bmatrix} 1 & 0 & 0 & \frac{12}{16} & -\frac{5}{16} & -\frac{6}{16} \\ 0 & 1 & 0 & \frac{4}{8} & -\frac{3}{8} & -\frac{2}{8} \\ 0 & 0 & 1 & -1 & 1 & 1 \end{bmatrix} = [I \quad A^{-1}].$$

At the last step, we divided through by the pivots. The coefficient matrix in the left half became the identity. Since A went to I, the same operations on the right half must have carried I into A^{-1}. Therefore we have computed the inverse.

A note for the future: You can see the determinant -16 appearing in the denominators of A^{-1}. It is the product of the pivots $(2)(-8)(1)$, and it enters at the end when the rows are divided by the pivots.

In spite of this tremendous success in computing A^{-1}, I do not recommend it. I admit that the inverse solves $Ax = b$ in one step instead of two:

$$x = A^{-1}b \quad \text{instead of} \quad Lc = b \quad \text{and} \quad Ux = c.$$

We could go further and write $c = L^{-1}b$ and $x = U^{-1}c = U^{-1}L^{-1}b$. But note that we did not explicitly form, and in actual computation *should not form*, these matrices L^{-1} and U^{-1}. It would be a waste of time, since it would still take the same $n^2/2$ operations to multiply c by U^{-1} or b by L^{-1}. A similar remark applies to A^{-1}; the multiplication $A^{-1}b$ would still take n^2 steps. *It is the solution that we want, and not all the entries in the inverse.*

Remark 1 Purely out of curiosity, we might count the number of operations required to find A^{-1}. The normal count for each new right-hand side is n^2, half in the forward direction and half in back-substitution. With n different right-hand sides this makes n^3, and after including the $n^3/3$ operations on A itself, the total would be $4n^3/3$.

This result is a little too high, however, because of the special form of the right sides e_j. In forward elimination, changes occur only below the 1 in the jth place. This part has only $n - j$ components, so the count for e_j is effectively changed to $(n - j)^2/2$. Summing over all j, the total for forward elimination is $n^3/6$. This is to be combined with the usual $n^3/3$ operations that are applied to A, and the $n(n^2/2)$ back-substitution steps that finally produce the x_j—whether done separately, or simultaneously in the Gauss-Jordan method. *The final operation count for computing A^{-1} is*

$$\frac{n^3}{6} + \frac{n^3}{3} + n\left(\frac{n^2}{2}\right) = n^3.$$

This count is remarkably low. In fact, since matrix multiplication already takes n^3 steps, it requires as many operations to compute A^2 as it does to compute A^{-1}. That fact seems almost unbelievable (and computing A^3 requires twice as many, as far as we can see). Nevertheless, if A^{-1} is not needed, it should not be computed.

Remark 2 In the Gauss-Jordan calculation we went all the way forward to U, before starting backward to produce zeros above the diagonal. That is like Gaussian elimination, but other orders are possible. We could have used the second pivot when we were there earlier, to create a zero above it as well as below it. This is not smart. At that time the second row is virtually full, whereas near the end it has zeros from the upward row operations that have already taken place.

Invertible = Nonsingular

Ultimately we want to know which matrices are invertible and which are not. This is so important a question that it has several answers. In fact, each of the first five chapters will give a different (but equivalent) test for invertibility. Sometimes the tests extend to rectangular matrices and one-sided inverses: Chapter 2 requires independent rows or independent columns, and Chapter 3 inverts AA^T or A^TA. The other chapters allow only square matrices, and look for **nonzero determinants** or **nonzero eigenvalues** or **nonzero pivots**. This last test is the one we meet through Gaussian elimination. We want to show (in a few more theoretical paragraphs) that it succeeds.

Suppose a matrix has a full set of n pivots—by definition it is nonsingular. (We mention again that pivots are nonzero.) The equation $AA^{-1} = I$ gives n separate systems $Ax_i = e_i$ for the columns of A^{-1}, and they can be solved—by elimination or by Gauss-Jordan. Row exchanges may be necessary, but the columns of A^{-1} are uniquely determined.

Strictly speaking, we have to show that the matrix A^{-1} with those columns is also a *left*-inverse. Solving $AA^{-1} = I$ has at the same time solved $A^{-1}A = I$, but why? That is not a trivial question. One argument looks ahead to the next chapter: a 1-sided inverse of a square matrix is automatically a 2-sided inverse. But we can do without Chapter 2, by a simple observation: *Every Gauss-Jordan step is a multiplication on the left by an elementary matrix.* We are allowing three types of elementary matrices:

1) E_{ij}, to subtract a multiple l of row j from row i
2) P_{ij}, to exchange rows i and j
3) D (or D^{-1}), to divide all rows by their pivots.

The Gauss-Jordan process is really a giant sequence of matrix multiplications:

$$(D^{-1} \cdots E \cdots P \cdots E)A = I. \tag{4}$$

That matrix in parentheses, to the left of A, is evidently a left-inverse! It exists, it must equal the right-inverse by Note 1, and thus **every nonsingular matrix is invertible.**

·The converse is also true: **Every invertible matrix is nonsingular.** If A is invertible, it has n pivots. In an extreme case that is clear: If A has an inverse, it cannot have a whole column of zeros. (The inverse could never multiply a column of zeros to produce a column of the identity.) In a less extreme case, suppose elimination starts on an invertible matrix A but breaks down at

$$A' = \begin{bmatrix} d_1 & x & x & x & x \\ 0 & d_2 & x & x & x \\ 0 & 0 & 0 & x & x \\ 0 & 0 & 0 & x & x \\ 0 & 0 & 0 & x & x \end{bmatrix}.$$

This matrix cannot have an inverse, no matter what the x's are. One proof is to use column operations (for the first time?) to make the whole third column zero. Subtracting the right multiple of column 2, and then of column 1, we reach a matrix that is certainly not invertible. Therefore the original A was not invertible, and we have a contradiction. When we start with an invertible matrix, elimination cannot break down.

Elimination gives a reliable test for the existence of A^{-1}: There must be n pivots.

A square matrix is invertible if and only if it is nonsingular.

Chapter 2 will use row operations on A', to reach a row of zeros and the same conclusion. For some right sides, the system $A'x = b$ has no solution. For an invertible matrix, there is always the unique solution $x = A^{-1}b$.

Transposes

We need one more matrix, and fortunately it is much simpler than the inverse. It is called the *transpose* of A, and is denoted by A^T. Its columns are taken directly from the rows of A—the ith *row of* A *becomes the* ith *column of* A^T—so it can be constructed without any calculations:

$$\text{If} \quad A = \begin{bmatrix} 2 & 1 & 4 \\ 0 & 0 & 3 \end{bmatrix} \quad \text{then} \quad A^T = \begin{bmatrix} 2 & 0 \\ 1 & 0 \\ 4 & 3 \end{bmatrix}.$$

At the same time the columns of A become the rows of A^T. If A is an m by n matrix, then A^T is n by m. The final effect is to flip the matrix across its main diagonal, and the entry in row i, column j of A^T comes from row j, column i of A:

$$(A^T)_{ij} = A_{ji}. \tag{5}$$

The transpose of a lower triangular matrix is upper triangular. The transpose of A^T brings us back to A.

Now suppose there are two matrices A and B. If we add them and then transpose, the result is the same as first transposing and then adding: $(A + B)^T$ is the same as $A^T + B^T$. But it is not so clear what is the transpose of a product AB or an inverse A^{-1}, and those are the essential formulas of this section:

1M   (i)   The transpose of AB is $(AB)^T = B^T A^T$.

       (ii)   The transpose of A^{-1} is $(A^{-1})^T = (A^T)^{-1}$.

Notice how the formula for $(AB)^T$ resembles the one for $(AB)^{-1}$. In both cases we reverse the order, giving $B^T A^T$ and $B^{-1}A^{-1}$. The proof for the inverse was easy, but this one requires an unnatural patience with matrix multiplication. The first row of $(AB)^T$ is the first column of AB, which means that the columns of A are weighted by the first column of B. Staying with rows, this amounts to the rows of A^T weighted by the first row of B^T. That is exactly the first row of $B^T A^T$.

The other rows of $(AB)^T$ and $B^T A^T$ also agree.

EXAMPLE

$$AB = \begin{bmatrix} 1 & 0 \\ 1 & 1 \end{bmatrix} \begin{bmatrix} 3 & 3 & 3 \\ 2 & 2 & 2 \end{bmatrix} = \begin{bmatrix} 3 & 3 & 3 \\ 5 & 5 & 5 \end{bmatrix}$$

$$B^T A^T = \begin{bmatrix} 3 & 2 \\ 3 & 2 \\ 3 & 2 \end{bmatrix} \begin{bmatrix} 1 & 1 \\ 0 & 1 \end{bmatrix} = \begin{bmatrix} 3 & 5 \\ 3 & 5 \\ 3 & 5 \end{bmatrix}.$$

To establish the formula for $(A^{-1})^T$, start from $AA^{-1} = I$ and $A^{-1}A = I$ and take transposes. On one side, $I^T = I$. On the other side, we know from part (i) the transpose of a product:

$$(A^{-1})^T A^T = I \qquad \text{and} \qquad A^T (A^{-1})^T = I. \tag{6}$$

This makes $(A^{-1})^T$ the inverse of A^T, proving (ii).

   With these rules established, we can introduce a special class of matrices, probably the most important class of all. *A symmetric matrix is a matrix which equals its own transpose*: $A^T = A$. The matrix is necessarily square, and each entry on one side of the diagonal equals its "mirror image" on the other side: $a_{ij} = a_{ji}$. Two simple examples are

$$A = \begin{bmatrix} 1 & 2 \\ 2 & 8 \end{bmatrix} \qquad \text{and} \qquad D = \begin{bmatrix} 3 & 0 \\ 0 & 5 \end{bmatrix}.$$

A symmetric matrix need not be invertible; it could even be a matrix of zeros. Nevertheless, *if A^{-1} exists it is also symmetric*. From formula (ii) above, the transpose of A^{-1} always equals $(A^T)^{-1}$; for a symmetric matrix this is just A^{-1}. Therefore, in this case A^{-1} equals its own transpose; it is symmetric whenever A is.

   Symmetric matrices appear in every subject whose laws are fair. "Each action has an equal and opposite reaction," and the entry which gives the action of i onto j is matched by the action of j onto i. We will see this symmetry in the next section, for differential equations. Here we stick to Gaussian elimination, and look to see what are the consequences of symmetry. Essentially, *it cuts the work in half*. The calculations above the diagonal duplicate the calculations below, and the upper triangular U is completely determined by the lower triangular L:

1N   If A is symmetric, and if it can be factored into $A = LDU$ without row exchanges to destroy the symmetry, then the upper triangular U is the transpose of the lower triangular L. *The factorization becomes $A = LDL^T$.*

For proof, take the transpose of $A = LDU$; the transposes come in reverse order to give $A^T = U^T D^T L^T$. Since A is symmetric it equals A^T, so we now have two factorizations of A into lower triangular times diagonal times upper triangular. (L^T is upper triangular with ones on the diagonal, exactly like U). According to 1I, such a factorization is unique. Therefore L^T must be identical to U, which completes the proof.

EXAMPLE   (with symmetric A, and $L^T = U$)

$$\begin{bmatrix} 1 & 2 \\ 2 & 8 \end{bmatrix} = \begin{bmatrix} 1 & 0 \\ 2 & 1 \end{bmatrix}\begin{bmatrix} 1 & 0 \\ 0 & 4 \end{bmatrix}\begin{bmatrix} 1 & 2 \\ 0 & 1 \end{bmatrix}.$$

After every stage of elimination, the matrix in the lower right corner remains symmetric—as it is after the first stage:

$$\begin{bmatrix} a & b & c \\ b & d & e \\ c & e & f \end{bmatrix} \rightarrow \begin{bmatrix} a & b & c \\ 0 & d - \dfrac{b^2}{a} & e - \dfrac{bc}{a} \\ 0 & e - \dfrac{bc}{a} & f - \dfrac{c^2}{a} \end{bmatrix}.$$

The work of elimination is cut essentially in half by symmetry, from $n^3/3$ to $n^3/6$. There is no need to store entries from both sides of the diagonal, or to store both L and U.

EXERCISES

1.6.1   Find the inverses (no special system required) of

$$A_1 = \begin{bmatrix} 0 & 2 \\ 3 & 0 \end{bmatrix}, \quad A_2 = \begin{bmatrix} 2 & 0 \\ 4 & 2 \end{bmatrix}, \quad A_3 = \begin{bmatrix} \cos\theta & -\sin\theta \\ \sin\theta & \cos\theta \end{bmatrix}.$$

1.6.2   (a)   Find the inverses of the permutation matrices

$$P = \begin{bmatrix} 0 & 0 & 1 \\ 0 & 1 & 0 \\ 1 & 0 & 0 \end{bmatrix} \quad \text{and} \quad P = \begin{bmatrix} 0 & 0 & 1 \\ 1 & 0 & 0 \\ 0 & 1 & 0 \end{bmatrix}.$$

(b)   Explain why for permutations P^{-1} is *always the same as* P^T, by using showing that the 1's are in the right places to give $PP^T = I$.

1.6.3   From $AB = C$ find a formula for A^{-1}. Do the same from $PA = LU$.

1.6.4   (a)   If A is invertible and $AB = AC$, prove quickly that $B = C$.
(b)   If $A = \begin{bmatrix} 1 & 0 \\ 0 & 0 \end{bmatrix}$, find an example with $AB = AC$ but $B \neq C$.

1.6.5   If the inverse of A^2 is B, show that the inverse of A is AB. (Thus A is invertible whenever A^2 is invertible.)

1.6.6   Use the Gauss-Jordan method to invert

$$A_1 = \begin{bmatrix} 1 & 0 & 0 \\ 1 & 1 & 1 \\ 0 & 0 & 1 \end{bmatrix}, \quad A_2 = \begin{bmatrix} 2 & -1 & 0 \\ -1 & 2 & -1 \\ 0 & -1 & 2 \end{bmatrix}, \quad A_3 = \begin{bmatrix} 0 & 0 & 1 \\ 0 & 1 & 1 \\ 1 & 1 & 1 \end{bmatrix}.$$

1.6.7 Find three 2 by 2 matrices, other than $A = I$ and $A = -I$, that are their own inverses: $A^2 = I$.

1.6.8 Show that $\begin{bmatrix} 1 & 1 \\ 3 & 3 \end{bmatrix}$ has no inverse by trying to solve

$$\begin{bmatrix} 1 & 1 \\ 3 & 3 \end{bmatrix} \begin{bmatrix} a & b \\ c & d \end{bmatrix} = \begin{bmatrix} 1 & 0 \\ 0 & 1 \end{bmatrix}.$$

1.6.9 When elimination fails for a singular matrix like

$$A = \begin{bmatrix} 2 & 1 & 4 & 6 \\ 0 & 3 & 8 & 5 \\ 0 & 0 & 0 & 7 \\ 0 & 0 & 0 & 9 \end{bmatrix}$$

show that A cannot be invertible. The third row of A^{-1}, multiplying A, should give the third row of $A^{-1}A = I$. Why is this impossible?

1.6.10 Find the inverses (in any legal way) of

$$A_1 = \begin{bmatrix} 0 & 0 & 0 & 1 \\ 0 & 0 & 2 & 0 \\ 0 & 3 & 0 & 0 \\ 4 & 0 & 0 & 0 \end{bmatrix}, \quad A_2 = \begin{bmatrix} 1 & 0 & 0 & 0 \\ -\frac{1}{2} & 1 & 0 & 0 \\ 0 & -\frac{2}{3} & 1 & 0 \\ 0 & 0 & -\frac{3}{4} & 1 \end{bmatrix}, \quad A_3 = \begin{bmatrix} a & b & 0 & 0 \\ c & d & 0 & 0 \\ 0 & 0 & a & b \\ 0 & 0 & c & d \end{bmatrix}.$$

1.6.11 Give examples of A and B such that

(i) $A + B$ is not invertible although A and B are invertible
(ii) $A + B$ is invertible although A and B are not invertible
(iii) all of A, B, and $A + B$ are invertible.

In the last case use $A^{-1}(A + B)B^{-1} = B^{-1} + A^{-1}$ to show that $B^{-1} + A^{-1}$ is also invertible—and find a formula for its inverse.

1.6.12 Which properties of a matrix A are preserved by its inverse (assuming A^{-1} exists)? (1) A is triangular (2) A is symmetric (3) A is tridiagonal (4) all entries are whole numbers (5) all entries are fractions (including whole numbers like $\frac{3}{1}$).

1.6.13 If $A = \begin{bmatrix} 3 \\ 1 \end{bmatrix}$ and $B = \begin{bmatrix} 2 \\ 2 \end{bmatrix}$, compute A^TB, B^TA, AB^T, and BA^T.

1.6.14 (Important)   Prove that even for rectangular matrices, AA^T and A^TA are always symmetric. Show by example that they may not be equal, even for square matrices.

1.6.15 Show that for any square matrix B, $A = B + B^T$ is always symmetric and $K = B - B^T$ is always *skew-symmetric*—which means that $K^T = -K$. Find these matrices when $B = \begin{bmatrix} 1 & 3 \\ 1 & 1 \end{bmatrix}$, and write B as the sum of a symmetric matrix and a skew-symmetric matrix.

1.6.16 (a)   How many entries can be chosen independently, in a symmetric matrix of order n?
(b)   How many entries can be chosen independently, in a skew-symmetric matrix of order n?

1.6.17 (a)   If $A = LDU$, with 1's on the diagonals of L and U, what is the corresponding factorization of A^T? Note that A and A^T (square matrices with no row exchanges) share the same pivots.
(b)   What triangular systems will give the solution to $A^T y = b$?

1.6.18 If $A = L_1 D_1 U_1$ and $A = L_2 D_2 U_2$, prove that $L_1 = L_2$, $D_1 = D_2$, and $U_1 = U_2$. If A is invertible, the factorization is unique.
(a)   Derive the equation $L_1^{-1} L_2 D_2 = D_1 U_1 U_2^{-1}$ and explain why one side is lower triangular and the other side is upper triangular.
(b)   Compare the main diagonals in that equation, and then compare the off-diagonals.

1.6.19 Under what conditions on its entries is A invertible, if

$$A = \begin{bmatrix} a & b & c \\ d & e & 0 \\ f & 0 & 0 \end{bmatrix} \quad \text{or} \quad A = \begin{bmatrix} a & b & 0 \\ c & d & 0 \\ 0 & 0 & e \end{bmatrix}?$$

1.6.20 If the 3 by 3 matrix A has row 1 + row 2 = row 3, show that it is impossible to solve $Ax = \begin{bmatrix} 1 & 2 & 4 \end{bmatrix}^T$. Can A be invertible?

1.6.21 Compute the symmetric LDL^T factorization of

$$A = \begin{bmatrix} 1 & 3 & 5 \\ 3 & 12 & 18 \\ 5 & 18 & 30 \end{bmatrix} \quad \text{and} \quad A = \begin{bmatrix} a & b \\ b & d \end{bmatrix}.$$

1.6.22 Find the inverse of

$$A = \begin{bmatrix} 1 & 0 & 0 & 0 \\ \frac{1}{4} & 1 & 0 & 0 \\ \frac{1}{3} & \frac{1}{3} & 1 & 0 \\ \frac{1}{2} & \frac{1}{2} & \frac{1}{2} & 1 \end{bmatrix}$$

1.6.23 If A and B are square matrices, show that $I - AB$ is invertible if $I - BA$ is invertible. Start from $B(I - AB) = (I - BA)B$.

1.7 ■ SPECIAL MATRICES AND APPLICATIONS

In this section we have two goals. The first is to explain one way in which large systems of linear equations can arise in practice. So far this book has not mentioned any applications, and the truth is that to describe a large and completely realistic problem in structural engineering or economics would lead us far afield. But there is one natural and important application that does not require a lot of preparation.

The other goal is to illustrate, by this same application, the special properties that coefficient matrices frequently have. It is unusual to meet large matrices that look as if they were constructed at random. Almost always there is a pattern, visible even at first sight—frequently a pattern of symmetry, and of very many zero entries. In the latter case, since a sparse matrix contains far fewer than n^2 pieces of information, the computations ought to work out much more simply than for a full matrix. We shall look particularly at *band matrices*, whose nonzero entries are concentrated near the main diagonal, to see how this property is reflected in the elimination process. In fact we look at one special band matrix.

The matrix itself can be seen in equation (6). The next paragraphs explain the application.

Our example comes from changing a continuous problem into a discrete one. The continuous problem will have infinitely many unknowns (it asks for $u(x)$ at every x), and it cannot be solved exactly on a computer. Therefore it has to be approximated by a discrete problem—the more unknowns we keep, the better will be the accuracy and the greater the expense. As a simple but still very typical continuous problem, our choice falls on the differential equation

$$-\frac{d^2u}{dx^2} = f(x), \quad 0 \le x \le 1. \tag{1}$$

This is a linear equation for the unknown function u, with inhomogeneous term f. There is some arbitrariness left in the problem, because any combination $C + Dx$ could be added to any solution. The sum would constitute another solution, since the second derivative of $C + Dx$ contributes nothing. Therefore the uncertainty left by these two arbitrary constants C and D will be removed by adding a "*boundary condition*" at each end of the interval:

$$u(0) = 0, \quad u(1) = 0. \tag{2}$$

The result is a *two-point boundary-value problem*, describing not a transient but a steady-state phenomenon—the temperature distribution in a rod, for example, with ends fixed at $0°$ and with a heat source $f(x)$.

Remember that our goal is to produce a problem that is discrete, or finite-dimensional—in other words, a problem in linear algebra. For that reason we cannot accept more than a finite amount of information about f, say its values at the equally spaced points $x = h$, $x = 2h, \ldots, x = nh$. And what we compute will be approximate values $u_1, \ldots, u_n$ for the true solution u at these same points. At the ends $x = 0$ and $x = 1 = (n + 1)h$, we are already given the correct boundary values $u_0 = 0$, $u_{n+1} = 0$.

The first question is, How do we replace the derivative d^2u/dx^2? Since every derivative is a limit of difference quotients, it can be approximated by stopping at a finite stepsize, and not permitting h (or Δx) to approach zero. For du/dx there are several alternatives:

$$\frac{du}{dx} \approx \frac{u(x+h)-u(x)}{h} \quad \text{or} \quad \frac{u(x)-u(x-h)}{h} \quad \text{or} \quad \frac{u(x+h)-u(x-h)}{2h}. \quad (3)$$

The last, because it is symmetric about x, is actually the most accurate. For the second derivative there is just one combination that uses only the values at x and $x \pm h$:

$$\frac{d^2u}{dx^2} \approx \frac{u(x+h)-2u(x)+u(x-h)}{h^2}. \quad (4)$$

It also has the merit of being symmetric about x. To repeat, the right side approaches the true value of d^2u/dx^2 as $h \to 0$, but we have to stop at a positive h.

At a typical meshpoint $x = jh$, the differential equation $-d^2u/dx^2 = f(x)$ is now replaced by this discrete analogue (4); after multiplying through by h^2,

$$-u_{j+1} + 2u_j - u_{j-1} = h^2 f(jh). \quad (5)$$

There are n equations of exactly this form, one for every value $j = 1, \ldots, n$. The first and last equations include the expressions u_0 and u_{n+1}, which are not unknowns—their values are the boundary conditions, and they are shifted to the right side of the equation and contribute to the inhomogeneous terms (or at least they might, if they were not known to equal zero). It is easy to understand (5) as a steady-state equation, in which the flows $(u_j - u_{j+1})$ coming from the right and $(u_j - u_{j-1})$ coming from the left are balanced by the loss of $h^2 f(jh)$ at the center.

The structure of the n equations (5) can be better visualized in matrix form $Au = b$. We shall choose $h = \frac{1}{6}$, or $n = 5$:

$$\begin{bmatrix} 2 & -1 & & & \\ -1 & 2 & -1 & & \\ & -1 & 2 & -1 & \\ & & -1 & 2 & -1 \\ & & & -1 & 2 \end{bmatrix} \begin{bmatrix} u_1 \\ u_2 \\ u_3 \\ u_4 \\ u_5 \end{bmatrix} = h^2 \begin{bmatrix} f(h) \\ f(2h) \\ f(3h) \\ f(4h) \\ f(5h) \end{bmatrix}. \quad (6)$$

From now on, *we will work with equation* (6), and it is not essential to look back at the source of the problem. What matters is that we have coefficient matrices whose order n can be very large, but which are obviously very far from random. The matrix A possesses many special properties, and three of those properties are fundamental:

(1) ***The matrix is tridiagonal.*** All its nonzero entries lie on the main diagonal and the two adjacent diagonals. Outside this band there is nothing: $a_{ij} = 0$ if $|i - j| > 1$. These zeros will bring a tremendous simplification to Gaussian elimination.

(2) **The matrix is symmetric.** Each entry a_{ij} equals its mirror image a_{ji}, so that $A^T = A$. Therefore the upper triangular U will be the transpose of the lower triangular L, and the final factorization will be $A = LDL^T$. This symmetry of A reflects the symmetry of the original differential equation. If there had been an odd derivative like d^3u/dx^3 or du/dx, A would not have been symmetric.

(3) **The matrix is positive definite.** This is an extra property to be verified as the pivots are computed; it says that *the pivots are positive*. Chapter 6 will give several equivalent definitions of positive-definiteness, most of them having nothing to do with elimination, but symmetry with positive pivots does have one immediate consequence: Row exchanges are unnecessary both in theory and in practice. This is in contrast to the matrix A' at the end of this section, which is not positive definite. Without a row exchange it is totally vulnerable to roundoff.

We return to the central fact, that A is tridiagonal. What effect does this have on elimination? To start, suppose we carry out the first stage of the elimination process and produce zeros below the first pivot:

$$
\begin{bmatrix}
2 & -1 & & & \\
-1 & 2 & -1 & & \\
& -1 & 2 & -1 & \\
& & -1 & 2 & -1 \\
& & & -1 & 2
\end{bmatrix}
\rightarrow
\begin{bmatrix}
2 & -1 & & & \\
0 & \frac{3}{2} & -1 & & \\
& -1 & 2 & -1 & \\
& & -1 & 2 & -1 \\
& & & -1 & 2
\end{bmatrix}.
$$

Compared with a general 5 by 5 matrix, there were two major simplifications:

(a) There was *only one nonzero entry* below the pivot.

(b) This one operation was carried out on a very short row. After the multiple $l_{21} = -\frac{1}{2}$ was determined, *only a single multiplication-subtraction* was required.

Thus the first step was very much simplified by the zeros in the first row and column. Furthermore, *the tridiagonal pattern is preserved during elimination* (in the absence of row exchanges!).

(c) The second stage of elimination, as well as every succeeding stage, also admits the simplifications (a) and (b).

We can summarize the final result in several ways. The most revealing is to look at the *LDU* factorization of A:

$$
A =
\begin{bmatrix}
1 & & & & \\
-\frac{1}{2} & 1 & & & \\
& -\frac{2}{3} & 1 & & \\
& & -\frac{3}{4} & 1 & \\
& & & -\frac{4}{5} & 1
\end{bmatrix}
\begin{bmatrix}
\frac{2}{1} & & & & \\
& \frac{3}{2} & & & \\
& & \frac{4}{3} & & \\
& & & \frac{5}{4} & \\
& & & & \frac{6}{5}
\end{bmatrix}
\begin{bmatrix}
1 & -\frac{1}{2} & & & \\
& 1 & -\frac{2}{3} & & \\
& & 1 & -\frac{3}{4} & \\
& & & 1 & -\frac{4}{5} \\
& & & & 1
\end{bmatrix}.
$$

The observations (a)—(c) can be expressed as follows: *The L and U factors of a tridiagonal matrix are bidiagonal.* These factors have more or less the same structure

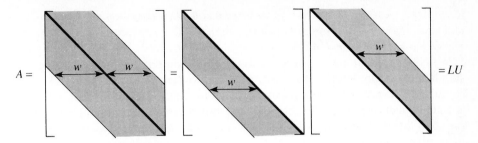

Fig. 1.7 A band matrix and its factors.

of zeros as A itself. Note too that L and U are transposes of one another, as was expected from the symmetry, and that the pivots d_i are all positive.† The pivots are obviously converging to a limiting value of $+1$, as n gets large. Such matrices make a computer very happy.

These simplifications lead to a complete change in the usual operation count. At each elimination stage only two operations are needed, and there are n such stages. Therefore *in place of $n^3/3$ operations we need only $2n$*; the computation is quicker by orders of magnitude. And the same is true of back-substitution; instead of $n^2/2$ operations we again need only $2n$. Thus *the number of operations for a tridiagonal system is proportional to n*, not to a higher power of n. Tridiagonal systems $Ax = b$ can be solved almost instantaneously.

Suppose, more generally, that A is a *band matrix*; its entries are all zero except within the band $|i - j| < w$ (Fig. 1.7). The "half bandwidth" is $w = 1$ for a diagonal matrix, $w = 2$ for a tridiagonal matrix, and $w = n$ for a full matrix. The first stage of elimination requires $w(w - 1)$ operations, and after this stage we still have bandwidth w. Since there are about n stages, *elimination on a band matrix must require about $w^2 n$ operations*.

The operation count is proportional to n, and now we see that it is proportional also to the square of w. As w approaches n, the matrix becomes full, and the count again is roughly n^3. A more exact count depends on the fact that in the lower right corner the bandwidth is no longer w; there is not room for that many bands. The precise number of divisions and multiplication-subtractions that produce L, D, and U (without assuming a symmetric A) is $P = \frac{1}{3}w(w - 1)(3n - 2w + 1)$. For a full matrix, which has $w = n$, we recover $P = \frac{1}{3}n(n - 1)(n + 1)$.†† To summarize: A band matrix A has triangular factors L and U that lie within the same band, and both elimination and back-substitution are very fast.

This is our last operation count, but we must emphasize the main point. For a finite difference matrix like A, the inverse is a full matrix. Therefore, in solving $Ax = b$, *we are actually much worse off knowing A^{-1} than knowing L and U*.

---

† The product of the pivots is the ***determinant*** of A: det $A = 6$.

†† We are happy to confirm that this P is a whole number; since $n - 1$, n, and $n + 1$ are consecutive integers, one of them must be divisible by 3.

Multiplying A^{-1} by b takes n^2 steps, whereas $4n$ are sufficient to solve $Lc = b$ and then $Ux = c$ — the forward elimination and back-substitution that produce $x = U^{-1}c = U^{-1}L^{-1}b = A^{-1}b$.

We hope this example has served two purposes: to reinforce the reader's understanding of the elimination sequence (which we now assume to be perfectly understood!) and to provide a genuine example of the kind of large linear system that is actually met in practice. In the next chapter we turn to the "theoretical" structure of a linear system $Ax = b$—the existence and the uniqueness of x.

Roundoff Error

In theory the nonsingular case is completed. Row exchanges may be necessary to achieve a full set of pivots; then back-substitution solves $Ax = b$. In practice, other row exchanges may be equally necessary—or the computed solution can easily become worthless. We will devote two pages (entirely optional in class) to making elimination more stable—why it is needed and how it is done.

Remember that for a system of moderate size, say 100 by 100, elimination involves a third of a million operations. With each operation we must expect a roundoff error. Normally, we keep a fixed number of significant digits (say three, for an extremely weak computer). Then adding two numbers of different sizes gives

$$.345 + .00123 \rightarrow .346$$

and the last digits in the smaller number are completely lost. The question is, how do all these individual roundoff errors contribute to the final error in the solution?

This is not an easy problem. It was attacked by John von Neumann, who was the leading mathematician at the time when computers suddenly made a million operations possible. In fact the combination of Gauss and von Neumann gives the simple elimination algorithm a remarkably distinguished history, although even von Neumann got a very complicated estimate of the roundoff error; it was Wilkinson who found the right way to answer the question, and his books are now classics.

Two simple examples, borrowed from the texts by Noble and by Forsythe and Moler, will illustrate three important points about roundoff error. The examples are

$$A = \begin{bmatrix} 1. & 1. \\ 1. & 1.0001 \end{bmatrix} \quad \text{and} \quad A' = \begin{bmatrix} .0001 & 1. \\ 1. & 1. \end{bmatrix}.$$

The first point is:

10 Some matrices are extremely sensitive to small changes, and others are not. The matrix A is ill-conditioned (that is, sensitive); A' is well-conditioned.

Qualitatively, A is nearly singular while A' is not. If we change the last entry of A to $a_{22} = 1$, it *is* singular and the two columns become the same. Consider two very close right-hand sides for $Ax = b$:

$$u + \qquad v = 2 \qquad\qquad u + \qquad v = 2$$
$$\text{and}$$
$$u + 1.0001v = 2 \qquad\qquad u + 1.0001v = 2.0001.$$

The solution to the first is $u = 2$, $v = 0$; the solution to the second is $u = v = 1$. *A change in the fifth digit of b was amplified to a change in the first digit of the solution. No numerical method can avoid this sensitivity to small perturbations.* The ill-conditioning can be shifted from one place to another, but it cannot be removed. The true solution is very sensitive, and the computed solution cannot be less so.

The second point is:

1P Even a well-conditioned matrix can be ruined by a poor algorithm.

We regret to say that for the matrix A', a straightforward Gaussian elimination is among the poor algorithms. Suppose .0001 is accepted as the first pivot, and 10,000 times the first row is subtracted from the second. The lower right entry becomes -9999, but roundoff to three places gives $-10,000$. Every trace has disappeared of the entry 1 which was originally there.

Consider the specific example

$$.0001u + v = 1$$
$$u + v = 2.$$

After elimination the second equation *should* read

$$-9999v = -9998, \qquad \text{or} \qquad v = .99990.$$

Roundoff will produce $-10,000v = -10,000$, or $v = 1$. So far the destruction of the second equation is not reflected in a poor solution; v is correct to three figures. As back-substitution continues, however, the first equation with the correct v should be

$$.0001u + .9999 = 1, \qquad \text{or} \qquad u = 1.$$

Instead, accepting the value $v = 1$ that is wrong only in the fourth place, we have

$$.0001u + 1 = 1, \qquad \text{or} \qquad u = 0.$$

The computed u is completely mistaken. Even though A' is well-conditioned, a straightforward elimination is violently unstable. The factors L, D, and U, whether exact or approximate, are completely out of scale with the original matrix:

$$A' = \begin{bmatrix} 1 & 0 \\ 10,000 & 1 \end{bmatrix} \begin{bmatrix} .0001 & 0 \\ 0 & -9999 \end{bmatrix} \begin{bmatrix} 1 & 10,000 \\ 0 & 1 \end{bmatrix}.$$

The small pivot .0001 brought instability, and the remedy is clear—*exchange rows*. This is our third point:

1Q   Just as a zero in the pivot position forced a theoretical change in elimination, so a small pivot forces a practical change. Unless it has special assurances to the contrary, a computer must compare each pivot with all the other possible pivots in the same column. Choosing the largest of these candidates, and exchanging the corresponding rows so as to make this largest value the pivot, is called *partial pivoting*.

In the matrix A', the possible pivot .0001 would be compared with the entry below it, which equals 1, and a row exchange would take place immediately. In matrix terms, this is just multiplication by a permutation matrix as before. The new matrix $A'' = PA'$ has the factorization

$$A'' = \begin{bmatrix} 1 & 1 \\ .0001 & 1 \end{bmatrix} = \begin{bmatrix} 1 & 0 \\ .0001 & 1 \end{bmatrix} \begin{bmatrix} 1 & 0 \\ 0 & .9999 \end{bmatrix} \begin{bmatrix} 1 & 1 \\ 0 & 1 \end{bmatrix}$$

The two pivots are now 1 and .9999, and they are perfectly in scale; previously they were .0001 and -9999.

Partial pivoting is distinguished from the still more conservative strategy of *complete pivoting*, which looks not only in the kth column but also in all later columns for the largest possible pivot. With complete pivoting, not only a row but also a column exchange is needed to move this largest value into the pivot. (In other words, there is a renumbering of the unknowns, or a *post*multiplication by a permutation matrix.) The difficulty with being so conservative is the expense; searching through all the remaining columns is time-consuming, and partial pivoting is quite adequate.

We have finally arrived at the fundamental algorithm of numerical linear algebra: *elimination with partial pivoting*. Some further refinements, such as watching to see whether a whole row or column needs to be rescaled, are still possible. But essentially the reader now knows what a computer does with a system of linear equations. Compared with the "theoretical" description—*find A^{-1}, and multiply $A^{-1}b$*—our description has consumed a lot of the reader's time (and patience). I wish there were an easier way to explain how x is actually found, but I do not think there is.

EXERCISES

1.7.1   Modify the example in the text by changing from $a_{11} = 2$ to $a_{11} = 1$, and find the *LDU* factorization of this new tridiagonal matrix.

1.7.2   Write down the 3 by 3 finite-difference matrix ($h = \frac{1}{4}$) for

$$-\frac{d^2u}{dx^2} + u = x, \qquad u(0) = u(1) = 0.$$

1.7.3   Find the 5 by 5 matrix A that approximates

$$-\frac{d^2u}{dx^2} = f(x), \qquad \frac{du}{dx}(0) = \frac{du}{dx}(1) = 0,$$

replacing the boundary conditions by $u_0 = u_1$ and $u_6 = u_5$. Check that your matrix, applied to the constant vector $(1, 1, 1, 1, 1)$, yields zero; A is *singular*. Analogously, show that if $u(x)$ is a solution of the continuous problem, then so is $u(x) + 1$. The two boundary conditions do not remove the uncertainty in the term $C + Dx$, and the solution is not unique.

1.7.4   With $h = \frac{1}{4}$ and $f(x) = 4\pi^2 \sin 2\pi x$, the difference equation (5) is

$$\begin{bmatrix} 2 & -1 & 0 \\ -1 & 2 & -1 \\ 0 & -1 & 2 \end{bmatrix} \begin{bmatrix} u_1 \\ u_2 \\ u_3 \end{bmatrix} = \frac{\pi^2}{4} \begin{bmatrix} 1 \\ 0 \\ -1 \end{bmatrix}.$$

Solve for u_1, u_2, u_3 and find their error in comparison with the true solution $u = \sin 2\pi x$ at $x = \frac{1}{4}$, $x = \frac{1}{2}$, and $x = \frac{3}{4}$.

1.7.5   What 5 by 5 system replaces (6) if the boundary conditions are changed to $u(0) = 1$, $u(1) = 0$?

1.7.6   (recommended)   Compute the inverse of the 3 by 3 Hilbert matrix

$$A = \begin{bmatrix} 1 & \frac{1}{2} & \frac{1}{3} \\ \frac{1}{2} & \frac{1}{3} & \frac{1}{4} \\ \frac{1}{3} & \frac{1}{4} & \frac{1}{5} \end{bmatrix}$$

in two ways using the ordinary Gauss-Jordan elimination sequence: (i) by exact computation, and (ii) by rounding off each number to three figures. *Note*: This is a case where pivoting does not help; A is ill-conditioned and incurable.

1.7.7   For the same matrix, compare the right sides of $Ax = b$ when the solutions are $x = (1, 1, 1)$ and $x = (0, 6, -3.6)$.

1.7.8   Solve $Ax = b = (1, 0, \ldots, 0)$ for the 10 by 10 Hilbert matrix with $a_{ij} = 1/(i + j - 1)$, using any computer code for linear equations. Then make a small change in an entry of A or b, and compare the solutions.

1.7.9   Compare the pivots in direct elimination to those with partial pivoting for

$$A = \begin{bmatrix} .001 & 0 \\ 1 & 1000 \end{bmatrix}.$$

(This is actually an example that needs rescaling before elimination.)

1.7.10   Explain why with partial pivoting all the multipliers l_{ij} in L satisfy $|l_{ij}| \leq 1$. Deduce that if the original entries of A satisfy $|a_{ij}| \leq 1$, then after producing zeros in the first column all entries are bounded by 2; after k stages they are bounded by 2^k. Can you construct a 3 by 3 example with all $|a_{ij}| \leq 1$ and $|l_{ij}| \leq 1$ whose last pivot is 4?

REVIEW EXERCISES: Chapter 1

1.1   (a)   Write down the 3 by 3 matrices with entries

$$a_{ij} = i - j \qquad \text{and} \qquad b_{ij} = \frac{i}{j}.$$

(b)   Compute the products AB and BA and A^2.

1.2   For the matrices

$$A = \begin{bmatrix} 1 & 0 \\ 2 & 1 \end{bmatrix} \qquad \text{and} \qquad B = \begin{bmatrix} 1 & 2 \\ 0 & 1 \end{bmatrix},$$

compute AB and BA and A^{-1} and B^{-1} and $(AB)^{-1}$.

1.3   Find examples of 2 by 2 matrices with $a_{12} = \frac{1}{2}$ for which

(a) $A^2 = I$ (b) $A^{-1} = A^{\mathrm{T}}$ (c) $A^2 = A$.

1.4   Solve by elimination and back-substitution:

$$\begin{aligned}
u \quad\;\; + w &= 4 \\
u + v \quad\;\; &= 3 \\
u + v + w &= 6
\end{aligned} \qquad \text{and} \qquad \begin{aligned}
v + w &= 0 \\
u \quad\;\; + w &= 0 \\
u + v \quad\;\; &= 6
\end{aligned}$$

1.5   Factor the preceding matrices into $A = LU$ or $PA = LU$.

1.6   (a)   There are sixteen 2 by 2 matrices whose entries are 1's and 0's. How many are invertible?
(b)   (Much harder!) If you put 1's and 0's at random into the entries of a 10 by 10 matrix, is it more likely to be invertible or singular?

1.7   There are sixteen 2 by 2 matrices whose entries are 1's and -1's. How many are invertible?

1.8   How are the rows of EA related to the rows of A, if

$$E = \begin{bmatrix} 1 & 0 & 0 \\ 0 & 2 & 0 \\ 4 & 0 & 1 \end{bmatrix} \quad \text{or} \quad E = \begin{bmatrix} 1 & 1 & 1 \\ 0 & 0 & 0 \end{bmatrix} \quad \text{or} \quad E = \begin{bmatrix} 0 & 0 & 1 \\ 0 & 1 & 0 \\ 1 & 0 & 0 \end{bmatrix}?$$

1.9   Write down a 2 by 2 system with infinitely many solutions.

1.10   Find inverses if they exist, by inspection or by Gauss-Jordan:

$$A = \begin{bmatrix} 1 & 0 & 1 \\ 1 & 1 & 0 \\ 0 & 1 & 1 \end{bmatrix} \quad \text{and} \quad A = \begin{bmatrix} 2 & 1 & 0 \\ 1 & 2 & 1 \\ 0 & 1 & 2 \end{bmatrix} \quad \text{and} \quad A = \begin{bmatrix} 1 & 1 & -2 \\ 1 & -2 & 1 \\ -2 & 1 & 1 \end{bmatrix}.$$

1.11   If E is 2 by 2 and it adds the first equation to the second, what are E^2 and E^8 and $8E$?

1.12   True or false, with *reason* if true or *counterexample* if false:
(1)   If A is invertible and its rows are in reverse order in B, then B is invertible.
(2)   If A and B are symmetric then AB is symmetric.
(3)   If A and B are invertible then BA is invertible.

(4) Every nonsingular matrix can be factored into the product $A = LU$ of a lower triangular L and an upper triangular U.

1.13 Solve $Ax = b$ by solving the triangular systems $Lc = b$ and $Ux = c$:

$$A = LU = \begin{bmatrix} 1 & 0 & 0 \\ 4 & 1 & 0 \\ 1 & 0 & 1 \end{bmatrix} \begin{bmatrix} 2 & 2 & 4 \\ 0 & 1 & 3 \\ 0 & 0 & 1 \end{bmatrix}, \quad b = \begin{bmatrix} 0 \\ 0 \\ 1 \end{bmatrix}.$$

What part of A^{-1} have you found, with this particular b?

1.14 If possible, find 3 by 3 matrices B such that
(a) $BA = 2A$ for every A
(b) $BA = 2B$ for every A
(c) BA has the first and last rows of A reversed
(d) BA has the first and last columns of A reversed.

1.15 Find the value for c in the following n by n inverse:

$$\text{if} \quad A = \begin{bmatrix} n & -1 & \cdot & -1 \\ -1 & n & \cdot & -1 \\ \cdot & \cdot & \cdot & -1 \\ -1 & -1 & -1 & n \end{bmatrix} \quad \text{then} \quad A^{-1} = \frac{1}{n+1} \begin{bmatrix} c & 1 & \cdot & 1 \\ 1 & c & \cdot & 1 \\ \cdot & \cdot & \cdot & 1 \\ 1 & 1 & 1 & c \end{bmatrix}.$$

1.16 For which values of k does

$$kx + y = 1$$
$$x + ky = 1$$

have no solution, one solution, or infinitely many solutions?

1.17 Find the symmetric factorization $A = LDL^T$ of

$$A = \begin{bmatrix} 1 & 2 & 0 \\ 2 & 6 & 4 \\ 0 & 4 & 11 \end{bmatrix} \quad \text{and} \quad A = \begin{bmatrix} a & b \\ b & c \end{bmatrix}.$$

1.18 Suppose A is the 4 by 4 identity matrix except for a vector v in column 2:

$$A = \begin{bmatrix} 1 & v_1 & 0 & 0 \\ 0 & v_2 & 0 & 0 \\ 0 & v_3 & 1 & 0 \\ 0 & v_4 & 0 & 1 \end{bmatrix}.$$

(a) Factor A into LU, assuming $v_2 \neq 0$.
(b) Find A^{-1}, which has the same form as A.

1.19 Solve by elimination, or show that there is no solution:

$$u + v + w = 0 \qquad\qquad u + v + w = 0$$
$$u + 2v + 3w = 0 \quad \text{and} \quad u + v + 3w = 0$$
$$3u + 5u + 7w = 1 \qquad\qquad 3u + 5v + 7w = 1.$$

1.20 The n by n permutation matrices are an important example of a "*group.*" If you multiply them you stay inside the group; they have inverses in the group; the identity is in the group; and the law $P_1(P_2P_3) = (P_1P_2)P_3$ is true—because it is true for all matrices.
(a) How many members belong to the groups of 4 by 4 and n by n permutation matrices?
(b) Find a power k so that all 3 by 3 permutation matrices satisfy $P^k = I$.

1.21 Describe the rows of DA and the columns of AD if $D = \begin{bmatrix} 2 & 0 \\ 0 & 5 \end{bmatrix}$.

1.22 (a) If A is invertible what is the inverse of A^T?
(b) If A is also symmetric what is the transpose of A^{-1}?
(c) Illustrate both formulas when $A = \begin{bmatrix} 2 & 1 \\ 1 & 1 \end{bmatrix}$.

1.23 By experiment with $n = 2$ and $n = 3$ find

$$\begin{bmatrix} 2 & 3 \\ 0 & 0 \end{bmatrix}^n, \quad \begin{bmatrix} 2 & 3 \\ 0 & 1 \end{bmatrix}^n, \quad \begin{bmatrix} 2 & 3 \\ 0 & 1 \end{bmatrix}^{-1}.$$

1.24 Starting with a first plane $u + 2v - w = 6$, find the equation for
(a) the parallel plane through the origin
(b) a second plane that also contains the points $(6, 0, 0)$ and $(2, 2, 0)$
(c) a third plane that meets the first and second in the point $(4, 1, 0)$.

1.25 What multiple of row 2 is subtracted from row 3 in forward elimination of

$$A = \begin{bmatrix} 1 & 0 & 0 \\ 2 & 1 & 0 \\ 0 & 5 & 1 \end{bmatrix} \begin{bmatrix} 1 & 2 & 0 \\ 0 & 1 & 5 \\ 0 & 0 & 1 \end{bmatrix} ?$$

How do you know (without multiplying those factors) that A is *invertible, symmetric,* and *tridiagonal*? What are its pivots?

1.26 (a) What vector x will make $Ax =$ column 1 + 2(column 3), for a 3 by 3 matrix A?
(b) Construct a matrix which has column 1 + 2(column 3) $= 0$. Check that A is singular (fewer than 3 pivots) and explain why that must happen.

1.27 True or false, with reason if true and counterexample if false:
(1) If $L_1U_1 = L_2U_2$ (upper triangular U's with nonzero diagonal, lower triangular L's with unit diagonal) then $L_1 = L_2$ and $U_1 = U_2$. The LU factorization is unique.
(2) If $A^2 + A = I$ then $A^{-1} = A + I$.
(3) If all diagonal entries of A are zero, then A is singular.

1.28 By experiment or the Gauss-Jordan method compute

$$\begin{bmatrix} 1 & 0 & 0 \\ l & 1 & 0 \\ m & 0 & 1 \end{bmatrix}^n, \quad \begin{bmatrix} 1 & 0 & 0 \\ l & 1 & 0 \\ m & 0 & 1 \end{bmatrix}^{-1}, \quad \begin{bmatrix} 1 & 0 & 0 \\ l & 1 & 0 \\ 0 & m & 1 \end{bmatrix}^{-1}.$$

1.29 Write down the 2 by 2 matrices which
(a) reverse the direction of every vector
(b) project every vector onto the x_2-axis
(c) turn every vector counterclockwise through $90°$
(d) reflect every vector through the $45°$ line $x_1 = x_2$.

2

VECTOR SPACES AND
LINEAR EQUATIONS

Elimination can simplify, one entry at a time, the linear system $Ax = b$. Fortunately it also simplifies the theory. The basic questions of *existence* and *uniqueness*—Is there one solution, or no solution, or an infinity of solutions?—are much easier to answer after elimination. We need to devote one more section to those questions; then that circle of ideas will be complete. But the mechanics of elimination produces only one kind of understanding of a linear system, and our chief object is to achieve a different and deeper understanding. This chapter may be more difficult than the first one. It goes to the heart of linear algebra.

First we need the concept of a ***vector space***. To introduce that idea we start immediately with the most important spaces. They are denoted by $\mathbf{R}^1, \mathbf{R}^2, \mathbf{R}^3, \ldots$; there is one for every positive integer. The space $\mathbf{R}^n$ consists of all column vectors with n components. (The components are real numbers.) The space $\mathbf{R}^2$ is represented by the usual x-y plane; the two components of the vector become the x and y coordinates of the corresponding point. $\mathbf{R}^3$ is equally familiar, with the three components giving a point in three-dimensional space. The one-dimensional space $\mathbf{R}^1$ is a line. The valuable thing for linear algebra is that the extension to n dimensions is so straightforward; for a vector in seven-dimensional space $\mathbf{R}^7$ we just need to know the seven components, even if the geometry is hard to visualize.

Within these spaces, and within all vector spaces, two operations are possible:

We can add any two vectors, and we can multiply vectors by scalars.

For the spaces $\mathbf{R}^n$ these operations are done a component at a time. If x is the vector in $\mathbf{R}^4$ with components 1, 0, 0, 3, then $2x$ is the vector with components

2, 0, 0, 6. A whole series of properties could be verified—the commutative law $x + y = y + x$, or the existence of a "zero vector" satisfying $0 + x = x$, or the existence of a vector "$-x$" satisfying $-x + x = 0$. Out of all such properties, eight (including those three) are fundamental; the full list is given in Exercise 2.1.5. Formally, *a real vector space is a set of* "vectors" *together with rules for vector addition and multiplication by real numbers*. The addition and multiplication must produce vectors that are *within the space*, and they must satisfy the eight conditions.

Normally our vectors belong to one of the spaces $\mathbf{R}^n$; they are ordinary column vectors. The formal definition, however, allows us to think of other things as vectors—provided that addition and scalar multiplication are all right. We give three examples:

(i)   *The infinite-dimensional space* $\mathbf{R}^\infty$. Its vectors have infinitely many components, as in $x = (1, 2, 1, 2, 1, \ldots)$, but the laws of addition and multiplication stay unchanged.

(ii)   *The space of 3 by 2 matrices.* In this case the "vectors" are matrices! We can add two matrices, and $A + B = B + A$, and there is a zero matrix, and so on. This space is almost the same as $\mathbf{R}^6$. (The six components are arranged in a rectangle instead of a column.) Any choice of m and n would give, as a similar example, the vector space of all m by n matrices.

(iii)   *The space of functions* $f(x)$. Here we admit all functions f that are defined on a fixed interval, say $0 \le x \le 1$. The space includes $f(x) = x^2$, $g(x) = \sin x$, their sum $(f + g)(x) = x^2 + \sin x$, and all multiples like $3x^2$ and $-\sin x$. The vectors are functions, and again the dimension is infinite—in fact, it is a larger infinity than for $\mathbf{R}^\infty$.

Other examples are given in the exercises, but the vector spaces we need most are somewhere else—they are *inside* the standard spaces $\mathbf{R}^n$. We want to describe them and explain why they are important. Geometrically, think of the usual three-dimensional $\mathbf{R}^3$ and choose any plane through the origin. *That plane is a vector space in its own right.* If we multiply a vector in the plane by 3, or -3, or any other scalar, we get a vector which lies in the same plane. If we add two vectors in the plane, their sum stays in the plane. This plane illustrates one of the most fundamental ideas in the theory of linear algebra; it is a *subspace* of the original space $\mathbf{R}^3$.

DEFINITION   A *subspace* of a vector space is a nonempty subset that satisfies two requirements:

(i)   If we add any vectors x and y in the subspace, their sum $x + y$ is in the subspace.

(ii)   If we multiply any vector x in the subspace by any scalar c, the multiple cx is still in the subspace.

In other words, a subspace is a subset which is "closed" under addition and scalar multiplication. Those operations follow the rules of the host space, without taking us outside the subspace. There is no need to verify the eight required properties, because they are satisfied in the larger space and will automatically be satisfied in every subspace. Notice in particular that *the zero vector will belong to every subspace.* That comes from rule (ii): Choose the scalar to be $c = 0$.

The most extreme possibility for a subspace is to contain only one vector, the zero vector. It is a "zero-dimensional space," containing only the point at the origin. Rules (i) and (ii) are satisfied, since addition and scalar multiplication are entirely permissible; the sum $0 + 0$ is in this one-point space, and so are all multiples $c0$. *This is the smallest possible vector space*: the empty set is not allowed. At the other extreme, the largest subspace is the whole of the original space—we can allow every vector into the subspace. If the original space is $\mathbf{R}^3$, then the possible subspaces are easy to describe: $\mathbf{R}^3$ itself, any plane through the origin, any line through the origin, or the origin (the zero vector) alone.

The distinction between a subset and a subspace is made clear by examples; we give some now and more later. In each case, the question to be answered is whether or not requirements (i) and (ii) are satisfied: Can you add vectors, and can you multiply by scalars, without leaving the space?

EXAMPLE 1 Consider all vectors whose components are positive or zero. If the original space is the x-y plane $\mathbf{R}^2$, then this subset is the first quadrant; the coordinates satisfy $x \geq 0$ and $y \geq 0$. It is not a subspace, even though it contains zero and addition does leave us within the subset. Rule (ii) is violated, since if the scalar is -1 and the vector is $\begin{bmatrix} 1 & 1 \end{bmatrix}$, the multiple $cx = \begin{bmatrix} -1 & -1 \end{bmatrix}$ is in the third quadrant instead of the first.

If we include the third quadrant along with the first, then scalar multiplication is all right; every multiple cx will stay in this subset, and rule (ii) is satisfied. However, rule (i) is now violated, since the addition of $\begin{bmatrix} 1 & 2 \end{bmatrix}$ and $\begin{bmatrix} -2 & -1 \end{bmatrix}$ gives a vector $\begin{bmatrix} -1 & 1 \end{bmatrix}$ which is not in either quadrant. The smallest subspace containing the first quadrant is the whole space $\mathbf{R}^2$.

EXAMPLE 2 If we start from the vector space of 3 by 3 matrices, then one possible subspace is the set of *lower triangular* matrices. Another is the set of symmetric matrices. In both cases, the sums $A + B$ and the multiples cA inherit the properties of A and B. They are lower triangular if A and B are lower triangular, and they are symmetric if A and B are symmetric. Of course, the zero matrix is in both subspaces.

We now come to the key examples of subspaces. They are tied directly to a matrix A, and they give information about the system $Ax = b$. In some cases they contain vectors with m components, like the columns of A; then they are subspaces of $\mathbf{R}^m$. In other cases the vectors have n components, like the rows of A (or like x itself); those are subspaces of $\mathbf{R}^n$. We illustrate by a system of three equations in two unknowns:

$$\begin{bmatrix} 1 & 0 \\ 5 & 4 \\ 2 & 4 \end{bmatrix} \begin{bmatrix} u \\ v \end{bmatrix} = \begin{bmatrix} b_1 \\ b_2 \\ b_3 \end{bmatrix}. \tag{1}$$

If there were more unknowns than equations, we might expect to find plenty of solutions (although that is not always so). In the present case there are more equations than unknowns—and we must expect that *usually there will be no solution.* A system with $m > n$ will be solvable only for certain right-hand sides, in fact, for a very "thin" subset of all possible three-dimensional vectors b. We want to find that subset of b's.

One way of describing this subset is so simple that it is easy to overlook.

2A The system $Ax = b$ is solvable if and only if the vector b can be expressed as a combination of the columns of A.

This description involves nothing more than a restatement of the system $Ax = b$, writing it in the following way:

$$u \begin{bmatrix} 1 \\ 5 \\ 2 \end{bmatrix} + v \begin{bmatrix} 0 \\ 4 \\ 4 \end{bmatrix} = \begin{bmatrix} b_1 \\ b_2 \\ b_3 \end{bmatrix}. \tag{2}$$

These are the same three equations in two unknowns. But now the problem is seen to be this: Find numbers u and v that multiply the first and second columns to produce the vector b. The system is solvable exactly when such coefficients exist, and the vector (u, v) is the solution x.

Thus, the subset of attainable right-hand sides b is *the set of all combinations of the columns of A.* One possible right side is the first column itself; the weights are $u = 1$ and $v = 0$. Another possibility is the second column: $u = 0$ and $v = 1$. A third is the right side $b = 0$; the weights are $u = 0$, $v = 0$ (and with that trivial choice, the vector $b = 0$ will be attainable no matter what the matrix is).

Now we have to consider all combinations of the two columns, and we describe the result geometrically: $Ax = b$ can be solved if and only if b lies in the plane that is spanned by the two column vectors (Fig. 2.1). This is the thin set of attainable b. If b lies off the plane, then it is not a combination of the two columns. In that case $Ax = b$ has no solution.

What is important is that this plane is not just a subset of $\mathbf{R}^3$; it is a subspace. It is called the **column space** of A. The column space of a matrix consists of **all combinations of the columns**. It is denoted by $\mathcal{R}(A)$. The equation $Ax = b$ can be solved if and only if b lies in the column space of A. For an m by n matrix this will be a subspace of $\mathbf{R}^m$, since the columns have m components, and the requirements (i) and (ii) for a subspace are easy to check:

(i) Suppose b and b' lie in the column space, so that $Ax = b$ for some x and $Ax' = b'$ for some x'; x and x' just give the combinations which produce b and b'. Then $A(x + x') = b + b'$, so that $b + b'$ is also a combination of the columns. If b is column 1 minus column 2, and b' is twice column 2, then $b + b'$ is column 1 plus column 2. The attainable vectors are closed under addition, and the first requirement for a subspace is met.

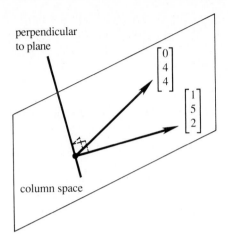

Fig. 2.1. The column space, a plane in three-dimensional space.

(ii) If b is in the column space, so is any multiple cb. If some combination of columns produces b (say $Ax = b$), then multiplying every coefficient in the combination by c will produce cb. In other words, $A(cx) = cb$.

Geometrically, the general case is like Fig. 2.1—except that the dimensions may be very different. We need not have a two-dimensional plane within three-dimensional space. Similarly, the perpendicular to the column space, which we drew in Fig. 2.1, may not always be a line. At one extreme, the smallest possible column space comes from the zero matrix $A = 0$. The only vector in its column space (the only combination of the columns) is $b = 0$, and no other choice of b allows us to solve $0x = b$. At the other extreme, suppose A is the 5 by 5 identity matrix. Then the column space is the whole of $\mathbf{R}^5$; the five columns of the identity matrix can combine to produce any five-dimensional vector b. This is not at all special to the identity matrix. *Any 5 by 5 matrix which is nonsingular will have the whole of* $\mathbf{R}^5$ *as its column space.* For such a matrix we can solve $Ax = b$ by Gaussian elimination; there are five pivots. Therefore every b is in the column space of a nonsingular matrix.

You can see how Chapter 1 is contained in this chapter. There we studied the most straightforward (and most common) case, an n by n matrix whose column space is $\mathbf{R}^n$. Now we allow also singular matrices, and rectangular matrices of any shape; the column space is somewhere between the zero space and the whole space. Together with its perpendicular space, it gives one of our two approaches to understanding $Ax = b$.

The Nullspace of A

The second approach to $Ax = b$ is "dual" to the first. We are concerned not only with which right sides b are attainable, but also with the set of solutions x

that attain them. The right side $b = 0$ always allows the particular solution $x = 0$, but there may be infinitely many other solutions. (There always are, if there are more unknowns than equations, $n > m$.) **The set of solutions to $Ax = 0$ is itself a vector space—the nullspace of A.**

> The nullspace of a matrix consists of all vectors x such that $Ax = 0$. It is denoted by $\mathcal{N}(A)$. It is a subspace of $\mathbf{R}^n$, just as the column space was a subspace of $\mathbf{R}^m$.

Requirement (i) holds: If $Ax = 0$ and $Ax' = 0$ then $A(x + x') = 0$. Requirement (ii) also holds: If $Ax = 0$ then $A(cx) = 0$. Both requirements fail if the right side is not zero. Only the solutions to a *homogeneous* equation ($b = 0$) form a subspace. The nullspace is easy to find for the example given above:

$$\begin{bmatrix} 1 & 0 \\ 5 & 4 \\ 2 & 4 \end{bmatrix} \begin{bmatrix} u \\ v \end{bmatrix} = \begin{bmatrix} 0 \\ 0 \\ 0 \end{bmatrix}.$$

The first equation gives $u = 0$, and the second equation then forces $v = 0$. In this case the nullspace contains only the zero vector; the only combination to produce zero on the right-hand side is $u = v = 0$.

The situation is changed when a third column is a combination of the first two:

$$B = \begin{bmatrix} 1 & 0 & 1 \\ 5 & 4 & 9 \\ 2 & 4 & 6 \end{bmatrix}.$$

The column space of B is the same as that of A. The new column lies in the plane of Fig. 2.1; it is just the sum of the two column vectors we started with. But the nullspace of this new matrix B contains the vector with components $1, 1, -1$, and it contains any multiple of that vector:

$$\begin{bmatrix} 1 & 0 & 1 \\ 5 & 4 & 9 \\ 2 & 4 & 6 \end{bmatrix} \begin{bmatrix} c \\ c \\ -c \end{bmatrix} = \begin{bmatrix} 0 \\ 0 \\ 0 \end{bmatrix}.$$

The nullspace of B is the line containing all points $x = c$, $y = c$, $z = -c$, where c ranges from $-\infty$ to ∞. (The line goes through the origin, as any subspace must.) This one-dimensional nullspace has a perpendicular space (a plane), which is directly related to the rows of the matrix, and is of special importance.

To summarize: We want to be able, for any system $Ax = b$, to find all attainable right-hand sides b and all solutions to $Ax = 0$. The vectors b are in the column space and the vectors x are in the nullspace. This means that we shall compute the dimensions of those subspaces and a convenient set of vectors to generate them. We hope to end up by understanding all *four* of the subspaces that are intimately related to each other and to A—the column space of A, the nullspace of A, and their two perpendicular spaces.

EXERCISES

2.1.1 Show that requirements (i) and (ii) for a vector space are genuinely independent by constructing:
(a) a subset of two-dimensional space closed under vector addition and even subtraction, but not under scalar multiplication;
(b) a subset of two-dimensional space (other than two opposite quadrants) closed under scalar multiplication but not under vector addition.

2.1.2 Which of the following subsets of $\mathbf{R}^3$ are actually subspaces?
(a) The plane of vectors with first component $b_1 = 0$.
(b) The plane of vectors b with $b_1 = 1$.
(c) The vectors b with $b_1 b_2 = 0$ (this is the union of two subspaces, the plane $b_1 = 0$ and the plane $b_2 = 0$).
(d) The solitary vector $b = (0, 0, 0)$.
(e) All combinations of two given vectors $x = (1, 1, 0)$ and $y = (2, 0, 1)$.
(f) The vectors (b_1, b_2, b_3) that satisfy $b_3 - b_2 + 3b_1 = 0$.

2.1.3 Describe the column space and the nullspace of the matrices

$$A = \begin{bmatrix} 1 & -1 \\ 0 & 0 \end{bmatrix} \quad \text{and} \quad B = \begin{bmatrix} 0 & 0 & 0 \\ 0 & 0 & 0 \end{bmatrix}.$$

2.1.4 What is the smallest subspace of 3 by 3 matrices which contains all symmetric matrices and all lower triangular matrices? What is the largest subspace which is contained in both of those subspaces?

2.1.5 In the definition of a vector space, addition and scalar multiplication are required to satisfy the following rules:

> 1. $x + y = y + x$
> 2. $x + (y + z) = (x + y) + z$
> 3. There is a unique "zero vector" such that $x + 0 = x$ for all x
> 4. For each x there is a unique vector $-x$ such that $x + (-x) = 0$
> 5. $1x = x$
> 6. $(c_1 c_2)x = c_1(c_2 x)$
> 7. $c(x + y) = cx + cy$
> 8. $(c_1 + c_2)x = c_1 x + c_2 x.$

(a) Suppose addition in $\mathbf{R}^2$ adds an extra one to each component, so that $(3, 1) + (5, 0)$ equals $(9, 2)$ instead of $(8, 1)$. With scalar multiplication unchanged, which rules are broken?
(b) Show that the set of all positive real numbers, with $x + y$ and cx redefined to equal the usual xy and x^c, respectively, is a vector space. What is the "zero vector?"

2.1.6 Let P be the plane in 3-space with equation $x + 2y + z = 6$. What is the equation of the plane P_0 through the origin parallel to P? Are P and P_0 subspaces of $\mathbf{R}^3$?

2.1.7   Which of the following are subspaces of $\mathbf{R}^\infty$?

    (a)   All sequences like $(1, 0, 1, 0, \ldots)$ which include infinitely many zeros.

    (b)   All sequences $(x_1, x_2, \ldots)$ with $x_j = 0$ from some point onward.

    (c)   All decreasing sequences: $x_{j+1} \le x_j$ for each j.

    (d)   All convergent sequences: the x_j have a limit as $j \to \infty$.

    (e)   All arithmetic progressions: $x_{j+1} - x_j$ is the same for all j.

    (f)   All geometric progressions $(x_1, kx_1, k^2x_1, \ldots)$ allowing all k and x_1.

2.1.8   Which descriptions are correct? The solutions x of

$$Ax = \begin{bmatrix} 1 & 1 & 1 \\ 1 & 0 & 2 \end{bmatrix} \begin{bmatrix} x_1 \\ x_2 \\ x_3 \end{bmatrix} = \begin{bmatrix} 0 \\ 0 \end{bmatrix}$$

form a plane, line, point, subspace, nullspace of A, column space of A.

2.1.9   Show that the set of nonsingular 2 by 2 matrices is not a vector space. Show also that the set of *singular* 2 by 2 matrices is not a vector space.

THE SOLUTION OF *m* EQUATIONS IN *n* UNKNOWNS ■ 2.2

The elimination process is by now very familiar for square matrices, and one example will be enough to illustrate the new possibilities when the matrix is rectangular. The elimination itself goes forward without major changes, but when it comes to reading off the solution by back-substitution, there are some differences.

Perhaps, even before the example, we should illustrate the possibilities by looking at the scalar equation $ax = b$. This is a "system" of only one equation in one unknown. It might be $3x = 4$ or $0x = 0$ or $0x = 4$, and those three examples display the three alternatives:

   (i)   If $a \neq 0$, then for any b there exists a solution $x = b/a$, and this solution is unique. This is the ***nonsingular*** case (of a 1 by 1 invertible matrix a).

   (ii)   If $a = 0$ and $b = 0$, there are infinitely many solutions; any x satisfies $0x = 0$. This is the ***underdetermined*** case; a solution exists, but it is not unique.

   (iii)   If $a = 0$ and $b \neq 0$, there is no solution to $0x = b$. This is the ***inconsistent*** case.

For square matrices all these alternatives may occur. We will replace "$a \neq 0$" by "A is invertible," but it still means that A^{-1} makes sense. With a rectangular matrix possibility (i) disappears; we cannot have existence and also uniqueness, one solution x for every b. There may be infinitely many solutions for every b; or infinitely many for some b and no solution for others; or one solution for some b and none for others.

We start with a 3 by 4 matrix, ignoring at first the right side b:

$$A = \begin{bmatrix} 1 & 3 & 3 & 2 \\ 2 & 6 & 9 & 5 \\ -1 & -3 & 3 & 0 \end{bmatrix}.$$

The pivot $a_{11} = 1$ is nonzero, and the usual elementary operations will produce zeros in the first column below this pivot:

$$A \rightarrow \begin{bmatrix} 1 & 3 & 3 & 2 \\ 0 & 0 & 3 & 1 \\ 0 & 0 & 6 & 2 \end{bmatrix}.$$

The candidate for the second pivot has become zero, and therefore we look below it for a nonzero entry—intending to carry out a row exchange. In this case *the entry below it is also zero.* If the original matrix were square, this would signal that the matrix was singular. With a rectangular matrix, we must expect trouble anyway, and there is no reason to terminate the elimination. All we can do is to *go on to the next column*, where the pivot entry is nonzero. Subtracting twice the second row from the third, we arrive at

$$U = \begin{bmatrix} 1 & 3 & 3 & 2 \\ 0 & 0 & 3 & 1 \\ 0 & 0 & 0 & 0 \end{bmatrix}.$$

Strictly speaking, we then proceed to the fourth column; there we meet another zero in the pivot position, and nothing can be done. The forward stage of elimination is complete.

The final form U is again upper triangular, but the pivots† are not necessarily on the main diagonal. The important thing is that the nonzero entries are confined to a "staircase pattern," or **echelon form**, which is indicated in a 5 by 9 case by Fig. 2.2. The pivots are circled, whereas the other starred entries may or may not be zero.

$$
U = \begin{bmatrix}
\circledast & * & * & * & * & * & * & * & * \\
0 & \circledast & * & * & * & * & * & * & * \\
0 & 0 & 0 & \circledast & * & * & * & * & * \\
0 & 0 & 0 & 0 & 0 & 0 & 0 & 0 & \circledast \\
0 & 0 & 0 & 0 & 0 & 0 & 0 & 0 & 0
\end{bmatrix}
$$

Fig. 2.2. The nonzero entries of a typical echelon matrix U.

We can summarize in words what the figure illustrates:

(i)   The nonzero rows come first—otherwise there would have been row exchanges—and the pivots are the first nonzero entries in those rows.

(ii)   Below each pivot is a column of zeros, obtained by elimination.

(iii)   Each pivot lies to the right of the pivot in the row above; this produces the staircase pattern.

Since we started with A and ended with U, the reader is certain to ask: Are these matrices connected by a lower triangular L as before? Is $A = LU$? There is no reason why not, since the elimination steps have not changed; each step still subtracts a multiple of one row from a row beneath it. The inverse of each step is also accomplished just as before, by adding back the multiple that was subtracted. These inverses still come in an order that permits us to record them directly in L:

$$
L = \begin{bmatrix}
1 & 0 & 0 \\
2 & 1 & 0 \\
-1 & 2 & 1
\end{bmatrix}.
$$

The reader can verify that $A = LU$, and should note that L is not rectangular but *square*. It is a matrix of the same order $m = 3$ as the number of rows in A and U.

The only operation not required by our example, but needed in general, is an exchange of rows. As in Chapter 1, this would introduce a permutation matrix P

---

† Remember that **pivots are nonzero**. During elimination we may find a zero in the pivot position, but this is only temporary; by exchanging rows or by giving up on a column and going to the next, we end up with a string of (nonzero) pivots and zeros beneath them.

and it can carry out row exchanges before elimination begins. Since we keep going to the next column when no pivots are available in a given column, there is no need to assume that A is nonsingular. Here is the main theorem:

2B To any m by n matrix A there correspond a permutation matrix P, a lower triangular matrix L with unit diagonal, and an m by n echelon matrix U, such that $PA = LU$.

Our goal is now to read off the solutions (if any) to $Ax = b$.

Suppose we start with the homogeneous case, $b = 0$. Then, since the row operations will have no effect on the zeros on the right side of the equation, $Ax = 0$ is simply reduced to $Ux = 0$:

$$Ux = \begin{bmatrix} 1 & 3 & 3 & 2 \\ 0 & 0 & 3 & 1 \\ 0 & 0 & 0 & 0 \end{bmatrix} \begin{bmatrix} u \\ v \\ w \\ y \end{bmatrix} = \begin{bmatrix} 0 \\ 0 \\ 0 \end{bmatrix}.$$

The unknowns u, v, w, and y go into two groups. One group is made up of the **basic variables**, those that correspond to **columns with pivots**. The first and third columns contain the pivots, so u and w are the basic variables. The other group is made up of the **free variables**, corresponding to **columns without pivots**; these are the second and fourth columns, so that v and y are free variables.

To find the most general solution to $Ux = 0$ (or equivalently, to $Ax = 0$) we may assign arbitrary values to the free variables. Suppose we call these values simply v and y. The basic variables are then completely determined, and can be computed in terms of the free variables by back-substitution. Proceeding upward,

$$3w + y = 0 \quad \text{yields} \quad w = -\tfrac{1}{3}y$$
$$u + 3v + 3w + 2y = 0 \quad \text{yields} \quad u = -3v - y.$$

There is a "double infinity" of solutions to the system, with two free and independent parameters v and y. The general solution is a combination

$$x = \begin{bmatrix} -3v - y \\ v \\ -\tfrac{1}{3}y \\ y \end{bmatrix} = v \begin{bmatrix} -3 \\ 1 \\ 0 \\ 0 \end{bmatrix} + y \begin{bmatrix} -1 \\ 0 \\ -\tfrac{1}{3} \\ 1 \end{bmatrix}. \tag{1}$$

Please look again at the last form of the solution to $Ax = 0$. The vector $(-3, 1, 0, 0)$ gives the solution when the free variables have the values $v = 1$, $y = 0$. The last vector is the solution when $v = 0$ and $y = 1$. *All solutions are linear combinations of these two.* Therefore a good way to find all solutions to $Ax = 0$ is

1. After elimination reaches $Ux = 0$, identify the basic and free variables.
2. Give one free variable the value one, set the other free variables to zero, and solve $Ux = 0$ for the basic variables.

3.   Every free variable produces its own solution by step 2, and the combinations of those solutions form the nullspace—the space of all solutions to $Ax = 0$.

Geometrically, the picture is this: Within the four-dimensional space of all possible vectors x, the solutions to $Ax = 0$ form a two-dimensional subspace—the *nullspace* of A. In the example it is generated by the two vectors $(-3, 1, 0, 0)$ and $(-1, 0, -\frac{1}{3}, 1)$. The combinations of these two vectors form a set that is closed under addition and scalar multiplication; these operations simply lead to more solutions to $Ax = 0$, and all these combinations comprise the nullspace.

This is the place to recognize one extremely important theorem. Suppose we start with a matrix that has more columns than rows, $n > m$. Then, since there can be at most m pivots (there are not rows enough to hold any more), *there must be at least $n - m$ free variables.* There will be even more free variables if some rows of U happen to reduce to zero; but no matter what, at least one of the variables must be free. This variable can be assigned an arbitrary value, leading to the following conclusion:

2C   If a homogeneous system $Ax = 0$ has more unknowns than equations ($n > m$), it has a nontrivial solution: There is a solution x other than the trivial solution $x = 0$.

There must actually be infinitely many solutions, since any multiple cx will also satisfy $A(cx) = 0$. The nullspace contains the line through x. And if there are additional free variables, the nullspace becomes more than just a line in n-dimensional space. *The nullspace is a subspace of the same "dimension" as the number of free variables.*

This central idea—the *dimension* of a subspace—is made precise in the next section. It is a count of the degrees of freedom.

The inhomogeneous case, $b \neq 0$, is quite different. We return to the original example $Ax = b$, and apply to both sides of the equation the operations that led from A to U. The result is an upper triangular system $Ux = c$:

$$\begin{bmatrix} 1 & 3 & 3 & 2 \\ 0 & 0 & 3 & 1 \\ 0 & 0 & 0 & 0 \end{bmatrix} \begin{bmatrix} u \\ v \\ w \\ y \end{bmatrix} = \begin{bmatrix} b_1 \\ b_2 - 2b_1 \\ b_3 - 2b_2 + 5b_1 \end{bmatrix} \tag{2}$$

The vector c on the right side, which appeared after the elimination steps, is just $L^{-1}b$ as in the previous chapter.

It is not clear that these equations have a solution. The third equation is very much in doubt. Its left side is zero, and *the equations are inconsistent unless* $b_3 - 2b_2 + 5b_1 = 0$. In other words, the *set of attainable vectors b is not the whole of three-dimensional space.* Even though there are more unknowns than equations, there may be no solution. We know, from Section 2.1, another way of considering the same question: $Ax = b$ can be solved if and only if b lies in the column space of A. This subspace is spanned by the four columns of A (not of U!):

$$\begin{bmatrix} 1 \\ 2 \\ -1 \end{bmatrix}, \quad \begin{bmatrix} 3 \\ 6 \\ -3 \end{bmatrix}, \quad \begin{bmatrix} 3 \\ 9 \\ 3 \end{bmatrix}, \quad \begin{bmatrix} 2 \\ 5 \\ 0 \end{bmatrix}.$$

Even though there are four vectors, their combinations only fill out a plane in three-dimensional space. The second column is three times the first, and the fourth column equals the first plus some fraction of the third. (Note that these dependent columns, the second and fourth, are exactly the ones without pivots.) The column space can now be described in two completely different ways. On the one hand, it is the plane generated by columns 1 and 3; the other columns lie in that plane, and contribute nothing new. Equivalently, it is the plane composed of all points (b_1, b_2, b_3) that satisfy $b_3 - 2b_2 + 5b_1 = 0$; this is the constraint that must be imposed if the system is to be solvable. Every column satisfies this constraint, so it is forced on b. Geometrically, we shall see that the vector $(5, -2, 1)$ is perpendicular to each column.

If b lies in this plane, and thus belongs to the column space, then the solutions of $Ax = b$ are easy to find. The last equation in the system amounts only to $0 = 0$. To the free variables v and y, we may assign arbitrary values as before. Then the basic variables are still determined by back-substitution. We take a specific example, in which the components of b are chosen as 1, 5, 5 (we were careful to make $b_3 - 2b_2 + 5b_1 = 0$). The system $Ax = b$ becomes

$$\begin{bmatrix} 1 & 3 & 3 & 2 \\ 2 & 6 & 9 & 5 \\ -1 & -3 & 3 & 0 \end{bmatrix} \begin{bmatrix} u \\ v \\ w \\ y \end{bmatrix} = \begin{bmatrix} 1 \\ 5 \\ 5 \end{bmatrix}.$$

Elimination converts this into

$$\begin{bmatrix} 1 & 3 & 3 & 2 \\ 0 & 0 & 3 & 1 \\ 0 & 0 & 0 & 0 \end{bmatrix} \begin{bmatrix} u \\ v \\ w \\ y \end{bmatrix} = \begin{bmatrix} 1 \\ 3 \\ 0 \end{bmatrix}.$$

The last equation is $0 = 0$, as expected, and the others give

$$3w + y = 3 \quad \text{or} \quad w = 1 - \tfrac{1}{3}y$$
$$u + 3v + 3w + 2y = 1 \quad \text{or} \quad u = -2 - 3v - y.$$

Again there is a double infinity of solutions. Looking at all four components together, the general solution can be written as

$$
x = \begin{bmatrix} u \\ v \\ w \\ y \end{bmatrix} = \begin{bmatrix} -2 \\ 0 \\ 1 \\ 0 \end{bmatrix} + v \begin{bmatrix} -3 \\ 1 \\ 0 \\ 0 \end{bmatrix} + y \begin{bmatrix} -1 \\ 0 \\ -\frac{1}{3} \\ 1 \end{bmatrix}. \tag{3}
$$

This is exactly like the solution to $Ax = 0$ in equation (1), except there is one new term. It is $(-2, 0, 1, 0)$, which is a *particular solution* to $Ax = b$. It solves the equation, and then the last two terms yield more solutions (because they satisfy $Ax = 0$). *Every solution to $Ax = b$ is the sum of one particular solution and a solution to $Ax = 0$:*

$$\boxed{x_{\text{general}} = x_{\text{particular}} + x_{\text{homogeneous}}}$$

The homogeneous part comes from the nullspace. The particular solution in (3) comes from solving the equation *with all free variables set to zero*. That is the only new part, since the nullspace is already computed. When you multiply the equation in the box by A, you get $Ax_{\text{general}} = b + 0$.

Geometrically, the general solutions again fill a two-dimensional surface—but it is not a subspace. It does not contain the origin. It is *parallel* to the nullspace we had before, but it is shifted by the particular solution. Thus the computations include one new step:

1.   Reduce $Ax = b$ to $Ux = c$.
2.   Set all free variables to zero and find a particular solution.
3.   Set the right side to zero and give each free variable, in turn, the value one. With the other free variables at zero, find a homogeneous solution (a vector x in the nullspace).

Previously step 2 was absent. When the equation was $Ax = 0$, the particular solution was the zero vector! It fits the pattern, but $x_{\text{particular}} = 0$ was not printed in equation (1). Now it is added to the homogeneous solutions, as in (3).

Elimination reveals the number of pivots and the number of free variables. *If there are r pivots, there are r basic variables and $n - r$ free variables.* That number r will be given a name — it is the *rank of the matrix* — and the whole elimination process can be summarized:

2D   Suppose elimination reduces $Ax = b$ to $Ux = c$. Let there be r pivots; the last $m - r$ rows of U are zero. Then there is a solution only if the last $m - r$ components of c are also zero. If $r = m$, there is always a solution.

The general solution is the sum of a particular solution (with all free variables zero) and a homogeneous solution (with the $n - r$ free variables as independent parameters). If $r = n$, there are no free variables and the nullspace contains only $x = 0$.

The number r is called the **rank** of the matrix A.

Note the two extreme cases, when the rank is as large as possible:

(1) If $r = n$, there are no free variables in x.

(2) If $r = m$, there are no zero rows in U.

With $r = n$ the nullspace contains only $x = 0$. The only solution is $x_{\text{particular}}$ .With $r = m$ there are no constraints on b, the column space is all of $\mathbf{R}^m$, and for every right-hand side the equation can be solved.

An optional remark   In many texts the elimination process does not stop at U, but continues until the matrix is in a still simpler "row-reduced echelon form." The difference is that all pivots are normalized to $+1$, by dividing each row by a constant, and zeros are produced not only below but also above every pivot. For the matrix A in the text, this form would be

$$\begin{bmatrix} 1 & 3 & 0 & 1 \\ 0 & 0 & 1 & \frac{1}{3} \\ 0 & 0 & 0 & 0 \end{bmatrix}.$$

If A is square and nonsingular we reach the identity matrix. It is an instance of *Gauss-Jordan elimination*, instead of the ordinary Gaussian reduction to $A = LU$. Just as Gauss-Jordan is slower in practical calculations with square matrices, and any band structure of the matrix is lost in A^{-1}, this special echelon form requires too many operations to be the first choice on a computer. It does, however, have some theoretical importance as a "canonical form" for A: Regardless of the choice of elementary operations, including row exchanges and row divisions, the final row-reduced echelon form of A is always the same.

EXERCISES

2.2.1   How many possible patterns can you find (like the one in Fig. 2.2) for 2 by 3 echelon matrices? Entries to the right of the pivots are irrelevant.

2.2.2   Construct the smallest system you can with more unknowns than equations, but no solution.

2.2.3   Compute an LU factorization for

$$A = \begin{bmatrix} 1 & 2 & 0 & 1 \\ 0 & 1 & 1 & 0 \\ 1 & 2 & 0 & 1 \end{bmatrix}.$$

Determine a set of basic variables and a set of free variables, and find the general solution to $Ax = 0$. Write it in a form similar to (1). What is the rank of A?

2.2.4   For the matrix

$$A = \begin{bmatrix} 0 & 1 & 4 & 0 \\ 0 & 2 & 8 & 0 \end{bmatrix},$$

determine the echelon form U, the basic variables, the free variables, and the general solution to $Ax = 0$. Then apply elimination to $Ax = b$, with components b_1 and b_2 on the right side; find the conditions for $Ax = b$ to be consistent (that is, to have a solution) and find the general solution in the same form as Equation (3). What is the rank of A?

2.2.5   Carry out the same steps, with b_1, b_2, b_3, b_4 on the right side, for the transposed matrix

$$A = \begin{bmatrix} 0 & 0 \\ 1 & 2 \\ 4 & 8 \\ 0 & 0 \end{bmatrix}.$$

2.2.6   Write the general solution to

$$\begin{bmatrix} 1 & 2 & 2 \\ 2 & 4 & 5 \end{bmatrix} \begin{bmatrix} u \\ v \\ w \end{bmatrix} = \begin{bmatrix} 1 \\ 4 \end{bmatrix}$$

as the sum of a particular solution to $Ax = b$ and the general solution to $Ax = 0$, as in (3).

2.2.7   Describe the set of attainable right sides b for

$$\begin{bmatrix} 1 & 0 \\ 0 & 1 \\ 2 & 3 \end{bmatrix} \begin{bmatrix} u \\ v \end{bmatrix} = \begin{bmatrix} b_1 \\ b_2 \\ b_3 \end{bmatrix},$$

by finding the constraints on b that turn the third equation into $0 = 0$ (after elimination). What is the rank? How many free variables, and how many solutions?

2.2.8   Find the value of c which makes it possible to solve

$$u + v + 2w = 2$$
$$2u + 3v - w = 5$$
$$3u + 4v + w = c.$$

2.2.9   Under what conditions on b_1 and b_2 (if any) does $Ax = b$ have a solution, if

$$A = \begin{bmatrix} 1 & 2 & 0 & 3 \\ 2 & 4 & 0 & 7 \end{bmatrix}, \qquad b = \begin{bmatrix} b_1 \\ b_2 \end{bmatrix}?$$

Find two vectors x in the nullspace of A, and the general solution to $Ax = b$.

2.2.10   (a) Find all solutions to

$$Ux = \begin{bmatrix} 1 & 2 & 3 & 4 \\ 0 & 0 & 1 & 2 \\ 0 & 0 & 0 & 0 \end{bmatrix} \begin{bmatrix} x_1 \\ x_2 \\ x_3 \\ x_4 \end{bmatrix} = \begin{bmatrix} 0 \\ 0 \\ 0 \end{bmatrix}.$$

(b) If the right side is changed from $(0, 0, 0)$ to $(a, b, 0)$, what are the solutions?

2.2.11 Suppose the only solution to $Ax = 0$ (m equations in n unknowns) is $x = 0$. What is the rank of A?

2.2.12 Find a 2 by 3 system $Ax = b$ whose general solution is

$$x = \begin{bmatrix} 1 \\ 1 \\ 0 \end{bmatrix} + w \begin{bmatrix} 1 \\ 2 \\ 1 \end{bmatrix}.$$

2.2.13 Find a 3 by 3 system with the same general solution as above, and with no solution when $b_1 + b_2 \neq b_3$.

2.2.14 Write down a 2 by 2 system $Ax = b$ in which there are many solutions $x_{homogeneous}$ but no solution $x_{particular}$. Therefore the system has no solution.

2.3 ■ LINEAR INDEPENDENCE, BASIS, AND DIMENSION

By themselves, the numbers m and n give an incomplete picture of the true size of a linear system. The matrix in our example had three rows and four columns, but the third row was only a combination of the first two. After elimination it became a zero row. It had no real effect on the homogeneous problem $Ax = 0$. The four columns also failed to be independent, and the column space degenerated into a two-dimensional plane; the second and fourth columns were simple combinations of the first and third.

The important number which is beginning to emerge is the **rank** r. The rank was introduced in a purely computational way, as the *number of pivots* in the elimination process—or equivalently, as the number of nonzero rows in the final matrix U. This definition is so mechanical that it could be given to a computer. But it would be wrong to leave it there because the rank has a simple and intuitive meaning: *The rank counts the number of genuinely independent rows in the matrix A.* We want to give this quantity, and others like it, a definition that is mathematical rather than computational.

The goal of this section is to explain and use four ideas:

1. Linear independence or dependence
2. Spanning a subspace
3. Basis for a subspace
4. Dimension of a subspace.

The first step is to define **linear independence**. Given a set of vectors $v_1, \ldots, v_k$, we look at their combinations $c_1 v_1 + c_2 v_2 + \cdots + c_k v_k$. The trivial combination, with all weights $c_i = 0$, obviously produces the zero vector: $0 v_1 + \cdots + 0 v_k = 0$. The question is whether this is the only way to produce zero. If so, the vectors are independent. If any other combination gives zero, they are *dependent*.

> **2E**   If only the trivial combination gives zero, so that
>
> $$c_1 v_1 + \cdots + c_k v_k = 0 \quad \text{only happens when} \quad c_1 = c_2 = \cdots = c_k = 0,$$
>
> then the vectors $v_1, \ldots, v_k$ are *linearly independent*. Otherwise they are linearly dependent, and one of them is a linear combination of the others.

Linear dependence is easy to visualize in three-dimensional space, when all vectors go out from the origin. Two vectors are dependent if they lie on the same line. Three vectors are dependent if they lie in the same plane. A random choice of three vectors, without any special accident, should produce linear independence. On the other hand, four vectors are always linearly dependent in $\mathbf{R}^3$.

EXAMPLE 1   If one of the vectors, say v_1, is already the zero vector, then the set is certain to be linearly dependent. We may choose $c_1 = 3$ and all other $c_i = 0$; this is a nontrivial combination that produces zero.

EXAMPLE 2   The columns of the matrix

$$A = \begin{bmatrix} 1 & 3 & 3 & 2 \\ 2 & 6 & 9 & 5 \\ -1 & -3 & 3 & 0 \end{bmatrix}$$

are linearly dependent, since the second column is three times the first. The combination of columns with weights $-3, 1, 0, 0$ gives a column of zeros.

The rows are also linearly dependent; row 3 is two times row 2 minus five times row 1. (This is the same as the combination of b_1, b_2, b_3, which had to vanish on the right side in order for $Ax = b$ to be consistent. Unless $b_3 - 2b_2 + 5b_1 = 0$, the third equation would not become $0 = 0$.)

EXAMPLE 3   The columns of the triangular matrix

$$A = \begin{bmatrix} 3 & 4 & 2 \\ 0 & 1 & 5 \\ 0 & 0 & 2 \end{bmatrix}$$

are linearly *independent*. This is automatic whenever the diagonal entries are nonzero. To see why, we look for a combination of the columns that makes zero:

$$c_1 \begin{bmatrix} 3 \\ 0 \\ 0 \end{bmatrix} + c_2 \begin{bmatrix} 4 \\ 1 \\ 0 \end{bmatrix} + c_3 \begin{bmatrix} 2 \\ 5 \\ 2 \end{bmatrix} = \begin{bmatrix} 0 \\ 0 \\ 0 \end{bmatrix}.$$

We have to show that c_1, c_2, c_3 are all forced to be zero. The last equation gives $c_3 = 0$. Then the next equation gives $c_2 = 0$, and substituting into the first equation forces $c_1 = 0$. The only combination to produce the zero vector is the trivial combination, and the vectors are linearly independent.

Written in matrix notation, this example looked at

$$\begin{bmatrix} 3 & 4 & 2 \\ 0 & 1 & 5 \\ 0 & 0 & 2 \end{bmatrix} \begin{bmatrix} c_1 \\ c_2 \\ c_3 \end{bmatrix} = \begin{bmatrix} 0 \\ 0 \\ 0 \end{bmatrix}.$$

We showed that *the nullspace contained only the zero vector $c_1 = c_2 = c_3 = 0$*. That is exactly the same as saying that *the columns are linearly independent*.

A similar reasoning applies to the rows of A, which are also independent. Suppose we had

$$c_1(3, 4, 2) + c_2(0, 1, 5) + c_3(0, 0, 2) = (0, 0, 0).$$

From the first components we find $3c_1 = 0$ or $c_1 = 0$. Then the second components give $c_2 = 0$, and finally $c_3 = 0$.

The same idea can be extended to any echelon matrix U; the nonzero rows must be independent. Furthermore, if we pick out the columns that contain the pivots, they also are linearly independent. In our earlier example, with

$$U = \begin{bmatrix} 1 & 3 & 3 & 2 \\ 0 & 0 & 3 & 1 \\ 0 & 0 & 0 & 0 \end{bmatrix},$$

columns 1 and 3 are independent. No set of three columns is independent, and certainly not all four. It is true that columns 1 and 4 are also independent, but if that last 1 was changed to 0 they would be dependent. *It is the columns with pivots that are guaranteed to be independent.* The general rule is this:

2F   The r nonzero rows of an echelon matrix U are linearly independent, and so are the r columns that contain pivots.

EXAMPLE 4   The columns of the n by n identity matrix

$$I = \begin{bmatrix} 1 & 0 & \cdot & 0 \\ 0 & 1 & \cdot & 0 \\ \cdot & \cdot & \cdot & 0 \\ 0 & 0 & 0 & 1 \end{bmatrix}$$

are linearly independent. These particular vectors have the special notation $e_1, \ldots, e_n$; they represent unit vectors in the coordinate directions. In $\mathbf{R}^4$ they are

$$e_1 = \begin{bmatrix} 1 \\ 0 \\ 0 \\ 0 \end{bmatrix}, \qquad e_2 = \begin{bmatrix} 0 \\ 1 \\ 0 \\ 0 \end{bmatrix}, \qquad e_3 = \begin{bmatrix} 0 \\ 0 \\ 1 \\ 0 \end{bmatrix}, \qquad e_4 = \begin{bmatrix} 0 \\ 0 \\ 0 \\ 1 \end{bmatrix}.$$

Most sets of four vectors in $\mathbf{R}^4$ are linearly independent, but this set is the easiest and safest.

To check any set of vectors $v_1, \ldots, v_n$ for linear independence, form the matrix A whose n columns are the given vectors. Then solve the system $Ac = 0$; the vectors are dependent if there is a solution other than $c = 0$. If there are no free variables (the rank is n) then there is no nullspace except $c = 0$; the vectors are independent. If the rank is less than n, at least one variable is free to be chosen nonzero and the columns are linearly dependent.

One case is of special importance. Let the vectors have m components, so that A is an m by n matrix. Suppose now that $n > m$. Then it will be impossible for the columns to be independent! There cannot be n pivots, since there are not enough rows to hold them. The rank must be less than n, and a homogeneous system $Ac = 0$ with more unknowns than equations always has solutions $c \neq 0$.

2G   A set of n vectors in $\mathbf{R}^m$ must be linearly dependent if $n > m$.

The reader will recognize this as a disguised form of 2C.

EXAMPLE 5 Consider the three columns of

$$A = \begin{bmatrix} 1 & 2 & 1 \\ 1 & 3 & 2 \end{bmatrix}.$$

There cannot be three independent vectors in $\mathbf{R}^2$, and to find the combination of the columns producing zero we solve $Ac = 0$:

$$A \to U = \begin{bmatrix} 1 & 2 & 1 \\ 0 & 1 & 1 \end{bmatrix}.$$

If we give the value 1 to the free variable c_3, then back-substitution in $Uc = 0$ gives $c_2 = -1$, $c_1 = 1$. With these three weights, the first column minus the second plus the third equals zero.

Spanning a Subspace

The next step in discussing vector spaces is to define what it means for a set of vectors to *span the space*. We used this term at the beginning of the chapter, when we spoke of the plane that was spanned by the two columns of the matrix, and called this plane the column space. The general definition is simply this:

2H If a vector space V consists of all linear combinations of the particular vectors $w_1, \ldots, w_l$, then these vectors *span* the space. In other words, every vector v in V can be expressed as some combination of the w's:

$$v = c_1 w_1 + \cdots + c_l w_l \quad \text{for some coefficients } c_i.$$

It is permitted that more than one set of coefficients c_i could give the same vector v. The coefficients need not be unique because the spanning set might be excessively large—it could even include the zero vector.

EXAMPLE 6 The vectors $w_1 = (1, 0, 0)$, $w_2 = (0, 1, 0)$, and $w_3 = (-2, 0, 0)$ span a plane (the x-y plane) within three-dimensional space. So would the first two vectors alone, whereas w_1 and w_3 span only a line.

EXAMPLE 7 The column space of a matrix is exactly *the space that is spanned by the columns*. The definition is made to order. Taking all combinations of the columns is the same as taking the space they span. Multiplying A by x gives a combination of the columns; it is a vector Ax in the column space.

If the columns are the coordinate vectors $e_1, e_2, \ldots, e_n$, coming from the identity matrix, then they span $\mathbf{R}^n$. Every vector $b = (b_1, b_2, \ldots, b_n)$ is a combination of those columns. In this example the weights are the components b_i themselves: $b = b_1 e_1 + \cdots + b_n e_n$. But it is not only the columns of the identity matrix that span $\mathbf{R}^n$!

To decide if b is a combination of the columns, we try to solve $Ax = b$. To decide if the columns are independent, we solve $Ax = 0$. *Spanning involves the column space, and independence involves the nullspace.* One should be large enough; the other should contain only the zero vector. For the coordinate vectors $e_1, \ldots, e_n$ both tests are easily passed: They span $\mathbf{R}^n$ and they are linearly independent. Roughly speaking, ***no vectors in that set are wasted.*** This leads to the idea of a ***basis***.

2I   A ***basis*** for a vector space is a set of vectors having two properties at once:

(1) It is linearly independent.
(2) It spans the space.

This combination of properties is absolutely fundamental to linear algebra. It means that every vector in the space is a combination of the basis vectors, because they span. It also means that the combination is unique: If $v = a_1 v_1 + \cdots + a_k v_k$ and also $v = b_1 v_1 + \cdots + b_k v_k$, then subtraction gives $0 = \sum (a_i - b_i) v_i$. Now independence plays its part; every coefficient $a_i - b_i$ must be zero. Therefore $a_i = b_i$, and *there is one and only one way to write v as a combination of the basis vectors.*

We had better say at once that the coordinate vectors $e_1, \ldots, e_n$ are not the only basis for $\mathbf{R}^n$. Some things in linear algebra are unique, but not this. A vector space has ***infinitely many different bases***. Whenever a square matrix is invertible, its columns are independent—and they are a basis for $\mathbf{R}^n$. The two columns of any nonsingular matrix like

$$A = \begin{bmatrix} 1 & 1 \\ 2 & 3 \end{bmatrix}$$

are a basis for $\mathbf{R}^2$. Every two-dimensional vector is a combination of those columns, and they are independent.

EXAMPLE 8   Consider the usual x-y plane (Fig. 2.3), which is just $\mathbf{R}^2$. The vector v_1 by itself is linearly independent, but fails to span $\mathbf{R}^2$. The three vectors v_1, v_2, v_3 certainly span $\mathbf{R}^2$, but are not independent. *Any two* of these vectors, say v_1 and v_2, have both properties—they span, and they are independent, so they form a basis. Notice again that *a vector space does not have a unique basis.*

EXAMPLE 9   Consider the 3 by 4 echelon matrix U:

$$U = \begin{bmatrix} 1 & 3 & 3 & 2 \\ 0 & 0 & 3 & 1 \\ 0 & 0 & 0 & 0 \end{bmatrix}.$$

The four columns span the column space, as always, but they are not independent. There are many possibilities for a basis, but we propose a specific choice: ***The columns that contain pivots*** (in this case the first and third, which correspond

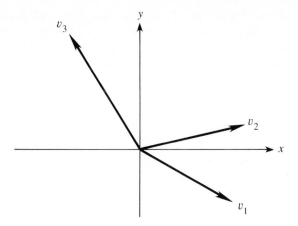

Fig. 2.3. A spanning set and a basis in $\mathbf{R}^2$.

to the basic variables) *are a basis for the column space*. We noted in 2F that these columns are independent, and it is easy to see that they span the space. In fact, the column space of U is just the x-y plane within $\mathbf{R}^3$. It is *not the same* as the column space of A.

To summarize: The columns of a matrix span its column space. If they are independent, they are a basis for the column space—whether the matrix is square or rectangular. If we are speaking about the whole space $\mathbf{R}^n$, and asking the columns to be a basis for that space, then the matrix must be square and invertible.

Dimension of a Vector Space

In spite of the fact that there is no unique choice of basis, and infinitely many different possibilities would do equally well, there *is* something common to all of these choices. It is a property that is intrinsic to the space itself:

2J Any two bases for a vector space V contain the same number of vectors. This number, which is shared by all bases and expresses the number of "degrees of freedom" of the space, is called the ***dimension*** of V.†

We have to prove this fact: All possible bases contain the same number of vectors. First we ask the reader to look at the examples, and notice their dimension.

---

† You must notice that the word "dimensional" is used in two different ways. We speak about a four-dimensional ***vector***, meaning a vector with four components—in other words, a member of $\mathbf{R}^4$. Now we have defined a four-dimensional ***subspace***; an example is the set of vectors in $\mathbf{R}^6$ whose first and last components are zero. The members of this four-dimensional subspace are six-dimensional vectors like (0, 5, 1, 3, 4, 0).

The x-y plane in Fig. 2.3 has two vectors in every basis; its dimension is 2. In three dimensions we need three vectors, along the x-y-z axes or in three other (linearly independent!) directions. *The dimension of the space $\mathbf{R}^n$ is n.* The column space of U in Example 9 had dimension 2; it was a "two-dimensional subspace of $\mathbf{R}^3$." The zero matrix is rather exceptional, because its column space contains only the zero vector. By convention, the empty set is a basis for that space, and its dimension is zero.

The statement 2J, on which the idea of dimension depends, is equivalent to

2K   Suppose that $v_1, \ldots, v_m$ and $w_1, \ldots, w_n$ are both bases for the same vector space V. Then $m = n$.

Proof   Suppose one set is smaller than the other, say $m < n$; we want to arrive at a contradiction. Since the v's form a basis, they must span the space, and every w_j can be written as a combination of the v's:

$$w_j = a_{1j}v_1 + \cdots + a_{mj}v_m = \sum_{i=1}^{m} a_{ij}v_i.$$

In matrix terms this is $W = VA$, if the w's are the columns of W and the v's are the columns of V. We have no way to know the coefficients a_{ij}, but we do know the important thing: A is m by n, with $m < n$. By 2C there must be a nontrivial solution to $Ac = 0$. Multiplying by V gives $VAc = 0$, or $Wc = 0$, which means that the vectors w_j combined with the coefficients c_j add to zero. Evidently, the vectors w_j are not linearly independent. Since this contradicts the hypothesis that they form a basis, we must give up the possibility that $m < n$.

This proof was the same as the one used earlier to show that every set of $m + 1$ vectors in $\mathbf{R}^m$ must be dependent. A similar proof extends to arbitrary vector spaces, not necessarily spaces of column vectors. In fact we can see that the general result is this: *In a subspace of dimension k, no set of more than k vectors can be linearly independent, and no set of fewer than k vectors can span the space.*

There are other "dual" theorems, of which we mention only one; it permits us to start with a set of vectors that is either too small or too big, and to end up with a basis:

2L   Any linearly independent set in V can be extended to a basis, by adding more vectors if necessary.

Any spanning set in V can be reduced to a basis, by discarding vectors if necessary.

The point is that a basis is a *maximal independent set*. It cannot be made larger without losing independence. A basis is also a *minimal spanning set*. It cannot be made smaller and still span the space.

One final note about the language of linear algebra. We never use the terms "basis of a matrix" or "rank of a space" or "dimension of a basis"; these phrases have no meaning. It is the *dimension of the column space* that equals the *rank of the matrix*, as we prove in the coming section.

EXERCISES

2.3.1   Decide whether or not the following vectors are linearly independent, by solving $c_1 v_1 + c_2 v_2 + c_3 v_3 + c_4 v_4 = 0$:

$$v_1 = \begin{bmatrix} 1 \\ 1 \\ 0 \\ 0 \end{bmatrix}, \quad v_2 = \begin{bmatrix} 1 \\ 0 \\ 1 \\ 0 \end{bmatrix}, \quad v_3 = \begin{bmatrix} 0 \\ 0 \\ 1 \\ 1 \end{bmatrix}, \quad v_4 = \begin{bmatrix} 0 \\ 1 \\ 0 \\ 1 \end{bmatrix}.$$

Decide also if they span $\mathbf{R}^4$, by trying to solve $c_1 v_1 + \cdots + c_4 v_4 = (0, 0, 0, 1)$.

2.3.2   Decide the dependence or independence of
(a) $(1, 1, 2), (1, 2, 1), (3, 1, 1)$;
(b) $v_1 - v_2, v_2 - v_3, v_3 - v_4, v_4 - v_1$ for any vectors v_1, v_2, v_3, v_4;
(c) $(1, 1, 0), (1, 0, 0), (0, 1, 1), (x, y, z)$ for any numbers x, y, z.

2.3.3   Prove that if any diagonal element of

$$T = \begin{bmatrix} a & b & c \\ 0 & d & e \\ 0 & 0 & f \end{bmatrix}$$

is zero, the rows are linearly dependent.

2.3.4   Is it true that if v_1, v_2, v_3 are linearly dependent, then also the vectors $w_1 = v_1 + v_2$, $w_2 = v_1 + v_3, w_3 = v_2 + v_3$ are linearly independent? (Hint: Assume some combination $c_1 w_1 + c_2 w_2 + c_3 w_3 = 0$, and find which c_i are possible.)

2.3.5   Suppose the vectors to be tested for independence are placed into the rows instead of the columns of A. How does the elimination process decide for or against independence? Apply this to the vectors in Ex. 2.3.1.

2.3.6   Describe geometrically the subspace of $\mathbf{R}^3$ spanned by
(a) $(0, 0, 0), (0, 1, 0), (0, 2, 0)$;
(b) $(0, 0, 1), (0, 1, 1), (0, 2, 1)$;
(c)   all six of these vectors. Which two form a basis?
(d)   all vectors with positive components.

2.3.7   To decide whether b is in the subspace spanned by $w_1, \ldots, w_l$, let the vectors w be the columns of A and try to solve $Ax = b$. What is the result for
(a) $w_1 = (1, 1, 0), w_2 = (2, 2, 1), w_3 = (0, 0, 2), b = (3, 4, 5)$;
(b) $w_1 = (1, 2, 0), w_2 = (2, 5, 0), w_3 = (0, 0, 2), w_4 = (0, 0, 0)$, and any b?

2.3.8   Describe, in words or in a sketch of the x-y plane, the column space of $A = \begin{bmatrix} 1 & 2 \\ 3 & 6 \end{bmatrix}$ and of A^2. Give a basis for the column space.

2.3.9   By locating the pivots, find a basis for the column space of

$$U = \begin{bmatrix} 0 & 1 & 4 & 3 \\ 0 & 0 & 2 & 2 \\ 0 & 0 & 0 & 0 \\ 0 & 0 & 0 & 0 \end{bmatrix}.$$

Express each column that is not in the basis as a combination of the basic columns. Find also a matrix A with this echelon form U, but a different column space.

2.3.10   Suppose we think of each 2 by 2 matrix as a "vector." Although these are not vectors in the usual sense, we do have rules for adding matrices and multiplying by scalars, and the set of matrices is closed under these operations. Find a basis for this vector space. What subspace is spanned by the set of all echelon matrices U?

2.3.11   Find two different bases for the subspace of all vectors in $\mathbf{R}^3$ whose first two components are equal.

2.3.12   Find a counterexample to the following statement: If v_1, v_2, v_3, v_4 is a basis for the vector space $\mathbf{R}^4$, and if W is a subspace, then some subset of the v's is a basis for W.

2.3.13   Find the dimensions of
(a)   the space of all vectors in $\mathbf{R}^4$ whose components add to zero;
(b)   the nullspace of the 4 by 4 identity matrix;
(c)   the space of all 4 by 4 matrices.

2.3.14   For the matrix $A = \begin{bmatrix} 1 & 2 & 1 \\ 0 & 0 & 4 \end{bmatrix}$, extend the set of rows to a basis for $\mathbf{R}^3$, and (separately) reduce the set of columns to a basis for $\mathbf{R}^2$.

2.3.15   Suppose V is known to have dimension k. Prove that

(i)   any k independent vectors in V form a basis;
(ii)   any k vectors that span V form a basis.

In other words, if the number of vectors is known to be right, either of the two properties of a basis implies the other.

2.3.16   Find the dimension of the space of 3 by 3 symmetric matrices, as well as a basis.

2.3.17   Prove that if V and W are three-dimensional subspaces of $\mathbf{R}^5$, then V and W must have a nonzero vector in common. *Hint*: Start with bases for the two subspaces, making six vectors in all.

2.3.18   *True or false*:   (a)   If the columns of A are linearly independent, then $Ax = b$ has exactly one solution for every b.
(b)   A 5 by 7 matrix never has linearly independent columns.

2.3.19   Suppose n vectors from $\mathbf{R}^m$ go into the columns of A. If they are linearly independent, what is the rank of A? If they span $\mathbf{R}^m$, what is the rank? If they are a basis for $\mathbf{R}^m$, what then?

2.3.20   In the space of 2 by 2 matrices, find a basis for the subspace of matrices whose row sums and column sums are all equal. (*Extra credit*: Find five linearly independent 3 by 3 matrices with this property.)

2.3.21   If A is a 64 by 17 matrix of rank 11, how many independent vectors satisfy $Ax = 0$? How many independent vectors satisfy $A^Ty = 0$?

2.3.22   Suppose V is a vector space of dimension 7 and W is a subspace of dimension 4. *True or false*:
(1)   Every basis for W can be extended to a basis for V by adding three more vectors;
(2)   Every basis for V can be reduced to a basis for W by removing three vectors.

2.3.23   Suppose $v_1, v_2, \ldots, v_9$ are nine vectors in $\mathbf{R}^7$.
(a)   Those vectors (are) (are not) (might be) linearly independent.
(b)   They (do) (do not) (might) span $\mathbf{R}^7$.
(c)   If those vectors are the columns of A, then $Ax = b$ (has) (does not have) (might not have) a solution.

2.4 ■ THE FOUR FUNDAMENTAL SUBSPACES

The previous section dealt with definitions rather than constructions. We know what a basis is, but not how to find one. Now, starting from an explicit description of a subspace, we would like to compute an explicit basis.

Subspaces are generally described in one of two ways. First, we may be given a set of vectors that span the space; this is the case for the column space, when the columns are specified. Second, we may be given a list of constraints on the subspace; we are told, not which vectors are in the space, but which conditions they must satisfy. The nullspace consists of all vectors which satisfy $Ax = 0$, and each equation in this system represents a constraint. In the first description, there may be useless columns; in the second there may be repeated constraints. In neither case is it possible to write down a basis by inspection, and a systematic procedure is necessary.

The reader can guess what that procedure will be. We shall show how to find, from the L and U (and P) which are produced by elimination, a basis for each of the subspaces associated with A. Then, even if it makes this section fuller than the others, we have to look at the extreme case:

> When the rank is as large as possible, $r = n$ or $r = m$ or $r = m = n$,
> the matrix has a left-inverse B or a right-inverse C or a two-sided A^{-1}.

To organize the whole discussion, we consider each of the four fundamental subspaces in turn. Two of them are familiar and two are new.

1. The **column space** of A, denoted by $\mathcal{R}(A)$.
2. The **nullspace** of A, denoted by $\mathcal{N}(A)$.
3. The **row space** of A, which is the *column space of A^T*. It is $\mathcal{R}(A^T)$, and it is spanned by the rows of A.
4. The **left nullspace** of A, which is the *nullspace of A^T*. It contains all vectors y such that $A^T y = 0$, and it is written $\mathcal{N}(A^T)$.

The point about the last two subspaces is that *they come from A^T*. If A is an m by n matrix, you can see which "host" spaces contain the four subspaces by looking at the number of components:

The nullspace $\mathcal{N}(A)$ and row space $\mathcal{R}(A^T)$ are subspaces of $\mathbf{R}^n$
The left nullspace $\mathcal{N}(A^T)$ and column space $\mathcal{R}(A)$ are subspaces of $\mathbf{R}^m$.

The rows have n components and the columns have m. For a simple matrix like

$$A = \begin{bmatrix} 1 & 0 & 0 \\ 0 & 0 & 0 \end{bmatrix},$$

the column space is the line through $\begin{bmatrix} 1 \\ 0 \end{bmatrix}$. The row space is the line through $[1 \quad 0 \quad 0]^T$. It is in $\mathbf{R}^3$. The nullspace is a plane in $\mathbf{R}^3$ and the left nullspace is a line in $\mathbf{R}^2$:

$$\mathcal{N}(A) \text{ contains } \begin{bmatrix} 0 \\ 1 \\ 0 \end{bmatrix} \text{ and } \begin{bmatrix} 0 \\ 0 \\ 1 \end{bmatrix}, \quad \mathcal{N}(A^{\mathrm{T}}) \text{ contains } \begin{bmatrix} 0 \\ 1 \end{bmatrix}.$$

Note that all vectors are column vectors. Even the rows are transposed, and the row space of A is the *column* space of A^{T}. Normally the simple matrix is U, after elimination. Our problem will be to connect the spaces for U to the spaces for A. Therefore we watch the echelon matrix at the same time as the original matrix:

$$U = \begin{bmatrix} 1 & 3 & 3 & 2 \\ 0 & 0 & 3 & 1 \\ 0 & 0 & 0 & 0 \end{bmatrix} \quad \text{and} \quad A = \begin{bmatrix} 1 & 3 & 3 & 2 \\ 2 & 6 & 9 & 5 \\ -1 & -3 & 3 & 0 \end{bmatrix}.$$

For novelty we take the four subspaces in a more interesting order.

3. The row space of A   For an echelon matrix like U, the row space is clear. It contains all combinations of the rows, as every row space does—but here the third row contributes nothing. The first two rows are a basis for the row space. A similar rule applies to every echelon matrix, with r pivots and r nonzero rows: *Its nonzero rows are independent, and its row space has dimension r.* Fortunately, it is equally easy to deal with the original matrix A. Its third row contributes nothing too.

2M   The row space of A has the same dimension r as the row space of U, and it has the same bases, because *the two row spaces are the same.*

The reason is that each elementary operation leaves the row space unchanged. The rows in U are combinations of the original rows in A. Therefore the row space of U contains nothing new. At the same time, because every step can be reversed, nothing is lost; the rows of A can be recovered from U. Row 2 of U came from rows 1 and 2 of A. The rows of A come from rows 1 and 2 of U. It is true that A and U have different rows, but the *combinations* of the rows are identical. It is those combinations that make up the row space.

Note that we did not start with the m rows of A, which span the row space, and discard $m - r$ of them to end up with a basis. According to 2L, we could have done so. But it might be hard to decide which rows to keep and which to discard, so it was easier just to take the nonzero rows of U.

2. The nullspace of A   Recall that the original purpose of elimination was to simplify a system of linear equations without changing any of the solutions. The system $Ax = 0$ is reduced to $Ux = 0$, and this process is reversible. ***Therefore the nullspace of A is the same as the nullspace of U.*** Of the m constraints apparently imposed by the m equations $Ax = 0$, only r are independent. They are specified by any r linearly independent rows of A, or (more clearly) by the r nonzero rows of U. If we choose the latter, it provides a definite way to find a basis for the nullspace:

2N   The nullspace $\mathcal{N}(A)$ is of dimension $n - r$. A basis can be constructed by reducing to $Ux = 0$, which has $n - r$ free variables—corresponding to the columns of U that do not contain pivots. Then, in turn, we give to each free variable the value 1, to the other free variables the value 0, and solve $Ux = 0$ by back-substitution for the remaining (basic) variables. The $n - r$ vectors produced in this way are a basis for $\mathcal{N}(A)$.

This is exactly the way we have been solving $Ux = 0$, with basic variables and free variables. The example has pivots in columns 1 and 3. Therefore its free variables are the second and fourth, v and y, and the basis for the nullspace is

$$
\begin{matrix} v = 1 \\ y = 0 \end{matrix}\ x_1 = \begin{bmatrix} -3 \\ 1 \\ 0 \\ 0 \end{bmatrix}; \qquad \begin{matrix} v = 0 \\ y = 1 \end{matrix}\ x_2 = \begin{bmatrix} -1 \\ 0 \\ -\frac{1}{3} \\ 1 \end{bmatrix}.
$$

It is easy to see, for this example or in general, that these vectors are independent. Any combination $c_1 x_1 + c_2 x_2$ has the value c_1 as its v component, and c_2 as its y component. The only way to have $c_1 x_1 + c_2 x_2 = 0$ is to have $c_1 = c_2 = 0$. These two vectors also span the nullspace; the general solution is a combination $v x_1 + y x_2$. Thus the $n - r = 4 - 2$ vectors are a basis.

The nullspace is also called the *kernel* of A, and its dimension $n - r$ is the *nullity*.

1. The column space of A   First another point of notation; the column space is often called the **range** of A (which accounts for the letter $\mathcal{R}$). This is consistent with the usual idea of the range of a function f, as the set of ll possible values $f(x)$; x is in the domain and $f(x)$ is in the range. In our case the function is $f(x) = Ax$. Its domain consists of all x in $\mathbf{R}^n$; its range is all possible vectors Ax. (In other words, all b for which $Ax = b$ can be solved.) We know that this is the same as all combinations of the columns; the range is the column space. We plan to keep the useful term *column space*, but also to adopt the shorthand notation $\mathcal{R}(A)$.†

Our problem is to find a basis for the column space $\mathcal{R}(U)$ and also the column space $\mathcal{R}(A)$. ***Those spaces are different*** (just look at the matrices!) but their dimensions are the same. That is the main point.

---

† It is a sad accident that *row space* also starts with the same letter. In this book, r stands for rank and $\mathcal{R}$ stands for column space.

The first and third columns of U are a basis for its column space. They are the columns *with pivots*. Every other column is a combination of those two. Furthermore, the same is true of the original matrix—even though the pivots are invisible and the columns are different. The first and third columns of A are a basis for *its* column space. Certainly the second column is not independent; it is three times the first. The fourth column equals (column 1) $+ \frac{1}{3}$(column 3). Whenever certain columns of U form a basis for the column space of U, the corresponding columns of A form a basis for the column space of A.

The reason is this: $Ax = 0$ exactly when $Ux = 0$. The two systems are equivalent and have the same solutions. The fourth column of U was also (column 1) $+ \frac{1}{3}$(column 3). Looking at matrix multiplication, $Ax = 0$ expresses a linear dependence among the columns of A, with coefficients given by x. That dependence is matched by a dependence $Ux = 0$ among the columns of U, with exactly the same coefficients. *If a set of columns of A is independent, then so are the corresponding columns of U, and vice versa.*†

Now, to find a basis for the column space $\mathcal{R}(A)$, we use what is already done for U. The r columns containing pivots are a basis for the column space of U. We will pick those same columns of A, as follows:

20   The dimension of the column space $\mathcal{R}(A)$ equals the rank r, which also equals the dimension of the row space: *The number of independent columns equals the number of independent rows.* A basis for $\mathcal{R}(A)$ is formed by the r columns of A which correspond, over in U, to the columns containing pivots.

This fact, that the row space and the column space have the same dimension r, is one of the most important theorems in linear algebra. It is often abbreviated as "*row rank = column rank*." It expresses a result that, for a random 10 by 12 matrix, is not at all obvious. It also says something about square matrices: *If the rows of a square matrix are linearly independent, then so are the columns* (and vice versa). Again that does not seem self-evident, at least not to the author.

To see once more that both the row and column spaces of U have dimension r, consider a typical situation with rank $r = 3$. The echelon matrix U certainly has three independent rows:

$$U = \begin{bmatrix} d_1 & * & * & * & * & * \\ 0 & 0 & 0 & d_2 & * & * \\ 0 & 0 & 0 & 0 & 0 & d_3 \\ 0 & 0 & 0 & 0 & 0 & 0 \end{bmatrix}.$$

We claim that there are also three independent columns, and no more. The columns have only three nonzero components. Therefore if we can show that the three basic

---

† I think this is the most subtle argument to appear so far in the book. Fortunately, it is not wasted: The conclusion 2O to which it leads is also the most subtle and most significant.

columns—the first, fourth, and sixth—are linearly independent, they must be a basis (for the column space of U, not A!). Suppose that some combination of these basic columns produced zero:

$$c_1 \begin{bmatrix} d_1 \\ 0 \\ 0 \\ 0 \end{bmatrix} + c_2 \begin{bmatrix} * \\ d_2 \\ 0 \\ 0 \end{bmatrix} + c_3 \begin{bmatrix} * \\ * \\ d_3 \\ 0 \end{bmatrix} = \begin{bmatrix} 0 \\ 0 \\ 0 \\ 0 \end{bmatrix}.$$

Working upward in the usual way, c_3 must be zero because the pivot $d_3 \neq 0$, then c_2 must be zero because $d_2 \neq 0$, and finally $c_1 = 0$. This establishes linear independence and completes the proof. Since $Ax = 0$ if and only if $Ux = 0$, we must find that the first, fourth, and sixth columns of A—whatever the original matrix A was, which we do not even know in this example—are a basis for $\mathcal{R}(A)$.

We emphasize that the row space and column space both became clear after elimination on A. We did not have to work with A^T. Certainly we could have transposed A, exchanging its columns for its rows (and its column space for its row space). Then we could have reduced A^T to its own echelon form (which is different from U^T). That leads to the right spaces, but it is not the right idea. There are many uses for the transpose; this is not one of them. The point is that $\mathcal{R}(A)$ and $\mathcal{R}(A^T)$ share the same dimension and can be found at the same time, from U.

Now comes the fourth fundamental subspace, which has been keeping quietly out of sight. Since the first three spaces were $\mathcal{R}(A)$, $\mathcal{N}(A)$, and $\mathcal{R}(A^T)$, the fourth must be $\mathcal{N}(A^T)$. It is the nullspace of the transpose, or the **left nullspace** of A— because $A^T y = 0$ means $y^T A = 0$, and the vector appears on the left side of A.

4. The left nullspace of A (= the nullspace of A^T)   If A is an m by n matrix, then A^T is n by m. Its nullspace is a subspace of $\mathbf{R}^m$; the vector y has m components. Written as $y^T A = 0$, those components multiply the *rows* of A to produce the zero row:

$$y^T A = [y_1 \cdots y_m] \begin{bmatrix} & & \\ & A & \\ & & \end{bmatrix} = [0 \cdots 0].$$

Such a row vector y^T is called a left nullvector of A.

The dimension of this nullspace $\mathcal{N}(A^T)$ is easy to find. For *any* matrix, the number of basic variables plus the number of free variables must match the total number of columns. For A that was $r + (n - r) = n$. In other words,

dimension of column space + dimension of nullspace = number of columns.

This rule applies equally to A^T, which has m columns and is just as good a matrix as A. But the dimension of its column space is also r, so

$$r + \dim \mathcal{N}(A^T) = m. \tag{1}$$

2P   The left nullspace $\mathcal{N}(A^T)$ has dimension $m - r$.

The vectors y are hiding somewhere in elimination, when the rows of A are combined to produce the $m - r$ *zero rows* of U. To find y we start from $PA = LU$, or $L^{-1}PA = U$. The last $m - r$ rows of $L^{-1}P$ must be a basis for the left nullspace—because they multiply A to give the zero rows in U. To repeat: The left nullspace contains the coefficients that make the rows of A combine to give zero.

In our 3 by 4 example, the zero row was row $3 - 2(\text{row } 2) + 5(\text{row } 1)$. It was the same combination as in $b_3 - 2b_2 + 5b_1$ on the right side, leading to $0 = 0$ as the final equation. Therefore the components of y are $5, -2, 1$. That vector is a basis for the left nullspace, which has dimension $m - r = 3 - 2 = 1$. It is the last row of $L^{-1}P$, and produces the zero row in U—and we can often see it without computing L^{-1}. If desperate, it is always possible just to solve $A^{T}y = 0$.

I realize that so far the book has given no reason to care about $\mathcal{N}(A^{T})$. It is correct but not convincing if I write in italics *the left nullspace is also important*. The next section does better, by finding a physical meaning for y.

Now we know the dimensions of the four spaces. We can summarize them in a table, and it even seems fair to advertise them as the

Fundamental Theorem of Linear Algebra, Part I

1. $\mathcal{R}(A)$ = column space of A; dimension r
2. $\mathcal{N}(A)$ = nullspace of A; dimension $n - r$
3. $\mathcal{R}(A^{T})$ = row space of A; dimension r
4. $\mathcal{N}(A^{T})$ = left nullspace of A; dimension $m - r$

EXAMPLE $A = \begin{bmatrix} 1 & 2 \\ 3 & 6 \end{bmatrix}$ $m = n = 2, r = 1$

1. The **column space** contains all multiples of $\begin{bmatrix} 1 \\ 3 \end{bmatrix}$. The second column is in the same direction and contributes nothing new.
2. The **nullspace** contains all multiples of $\begin{bmatrix} -2 \\ 1 \end{bmatrix}$. The vector satisfies $Ax = 0$ and so do its multiples.
3. The **row space** contains all multiples of $\begin{bmatrix} 1 \\ 2 \end{bmatrix}$. I write it as a column vector, since strictly speaking it is in the column space of A^{T}.
4. The **left nullspace** contains all multiples of $\begin{bmatrix} -3 \\ 1 \end{bmatrix}$. That vector satisfies $A^{T}y = 0$—and the rows of A with coefficients -3 and 1 add to zero.

In this example *all four subspaces were lines!* That is an accident, coming from $r = 1$ and $n - r = 1$ and $m - r = 1$. The exercises will be more varied.

Note that if you change the last entry of A from 6 to 7, all the dimensions are different. The column space and row space have dimension $r = 2$. The nullspace and left nullspace contain only the vectors $x = 0$ and $y = 0$. *The matrix is invertible.*

Existence of Inverses

We know that if A has a left-inverse ($BA = I$) and a right-inverse ($AC = I$), then the two inverses are equal: $B = B(AC) = (BA)C = C$. Now, from the rank of a matrix, it is easy to decide which matrices actually have these inverses. Roughly speaking, *an inverse exists only when the rank is as large as possible.*

The rank always satisfies $r \leq m$ and also $r \leq n$. An m by n matrix cannot have more than m independent rows or n independent columns. There is not space for more than m pivots, or more than n. We want to prove that when $r = m$ there is a right-inverse, and when $r = n$ there is a left-inverse. In the first case $Ax = b$ always has a solution. In the second case the solution (*if it exists*) is unique. Only a square matrix can have both $r = m$ and $r = n$, and therefore only a square matrix can achieve both existence and uniqueness. Only a square matrix has a two-sided inverse.

2Q   EXISTENCE: The system $Ax = b$ has *at least* one solution x for every b if and only if the columns span $\mathbf{R}^m$; then $r = m$. In this case there exists an n by m right-inverse C such that $AC = I_m$, the identity matrix of order m. This is possible only if $m \leq n$.

UNIQUENESS: The system $Ax = b$ has *at most* one solution x for every b if and only if the columns are linearly independent; then $r = n$. In this case there exists an n by m left-inverse B such that $BA = I_n$, the identity matrix of order n. This is possible only if $m \geq n$.

In the first case, one possible solution is $x = Cb$, since then $Ax = ACb = b$. But there will be other solutions if there are other right-inverses.

In the second case, if there is a solution to $Ax = b$, it has to be $x = BAx = Bb$. But there may be no solution.†

There are simple formulas for left and right inverses, if they exist:

$$B = (A^{\mathrm{T}}A)^{-1}A^{\mathrm{T}} \quad \text{and} \quad C = A^{\mathrm{T}}(AA^{\mathrm{T}})^{-1}.$$

Certainly $BA = I$ and $AC = I$. What is not so certain is that $A^{\mathrm{T}}A$ and AA^{T} are actually invertible. We show in Chapter 3 that $A^{\mathrm{T}}A$ does have an inverse if the rank is n, and AA^{T} has an inverse when the rank is m. Thus the formulas make sense exactly when the rank is as large as possible, and the one-sided inverses are found.

There is also a more basic approach. We can look, a column at a time, for a matrix C such that

$$AC = I \quad \text{or} \quad A[x_1 \quad x_2 \quad \cdots \quad x_m] = [e_1 \quad e_2 \quad \cdots \quad e_m].$$

Each column of C, when multiplied by A, gives a column of the identity matrix. To solve $Ax_i = e_i$ we need the coordinate vectors e_i to be in the column space. If it contains all those vectors, the column space must be all of $\mathbf{R}^m$! Its dimension (the rank) must be $r = m$. This is the "existence case," when the columns span $\mathbf{R}^m$.

---

† The number of solutions in the "uniqueness case" is 0 or 1, whereas in the "existence case" it is 1 or ∞.

EXAMPLE   Consider a simple 2 by 3 matrix of rank 2:

$$A = \begin{bmatrix} 4 & 0 & 0 \\ 0 & 5 & 0 \end{bmatrix}.$$

Since $r = m = 2$, the theorem guarantees a right-inverse C:

$$AC = \begin{bmatrix} 4 & 0 & 0 \\ 0 & 5 & 0 \end{bmatrix} \begin{bmatrix} \frac{1}{4} & 0 \\ 0 & \frac{1}{5} \\ c_{31} & c_{32} \end{bmatrix} = \begin{bmatrix} 1 & 0 \\ 0 & 1 \end{bmatrix}.$$

In fact, there are many right-inverses; the last row of C is completely arbitrary. This is a case of existence but no uniqueness. The matrix A has no left-inverse because the last column of BA is certain to be zero, and cannot agree with the 3 by 3 identity matrix.

For this example the formula $C = A^{T}(AA^{T})^{-1}$ gives the specific choice

$$C = \begin{bmatrix} 4 & 0 \\ 0 & 5 \\ 0 & 0 \end{bmatrix} \begin{bmatrix} \frac{1}{16} & 0 \\ 0 & \frac{1}{25} \end{bmatrix} = \begin{bmatrix} \frac{1}{4} & 0 \\ 0 & \frac{1}{5} \\ 0 & 0 \end{bmatrix}.$$

The formula chooses the arbitrary values c_{31} and c_{32} to be zero. This is an instance of the "*pseudoinverse*"—a way of deciding on a particular inverse when there is no normal way to decide. It is developed in the first Appendix.

The transpose of A yields an example in the opposite direction, with infinitely many *left*-inverses:

$$\begin{bmatrix} \frac{1}{4} & 0 & b_{13} \\ 0 & \frac{1}{5} & b_{23} \end{bmatrix} \begin{bmatrix} 4 & 0 \\ 0 & 5 \\ 0 & 0 \end{bmatrix} = \begin{bmatrix} 1 & 0 \\ 0 & 1 \end{bmatrix}.$$

Now it is the last column of B that is completely arbitrary. This is typical of the "uniqueness case," when the n columns of A are linearly independent. The rank is $r = n$. There are no free variables, since $n - r = 0$, so if there is a solution it will be the only one. You can see when this example has a solution:

$$\begin{bmatrix} 4 & 0 \\ 0 & 5 \\ 0 & 0 \end{bmatrix} \begin{bmatrix} x_1 \\ x_2 \end{bmatrix} = \begin{bmatrix} b_1 \\ b_2 \\ b_3 \end{bmatrix} \quad \text{is solvable if} \quad b_3 = 0.$$

When b_3 is zero, the solution (unique!) is $x_1 = \frac{1}{4}b_1$, $x_2 = \frac{1}{5}b_2$.

For a rectangular matrix, it is not possible to have both existence and uniqueness. If m is different from n, we cannot have $r = m$ and $r = n$. A square matrix is the opposite. If $m = n$, we cannot have one property *without* the other. A square matrix has a left-inverse if and only if it has a right-inverse. There is only one inverse, namely $B = C = A^{-1}$. *Existence implies uniqueness and uniqueness implies existence, when the matrix is square.* The condition for this invertibility is that the rank must be as large as possible: $r = m = n$. We can say this in another way: For a square matrix A of order n to be nonsingular, each of the following conditions is a necessary and sufficient test:

(1)   The columns span $\mathbf{R}^n$, so $Ax = b$ has at least one solution for every b.
(2)   The columns are independent, so $Ax = 0$ has only the solution $x = 0$.

This list can be made much longer, especially if we look ahead to later chapters; every condition in the list is equivalent to every other, and ensures that A is non-singular.

(3)   The rows of A span $\mathbf{R}^n$.
(4)   The rows are linearly independent.
(5)   Elimination can be completed: $PA = LDU$, with all $d_i \neq 0$.
(6)   There exists a matrix A^{-1} such that $AA^{-1} = A^{-1}A = I$.
(7)   The determinant of A is not zero.
(8)   Zero is not an eigenvalue of A.
(9) $A^T A$ is positive definite.

Here is a typical application. Consider all polynomials $P(t)$ of degree $n - 1$. The only such polynomial that vanishes at n given points $t_1, \ldots, t_n$ is $P(t) \equiv 0$. No other polynomial of degree $n - 1$ can have n roots. This is a statement of uniqueness, and it implies a statement of existence: Given any values $b_1, \ldots, b_n$, there exists a polynomial of degree $n - 1$ interpolating these values: $P(t_i) = b_i$, $i = 1, \ldots, n$. The point is that we are dealing with a square matrix; the number of coefficients in $P(t)$ (which is n) matches the number of equations. In fact the equations $P(t_i) = b_i$ are the same as

$$
\begin{bmatrix}
1 & t_1 & t_1^2 & \cdots & t_1^{n-1} \\
1 & t_2 & t_2^2 & \cdots & t_2^{n-1} \\
\vdots & \vdots & \vdots & \vdots & \vdots \\
1 & t_n & t_n^2 & \cdots & t_n^{n-1}
\end{bmatrix}
\begin{bmatrix}
x_1 \\
x_2 \\
\vdots \\
x_n
\end{bmatrix}
=
\begin{bmatrix}
b_1 \\
b_2 \\
\vdots \\
b_n
\end{bmatrix}.
$$

The coefficient matrix A is n by n, and is known as *Vandermonde's matrix*. To repeat the argument: Since $Ax = 0$ has only the solution $x = 0$ (in other words $P(t_i) = 0$ is only possible if $P \equiv 0$), it follows that A is nonsingular. Thus $Ax = b$ always has a solution—a polynomial can be passed through any n values b_i at distinct points t_i. Later we shall actually find the determinant of A; it is not zero.

Matrices of Rank One

Finally comes the easiest case, when the rank is as *small* as possible (except for the zero matrix with rank zero). One of the basic themes of mathematics is, given something complicated, to show how it can be broken into simple pieces. For linear algebra the simple pieces are matrices of **rank one**, $r = 1$. The following example is typical:

$$A = \begin{bmatrix} 2 & 1 & 1 \\ 4 & 2 & 2 \\ 8 & 4 & 4 \\ -2 & -1 & -1 \end{bmatrix}.$$

Every row is a multiple of the first row, so the row space is one-dimensional. In fact, we can write the whole matrix in the following special way, as *the product of a column vector and a row vector*:

$$A = \begin{bmatrix} 2 & 1 & 1 \\ 4 & 2 & 2 \\ 8 & 4 & 4 \\ -2 & -1 & -1 \end{bmatrix} = \begin{bmatrix} 1 \\ 2 \\ 4 \\ -1 \end{bmatrix} \begin{bmatrix} 2 & 1 & 1 \end{bmatrix}.$$

The product of a 4 by 1 matrix and a 1 by 3 matrix is a 4 by 3 matrix, and this product has rank one. Note that, at the same time, the columns are all multiples of the same column vector; the column space shares the dimension $r = 1$ and reduces to a line.

The same thing will happen for any other matrix of rank one:

Every matrix of rank one has the simple form $A = uv^{\mathrm{T}}$.

The rows are all multiples of the same vector v^{T}, and the columns are all multiples of the same vector u. The row space and column space are lines.

EXERCISES

2.4.1 True or false: If $m = n$, then the row space of A equals the column space.

2.4.2 Find the dimension and construct a basis for the four subspaces associated with each of the matrices

$$A = \begin{bmatrix} 0 & 1 & 4 & 0 \\ 0 & 2 & 8 & 0 \end{bmatrix} \quad \text{and} \quad U = \begin{bmatrix} 0 & 1 & 4 & 0 \\ 0 & 0 & 0 & 0 \end{bmatrix}.$$

2.4.3 Find the dimension and a basis for the four fundamental subspaces for both

$$A = \begin{bmatrix} 1 & 2 & 0 & 1 \\ 0 & 1 & 1 & 0 \\ 1 & 2 & 0 & 1 \end{bmatrix} \quad \text{and} \quad U = \begin{bmatrix} 1 & 2 & 0 & 1 \\ 0 & 1 & 1 & 0 \\ 0 & 0 & 0 & 0 \end{bmatrix}.$$

2.4.4 Describe the four subspaces in 3-dimensional space associated with

$$A = \begin{bmatrix} 0 & 1 & 0 \\ 0 & 0 & 1 \\ 0 & 0 & 0 \end{bmatrix}.$$

2.4.5 If the product of two matrices is the zero matrix, $AB = 0$, show that the column space of B is contained in the nullspace of A. (Also the row space of A is in the left nullspace of B, since each row of A multiplies B to give a zero row.)

2.4.6 Explain why $Ax = b$ is solvable if and only if rank $A = $ rank A', where A' is formed from A by adding b as an extra column. *Hint*:   The rank is the dimension of the column space; when does adding an extra column leave the dimension unchanged?

2.4.7 Suppose A is an m by n matrix of rank r. Under what conditions on those numbers does
(a) A have a two-sided inverse: $AA^{-1} = A^{-1}A = I$?
(b) $Ax = b$ have *infinitely many* solutions for *every* b?

2.4.8 Why is there no matrix whose row space and nullspace both contain the vector $\begin{bmatrix} 1 & 1 & 1 \end{bmatrix}^T$?

2.4.9 Suppose the only solution to $Ax = 0$ (m equations in n unknowns) is $x = 0$. What is the rank and why?

2.4.10 Find a 1 by 3 matrix whose nullspace consists of all vectors in $\mathbf{R}^3$ such that $x_1 + 2x_2 + 4x_3 = 0$. Find a 3 by 3 matrix with that same nullspace.

2.4.11 If $Ax = b$ always has at least one solution, show that the only solution to $A^Ty = 0$ is $y = 0$. *Hint*:   What is the rank?

2.4.12 If $Ax = 0$ has a nonzero solution, show that $A^Ty = f$ fails to be solvable for some right sides f. Construct an example of A and f.

2.4.13 Find the rank of A and write the matrix as $A = uv^T$:

$$A = \begin{bmatrix} 1 & 0 & 0 & 3 \\ 0 & 0 & 0 & 0 \\ 2 & 0 & 0 & 6 \end{bmatrix} \quad \text{and} \quad A = \begin{bmatrix} 2 & -2 \\ 2 & -2 \end{bmatrix}.$$

2.4.14 If a, b, and c are given with $a \neq 0$, how must d be chosen so that

$$A = \begin{bmatrix} a & b \\ c & d \end{bmatrix}$$

has rank one? With this choice of d, factor A into uv^T.

2.4.15 Find a left-inverse and/or a right-inverse (when they exist) for

$$A = \begin{bmatrix} 1 & 1 & 0 \\ 0 & 1 & 1 \end{bmatrix} \quad \text{and} \quad M = \begin{bmatrix} 1 & 0 \\ 1 & 1 \\ 0 & 1 \end{bmatrix} \quad \text{and} \quad T = \begin{bmatrix} a & b \\ 0 & a \end{bmatrix}.$$

2.4.16 If the columns of A are linearly independent (A is m by n) then the rank is _____ and the nullspace is _____ and the row space is _____ and there exists a _____-inverse.

2.4.17 (*A paradox*) Suppose we look for a right-inverse of A. Then $AB = I$ leads to $A^T AB = A^T$ or $B = (A^TA)^{-1}A^T$. But that satisfies $BA = I$; it is a *left*-inverse. What step is not justified?

2.4.18 If V is the subspace spanned by

$$\begin{bmatrix} 1 \\ 1 \\ 0 \end{bmatrix}, \quad \begin{bmatrix} 1 \\ 2 \\ 0 \end{bmatrix}, \quad \begin{bmatrix} 1 \\ 5 \\ 0 \end{bmatrix}$$

find a matrix A that has V as its row space and a matrix B that has V as its nullspace.

2.4.19 Find a basis for each of the four subspaces of

$$A = \begin{bmatrix} 0 & 1 & 2 & 3 & 4 \\ 0 & 1 & 2 & 4 & 6 \\ 0 & 0 & 0 & 1 & 2 \end{bmatrix} = \begin{bmatrix} 1 & 0 & 0 \\ 1 & 1 & 0 \\ 0 & 1 & 1 \end{bmatrix} \begin{bmatrix} 0 & 1 & 2 & 3 & 4 \\ 0 & 0 & 0 & 1 & 2 \\ 0 & 0 & 0 & 0 & 0 \end{bmatrix}.$$

2.4.20 Write down a matrix with the required property or explain why no such matrix exists.

(a) Column space contains $\begin{bmatrix} 1 \\ 0 \\ 0 \end{bmatrix}, \begin{bmatrix} 0 \\ 0 \\ 1 \end{bmatrix}$, row space contains $\begin{bmatrix} 1 \\ 1 \end{bmatrix}, \begin{bmatrix} 1 \\ 2 \end{bmatrix}$

(b) Column space has basis $\begin{bmatrix} 1 \\ 1 \\ 1 \end{bmatrix}$, nullspace has basis $\begin{bmatrix} 1 \\ 2 \\ 1 \end{bmatrix}$

(c) Column space $= \mathbf{R}^4$, row space $= \mathbf{R}^3$.

2.4.21 If A has the same four fundamental subspaces as B, does $A = B$?

2.5 ■ GRAPHS AND NETWORKS

I am not entirely happy with the 3 by 4 matrix in the previous section. From a theoretical point of view it was very satisfactory; the four subspaces were computable and not trivial. All of their dimensions r, $n - r$, r, $m - r$ were nonzero. But it was invented artificially, rather than produced by a genuine application, and therefore it did not show how fundamental those subspaces really are.

This section introduces a class of rectangular matrices with two advantages. They are simple, and they are important. They are known as ***incidence matrices***, and every entry is 1, -1, or 0. What is remarkable is that the same is true of L and U and the basis vectors for the four subspaces. Those subspaces play a central role in network theory and graph theory. The incidence matrix comes directly from a graph, and we begin with a specific example—after emphasizing that the word "graph" does not refer to the graph of a function (like a parabola for $y = x^2$). There is a second meaning, completely different, which is closer to computer science than to calculus—and it is easy to explain. *This section is optional*, but it gives a chance to see rectangular matrices in action—and to see how the square symmetric matrix $A^T A$ turns up in the end.

A ***graph*** has two ingredients: a set of vertices or "*nodes*," and a set of arcs or "*edges*" that connect them. The graph in Fig. 2.4 has 4 nodes and 5 edges. It does not have an edge between every pair of nodes; that is not required (and edges from a node to itself are forbidden). It is like a road map, with cities as nodes and roads as edges. Ours is a *directed graph*, because each edge has an arrow to indicate its direction.

The ***edge-node incidence matrix*** is 5 by 4; we denote it by A. It has a row for every edge, to indicate the two nodes connected by the edge. ***If the edge goes from node j to node k, then that row has -1 in column j and $+1$ in column k.*** The incidence matrix is printed next to the graph.

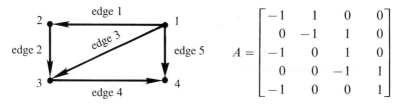

$$A = \begin{bmatrix} -1 & 1 & 0 & 0 \\ 0 & -1 & 1 & 0 \\ -1 & 0 & 1 & 0 \\ 0 & 0 & -1 & 1 \\ -1 & 0 & 0 & 1 \end{bmatrix}$$

Fig. 2.4. A directed graph and its edge-node incidence matrix.

Row 1 shows the edge from node 1 to node 2. Row 5 comes from the fifth edge, from node 1 to node 4.

Notice what happens to the columns. The third column gives information about node 3—it tells which edges enter and leave. Edges 2 and 3 go in, edge 4 goes out. A is sometimes called the *connectivity* matrix, or the *topology* matrix, and it normally has more rows than columns. When the graph has m edges and n nodes, A is m by n. Its transpose is the "node-edge" incidence matrix.

We start with the nullspace of A. Is there a combination of the columns that gives zero? Normally the answer comes from elimination, but here it comes at a glance. *The columns add up to the zero column.* Therefore the nullspace contains the vector of 1's; if $x = (1, 1, 1, 1)$ then $Ax = 0$. The equation $Ax = b$ does not have a unique solution (if it has a solution at all). Any "constant vector" $x = (c, c, c, c)$ can be added to any solution of $Ax = b$, and we still have a solution.

This has a meaning if we think of the components x_1, x_2, x_3, x_4 as the *potentials at the nodes*. The vector Ax then gives the potential differences. There are five components of Ax (the first is $x_2 - x_1$, from the ± 1 in the first row of A) and they give the *differences* in potential across the five edges. The equation $Ax = b$ therefore asks: Given the differences $b_1, \ldots, b_5$ find the actual potentials $x_1, \ldots, x_4$. But that is impossible to do! We can raise or lower all the potentials by the same constant c, and the differences will not change—confirming that $x = (c, c, c, c)$ is in the nullspace of A. In fact those are the only vectors in the nullspace, since $Ax = 0$ means equal potentials across every edge. The nullspace of this incidence matrix is 1-dimensional. Now we determine the other three subspaces.

Column space: For which differences $b_1, \ldots, b_5$ can we solve $Ax = b$? To find a direct test, look back at the matrix. The sum of rows 1 and 2 is row 3. On the right side we need $b_1 + b_2 = b_3$, or no solution is possible. Similarly the sum of rows 3 and 4 is row 5. Therefore the right side must satisfy $b_3 + b_4 = b_5$, in order for elimination to arrive at $0 = 0$. To repeat, if b is in the column space then

$$b_1 + b_2 - b_3 = 0 \qquad \text{and} \qquad b_3 + b_4 - b_5 = 0. \qquad (1)$$

Continuing the search, we also find that rows 1, 2, and 4 add to row 5. But this is nothing new; adding the equations in (1) already produces $b_1 + b_2 + b_4 = b_5$. There are two conditions on the five components, because the column space has dimension $3 = 5 - 2$. Those conditions would be found more systematically by elimination, but here they must have a meaning on the graph.

The rule is that *potential differences around a loop must add to zero.* The differences around the upper loop are b_1, b_2, and $-b_3$ (the minus sign is required by the direction of the arrow). To circle the loop and arrive back at the same potential, we need $b_1 + b_2 - b_3 = 0$. Equivalently, the potential differences must satisfy $(x_2 - x_1) + (x_1 - x_3) = (x_2 - x_3)$. Similarly the requirement $b_3 + b_4 - b_5 = 0$ comes from the lower loop. Notice that the columns of A satisfy these two requirements—they must, because $Ax = b$ is solvable exactly when b is in the column space. There are three independent columns and the rank is $r = 3$.

Left nullspace: What combinations of the rows give a zero row? That is also answered by the loops! The vectors that satisfy $y^T A = 0$ are

$$y_1^T = [1 \quad 1 \quad -1 \quad 0 \quad 0] \qquad \text{and} \qquad y_2^T = [0 \quad 0 \quad 1 \quad 1 \quad -1].$$

Each loop produces a vector y in the left nullspace. The component $+1$ or -1 indicates whether the edge arrow has the same direction as the loop arrow. The combinations of y_1 and y_2 are also in the left nullspace—in fact $y_1 + y_2 = (1, 1, 0, 1, -1)$ gives the loop around the outside of the graph.

You see that the column space and left nullspace are closely related. When the left nullspace contains $y_1 = (1, 1, -1, 0, 0)$, the vectors in the column space satisfy

$b_1 + b_2 - b_3 = 0$. This illustrates the rule $y^T b = 0$, soon to become part two of the "fundamental theorem of linear algebra." We hold back on the general case, and identify this specific case as a law of network theory—known as Kirchhoff's voltage law.

> **2R**   The vectors in the left nullspace correspond to loops in the graph. The test for b to be in the column space is *Kirchhoff's Voltage Law*:
>
> **The sum of potential differences around a loop must be zero.**

Row space: That leaves one more subspace to be given a meaning in terms of the graph. The row space contains vectors in 4-dimensional space, but not all vectors; its dimension is only $r = 3$. We could look to elimination to find three independent rows, or we could look to the graph. The first three rows are *dependent* (because row 1 + row 2 = row 3) but rows 1, 2, 4 are *independent*. Rows correspond to edges, and the rows are independent provided the edges contain no loops.

Rows 1, 2, 4 are a basis, but what do their combinations look like? In each row the entries *add to zero*. Therefore any combination will have that same property. If $f = (f_1, f_2, f_3, f_4)$ is a linear combination of the rows, then

$$f_1 + f_2 + f_3 + f_4 = 0. \tag{2}$$

That is the test for f to be in the row space. Looking back, there has to be a connection with the vector $x = (1, 1, 1, 1)$ in the nullspace. Those four 1's in equation (2) cannot be a coincidence:

If f is in the row space and x is in the nullspace then $f^T x = 0$.

Again that illustrates the fundamental theorem of linear algebra (Part 2). And again it comes from a basic law of network theory—which now is Kirchhoff's *current* law. *The total flow into every node is zero.* The numbers f_1, f_2, f_3, f_4 are "current sources" at the nodes. The source f_1 must balance $-y_1 - y_3 - y_5$, which is the flow leaving node 1 along edges 1, 3, 5. That is the first equation in $A^T y = f$. Similarly at the other three nodes—conservation of charge requires that "flow in = flow out." The beautiful thing is that *the transpose of A is exactly the right matrix for the current law.*

The system $A^T y = f$ is solvable when f is in the column space of A^T, which is the row space of A:

> **2S**   The four equations $A^T y = f$, from the four nodes of the graph, express *Kirchhoff's Current Law*:
>
> **The net current into every node is zero.**
>
> This law can only be satisfied if the total current entering the nodes from outside is $f_1 + f_2 + f_3 + f_4 = 0$.

If $f = 0$ then Kirchhoff's current law is $A^T y = 0$. It is satisfied by *any current that goes around a loop*. Thus the loops give the vectors y in the nullspace of A^T.

Spanning Trees and Independent Rows

It is remarkable that every entry of the nullvectors x and y is 1 or -1 or 0. The same is true of all the factors in $PA = LDU$, coming from elimination. That may not seem so amazing, since it was true of the incidence matrix that we started with. But ± 1's should not be regarded as automatic; they may not be inherited by L and U. If we begin with $A = \begin{bmatrix} 1 & -1 \\ 1 & 1 \end{bmatrix}$, then elimination will produce 2 as the second pivot (and also as the determinant). This matrix A is not an incidence matrix.

For incidence matrices, every elimination step has a meaning for the graph—and we carry out those steps on an example:

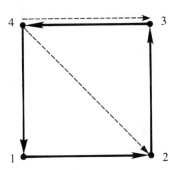

$$A = \begin{bmatrix} -1 & 1 & 0 & 0 \\ 0 & -1 & 1 & 0 \\ 0 & 0 & -1 & 1 \\ 1 & 0 & 0 & -1 \end{bmatrix}$$

The first step adds row 1 to row 4, to put a zero in the lower left corner. It produces the new fourth row 0, 1, 0, -1. That row still contains ± 1, and the matrix is still an incidence matrix. The new row corresponds to the dotted edge in the graph, from 4 to 2. The old edge from 4 to 1 is eliminated in favor of this new edge.

The next stage of elimination, using row 2 as pivot row, will be similar. Adding row 2 to the new row 4 produces 0, 0, 1, -1—which is a new edge from 4 to 3. The dotted edge should be removed, and replaced by this new edge (along top). It happens to run opposite to the existing edge from 3 to 4, since the arrows on 4–2 and 2–3 combine to give 4–3.

The last elimination step swallows up that new edge, and leaves zero in row 4. Therefore U is the same as A, except for the last row of zeros. The first three rows of A were linearly independent.

This leads back to the general question: Which rows of an incidence matrix are independent? The answer is:

Rows are independent if the corresponding edges are without a loop.

There is a name in graph theory for a set of edges without loops. It is called a ***tree***. The four edges in our square graph do not form a tree, and the four rows of A are not independent. But the first three edges (in fact *any* three edges) in the original graph do form a tree. So do any two edges, or any edge by itself; a tree can be small. But it is natural to look for the largest tree.

A tree that touches every node of the graph is a *spanning tree*. Its edges span the graph, and its rows span the row space. In fact those rows are a *basis* for the row space of A; adding another row (another edge) would close a loop. A spanning tree is as large a tree as possible. If a connected graph has n nodes, then *every spanning tree has $n - 1$ edges*. That is the number of independent rows in A, and it is the rank of the matrix.

There must also be $n - 1$ independent columns. There are n columns altogether, but they add up to the zero column. The nullspace of A is a line, passing through the nullvector $x = (1, 1, \ldots, 1)$. The dimensions add to $(n - 1) + 1 = n$, as required by the fundamental theorem of linear algebra.

That theorem also gives the number of independent loops—which is the dimension of the left nullspace. It is $m - r$, or $m - n + 1$.† If the graph lies in a plane, we can look immediately at the "mesh loops"—there were two of those small loops in Fig. 2.4, and the large loop around the outside was not independent. Even if the graph goes outside a plane—as long as it is connected—it still has $m - n + 1$ independent loops. Every node of a connected graph can be reached from every other node—there is a path of edges between them—and we summarize the properties of the incidence matrix:

Nullspace: dimension 1, contains $x = (1, \ldots, 1)$
Column space: dimension $n - 1$, any $n - 1$ columns are independent
Row space: dimension $n - 1$, independent rows from any spanning tree
Left nullspace: dimension $m - n + 1$, contains y's from the loops.

Every vector f in the row space has $x^T f = f_1 + \cdots + f_n = 0$—the currents from outside add to zero. Every vector b in the column space has $y^T b = 0$—the potential differences b_i add to zero around all loops. Those follow from Kirchhoff's laws, and in a moment we introduce a third law (*Ohm's law*). That law is a property of the material, not a property of the incidence matrix, and it will link x to y. First we stay with the matrix A, for an application that seems frivolous but is not.

The Ranking of Football Teams

At the end of the season, the polls rank college football teams. It is a subjective judgement, mostly an average of opinions, and it becomes pretty vague after the top dozen colleges. We want to rank all teams on a more mathematical basis.

The first step is to recognize the graph. If team j played team k, there is an edge between them. The *teams* are the *nodes*, and the *games* are the *edges*. Thus there are a few hundred nodes and a few thousand edges—which will be given a direction by

---

† That is *Euler's formula*, which now has a linear algebra proof: $m - n + 1$ loops $\Rightarrow$

(number of nodes) − (number of edges) + (number of loops) = 1.

an arrow from the visiting team to the home team. The figure shows part of the Ivy League, and some serious teams, and also a college that is not famous for big time football. Fortunately for that college (from which I am writing these words) the graph is not connected. Mathematically speaking, we cannot prove that MIT is not number 1 (unless it happens to play a game against somebody).

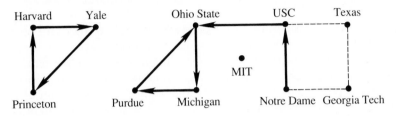

Fig. 2.5. The graph for football.

If football were perfectly consistent, we could assign a "potential" x_j to every team. Then if team v played team h, the one with higher potential would win. In the ideal case, the difference b in the score (home team minus visiting team) would exactly equal the difference $x_h - x_v$ in their potentials. They wouldn't even have to play the game! In that case there would be complete agreement that the team with highest potential was the best.

This method has two difficulties (at least). We are trying to find a number x for every team, so that $x_h - x_v = b_i$ for every game. That means a few thousand equations and only a few hundred unknowns. The equations $x_h - x_v = b_i$ go into a linear system $Ax = b$, in which A is an *incidence matrix*. Every game has a row, with $+1$ in column h and -1 in column v—to indicate which teams are in that game.

First difficulty: If b is not in the column space there is no solution. The scores must fit perfectly or exact potentials cannot be found. Second difficulty: If A has nonzero vectors in its nullspace, the potentials x are not well determined. In the first case x does not exist; in the second case it is not unique. Probably both difficulties are present.

The nullspace is easy, but it brings out an important point. It always contains the vector of 1's, since A looks only at the *differences* $x_h - x_v$. To determine the potentials we can arbitrarily assign zero potential to Harvard. That is absolutely justified (I am speaking mathematically). But if the graph is not connected, that is not enough. Every separate piece of the graph contributes a vector to the nullspace. There is even the vector with $x_{\text{MIT}} = 1$ and all other $x_j = 0$. Therefore we have to ground not only Harvard but one team in each piece. (There is nothing unfair in assigning zero potential; if all other potentials are below zero then the grounded team is ranked first.) The dimension of the nullspace is the *number of pieces* of the graph—it equals the number of degrees of freedom in x. That freedom is removed by fixing one of the potentials in every piece, and there will be no way to rank one piece against another.

The column space looks harder to describe. Which scores fit perfectly with a set of potentials? Certainly $Ax = b$ is unsolvable if Harvard beats Yale, Yale beats Princeton, and Princeton beats Harvard. But more than that, the score differences *have to add to zero around a loop*:

$$b_{HY} + b_{YP} + b_{PH} = 0.$$

This is Kirchhoff's voltage law!—the differences around loops must add to zero. It is also a law of linear algebra—the equation $Ax = b$ can be solved exactly when the vector b satisfies the same linear dependencies as the rows on the left side. Then elimination leads to $0 = 0$, and solutions can be found.

In reality b is almost certainly not in the column space. Football scores are not that consistent. The right way to obtain an actual ranking is *least squares*: Make Ax as close as possible to b. That is in Chapter 3, and we mention only one other adjustment. The winner gets a bonus of 50 or even 100 points on top of the score difference. Otherwise winning by 1 is too close to losing by 1. This brings the computed rankings very close to the polls.†

Note added in proof.   After writing that section I found the following in the *New York Times*:

> "In its final rankings for 1985, the computer placed Miami (10-2) in the seventh spot above Tennessee (9-1-2). A few days after publication, packages containing oranges and angry letters from disgruntled Tennessee fans began arriving at the *Times* sports department. The irritation stems from the fact that Tennessee thumped Miami 35-7 in the Sugar Bowl. Final AP and UPI polls ranked Tennessee fourth, with Miami significantly lower.
>
> Yesterday morning nine cartons of oranges arrived at the loading dock. They were sent to Bellevue Hospital with a warning that the quality and contents of the oranges were uncertain."

So much for that application of linear algebra.

Networks and Discrete Applied Mathematics

A graph becomes a *network* when numbers $c_1, \ldots, c_m$ are assigned to the edges. The number c_i can be the length of edge i, or its capacity, or its stiffness (if it contains a spring), or its conductance (if it contains a resistor). Those numbers go into a diagonal matrix C, which is m by m. It reflects "material properties," in contrast to the incidence matrix A—which gives information about the connections. Combined, those two matrices C and A enter the fundamental equations of network theory, and we want to explain those equations.

---

† Dr. Leake (Notre Dame) gave a full analysis in Management Science in Sports (1976).

Our description will be in electrical terms. On edge i, the conductance is c_i and the resistance is $1/c_i$. Ohm's law says that the current through the resistor is

$$y_i = c_i e_i, \quad \text{or} \quad \text{(current)} = \text{(conductance)(voltage drop).}$$

This is also written $E = IR$, voltage drop equals current times resistance. As a vector equation on all edges at once, *Ohm's law is* $y = Ce$.

We need Kirchhoff's voltage law and current law to complete the framework:

> **KVL:** The voltage drops around a loop add to zero
> **KCL:** The currents into a node add to zero.

The voltage law allows us to assign potentials $x_1, \ldots, x_n$ to the nodes. Then the differences around a loop give a sum like $(x_2 - x_1) + (x_3 - x_2) + (x_1 - x_3) = 0$, in which everything cancels. The current law asks us to add the currents into node j, which is represented in A by column j. It has $+1$ for edges that go into the node, and -1 for edges that go out. The multiplication $A^T y$ adds the currents with their correct signs (consistent with the arrows on the edges). This vector $A^T y$ of total currents into nodes is zero, if there are no external sources of current. In that case *Kirchhoff's current law is* $A^T y = 0$.

In general we allow for a source term. If external currents $f_1, \ldots, f_n$ are sent into the nodes, the law becomes $A^T y = f$.

The other equation is Ohm's law, but we need to find e—which is the voltage drop *across the resistor*. The multiplication Ax gave the potential difference between the nodes. Reversing the signs, $-Ax$ gives the potential *drop*. Part of that drop may be due to a *battery* in the edge, of strength b_i. The rest of the drop is across the resistor, and it is given by the difference $e = b - Ax$. Then Ohm's law $y = Ce$ is

$$y = C(b - Ax) \quad \text{or} \quad C^{-1}y + Ax = b. \tag{3}$$

It connects x to y. We are no longer trying to solve $Ax = b$ (which was hard to do, because there were more equations than unknowns). There is a new term $C^{-1}y$. In fact the special case when $Ax = b$ did accidentally have a solution is also the special case in which *no current flows*. In that case the football score differences or the batteries add to zero around loops—and there is no need for current.

We emphasize the *fundamental equations of equilibrium*, which combine Ohm's law with both of Kirchhoff's laws:

$$\boxed{\begin{aligned} C^{-1}y + Ax &= b \\ A^T y \quad\;\; &= f \end{aligned}} \tag{4}$$

That is a symmetric system, from which e has disappeared. The unknowns are the currents y and the potentials x. It is a linear system, and we can write it in "block form" as

$$\begin{bmatrix} C^{-1} & A \\ A^{\mathrm{T}} & 0 \end{bmatrix} \begin{bmatrix} y \\ x \end{bmatrix} = \begin{bmatrix} b \\ f \end{bmatrix}. \tag{5}$$

We can even do elimination on this block form. The pivot is C^{-1}, the multiplier is $A^{\mathrm{T}}C$, and subtraction knocks out A^{T} below the pivot. The result is

$$\begin{bmatrix} C^{-1} & A \\ 0 & -A^{\mathrm{T}}CA \end{bmatrix} \begin{bmatrix} y \\ x \end{bmatrix} = \begin{bmatrix} b \\ f - A^{\mathrm{T}}Cb \end{bmatrix}$$

The equation to be solved for x is in the bottom row:

$$\boxed{A^{\mathrm{T}}CAx = A^{\mathrm{T}}Cb - f.} \tag{6}$$

Then back-substitution in the first equation produces y.

There is nothing mysterious about (6). Substituting $y = C(b - Ax)$ into $A^{\mathrm{T}}y = f$, we reach that equation. The currents y are eliminated to leave an equation for x.

Important remark   Throughout those equations it is essential that one potential is fixed in advance: $x_n = 0$. The nth node is grounded, and the nth column of the original incidence matrix is removed. The resulting matrix is what we mean by A; it is m by $n - 1$, and its columns are independent. The square matrix $A^{\mathrm{T}}CA$, which is the key to solving equation (6) for x, is an invertible matrix of order $n - 1$:

$$\underset{n-1 \text{ by } m}{\begin{bmatrix} A^{\mathrm{T}} \end{bmatrix}} \underset{m \text{ by } m}{\begin{bmatrix} C \end{bmatrix}} \underset{m \text{ by } n-1}{\begin{bmatrix} A \end{bmatrix}} = \underset{n-1 \text{ by } n-1}{\begin{bmatrix} A^{\mathrm{T}}CA \end{bmatrix}}$$

EXAMPLE   Suppose a battery and a current source (and five resistors) are added to the network discussed earlier:

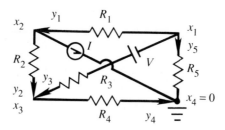

The first thing to check is the current law $A^{\mathrm{T}}y = f$ at nodes 1, 2, 3:

$$\begin{aligned} -y_1 - y_3 - y_5 &= -I \\ y_1 - y_2 &= 0 \\ y_2 + y_3 - y_4 &= I \end{aligned} \quad \text{has } A^{\mathrm{T}} = \begin{bmatrix} -1 & 0 & -1 & 0 & -1 \\ 1 & -1 & 0 & 0 & 0 \\ 0 & 1 & 1 & -1 & 0 \end{bmatrix} \text{ and } f = \begin{bmatrix} -I \\ 0 \\ I \end{bmatrix}.$$

No equation is written for node 4. At that node the current law would be $y_4 + y_5 = 0$. This follows from the other three equations, whose sum is exactly $-y_4 - y_5 = 0$.

The other equation is $C^{-1}y + Ax = b$. The potential x is connected to the current y by Ohm's law. The diagonal matrix C contains the five conductances $c_i = 1/R_i$. The right side accounts for the battery of strength $b_3 = V$, and the block form has $C^{-1}y + Ax = b$ above $A^T y = f$:

$$
\begin{bmatrix} C^{-1} & A \\ A^T & 0 \end{bmatrix} \begin{bmatrix} y \\ x \end{bmatrix} =
\begin{bmatrix}
R_1 & & & & & -1 & 1 & 0 \\
& R_2 & & & & 0 & -1 & 1 \\
& & R_3 & & & -1 & 0 & 1 \\
& & & R_4 & & 0 & 0 & -1 \\
& & & & R_5 & -1 & 0 & 0 \\
-1 & 0 & -1 & 0 & -1 & & & \\
1 & -1 & 0 & 0 & 0 & & & \\
0 & 1 & 1 & -1 & 0 & & &
\end{bmatrix}
\begin{bmatrix} y_1 \\ y_2 \\ y_3 \\ y_4 \\ y_5 \\ x_1 \\ x_2 \\ x_3 \end{bmatrix} =
\begin{bmatrix} 0 \\ 0 \\ V \\ 0 \\ 0 \\ -I \\ 0 \\ I \end{bmatrix}
$$

The system is 8 by 8, with five currents and three potentials. Elimination reduces it to the 3 by 3 system $A^T C A x = A^T C b - f$. The matrix in that system contains the reciprocals $c_i = 1/R_i$ (because in elimination you divide by the pivots). This matrix is $A^T C A$, and it is worth looking at—with the fourth row and column, from the grounded node, included too:

$$
A^T C A = \begin{bmatrix}
c_1 + c_3 + c_5 & -c_1 & -c_3 \\
-c_1 & c_1 + c_2 & -c_2 \\
-c_3 & -c_2 & c_2 + c_3 + c_4 \\
-c_5 & 0 & -c_4
\end{bmatrix}
\begin{matrix} -c_5 & \text{(node 1)} \\ 0 & \text{(node 2)} \\ -c_4 & \text{(node 3)} \\ c_4 + c_5 & \text{(node 4)} \end{matrix}
$$

You can almost see this matrix by multiplying A^T and A—the lower left corner and the upper right corner of the 8 by 8 matrix above. The first entry is $1 + 1 + 1$, or $c_1 + c_3 + c_5$ when C is included; edges 1, 3, 5 touch node 1. The next diagonal entry is $1 + 1$ or $c_1 + c_2$, from the edges touching node 2. Similarly $c_2 + c_3 + c_4$ comes from node 3. Off the diagonal the c's appear with minus signs, *but not the edges to the grounded node* 4. Those belong in the *fourth* row and column, which are deleted when column 4 is removed from A. By grounding the last node we reduce to a system of order $n - 1$—and more important, to a matrix $A^T C A$ that is invertible. The 4 by 4 matrix would have all rows and columns adding to zero, and $(1, 1, 1, 1)$ would be in its nullspace.

Notice that $A^T C A$ is symmetric. Its transpose is $(A^T C A)^T = A^T C^T A^{TT}$, which is again $A^T C A$. It also has positive pivots, but that is left for Chapter 6. It comes from the basic framework of applied mathematics, which is illustrated in the figure:

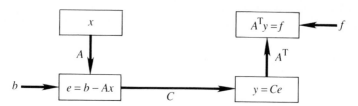

Fig. 2.6. The framework for equilibrium: sources b and f, matrix A^TCA.

For electrical networks x contained potentials and y contained currents. In mechanics x and y become displacements and stresses. In fluids they are pressure and flow rate.† In statistics e is the error and the equations give the best least squares fit to the data. The triple product of A^T, C, and A combines the three steps of the framework into the single matrix that governs equilibrium.

We end this chapter at that high point—the *formulation* of a fundamental problem in applied mathematics. Often that requires more insight than the *solution* of the problem. We solved linear equations in Chapter 1, as the first step in linear algebra, but to set them up has required the deeper insight of Chapter 2. The contribution of mathematics, and of people, is not computation but intelligence.

A Look Ahead

We introduced the column space as the set of vectors Ax, and the left nullspace as the solutions to $A^Ty = 0$, because those mathematical abstractions are needed in application. For networks Ax gives the potential differences, satisfying the voltage law; y satisfies the current law. With unit resistors $(C = I)$ the equilibrium equations (4) are

$$y + Ax = b$$
$$A^Ty \quad\;\; = 0. \tag{7}$$

Linear algebra (or just direct substitution) leads to $A^T(b - Ax) = 0$, and the computer solves $A^TAx = A^Tb$. But there is one more source of insight still to be heard from.

That final source is geometry. It goes together with algebra, but it is different from algebra. The spatial orientation of vectors is crucial, even if calculations are done on their separate components. In this problem the orientation is nothing short of sensational: Ax is *perpendicular* to y. The voltage differences are perpen-

---

† These matrix equations and the corresponding differential equations are studied in our textbook *Introduction to Applied Mathematics* (Wellesley–Cambridge Press, Box 157, Wellesley MA 02181).

dicular to the currents! Their sum is b, and therefore that vector b is split into two perpendicular pieces—its projection Ax onto the column space, and its projection y onto the left nullspace.

That will be the contribution of Chapter 3. It adds geometry to the algebra of bases and subspaces, in order to reach orthonormal bases and orthogonal subspaces. It also does what algebra could not do unaided—it gives an answer to $Ax = b$ **when b is not in the column space.** The equation as it stands has no solution. To solve it we have to remove the part of b that lies outside the column space and makes the solution impossible. What remains is equation (7), or $A^T A x = A^T b$, which leads—through geometry—to the best possible x.

EXERCISES

2.5.1 For the 3-node triangular graph in the figure below, write down the 3 by 3 incidence matrix A. Find a solution to $Ax = 0$ and describe all other vectors in the nullspace of A. Find a solution to $A^T y = 0$ and describe all other vectors in the left nullspace of A.

2.5.2 For the same 3 by 3 matrix, show directly from the columns that every vector b in the column space will satisfy $b_1 + b_2 - b_3 = 0$. Derive the same thing from the three rows—the equations in the system $Ax = b$. What does that mean about potential differences around a loop?

2.5.3 Show directly from the rows that every vector f in the row space will satisfy $f_1 + f_2 + f_3 = 0$. Derive the same thing from the three equations $A^T y = f$. What does that mean when the f's are currents into the nodes?

2.5.4 Compute the 3 by 3 matrix $A^T A$, and show it is symmetric but singular—what vectors are in its nullspace? Removing the last column of A (and last row of A^T) leaves the 2 by 2 matrix in the upper left corner; show that it is *not* singular.

2.5.5 Put the diagonal matrix C with entries c_1, c_2, c_3 in the middle and compute $A^T C A$. Show again that the 2 by 2 matrix in the upper left corner is invertible.

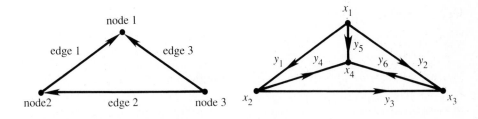

2.5.6 Write down the 6 by 4 incidence matrix A for the second graph in the figure. The vector $(1, 1, 1, 1)$ is in the nullspace of A, but now there will be $m - n + 1 = 3$ independent vectors that satisfy $A^T y = 0$. Find three vectors y and connect them to the loops in the graph.

2.5.7   If that graph represents six games between four teams, and the score differences are $b_1, \ldots, b_6$, when is it possible to assign potentials to the teams so that the potential differences agree exactly with the b's? In other words, find (from Kirchhoff or from elimination) the conditions on b that make $Ax = b$ solvable.

2.5.8   Write down the dimensions of the four fundamental subspaces for this 6 by 4 incidence matrix, and a basis for each subspace.

2.5.9   Compute $A^T A$ and $A^T C A$, where the 6 by 6 diagonal matrix C has entries $c_1, \ldots, c_6$. What is the pattern for the main diagonal of $A^T C A$—how can you tell from the graph which c's will appear in row j?

2.5.10   Draw a graph with numbered and directed edges (and numbered nodes) whose incidence matrix is

$$
A = \begin{bmatrix} -1 & 1 & 0 & 0 \\ -1 & 0 & 1 & 0 \\ 0 & 1 & 0 & -1 \\ 0 & 0 & -1 & 1 \end{bmatrix}.
$$

Is this graph a tree? (Are the rows of A independent?) Show that removing the last edge produces a spanning tree. Then the remaining rows are a basis for _____?

2.5.11   With the last column removed from the preceding A, and with the numbers 1, 2, 2, 1 on the diagonal of C, write out the 7 by 7 system

$$
\begin{aligned}
C^{-1}y + Ax &= 0 \\
A^T y \quad\; &= f.
\end{aligned}
$$

Eliminating y_1, y_2, y_3, y_4 leaves three equations $A^T C A x = -f$ for x_1, x_2, x_3. Solve the equations when $f = (1, 1, 6)$. With those currents entering nodes 1, 2, 3 of the network what are the potentials at the nodes and currents on the edges?

2.5.12   If A is a 12 by 7 incidence matrix from a connected graph, what is its rank? How many free variables in the solution to $Ax = b$? How many free variables in the solution to $A^T y = f$? How many edges must be removed to leave a spanning tree?

2.5.13   In a graph with 4 nodes and 6 edges, find all 16 spanning trees.

2.5.14   If E and H are square, what is the product of the block matrices

$$
\begin{matrix} m_1 \text{ rows} \\ m_2 \text{ rows} \end{matrix} \begin{bmatrix} A & B \\ C & D \end{bmatrix} \begin{bmatrix} E & F \\ G & H \end{bmatrix} \begin{matrix} n_1 \text{ rows} \\ n_2 \text{ rows} \end{matrix}
$$

and what will be the shapes of the blocks in the product?

2.5.15   If MIT beats Harvard 35-0 and Yale ties Harvard and Princeton beats Yale 7-6, what score differences in the other 3 games (H-P, MIT-P, MIT-Y) will allow potential differences that agree with the score differences? If the score differences are known for the games in a spanning tree, they are known for all games.

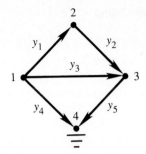

2.5.16 (a) What are the three current laws $A^\mathsf{T} y = 0$ at the ungrounded nodes above?
(b) How does the current law at the grounded node follow from those three equations?
(c) What is the rank of A^T?
(d) Describe the solutions of $A^\mathsf{T} y = 0$ in terms of loops in the network.

2.5.17 In our method for football rankings, should the strength of the opposition be considered—or is that already built in?

2.5.18 If there is an edge between every pair of nodes (a complete graph) how many edges are there? The graph has n nodes, and edges from a node to itself are not allowed.

2.6 ■ LINEAR TRANSFORMATIONS

At this point we know how a matrix moves subspaces around. The nullspace goes into the zero vector, when we multiply by A. All vectors go into the column space, since Ax is in all cases a combination of the columns. You will soon see something beautiful—that A takes its row space into its column space, and on those spaces of dimension r it is 100% invertible. That is the real action of a matrix. It is partly hidden by nullspaces and left nullspaces, which lie at right angles and go their own way (toward zero)—but when A is square and invertible those are insignificant. What matters is what happens *inside* the space—which means inside n-dimensional space, if A is n by n. That demands a closer look.

Suppose x is an n-dimensional vector. When A multiplies x, we can think of it as ***transforming*** that vector into a new vector Ax. This happens at every point x of the n-dimensional space $\mathbf{R}^n$. The whole space is transformed, or "mapped into itself," by the matrix A. We give four examples of the transformations that come from matrices:

$$A = \begin{bmatrix} c & 0 \\ 0 & c \end{bmatrix}$$

1. A multiple of the identity matrix, $A = cI$, ***stretches*** every vector by the same factor c. The whole space expands or contracts (or somehow goes through the origin and out the opposite side, when c is negative).

$$A = \begin{bmatrix} 0 & -1 \\ 1 & 0 \end{bmatrix}$$

2. A ***rotation*** matrix turns the whole space around the origin. This example turns all vectors through $90°$, transforming $(1, 0)$ on the x-axis to $(0, 1)$, and sending $(0, 1)$ on the y-axis to $(-1, 0)$.

$$A = \begin{bmatrix} 0 & 1 \\ 1 & 0 \end{bmatrix}$$

3. A ***reflection*** matrix transforms every vector into its image on the opposite side of a mirror. In this example the mirror is the $45°$ line $y = x$, and a point like $(2, 2)$ is unchanged. A point like $(2, -2)$ is reversed to $(-2, 2)$. On a combination like $(2, 2) + (2, -2) = (4, 0)$, the matrix leaves one part and reverses the other part. The result is to exchange y and x, and produce $(0, 4)$:

$$\begin{bmatrix} 0 & 1 \\ 1 & 0 \end{bmatrix}\begin{bmatrix} 4 \\ 0 \end{bmatrix} = \begin{bmatrix} 0 \\ 4 \end{bmatrix}, \text{ or } A\left(\begin{bmatrix} 2 \\ 2 \end{bmatrix} + \begin{bmatrix} 2 \\ -2 \end{bmatrix}\right) = \left(\begin{bmatrix} 2 \\ 2 \end{bmatrix} + \begin{bmatrix} -2 \\ 2 \end{bmatrix}\right).$$

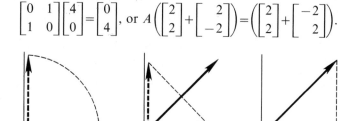

stretching   90° rotation   reflection   projection

Fig. 2.7. Transformations of the plane by four matrices.

That reflection matrix is also a permutation matrix! It is algebraically so simple, sending (x, y) to (y, x), that the geometric picture was concealed. The fourth example is simple in both respects:

$$A = \begin{bmatrix} 1 & 0 \\ 0 & 0 \end{bmatrix}$$

4. A *projection* matrix takes the whole space onto a lower-dimensional subspace (and therefore fails to be invertible). The example transforms each vector (x, y) in the plane to the nearest point $(x, 0)$ on the horizontal axis. That axis is the column space of A, and the vertical axis (which projects onto the origin) is the nullspace.

Those examples could easily be lifted into three dimensions. There are matrices to stretch the earth or spin it or reflect it across the plane of the equator (north pole transforming to south pole). There is a matrix that projects everything onto that plane (both poles to the center). Other examples are certainly possible and necessary. But it is also important to recognize that matrices cannot do everything, and some transformations are *not possible* with matrices:

(i)   It is impossible to move the origin, since $A0 = 0$ for every matrix.

(ii) If the vector x goes to x', then $2x$ must go to $2x'$. In general cx must go to cx', since $A(cx) = c(Ax)$.

(iii) If the vectors x and y go to x' and y', then their sum $x + y$ must go to $x' + y'$—since $A(x + y) = Ax + Ay$.

Matrix multiplication imposes those rules on the transformation of the space. The first two rules are easy, and the second one contains the first (just take $c = 0$). We saw rule (iii) in action when the vector $(4, 0)$ was reflected across the $45°$ line. It was split into $(2, 2) + (2, -2)$ and the two parts were reflected separately. The same could be done for projections: split, project separately, and add the projections. These rules apply to *any transformation that comes from a matrix*. Their importance has earned them a name: Transformations that obey rules (i)–(iii) are called *linear transformations*.

Those conditions can be combined into a single requirement:

2T For all numbers c and d and all vectors x and y, matrix multiplication satisfies the rule of linearity:

$$A(cx + dy) = c(Ax) + d(Ay). \tag{1}$$

Every transformation that meets this requirement is a *linear transformation*.

Any matrix leads immediately to a linear transformation. The more interesting question is in the opposite direction: *Does every linear transformation lead to a matrix?* The object of this section is to answer that question (affirmatively, in n dimensions). This theory is the foundation of an approach to linear algebra—starting with property (1) and developing its consequences—which is much more abstract than the main approach in this book. We preferred to begin directly with matrices, and now we see how they represent linear transformations.

We must emphasize that a transformation need not go from $\mathbf{R}^n$ to the same space $\mathbf{R}^n$. It is absolutely permitted to transform vectors in $\mathbf{R}^n$ to vectors in a different space $\mathbf{R}^m$. That is exactly what is done by an m by n matrix! The original vector x has n components, and the transformed vector Ax has m components. The rule of linearity is equally satisfied by rectangular matrices, so they also produce linear transformations.

Having gone that far, there is no reason to stop. The operations in the linearity condition (1) are addition and scalar multiplication, but x and y need not be column vectors in $\mathbf{R}^n$. That space was expected, but it is not the only one. By definition, *any vector space allows the combinations $cx + dy$*—the "vectors" are x and y, but they may actually be polynomials or matrices or functions $x(t)$ and $y(t)$. As long as a transformation between such spaces satisfies (1), it is linear.

We take as examples the spaces P_n, in which the vectors are polynomials of degree n. They look like $p = a_0 + a_1 t + \cdots + a_n t^n$, and the dimension of the vector space is $n + 1$ (because with the constant term, there are $n + 1$ coefficients).

EXAMPLE 1   The operation of *differentiation*, $A = d/dt$, is linear:

$$Ap = \frac{d}{dt}(a_0 + a_1 t + \cdots + a_n t^n) = a_1 + \cdots + na_n t^{n-1}. \tag{2}$$

Its nullspace is the one-dimensional space of constant polynomials: $da_0/dt = 0$. Its column space is the n-dimensional space P_{n-1}; the right side of (2) is always in that space. The sum of nullity $(= 1)$ and rank $(= n)$ is the dimension of the original space P_n.

EXAMPLE 2   *Integration* from 0 to t is also linear (*it takes P_n to P_{n+1}*):

$$Ap = \int_0^t (a_0 + \cdots + a_n t^n)\, dt = a_0 t + \cdots + \frac{a_n}{n+1} t^{n+1}. \tag{3}$$

This time there is no nullspace (except for the zero vector, as always!) but integration does not produce all polynomials in P_{n+1}. The right side of (3) has no constant term. Probably the constant polynomials will be the left nullspace.

EXAMPLE 3   *Multiplication* by a fixed polynomial like $2 + 3t$ is linear:

$$Ap = (2 + 3t)(a_0 + \cdots + a_n t^n) = 2a_0 + \cdots + 3a_n t^{n+1}.$$

Again this transforms P_n to P_{n+1}, with no nullspace except $p = 0$.

In these examples and in almost all examples, linearity is not difficult to verify. It hardly even seems interesting. If it is there, it is practically impossible to miss.

Nevertheless it is the most important property a transformation can have.† Of course most transformations are not linear—for example to square the polynomial $(Ap = p^2)$, or to add 1 $(Ap = p + 1)$, or to keep the positive coefficients $(A(t - t^2) = t)$. It will be linear transformations, and only those, that lead us back to matrices.

Transformations Represented by Matrices

Linearity has a crucial consequence: If we know Ax for each vector *in a basis*, then we know Ax for each vector *in the entire space*. Suppose the basis consists of the n vectors $x_1, \ldots, x_n$. Every other vector x is a combination of those particular vectors (they span the space). Then linearity determines Ax:

$$\text{if} \quad x = c_1 x_1 + \cdots + c_n x_n \quad \text{then} \quad Ax = c_1(Ax_1) + \cdots + c_n(Ax_n). \qquad (4)$$

The transformation A has no freedom left, after it has decided what to do with the basis vectors. The rest of the transformation is determined by linearity. The requirement (1) for two vectors x and y leads to (4) for n vectors $x_1, \ldots, x_n$. The transformation does have a free hand with the vectors in the basis (they are independent). When those are settled, the whole transformation is settled.

EXAMPLE 4 Question: What linear transformation takes

$$x_1 = \begin{bmatrix} 1 \\ 0 \end{bmatrix} \quad \text{to} \quad Ax_1 = \begin{bmatrix} 2 \\ 3 \\ 4 \end{bmatrix} \quad \text{and} \quad x_2 = \begin{bmatrix} 0 \\ 1 \end{bmatrix} \quad \text{to} \quad Ax_2 = \begin{bmatrix} 4 \\ 6 \\ 8 \end{bmatrix} ?$$

It must be multiplication by the matrix

$$A = \begin{bmatrix} 2 & 4 \\ 3 & 6 \\ 4 & 8 \end{bmatrix}.$$

Starting with a different basis (1, 1) and (2, −1), this is also the only linear transformation with

$$A \begin{bmatrix} 1 \\ 1 \end{bmatrix} = \begin{bmatrix} 6 \\ 9 \\ 12 \end{bmatrix} \quad \text{and} \quad A \begin{bmatrix} 2 \\ -1 \end{bmatrix} = \begin{bmatrix} 0 \\ 0 \\ 0 \end{bmatrix}.$$

Next we try a new problem—to find a matrix that represents differentiation, and a matrix that represents integration. That can be done as soon as we decide on a basis. For the polynomials of degree 3 (the space P_3 whose dimension is 4) there is a natural choice for the four basis vectors:

$$p_1 = 1, \quad p_2 = t, \quad p_3 = t^2, \quad p_4 = t^3.$$

---

† Invertibility is perhaps in second place.

That basis is not unique (it never is), but some choice is necessary and this is the most convenient. We look to see what differentiation does to those four basis vectors. Their derivatives are $0, 1, 2t, 3t^2$, or

$$Ap_1 = 0, \quad Ap_2 = p_1, \quad Ap_3 = 2p_2, \quad Ap_4 = 3p_3. \tag{5}$$

A is acting exactly like a matrix, but which matrix? Suppose we were in the usual 4-dimensional space with the usual basis—the coordinate vectors $p_1 = (1, 0, 0, 0)$, $p_2 = (0, 1, 0, 0)$, $p_3 = (0, 0, 1, 0)$, $p_4 = (0, 0, 0, 1)$. Then the matrix corresponding to (5) would be

$$A = \begin{bmatrix} 0 & 1 & 0 & 0 \\ 0 & 0 & 2 & 0 \\ 0 & 0 & 0 & 3 \\ 0 & 0 & 0 & 0 \end{bmatrix}.$$

This is the "differentiation matrix." Ap_1 is its first column, which is zero. Ap_2 is the second column, which is p_1. Ap_3 is $2p_2$, and Ap_4 is $3p_3$. The nullspace contains p_1 (the derivative of a constant is zero). The column space contains p_1, p_2, p_3 (the derivative of a cubic is a quadratic). The derivative of any other combination like $p = 2 + t - t^2 - t^3$ is decided by linearity, and there is nothing new about that—it is the only way to differentiate. It would be crazy to memorize the derivative of every polynomial.

The matrix can differentiate that polynomial:

$$\frac{dp}{dt} = Ap \rightarrow \begin{bmatrix} 0 & 1 & 0 & 0 \\ 0 & 0 & 2 & 0 \\ 0 & 0 & 0 & 3 \\ 0 & 0 & 0 & 0 \end{bmatrix} \begin{bmatrix} 2 \\ 1 \\ -1 \\ -1 \end{bmatrix} = \begin{bmatrix} 1 \\ -2 \\ -3 \\ 0 \end{bmatrix} \rightarrow 1 - 2t - 3t^2.$$

In short, *the matrix carries all the essential information*. If the basis is known, and the matrix is known, then the linear transformation is known.

The coding of the information is simple. For transformations from a space to itself one basis is enough. A transformation from one space to another requires a basis for each.

2U Suppose the vectors $x_1, \ldots, x_n$ are a basis for the space V, and $y_1, \ldots, y_m$ are a basis for W. Then each linear transformation A from V to W is represented by a matrix. The jth column is found by applying A to the jth basis vector; the result Ax_j is a combination of the y's and the coefficients in that combination go into column j:

$$Ax_j = a_{1j}y_1 + a_{2j}y_2 + \cdots + a_{mj}y_m. \tag{6}$$

For the differentiation matrix, column 1 came from the first basis vector $p_1 = 1$. Its derivative was zero, so column 1 was zero. The last column came from $(d/dt)t^3 =$

$3t^2$. Since $3t^2 = 0p_1 + 0p_2 + 3p_3 + 0p_4$, the last column contained 0, 0, 3, 0. The rule (6) constructed the matrix.

We do the same for integration. That goes from cubics to quartics, transforming $V = P_3$ into $W = P_4$, so for W we need a basis. The natural choice is $y_1 = 1$, $y_2 = t$, $y_3 = t^2$, $y_4 = t^3$, $y_5 = t^4$, spanning the polynomials of degree 4. The matrix will be m by n, or 5 by 4, and it comes from applying integration to each basis vector of V:

$$\int_0^t 1\,dt = t \quad \text{or} \quad Ax_1 = y_2, \quad \dots, \quad \int_0^t t^3\,dt = \tfrac{1}{4}t^4 \quad \text{or} \quad Ax_4 = \tfrac{1}{4}y_5.$$

Thus the matrix that represents integration is

$$A_{\text{int}} = \begin{bmatrix} 0 & 0 & 0 & 0 \\ 1 & 0 & 0 & 0 \\ 0 & \tfrac{1}{2} & 0 & 0 \\ 0 & 0 & \tfrac{1}{3} & 0 \\ 0 & 0 & 0 & \tfrac{1}{4} \end{bmatrix}.$$

Remark We think of differentiation and integration as *inverse operations*. Or at least integration *followed* by differentiation leads back to the original function. To make that happen for matrices, we need the differentiation matrix from quartics down to cubics, which is 4 by 5:

$$A_{\text{diff}} = \begin{bmatrix} 0 & 1 & 0 & 0 & 0 \\ 0 & 0 & 2 & 0 & 0 \\ 0 & 0 & 0 & 3 & 0 \\ 0 & 0 & 0 & 0 & 4 \end{bmatrix} \quad \text{and} \quad A_{\text{diff}}A_{\text{int}} = \begin{bmatrix} 1 & & & \\ & 1 & & \\ & & 1 & \\ & & & 1 \end{bmatrix}.$$

Differentiation is a **left-inverse** of integration. But rectangular matrices cannot have two-sided inverses! In the opposite order, it cannot be true that $A_{\text{int}}A_{\text{diff}} = I$. This fails in the first column, where the 5 by 5 product has zeros. The derivative of a constant is zero. In the other columns $A_{\text{int}}A_{\text{diff}}$ is the identity and the integral of the derivative of t^n is t^n.

Rotations Q, Projections P, and Reflections H

This section began with 90° rotations, and projections onto the x-axis, and reflections through the 45° line. Their matrices were especially simple:

$$Q = \begin{bmatrix} 0 & -1 \\ 1 & 0 \end{bmatrix}, \qquad P = \begin{bmatrix} 1 & 0 \\ 0 & 0 \end{bmatrix}, \qquad H = \begin{bmatrix} 0 & 1 \\ 1 & 0 \end{bmatrix}.$$
$$\text{(rotation)} \qquad\qquad \text{(projection)} \qquad\quad \text{(reflection)}$$

Of course the underlying linear transformations of the x-y plane are also simple. But it seems to me that rotations through other angles, and projections onto other lines, and reflections in other mirrors, are almost as easy to visualize. They are still linear transformations, provided the origin is fixed: $A0 = 0$. They must be represented by matrices. Using the natural basis $\begin{bmatrix} 1 \\ 0 \end{bmatrix}$ and $\begin{bmatrix} 0 \\ 1 \end{bmatrix}$, we want to discover what those matrices are.

1. Rotation   Figure 2.8 shows rotation through an angle θ. It also shows the effect on the two basis vectors. The first one goes to $(\cos\theta, \sin\theta)$, whose length is still one; it lies on the "θ-line." The second basis vector $(0, 1)$ rotates into $(-\sin\theta, \cos\theta)$. By rule (6) those numbers go into the columns of the matrix, and we introduce the abbreviations c and s for the cosine and sine.

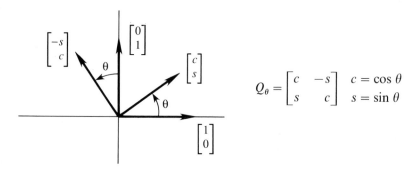

$$Q_\theta = \begin{bmatrix} c & -s \\ s & c \end{bmatrix} \qquad \begin{array}{l} c = \cos\theta \\ s = \sin\theta \end{array}$$

Fig. 2.8. Rotation through θ: the geometry and the matrix.

This family of rotations Q_θ is a perfect chance to test the correspondence between transformations and matrices:

Does the inverse of Q_θ equal $Q_{-\theta}$ (rotation backward through θ)? Yes.

$$Q_\theta Q_{-\theta} = \begin{bmatrix} c & -s \\ s & c \end{bmatrix} \begin{bmatrix} c & s \\ -s & c \end{bmatrix} = \begin{bmatrix} 1 & 0 \\ 0 & 1 \end{bmatrix}.$$

Does the square of Q_θ equal $Q_{2\theta}$ (rotation through a double angle)? Yes.

$$Q_\theta^2 = \begin{bmatrix} c & -s \\ s & c \end{bmatrix} \begin{bmatrix} c & -s \\ s & c \end{bmatrix} = \begin{bmatrix} c^2 - s^2 & -2cs \\ 2cs & c^2 - s^2 \end{bmatrix} = \begin{bmatrix} \cos 2\theta & -\sin 2\theta \\ \sin 2\theta & \cos 2\theta \end{bmatrix}.$$

Does the product of Q_θ and Q_φ equal $Q_{\theta+\varphi}$ (rotation through θ then φ)? Yes.

$$Q_\theta Q_\varphi = \begin{bmatrix} \cos\theta\cos\varphi - \sin\theta\sin\varphi & \underline{\qquad} \\ \sin\theta\cos\varphi + \cos\theta\sin\varphi & \underline{\qquad} \end{bmatrix}$$

$$= \begin{bmatrix} \cos(\theta+\varphi) & -\sin(\theta+\varphi) \\ \sin(\theta+\varphi) & \cos(\theta+\varphi) \end{bmatrix} = Q_{\theta+\varphi}.$$

The last case contains the first two. The inverse appears when φ is $-\theta$, and the square appears when φ is $+\theta$. All three questions were decided by trigonometric identities (and they give a new way to remember those identities). Of course it was no accident that all the answers were yes. *Matrix multiplication was defined exactly so that **the product of the matrices corresponds to the product of the transformations**.*

2V   Suppose A and B are linear transformations from V to W and from U to V. Their product AB starts with a vector u in U, goes to Bu in V, and finishes with ABu in W. This "composition" AB is again a linear transformation (from U to W). The matrix that represents it is the product of the individual matrices representing A and B.

That was tested earlier for $A_{\text{diff}}A_{\text{int}}$, where the composite transformation was the identity. (And also for $A_{\text{int}}A_{\text{diff}}$, which annihilated all constants.) For rotations it happens that the order of multiplication does not matter—U and V and W were all the x-y plane, and $Q_\theta Q_\varphi$ is the same as $Q_\varphi Q_\theta$. For a rotation and a reflection, the order makes a difference.

Technical note: To construct the matrices we need bases for V and W, and then for U and V. By keeping the same basis for V, the product matrix goes correctly from the basis in U to the basis in W. If we distinguish the transformation A from its matrix (call that $[A]$), then the product rule 2V becomes extremely concise: $[AB] = [A][B]$. To repeat, the rule for multiplying matrices in Chapter 1 was totally determined by this requirement—that it must match the product of linear transformations.
    We come back to concrete examples, with new matrices.

2. Projection   Figure 2.9 shows the projection of $(1, 0)$ onto the θ-line. The length of the projection is $c = \cos \theta$. Notice that the *point* of projection is not (c, s), as I mistakenly thought; that vector has length 1 (it is the rotation). The projection of $(1, 0)$ is c times that unit vector, or (c^2, cs). Similarly the projection of $(0, 1)$ has length s, and falls at (cs, s^2). That gives the second column of the projection matrix.

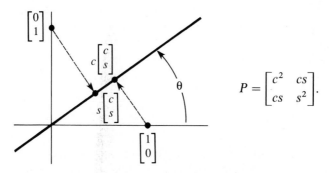

Fig. 2.9. Projection onto the θ-line: the geometry and the matrix.

This matrix has no inverse, because the transformation has no inverse. Points like $(-s, c)$ on the perpendicular line are projected onto the origin; that line is the nullspace of P. At the same time, points on the θ-line are projected to themselves! In other words, projecting twice is the same as projecting once, and $P^2 = P$:

$$P^2 = \begin{bmatrix} c^2 & cs \\ cs & s^2 \end{bmatrix}^2 = \begin{bmatrix} c^2(c^2 + s^2) & cs(c^2 + s^2) \\ cs(c^2 + s^2) & s^2(c^2 + s^2) \end{bmatrix} = P.$$

Of course $c^2 + s^2 = \cos^2 \theta + \sin^2 \theta = 1$. *A projection matrix equals its own square.* It is also symmetric.

3. Reflection   Figure 2.10 shows the reflection of $(1, 0)$ in the θ-line. The length of the reflection equals the length of the original, as it did after rotation—but those transformations are very different. Here the θ-line stays where it is. The perpendicular line reverses direction; all points go straight through the mirror. Linearity decides the rest.

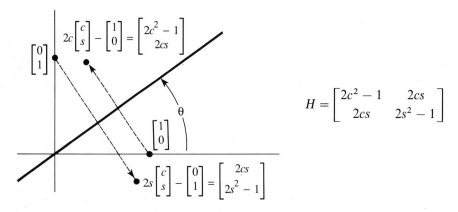

$$H = \begin{bmatrix} 2c^2 - 1 & 2cs \\ 2cs & 2s^2 - 1 \end{bmatrix}$$

Fig. 2.10. Reflection through the θ-line: the geometry and the matrix.

This matrix H has the remarkable property $H^2 = I$. *Two reflections bring back the original.* Thus a reflection is its own inverse, $H = H^{-1}$, which is clear from the geometry but less clear from the matrix. One approach is through the relationship of reflections to projections: $H = 2P - I$. This means that $Hx + x = 2Px$—the image plus the original equals twice the projection. It also confirms that

$$H^2 = (2P - I)^2 = 4P^2 - 4P + I = I,$$

since all projections satisfy $P^2 = P$.

Those three transformations either leave lengths unchanged (rotations and reflections), or reduce the length (projections). Other transformations can increase the length; stretching and shearing are in the exercises. Each example has a matrix to represent it—which is the main point of this section. But there is also the question of choosing a basis, and we emphasize that *the matrix depends on the choice of basis.* For example:

(i)   For projections, suppose the first basis vector is *on the θ-line* and the second basis vector is perpendicular. Then the projection matrix is back to $P = \begin{bmatrix} 1 & 0 \\ 0 & 0 \end{bmatrix}$. This matrix is constructed as always: its first column comes from the first basis vector (which is projected to itself), and the second column comes from the basis vector which is projected onto zero.

(ii)   For reflections, that same basis gives $H = \begin{bmatrix} 1 & 0 \\ 0 & -1 \end{bmatrix}$. The second basis vector is reflected onto its negative, to produce this second column. The matrix H is still $2P - I$, when the same basis is used for H and P.

(iii)   For rotations, we could again choose unit vectors along the θ-line and its perpendicular. But the matrix would not be changed. Those lines are still rotated through θ, and $Q = \begin{bmatrix} c & -s \\ s & c \end{bmatrix}$ as before.

The whole question of choosing the best basis is absolutely central, and we come back to it in Chapter 5. The goal is to make the matrix diagonal, as achieved for P and H. To make Q diagonal requires complex vectors, since all real vectors are rotated.

We mention here the effect on the matrix of a *change of basis*, while the linear transformation stays the same. The matrix A (or Q or P or H) **is altered to** $S^{-1}AS$. Thus a single transformation is represented by different matrices (via different bases, accounted for by S). The theory of eigenvectors will lead to this formula $S^{-1}AS$, and to the best basis.

EXERCISES

2.6.1   What matrix has the effect of rotating every vector through 90° and then projecting the result onto the *x*-axis?

2.6.2   What matrix represents projection onto the *x*-axis followed by projection onto the *y*-axis?

2.6.3   Does the product of 5 reflections and 8 rotations of the *x-y* plane produce a rotation or a reflection?

2.6.4   The matrix $A = \begin{bmatrix} 2 & 0 \\ 0 & 1 \end{bmatrix}$ produces a **stretching** in the *x*-direction. Draw the circle $x^2 + y^2 = 1$ and sketch around it the points $(2x, y)$ that result from multiplication by A. What shape is that curve?

2.6.5   Every straight line remains straight after a linear transformation. If z is halfway between x and y, show that Az is halfway between Ax and Ay.

2.6.6   The matrix $A = \begin{bmatrix} 1 & 0 \\ 3 & 1 \end{bmatrix}$ yields a **shearing** transformation, which leaves the *y*-axis unchanged. Sketch its effect on the *x*-axis, by indicating what happens to $(1, 0)$ and $(2, 0)$ and $(-1, 0)$—and how the whole axis is transformed.

2.6.7   What 3 by 3 matrices represent the transformations that
   i)   project every vector onto the *x–y* plane?
   ii)   reflect every vector through the *x–y* plane?

iii) rotate the x–y plane through $90°$, leaving the z-axis alone?

iv) rotate the x–y plane, then the x–z plane, then the y–z plane, all through $90°$?

v) carry out the same three rotations, but through $180°$?

2.6.8 On the space P_3 of cubic polynomials, what matrix represents d^2/dt^2? Construct the 4 by 4 matrix from the standard basis 1, t, t^2, t^3. What is its nullspace, what is its column space, and what do they mean in terms of polynomials?

2.6.9 From the cubics P_3 to the fourth degree polynomials P_4, what matrix represents multiplication by $2 + 3t$? The columns of the 5 by 4 matrix A come from applying the transformation to each basis vector $x_1 = 1$, $x_2 = t$, $x_3 = t^2$, $x_4 = t^3$.

2.6.10 The solutions to the linear differential equation $d^2u/dt^2 = u$ form a vector space (since combinations of solutions are still solutions). Find two independent solutions, to give a basis for that space.

2.6.11 With initial values $u = x$ and $du/dt = y$ at $t = 0$, what combination of basis vectors in Ex. 2.6.10 solves the equation? This transformation from the initial values to the solution is linear; what is its 2 by 2 matrix (using $x = 1$, $y = 0$ and $x = 0$, $y = 1$ as basis for V, and your basis for W)?

2.6.12 Verify directly from $c^2 + s^2 = 1$ that the reflection matrices satisfy $H^2 = I$.

2.6.13 Suppose A is a linear transformation from the x–y plane to itself. Show that A^{-1} is also a linear transformation (if it exists). If A is represented by the matrix M, explain why A^{-1} is represented by M^{-1}.

2.6.14 The product $(AB)C$ of linear transformations starts with a vector x, produces a vector $u = Cx$, and then follows the shaded rule 2V in applying AB to u. It reaches $(AB)Cx$.

i) Is the result the same as separately applying C then B then A?

ii) Is the result the same as applying BC followed by A? If so, parentheses are unnecessary and the associative law $(AB)C = A(BC)$ holds for linear transformations. Combined with the product rule 2V, this is the best proof of the same law for matrices.

2.6.15 Prove that A^2 is a linear transformation if A is (say from $\mathbf{R}^3$ to $\mathbf{R}^3$).

2.6.16 The space of all 2 by 2 matrices has the four basis "vectors"

$$\begin{bmatrix} 1 & 0 \\ 0 & 0 \end{bmatrix}, \quad \begin{bmatrix} 0 & 1 \\ 0 & 0 \end{bmatrix}, \quad \begin{bmatrix} 0 & 0 \\ 1 & 0 \end{bmatrix}, \quad \begin{bmatrix} 0 & 0 \\ 0 & 1 \end{bmatrix}.$$

Consider the linear transformation of *transposing* every 2 by 2 matrix, and find its matrix A with respect to this basis. Why is $A^2 = I$?

2.6.17 Find the 4 by 4 matrix that represents a cyclic permutation: each vector (x_1, x_2, x_3, x_4) is transformed to (x_2, x_3, x_4, x_1). What is the effect of A^2? Show that $A^3 = A^{-1}$.

2.6.18 Find the 4 by 3 matrix A that represents a *right shift*: each vector (x_1, x_2, x_3) is transformed to $(0, x_1, x_2, x_3)$. Find also the *left shift* matrix B from $\mathbf{R}^4$ back to $\mathbf{R}^3$, transforming (x_1, x_2, x_3, x_4) to (x_2, x_3, x_4). What are the products AB and BA?

2.6.19 In the vector space V of all cubic polynomials $P = a_0 + a_1x + a_2x^2 + a_3x^3$, let S be the subset of polynomials with $\int_0^1 p(x)\,dx = 0$. Verify that S is a subspace and find a basis.

2.6.20 A *nonlinear* transformation is invertible if there is existence and uniqueness; $f(x) = b$ has exactly one solution for every b. The example $f(x) = x^2$ is not invertible because $x^2 = b$ has two solutions for positive b and no solution for negative b. Which of the following transformations (from the real numbers $\mathbf{R}^1$ to the real numbers $\mathbf{R}^1$) are invertible? None are linear, not even (c).
   (a) $f(x) = x^3$ (b) $f(x) = e^x$ (c) $f(x) = x + 11$ (d) $f(x) = \cos x$.

2.6.21 What is the axis of rotation, and the angle of rotation, of the transformation that takes (x_1, x_2, x_3) into (x_2, x_3, x_1)?

REVIEW EXERCISES: Chapter 2

2.1   Find a basis for the following subspaces of $\mathbf{R}^4$:
   (a)   The vectors for which $x_1 = 2x_4$
   (b)   The vectors for which $x_1 + x_2 + x_3 = 0$ and $x_3 + x_4 = 0$
   (c)   The subspace spanned by $(1, 1, 1, 1)$, $(1, 2, 3, 4)$, and $(2, 3, 4, 5)$.

2.2   By giving a basis, describe a two-dimensional subspace of $\mathbf{R}^3$ that contains none of the coordinate vectors $(1, 0, 0)$, $(0, 1, 0)$, $(0, 0, 1)$.

2.3   True or false, with counterexample if false:
   (i)   If the vectors $x_1, \ldots, x_m$ span a subspace S, then dim $S = m$.
   (ii)   The intersection of two subspaces of a vector space cannot be empty.
   (iii) If $Ax = Ay$, then $x = y$.
   (iv)   The row space of A has a unique basis that can be computed by reducing A to echelon form.
   (v)   If a square matrix A has independent columns, so does A^2.

2.4   What is the echelon form U of

$$A = \begin{bmatrix} 1 & 2 & 0 & 2 & 1 \\ -1 & -2 & 1 & 1 & 0 \\ 1 & 2 & -3 & -7 & -2 \end{bmatrix}?$$

What are the dimensions of its four fundamental subspaces?

2.5   Find the rank and the nullspace of

$$A = \begin{bmatrix} 0 & 0 & 1 \\ 0 & 0 & 1 \\ 1 & 1 & 1 \end{bmatrix} \quad \text{and} \quad B = \begin{bmatrix} 0 & 0 & 1 & 2 \\ 0 & 0 & 1 & 2 \\ 1 & 1 & 1 & 0 \end{bmatrix}.$$

2.6   Find bases for the four fundamental subspaces associated with

$$A = \begin{bmatrix} 1 & 2 \\ 3 & 6 \end{bmatrix}, \quad B = \begin{bmatrix} 0 & 0 \\ 1 & 2 \end{bmatrix}, \quad C = \begin{bmatrix} 1 & 1 & 0 & 0 \\ 0 & 1 & 0 & 1 \end{bmatrix}.$$

2.7   What is the most general solution to $u + v + w = 1$, $u - w = 2$?

2.8   (a)   Construct a matrix whose nullspace contains the vector $x = (1, 1, 2)$.
   (b)   Construct a matrix whose left nullspace contains $y = (1, 5)$.
   (c)   Construct a matrix whose column space is spanned by $(1, 1, 2)$ and whose row space is spanned by $(1, 5)$.
   (d)   If you are given any three vectors in $\mathbf{R}^6$ and any three vectors in $\mathbf{R}^5$, is there a 6 by 5 matrix whose column space is spanned by the first three and whose row space is spanned by the second three?

2.9   In the vector space of 2 by 2 matrices,
   (a)   is the set of rank-one matrices a subspace?
   (b)   what subspace is spanned by the permutation matrices?
   (c)   what subspace is spanned by the positive matrices (all $a_{ij} > 0$)?
   (d)   what subspace is spanned by the invertible matrices?

2.10 Invent a vector space that contains all linear transformations from $\mathbf{R}^n$ to $\mathbf{R}^n$. You have to decide on a rule for addition. What is its dimension?

2.11 (a) Find the rank of A, and give a basis for its nullspace.

$$A = LU = \begin{bmatrix} 1 & & & \\ 2 & 1 & & \\ 2 & 1 & 1 & \\ 3 & 2 & 4 & 1 \end{bmatrix} \begin{bmatrix} 1 & 2 & 0 & 1 & 2 & 1 \\ 0 & 0 & 2 & 2 & 0 & 0 \\ 0 & 0 & 0 & 0 & 0 & 1 \\ 0 & 0 & 0 & 0 & 0 & 0 \end{bmatrix}$$

(b) T F   The first 3 rows of U are a basis for the row space of A
     T F   Columns 1, 3, 6 of U are a basis for the column space of A
     T F   The four rows of A are a basis for the row space of A
(c) Find as many linearly independent vectors b as possible for which $Ax = b$ has a solution.
(d) In elimination on A, what multiple of the third row is subtracted to knock out the fourth row?

2.12 If A is an n by $n - 1$ matrix, and its rank is $n - 2$, what is the dimension of its nullspace?

2.13 Use elimination to find the triangular factors in $A = LU$, if

$$A = \begin{bmatrix} a & a & a & a \\ a & b & b & b \\ a & b & c & c \\ a & b & c & d \end{bmatrix}.$$

Under what conditions on the numbers a, b, c, d are the columns linearly independent?

2.14 Do the vectors $(1, 1, 3)$, $(2, 3, 6)$, and $(1, 4, 3)$ form a basis for $\mathbf{R}^3$?

2.15 Give examples of matrices A for which the number of solutions to $Ax = b$ is

(i)   0 or 1, depending on b;
(ii) ∞, independent of b;
(iii) 0 or ∞, depending on b;
(iv)   1, regardless of b.

2.16 In the previous exercise, how is r related to m and n in each example?

2.17 If x is a vector in $\mathbf{R}^n$, and $x^T y = 0$ for every y, prove that $x = 0$.

2.18 If A is an n by n matrix such that $A^2 = A$ and rank $A = n$, prove that $A = I$.

2.19 What subspace of 3 by 3 matrices is spanned by the elementary matrices E_{ij}, with ones on the diagonal and at most one nonzero entry below?

2.20 How many 5 by 5 permutation matrices are there? Are they linearly independent? Do they span the space of all 5 by 5 matrices? No need to write them all down.

2.21 What is the rank of the n by n matrix with every entry equal to one? How about the "checkerboard matrix," with $a_{ij} = 0$ when $i + j$ is even, $a_{ij} = 1$ when $i + j$ is odd?

2.22 (a) $Ax = b$ has a solution under what conditions on b, if

$$A = \begin{bmatrix} 1 & 2 & 0 & 3 \\ 0 & 0 & 0 & 0 \\ 2 & 4 & 0 & 1 \end{bmatrix} \quad \text{and} \quad b = \begin{bmatrix} b_1 \\ b_2 \\ b_3 \end{bmatrix}?$$

(b) Find a basis for the nullspace of A.
(c) Find the general solution to $Ax = b$, when a solution exists.
(d) Find a basis for the column space of A.
(e) What is the rank of A^T?

2.23 How can you construct a matrix which transforms the coordinate vectors e_1, e_2, e_3 into three given vectors v_1, v_2, v_3? When will that matrix be invertible?

2.24 If e_1, e_2, e_3 are in the column space of a 3 by 5 matrix, does it have a left-inverse? Does it have a right-inverse?

2.25 Suppose T is the linear transformation on $\mathbf{R}^3$ that takes each point (u, v, w) to $(u + v + w, u + v, u)$. Describe what T^{-1} does to the point (x, y, z).

2.26 *True or false:* (a) Every subspace of $\mathbf{R}^4$ is the nullspace of some matrix.
(b) If A has the same nullspace as A^T, the matrix must be square.
(c) The transformation that takes x to $mx + b$ is linear (from $\mathbf{R}^1$ to $\mathbf{R}^1$).

2.27 Find bases for the four fundamental subspaces of

$$A_1 = \begin{bmatrix} 1 & 2 & 0 & 3 \\ 0 & 2 & 2 & 2 \\ 0 & 0 & 0 & 0 \\ 0 & 0 & 0 & 4 \end{bmatrix} \quad \text{and} \quad A_2 = \begin{bmatrix} 1 \\ 1 \\ 1 \end{bmatrix} \begin{bmatrix} 1 & 4 \end{bmatrix}.$$

2.28 (a) If the rows of A are linearly independent (A is m by n) then the rank is _____ and the column space is _____ and the left nullspace is _____.
(b) If A is 8 by 10 with a 2-dimensional nullspace, show that $Ax = b$ can be solved for every b.

2.29 Describe the linear transformations of the x–y plane that are represented with standard basis $(1, 0)$ and $(0, 1)$ by the matrices

$$A_1 = \begin{bmatrix} 1 & 0 \\ 0 & -1 \end{bmatrix}, \quad A_2 = \begin{bmatrix} 1 & 0 \\ 2 & 1 \end{bmatrix}, \quad A_3 = \begin{bmatrix} 0 & 1 \\ -1 & 0 \end{bmatrix}.$$

2.30 (a) If A is square, show that the nullspace of A^2 contains the nullspace of A.
(b) Show also that the column space of A^2 is contained in the column space of A.

2.31 When does the rank-one matrix $A = uv^T$ have $A^2 = 0$?

2.32 (a) Find a basis for the space of all vectors in $\mathbf{R}^6$ with $x_1 + x_2 = x_3 + x_4 = x_5 + x_6$.
(b) Find a matrix with that subspace as its nullspace.
(c) Find a matrix with that subspace as its column space.

2.33 Suppose the matrices in $PA = LU$ are

$$\begin{bmatrix} 0 & 1 & 0 & 0 \\ 1 & 0 & 0 & 0 \\ 0 & 0 & 0 & 1 \\ 0 & 0 & 1 & 0 \end{bmatrix} \begin{bmatrix} 0 & 0 & 1 & -3 & 2 \\ 2 & -1 & 4 & 2 & 1 \\ 4 & -2 & 9 & 1 & 4 \\ 2 & -1 & 5 & -1 & 5 \end{bmatrix}$$

$$= \begin{bmatrix} 1 & 0 & 0 & 0 \\ 0 & 1 & 0 & 0 \\ 1 & 1 & 1 & 0 \\ 2 & 1 & 0 & 1 \end{bmatrix} \begin{bmatrix} 2 & -1 & 4 & 2 & 1 \\ 0 & 0 & 1 & -3 & 2 \\ 0 & 0 & 0 & 0 & 2 \\ 0 & 0 & 0 & 0 & 0 \end{bmatrix}.$$

(a) What is the rank of A?
(b) What is a basis for the row space of A?
(c) *True or false*: Rows 1, 2, 3 of A are linearly independent.
(d) What is a basis for the column space of A?
(e) What is the dimension of the left nullspace of A?
(f) What is the general solution to $Ax = 0$?

3

ORTHOGONALITY

3.1 ■ PERPENDICULAR VECTORS AND ORTHOGONAL SUBSPACES

We know from the last chapter what a basis is. Algebraically, it is a set of independent vectors that span the space. Geometrically, it is a set of coordinate axes. A vector space is defined without those axes, but every time I think of the x-y plane or three-dimensional space or $\mathbf{R}^n$, the axes are there. Furthermore, they are usually perpendicular! *The coordinate axes that the imagination constructs are practically always orthogonal.* In choosing a basis, we tend to choose an orthogonal basis.

If the idea of a basis is one of the foundations of linear algebra, then the specialization to an orthogonal basis is not far behind. We need a basis to convert geometric constructions into algebraic calculations, and we need an orthogonal basis to make those calculations simple. There is even a further specialization, which makes the basis just about optimal: The vectors should have *length one*. That can be achieved, but to do it we have to know

(1)   the length of a vector
(2)   the test for perpendicular vectors
(3)   how to create perpendicular vectors from linearly independent vectors.

More than that, subspaces must enter the picture. They also can be perpendicular. We will discover, so beautifully and simply that it is a delight to see, that *the fundamental subspaces are at right angles to each other.* They are perpendicular in pairs, two in $\mathbf{R}^m$ and two in $\mathbf{R}^n$. That will complete the fundamental theorem of linear algebra.

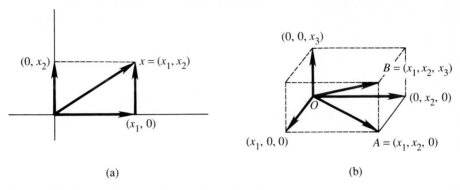

Fig. 3.1. The length of two- and three-dimensional vectors.

The first step is to find the **length of a vector**. It is denoted by $\|x\|$, and in two dimensions it comes from the hypotenuse of a right triangle (Fig. 3.1a). The square of the length was given a long time ago by Pythagoras: $\|x\|^2 = x_1^2 + x_2^2$.

In three-dimensional space, the vector $x = (x_1, x_2, x_3)$ is the diagonal of a box (Fig. 3.1b) and its length comes from *two* applications of the Pythagoras formula. The two-dimensional case takes care of the diagonal $OA = (x_1, x_2, 0)$ which runs across the base, and gives $\overline{OA}^2 = x_1^2 + x_2^2$. This forms a right angle with the vertical side $(0, 0, x_3)$, so we may appeal to Pythagoras again (in the plane of OA and AB). The hypotenuse of the triangle OAB is the length $\|x\|$ we want, and it is given by

$$\|x\|^2 = \overline{OA}^2 + \overline{AB}^2 = x_1^2 + x_2^2 + x_3^2.$$

The generalization to a vector in n dimensions, $x = (x_1, \ldots, x_n)$, is immediate. **The length $\|x\|$ of a vector in $\mathbf{R}^n$ is the positive square root of**

$$\|x\|^2 = x_1^2 + x_2^2 + \cdots + x_n^2 = x^Tx. \tag{1}$$

Geometrically, this amounts to applying the Pythagoras formula $n - 1$ times, adding one more dimension at each step. The sum of squares agrees with the multiplication x^Tx—and the length of $x = (1, 2, -3)$ is $\sqrt{14}$:

$$x^Tx = \begin{bmatrix} 1 & 2 & -3 \end{bmatrix} \begin{bmatrix} 1 \\ 2 \\ -3 \end{bmatrix} = 14.$$

Now suppose we are given two vectors x and y (Fig. 3.2). How can we decide whether they are perpendicular? In other words, what is the test for orthogonality? This is a question that can be answered in the two-dimensional plane by trigonometry. Here we need the generalization to $\mathbf{R}^n$, but still we can stay in the plane spanned by x and y. Within this plane, x is orthogonal to y provided they form

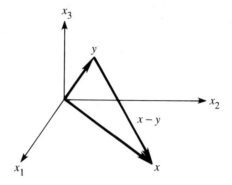

Fig. 3.2. The plane triangle with sides x, y, and $x - y$.

a right triangle. We go back to the Pythagoras formula as a test: *x and y are perpendicular if*

$$\|x\|^2 + \|y\|^2 = \|x - y\|^2. \tag{2}$$

Applying the length formula (1), this becomes

$$(x_1^2 + \cdots + x_n^2) + (y_1^2 + \cdots + y_n^2) = (x_1 - y_1)^2 + \cdots + (x_n - y_n)^2.$$

The left side cancels part of the right side, which expands into

$$(x_1^2 + \cdots + x_n^2) - 2(x_1 y_1 + \cdots + x_n y_n) + (y_1^2 + \cdots + y_n^2).$$

Thus equality holds in (2) *when the "cross-product terms" give zero: x **and** y **are** orthogonal if*

$$\boxed{x_1 y_1 + \cdots + x_n y_n = 0.} \tag{3}$$

Notice that this quantity is exactly the same as $x^{\mathrm{T}}y$, the product of a 1 by n matrix (the row vector x^{T}) with an n by 1 matrix (the column vector y):

$$x^{\mathrm{T}}y = [x_1 \ldots x_n] \begin{bmatrix} y_1 \\ \vdots \\ y_n \end{bmatrix} = x_1 y_1 + \cdots + x_n y_n. \tag{4}$$

Using the notation for summation, it is $\sum x_i y_i$. (*It is also* $y^{\mathrm{T}}x$.) This combination appears in every discussion of the geometry of n-dimensional space. It is sometimes called the scalar product or dot product of the two vectors, and denoted by (x, y) or $x \cdot y$, but we prefer to call it the **inner product** and to keep the notation $x^{\mathrm{T}}y$:

3A   The quantity $x^{\mathrm{T}}y$ is the inner product of the (column) vectors x and y in $\mathbf{R}^n$. It is zero if and only if x and y are orthogonal.

The ideas of length and inner product are connected by $x^Tx = x_1^2 + \cdots + x_n^2 = \|x\|^2$. The only vector with length zero—in other words, the only vector orthogonal to itself—is the zero vector. This vector $x = 0$ is orthogonal to every vector in $\mathbf{R}^n$.

EXAMPLE　$x = (2, 2, -1)$ is orthogonal to $y = (-1, 2, 2)$. Both have length $\sqrt{4 + 4 + 1} = 3$.

The next section studies vectors that are not orthogonal; the inner product x^Ty determines the angle between them. It gives a natural definition of the *cosine* in n-dimensional space; perpendicular vectors have $\cos \theta = 0$. In this section we stay with those right angles. The goal is still to understand the four fundamental subspaces—and the property we are after is orthogonality.

First, there is a simple connection to linear independence: *If the nonzero vectors $v_1, \ldots, v_k$ are mutually orthogonal* (every vector is perpendicular to every other), *then they are linearly independent.*

Proof　Suppose $c_1v_1 + \cdots + c_kv_k = 0$. To show that c_1 must be zero, take the inner product of both sides with v_1:

$$v_1^T(c_1v_1 + \cdots + c_kv_k) = v_1^T 0 = 0. \tag{5}$$

The orthogonality of the v's leaves only one term in (5), $c_1v_1^Tv_1 = 0$. Because the vectors were assumed nonzero, $v_1^Tv_1 \neq 0$ and therefore $c_1 = 0$. The same is true of every c_i. Thus the only combination of the v's producing zero is the trivial one with all $c_i = 0$, and the vectors are independent.

The most important example of mutually orthogonal vectors is the set of co-ordinate vectors $e_1, \ldots, e_n$ in $\mathbf{R}^n$. Those are the columns of the identity matrix. They form the simplest basis for $\mathbf{R}^n$, and they are *unit vectors*—each of them has length $\|e_i\| = 1$. They point in the direction of the coordinate axes. If this system of vectors is rotated, the result is a new "*orthonormal basis,*" that is, a new system of mutually orthogonal unit vectors. In the plane, this rotation produces

$$v_1 = (\cos \theta, \sin \theta), \qquad v_2 = (-\sin \theta, \cos \theta).$$

That gives another example of length ($v_1^Tv_1 = 1$ and $v_2^Tv_2 = 1$) and orthogonality ($v_1^Tv_2 = 0$).

Orthogonal Subspaces

We come to the orthogonality of two subspaces. That requires *every vector* in one subspace to be orthogonal to *every vector* in the other. In three-dimensional space, subspaces are represented by lines or planes through the origin—and in the extreme cases, by the origin alone or the whole space. The subspaces can have dimension 0, 1, 2, or 3. The subspace $\{0\}$ is orthogonal to all subspaces. A line can be orthogonal to another line, or it can be orthogonal to a plane, but *a plane cannot be orthogonal to a plane.*† The full space $\mathbf{R}^3$ is orthogonal only to $\{0\}$. In n dimensions the definition is this:

3B Two subspaces V and W of the same space $\mathbf{R}^n$ are *orthogonal* if every vector v in V is orthogonal to every vector w in W: $v^{\mathsf{T}}w = 0$ for all v and w.

EXAMPLE Suppose V is the plane spanned by $v_1 = (1, 0, 0, 0)$ and $v_2 = (1, 1, 0, 0)$, and W is the line spanned by $w = (0, 0, 4, 5)$. Then since w is orthogonal to both v's, the line W will be orthogonal to the whole plane V.

In this case, with subspaces of dimension 2 and 1 in $\mathbf{R}^4$, there is room for a third subspace. It is the line L through $z = (0, 0, 5, -4)$, perpendicular to V and W. Then the dimensions add to $2 + 1 + 1 = 4$, and only the zero vector is perpendicular to all three of V, W, and L.

Now we explain our interest in orthogonal subspaces. The important ones don't come by accident, and they come two at a time. In fact orthogonal subspaces are unavoidable: *They are the fundamental subspaces!* There are four subspaces, and they come in pairs.

The first pair is the *nullspace* and *row space*. Those are subspaces of $\mathbf{R}^n$—the rows have n components and so does the vector in $Ax = 0$. We have to show, using only the equation $Ax = 0$, that *the rows are orthogonal to the vector x*.

3C The row space is orthogonal to the nullspace (in $\mathbf{R}^n$) and the column space is orthogonal to the left nullspace (in $\mathbf{R}^m$).

First proof Suppose x is a vector in the nullspace. Then $Ax = 0$, and this system of m equations can be written out more fully as

$$Ax = \begin{bmatrix} \cdots & \text{row 1} & \cdots \\ \cdots & \text{row 2} & \cdots \\ & \vdots & \\ \cdots & \text{row } m & \cdots \end{bmatrix} \begin{bmatrix} x_1 \\ x_2 \\ \vdots \\ x_n \end{bmatrix} = \begin{bmatrix} 0 \\ 0 \\ \vdots \\ 0 \end{bmatrix}. \tag{6}$$

The main point is already in the first equation: *row 1 is orthogonal to x*. Their inner product is zero; that is the equation. The second equation says the same thing for row 2. Because of the zeros on the right side, x is orthogonal to every

---

† I have to admit that the front wall and side wall of a room look like perpendicular planes in $\mathbf{R}^3$. But by our definition, that is not so! There are lines v and w in the two walls that do not meet at a right angle. In fact the line along the corner is in *both* walls, and it is certainly not orthogonal to itself.

row. Therefore it is orthogonal to every combination of the rows. Each x in the nullspace is orthogonal to each vector in the row space, so $\mathcal{N}(A) \perp \mathcal{R}(A^T)$.

The other orthogonal subspaces come from $A^T y = 0$, or $y^T A = 0$:

$$y^T A = \begin{bmatrix} y_1 & \cdots & y_m \end{bmatrix} \begin{bmatrix} c & & c \\ o & & o \\ 1 & & 1 \\ u & \cdots & u \\ m & & m \\ n & & n \\ 1 & & n \end{bmatrix} = \begin{bmatrix} 0 & \cdots & 0 \end{bmatrix}. \tag{7}$$

The vector y is orthogonal to every column. The equation says so, from the zeros on the right. Therefore y is orthogonal to every combination of the columns. It is orthogonal to the column space, and it is a typical vector in the left nullspace: $\mathcal{N}(A^T) \perp \mathcal{R}(A)$. This is the same as the first half of the theorem, with A replaced by A^T.

Second proof   We want to establish the same result by a more coordinate-free argument. The contrast between the two proofs should be useful to the reader, as a specific example of an "abstract" versus a "concrete" method of reasoning. I wish I were sure which is clearer, and more permanently understood.

Suppose x is in the nullspace and v is in the row space. Then $Ax = 0$ and $v = A^T z$ for some vector z. (The vector v is a combination of the rows, since it is in the row space.) One line will prove that they are orthogonal:

$$v^T x = (A^T z)^T x = z^T A x = z^T 0 = 0. \tag{8}$$

EXAMPLE   Suppose A has rank one, so its row space and column space are lines:

$$A = \begin{bmatrix} 1 & 3 \\ 2 & 6 \\ 3 & 9 \end{bmatrix}.$$

The rows are multiples of $(1, 3)$. The nullspace contains $(-3, 1)$, which is orthogonal to the rows. In fact the nullspace is just the perpendicular line in $\mathbf{R}^2$, satisfying

$$\begin{bmatrix} 1 & 3 \end{bmatrix} \begin{bmatrix} x_1 \\ x_2 \end{bmatrix} = 0 \quad \text{and} \quad \begin{bmatrix} 2 & 6 \end{bmatrix} \begin{bmatrix} x_1 \\ x_2 \end{bmatrix} = 0 \quad \text{and} \quad \begin{bmatrix} 3 & 9 \end{bmatrix} \begin{bmatrix} x_1 \\ x_2 \end{bmatrix} = 0.$$

In contrast, the other pair of subspaces is in $\mathbf{R}^3$. The column space is the line through $(1, 2, 3)$, and the left nullspace is a plane. It must be the *perpendicular plane* $y_1 + 2y_2 + 3y_3 = 0$. That is exactly the content of $y^T A = 0$.

Notice that the first pair (the two lines) had dimensions $1 + 1 = 2$ in the space $\mathbf{R}^2$. The second pair (line and plane) had dimensions $1 + 2 = 3$ in the space $\mathbf{R}^3$. In general the row space and nullspace have dimensions that add to $r + (n - r) = n$.

The other pair adds to $r + (m - r) = m$. Something more than just orthogonality is occurring, and I have to ask your patience about that one further point.

It is certainly the truth that the nullspace is perpendicular to the row space—but it is not the whole truth. $\mathcal{N}(A)$ does not contain just some of the vectors orthogonal to the row space, *it contains every such vector*. The nullspace was formed from *all* solutions to $Ax = 0$.

DEFINITION   Given a subspace V of $\mathbf{R}^n$, the space of all vectors orthogonal to V is called the *orthogonal complement* of V, and denoted by $V^\perp$.†

Using this terminology, the nullspace is the orthogonal complement of the row space: $\mathcal{N}(A) = (\mathcal{R}(A^T))^\perp$. At the same time, the opposite relation also holds: The row space contains all vectors that are orthogonal to the nullspace. This is not so obvious, since in solving $Ax = 0$ we started with the row space and found all x that were orthogonal to it; now we are going in the opposite direction. Suppose, however, that some vector z is orthogonal to the nullspace but is outside the row space. Then adding z as an extra row of A would enlarge the row space without changing the nullspace. But we know that there is a fixed formula $r + (n - r) = n$:

$$\text{dim(row space)} + \text{dim(nullspace)} = \text{number of columns.}$$

Since the last two numbers are unchanged when the new row z is added, it is impossible for the first one to change either. We conclude that every vector orthogonal to the nullspace is already in the row space: $\mathcal{R}(A^T) = (\mathcal{N}(A))^\perp$.

The same reasoning applied to A^T produces the dual result: *The left nullspace $\mathcal{N}(A^T)$ and the column space $\mathcal{R}(A)$ are orthogonal complements.* Their dimensions add up to $(m - r) + r = m$. This completes the second half of the fundamental theorem of linear algebra. The first half gave the dimensions of the four subspaces, including the fact that row rank = column rank. Now we know that those subspaces are perpendicular, and more than that they are orthogonal complements.

3D   *Fundamental Theorem of Linear Algebra*, Part 2
The nullspace is the orthogonal complement of the row space in $\mathbf{R}^n$.
The left nullspace is the orthogonal complement of the column space in $\mathbf{R}^m$.

To repeat, those statements are reversible. The row space contains everything orthogonal to the nullspace. The column space contains everything orthogonal to the left nullspace. That is just a sentence, hidden in the middle of the book, but it

---

† Suggested pronunciation: "V perp."

decides exactly which equations can be solved! Looked at directly, $Ax = b$ requires b to be in the column space. Looked at indirectly, it requires b to be perpendicular to the left nullspace.

3E The equation $Ax = b$ is solvable if and only if $b^T y = 0$ whenever $A^T y = 0$.

The direct approach was "b must be a combination of the columns." The indirect approach is "b must be orthogonal to every vector that is orthogonal to the columns." That doesn't sound like an improvement (to put it mildly). But if there are many columns, and only one or two vectors are orthogonal to them, it is much easier to check those one or two conditions $b^T y = 0$. A good example is Kirchhoff's voltage law in Section 2.5. Testing for zero around loops is much easier than recognizing combinations of the columns.

EXAMPLE The left sides of the following equations add to zero:

$$\begin{array}{l} x_1 - x_2 = b_1 \\ x_2 - x_3 = b_2 \\ x_3 - x_1 = b_3 \end{array} \quad \text{or} \quad \begin{bmatrix} 1 & -1 & 0 \\ 0 & 1 & -1 \\ -1 & 0 & 1 \end{bmatrix} \begin{bmatrix} x_1 \\ x_2 \\ x_3 \end{bmatrix} = \begin{bmatrix} b_1 \\ b_2 \\ b_3 \end{bmatrix}.$$

They are solvable *if and only if the right sides add to zero.* It is easier to check $b_1 + b_2 + b_3 = 0$—which makes b orthogonal to $y = (1, 1, 1)$ in the left nullspace—than to check whether b is a combination of the columns. By the fundamental theorem, it is the same thing.

The Matrix and the Subspaces

We emphasize that V and W can be orthogonal without being complements, when their dimensions are too small. The line V spanned by $(1, 0, 0)$ is orthogonal to the line W spanned by $(0, 0, 1)$, but in three dimensions V is not $W^\perp$. The orthogonal complement of W is two-dimensional. It is a plane, and the line V is only part of it. If the dimensions are right, then orthogonal subspaces *are* necessarily orthogonal complements. That was the case for the row space and nullspace, and the proof can be applied in general:

$$\text{if} \quad W = V^\perp \quad \text{then} \quad V = W^\perp.$$

In other words $V^{\perp\perp} = V$! The dimensions of V and W are right, and the space is being decomposed into two perpendicular parts (Fig. 3.3).

When the space is split into orthogonal parts, so is every vector: $x = v + w$. The vector v is the projection onto the subspace V. The orthogonal component w is the projection of x onto W. The next sections show how to find those projections; here we want to use them. They lead to what is probably the most important figure in the book (Fig. 3.4).

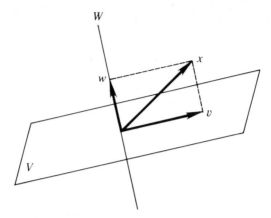

Fig. 3.3. Orthogonal complements in $\mathbf{R}^3$: a plane and a line.

Figure 3.4 summarizes the fundamental theorem of linear algebra. It illustrates the true effect of a matrix—what is happening below the surface of the multiplication Ax. One part of the theorem determined the *dimensions* of the subspaces. The key was that the row space and column space share the same dimension r (the rank). Now we also know the *orientation* of the four spaces. Two subspaces are orthogonal complements in $\mathbf{R}^n$, and the other two in $\mathbf{R}^m$. The nullspace is carried to the zero vector. Nothing is carried to the left nullspace. The real action is between the row space and column space, and you see it by looking at a typical vector x. It has a "row space component" and a "nullspace component," $x = x_r + x_n$. When multiplied by A, this is $Ax = Ax_r + Ax_n$:

the nullspace component goes to zero: $Ax_n = 0$
the row space component goes to the column space: $Ax_r = Ax$.

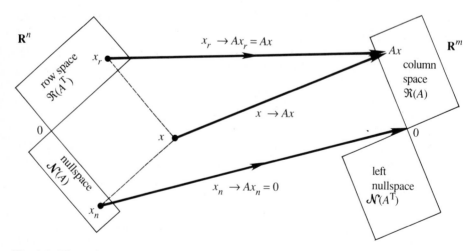

Fig. 3.4. The action of a matrix A.

Of course everything goes to the column space—the matrix cannot do anything else—and the figure shows how it happens.†

3F The mapping from row space to column space is actually invertible. Every vector b in the column space comes from one and only one vector x_r in the row space.

Proof If b is in the column space, it is a combination Ax of the columns. In fact it is Ax_r, with x_r in the row space, since the nullspace component gives $Ax_n = 0$. If another vector x'_r in the row space gives $Ax'_r = b$, then $A(x_r - x'_r) = b - b = 0$. This puts $x_r - x'_r$ in both the nullspace and the row space, which makes it orthogonal to itself. Therefore it is zero, and $x_r = x'_r$. Exactly one vector in the row space is carried to b.

Every matrix transforms its row space to its column space.

On those r-dimensional spaces A is invertible; on its nullspace it is zero. That is easy to see when A is diagonal; the submatrix holding the r nonzeros is invertible. Now we know it is always true. Furthermore A^T goes in the opposite direction, from $\mathbf{R}^m$ back to $\mathbf{R}^n$ and from $\mathcal{R}(A)$ back to $\mathcal{R}(A^T)$. Of course the transpose is not the inverse! A^T moves the spaces correctly, but not the individual vectors. That honor belongs to A^{-1} if it exists—and it only exists if $r = m = n$. Otherwise we are asking it to bring back a whole nullspace out of the single vector zero, which no matrix can do.

When A^{-1} fails to exist, you can see a natural substitute. It is called the ***pseudoinverse***, and denoted by A^+. It inverts A where that is possible: $A^+Ax = x$ for x in the row space. On the left nullspace nothing can be done: $A^+y = 0$. Thus A^+ inverts A where it is invertible, and it is computed in the Appendix. That computation depends on one of the great factorizations of a matrix—the ***singular value decomposition***—for which we first need to know about eigenvalues.

EXERCISES

3.1.1   Find the lengths and the inner product of $x = (1, 4, 0, 2)$ and $y = (2, -2, 1, 3)$.

3.1.2   Give an example in $\mathbf{R}^2$ of linearly independent vectors that are not mutually orthogonal. Also, give an example of mutually orthogonal vectors that are not independent.

---

† We did not really know how to draw two orthogonal subspaces of dimension r and $n - r$. If you understand these dimensions, and the orthogonality, do not allow Fig. 3.4 to confuse you!

3.1.3   According to analytic geometry, two lines in the plane are perpendicular when the product of their slopes is -1. Apply this to the vectors $x = (x_1, x_2)$ and $y = (y_1, y_2)$, whose slopes are x_2/x_1 and y_2/y_1, to derive again the orthogonality condition $x^T y = 0$.

3.1.4   How do we know that the ith row of an invertible matrix B is orthogonal to the jth column of B^{-1}, if $i \neq j$?

3.1.5   Which pairs are orthogonal among the vectors

$$v_1 = \begin{bmatrix} 1 \\ 2 \\ -2 \\ 1 \end{bmatrix}, \qquad v_2 = \begin{bmatrix} 4 \\ 0 \\ 4 \\ 0 \end{bmatrix}, \qquad v_3 = \begin{bmatrix} 1 \\ -1 \\ -1 \\ -1 \end{bmatrix}?$$

3.1.6   In $\mathbf{R}^3$ find all vectors that are orthogonal to $(1, 1, 1)$ and $(1, -1, 0)$. Produce from these vectors a mutually orthogonal system of unit vectors (an orthonormal system) in $\mathbf{R}^3$.

3.1.7   Find a vector x orthogonal to the row space, and a vector y orthogonal to the column space, of

$$A = \begin{bmatrix} 1 & 2 & 1 \\ 2 & 4 & 3 \\ 3 & 6 & 4 \end{bmatrix}.$$

3.1.8   If V and W are orthogonal subspaces, show that the only vector they have in common is the zero vector: $V \cap W = \{0\}$.

3.1.9   Find the orthogonal complement of the plane spanned by the vectors $(1, 1, 2)$ and $(1, 2, 3)$, by taking these to be the rows of A and solving $Ax = 0$. Remember that the complement is a whole line.

3.1.10   Construct a homogeneous equation in three unknowns whose solutions are the linear combinations of the vectors $(1, 1, 2)$ and $(1, 2, 3)$. This is the reverse of the previous exercise, but of course the two problems are really the same.

3.1.11   The fundamental theorem of linear algebra is often stated in the form of *Fredholm's alternative*: For any A and b, one and only one of the following systems has a solution:

$$(1) \quad Ax = b \qquad (2) \quad A^T y = 0, \; y^T b \neq 0.$$

In other words, either b is in the column space $\mathscr{R}(A)$ or there is a y in $\mathscr{N}(A^T)$ such that $y^T b \neq 0$. Show that it is contradictory for (1) and (2) both to have solutions.

3.1.12   Find a basis for the nullspace of

$$A = \begin{bmatrix} 1 & 0 & 2 \\ 1 & 1 & 4 \end{bmatrix},$$

and verify that it is orthogonal to the row space. Given $x = (3, 3, 3)$, split it into a row space component x_r and a nullspace component x_n.

3.1.13 Illustrate the action of A^T by a picture corresponding to Fig. 3.4, sending $\mathscr{R}(A)$ back to the row space and the left nullspace to zero.

3.1.14 Show that $x - y$ is orthogonal to $x + y$ if and only if $\|x\| = \|y\|$.

3.1.15 Find a matrix whose row space contains $(1, 2, 1)$ and whose nullspace contains $(1, -2, 1)$, or prove that there is no such matrix.

3.1.16 Find all vectors which are perpendicular to $(1, 4, 4, 1)$ and $(2, 9, 8, 2)$.

3.1.17 If V is the orthogonal complement of W in $\mathbf{R}^n$, is there a matrix with row space V and nullspace W? Starting with a basis for V show how such a matrix can be constructed.

3.1.18 If $S = \{0\}$ is the subspace of $\mathbf{R}^4$ containing only the origin, what is $S^\perp$? If S is spanned by $(0, 0, 0, 1)$, what is $S^\perp$?

3.1.19 *True or false*: (a) If V is orthogonal to W, then $V^\perp$ is orthogonal to $W^\perp$.
(b) If V is orthogonal to W and W is orthogonal to Z, then V is orthogonal to Z.

3.1.20 Let S be a subspace of $\mathbf{R}^n$. Explain what $(S^\perp)^\perp = S$ means and why it is true.

3.1.21 Let P be the plane (not a subspace) in 3-space with equation $x + 2y - z = 6$. Find the equation of a plane P' parallel to P but going through the origin. Find also a vector perpendicular to those planes. What matrix has the plane P' as its nullspace, and what matrix has P' as its row space?

3.1.22 Let S be the subspace of $\mathbf{R}^4$ containing all vectors with $x_1 + x_2 + x_3 + x_4 = 0$. Find a basis for the space $S^\perp$, containing all vectors orthogonal to S.

3.2 ■ INNER PRODUCTS AND PROJECTIONS ONTO LINES

We know that the inner product of two vectors x and y is the number $x^T y$. So far we have been interested only in whether that inner product is zero—in other words, whether the two vectors are orthogonal. Now we allow also the possibility of inner products that are *not* zero, and angles that are not right angles. We want to understand the relation of the inner product to the angle and also the connection between inner products and transposes. In Chapter 1 the transpose was constructed by flipping over a matrix as if it were some kind of pancake. We have to do better than that.

   If we try to summarize the rest of the chapter, there is no way to avoid the fact that *the orthogonal case is by far the most important.* Suppose we are given a point b in n-dimensional space, and we want to find its distance to a given line—say the line in the direction of the vector a. We are looking along that line for the point p closest to b. The key is in the geometry: ***The line connecting b to p*** (the dotted line in Fig. 3.5) ***is perpendicular to the vector*** a. *This fact will allow us to find the closest point p*, and to compute its distance from b. Even though the given vectors a and b are not orthogonal, the solution to the problem automatically brings in orthogonality.

   The situation is the same when, instead of a line in the direction of a, we are given a plane—or more generally any subspace S of $\mathbf{R}^n$. Again the problem is to find the point p on that subspace that is closest to b. *This point p is the projection of b onto the subspace.* When we draw a perpendicular from b to S, p is the point where the perpendicular meets the subspace. Geometrically speaking, that is a simple solution to a natural problem about distances between points b and subspaces S. But there are two questions that need to be asked:

(1)   Does this problem actually arise in practical applications?
(2)   If we have a basis for the subspace, is there a formula for the projection p?

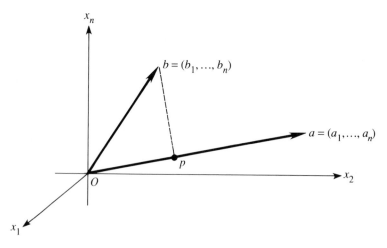

Fig. 3.5. A one-dimensional projection in n-dimensional space.

The answers are certainly yes. Our problem, described so far only in geometrical terms, is exactly the problem of the **least squares solution to an overdetermined system**. The vector b represents the data, given by experiments or questionnaires, and it contains too many errors to be found in the given subspace. When we try to write b as a combination of the basis vectors in the subspace, it cannot be done—the equations are inconsistent, and have no solution. The least squares method selects the point p as the best choice possible. There can be no doubt of the importance of this application.†

The second question, to find a formula for p, is easy when the subspace is a line. We will project one vector onto another in several different ways, in this section and the next, and relate this projection to inner products and angles. Fortunately, the formula for p remains fairly simple when we project onto a higher dimensional subspace, provided we are given a basis. This is by far the most important case; it corresponds to a least squares problem with several parameters, and it is solved in Section 3.3. Then it remains to make the formulas even simpler, by getting back to orthogonal vectors.

Inner Products and the Schwarz Inequality

We pick up the discussion of inner products and angles. You will soon see that it is not the angle, but **the cosine of the angle**, that is directly related to inner products. Therefore we first look back to trigonometry, that is to the two-dimensional case, in order to find that relationship. Suppose α is the angle that the vector a makes with the x axis (Fig. 3.6). Remembering that $\|a\|$ is the length of the vector, which is the hypotenuse in the triangle OaQ, the sine and cosine of

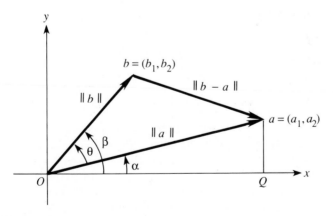

Fig. 3.6. The cosine of the angle $\theta = \beta - \alpha$.

---

† In economics and statistics, least squares enters *regression analysis*. In geodesy, the U.S. mapping survey plans to solve the largest system ever attempted, now 2.5 million equations in 400,000 unknowns.

α are

$$\sin \alpha = \frac{a_2}{\|a\|}, \qquad \cos \alpha = \frac{a_1}{\|a\|}.$$

The same is true for b and its corresponding angle β: the sine is $b_2/\|b\|$, and the cosine is $b_1/\|b\|$. Now, since θ is just $\beta - \alpha$, its cosine comes from a trigonometric identity which no one could forget:

$$\cos \theta = \cos \beta \cos \alpha + \sin \beta \sin \alpha = \frac{a_1 b_1 + a_2 b_2}{\|a\| \|b\|}. \tag{1}$$

The numerator in this formula is exactly the inner product of b and a, and gives the relationship we are looking for:

3G   The cosine of the angle between any two vectors a and b is

$$\cos \theta = \frac{a^{\mathrm{T}} b}{\|a\| \|b\|}. \tag{2}$$

Notice that the formula is dimensionally correct; if we double the length of b, then both numerator and denominator are doubled, and the cosine is unchanged. Reversing the sign of b, on the other hand, reverses the sign of $\cos \theta$—and changes the angle by 180°.

Remark   There is another law of trigonometry, the law of cosines, that leads directly to the same result. It is not quite so unforgettable as the formula in (1), but it relates the lengths of the sides of any triangle:

$$\|b - a\|^2 = \|b\|^2 + \|a\|^2 - 2\|b\| \|a\| \cos \theta. \tag{3}$$

When θ is a right angle, we are back to Pythagoras. But regardless of θ, the expression $\|b - a\|^2$ can be expanded as $(b - a)^{\mathrm{T}}(b - a)$, and (3) becomes

$$b^{\mathrm{T}} b - 2a^{\mathrm{T}} b + a^{\mathrm{T}} a = b^{\mathrm{T}} b + a^{\mathrm{T}} a - 2\|b\| \|a\| \cos \theta.$$

Canceling $b^{\mathrm{T}} b$ and $a^{\mathrm{T}} a$ on both sides of this equation, you recognize formula (2) for the cosine. In fact this proves the cosine formula in n dimensions, since we only have to worry about the plane triangle Oab.

Now we want to find the projection point p. This point must be some multiple $p = \bar{x} a$ of the given vector a—every point on the line is a multiple of a—and the problem is to compute the coefficient $\bar{x}$. All that we need for this computation is the geometrical fact that **the line from b to the closest point $p = \bar{x} a$ is perpendicular to the vector** a:

$$(b - \bar{x} a) \perp a, \quad \text{or} \quad a^{\mathrm{T}}(b - \bar{x} a) = 0, \quad \text{or} \quad \bar{x} = \frac{a^{\mathrm{T}} b}{a^{\mathrm{T}} a}. \tag{4}$$

That gives the formula for $\bar{x}$ and p:

3H The projection of b onto the line through O and a is

$$p = \bar{x}a = \frac{a^{\mathsf{T}}b}{a^{\mathsf{T}}a}\,a. \tag{5}$$

This allows us to redraw Fig. 3.5 with a correct formula for p.

This formula has a remarkable corollary, which is probably the most important inequality in mathematics. It includes as a special case the fact that arithmetic means $\frac{1}{2}(x + y)$ are larger than geometric means $\sqrt{xy}$. (It is also equivalent—see Exercise 3.2.1—to the triangle inequality for vectors.) The result seems to come almost accidentally from the statement that the squared distance $\|b - p\|^2$ in Fig. 3.7 cannot be negative:

$$\left\| b - \frac{a^{\mathsf{T}}b}{a^{\mathsf{T}}a}\,a \right\|^2 = b^{\mathsf{T}}b - 2\frac{(a^{\mathsf{T}}b)^2}{a^{\mathsf{T}}a} + \left(\frac{a^{\mathsf{T}}b}{a^{\mathsf{T}}a}\right)^2 a^{\mathsf{T}}a = \frac{(b^{\mathsf{T}}b)(a^{\mathsf{T}}a) - (a^{\mathsf{T}}b)^2}{(a^{\mathsf{T}}a)} \geq 0.$$

Since the last numerator is never negative, we have $(b^{\mathsf{T}}b)(a^{\mathsf{T}}a) \geq (a^{\mathsf{T}}b)^2$—and then we take square roots:

3I Any two vectors satisfy the *Schwarz inequality*

$$\left| a^{\mathsf{T}}b \right| \leq \|a\|\,\|b\|. \tag{6}$$

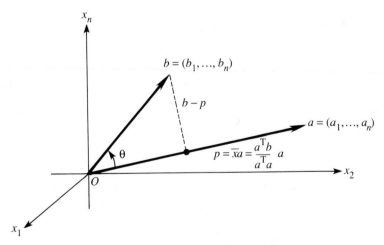

Fig. 3.7. The projection of b onto a, with $\cos\theta = \dfrac{Op}{Ob} = \dfrac{a^{\mathsf{T}}b}{\|a\|\,\|b\|}$.

Remark   According to formula (2), the ratio between the two sides of the Schwarz inequality is exactly $|\cos\theta|$. Since all cosines lie in the interval $-1 \le \cos\theta \le 1$, this gives another proof of (6): *the Schwarz inequality is the same as* $|\cos\theta| \le 1$. In some ways that is a more easily understood proof, because cosines are so familiar. Either proof is all right in $\mathbf{R}^n$, but notice that ours simply amounted to doing the mechanical calculation of $\|b - p\|^2$. This is nonnegative, and it will stay nonnegative when we later introduce new possibilities for the lengths and inner products. Therefore, without any appeal to trigonometry, the Schwarz inequality is proved.†

One final observation about $|a^{\mathrm{T}}b| \le \|a\|\,\|b\|$. *Equality holds if and only if b is a multiple of a.* The angle is $\theta = 0°$ or $\theta = 180°$ and the cosine is 1 or -1. In this case b is identical with its projection p, and the distance between b and the line is zero.

EXAMPLE   Project $b = (1, 2, 3)$ onto the line through $a = (1, 1, 1)$:

$$\bar{x} = \frac{a^{\mathrm{T}}b}{a^{\mathrm{T}}a} = \frac{6}{3} = 2.$$

The projection is $p = 2a = (2, 2, 2)$. The cosine is

$$\cos\theta = \frac{\|p\|}{\|b\|} = \frac{\sqrt{12}}{\sqrt{14}}.$$

The Schwarz inequality is $|a^{\mathrm{T}}b| \le \|a\|\,\|b\|$, or $6 \le \sqrt{3}\sqrt{14}$. If we write 6 as $\sqrt{36}$, that is the same as $\sqrt{12} \le \sqrt{14}$; the cosine is less than 1. Inequality holds, because b is not parallel to a.

Projections of Rank One

The projection of b onto the line through a lies at $p = a(a^{\mathrm{T}}b/a^{\mathrm{T}}a)$. That is our formula $p = \bar{x}a$, but it is written with a slight twist: The vector a is put before the number $\bar{x} = a^{\mathrm{T}}b/a^{\mathrm{T}}a$. There is a reason behind that apparently trivial change. Projection onto a line is carried out by a "***projection matrix***" P, and written in this new order we can see what it is. *It is the matrix that multiplies b and produces p*:

$$\boxed{P = \frac{aa^{\mathrm{T}}}{a^{\mathrm{T}}a}.} \tag{7}$$

That is a column times a row—a square matrix—divided by the number $a^{\mathrm{T}}a$.

---

† The name of Cauchy is also attached to this inequality $|a^{\mathrm{T}}b| \le \|a\|\,\|b\|$, and the Russians even refer to it as the Cauchy-Schwarz-Buniakowsky inequality! Mathematical historians seem to agree that Buniakowsky's claim is genuine.

EXAMPLE The matrix that projects onto the line through $a = (1, 1, 1)$ is

$$P = \frac{aa^T}{a^Ta} = \frac{1}{3} \begin{bmatrix} 1 \\ 1 \\ 1 \end{bmatrix} \begin{bmatrix} 1 & 1 & 1 \end{bmatrix} = \begin{bmatrix} \frac{1}{3} & \frac{1}{3} & \frac{1}{3} \\ \frac{1}{3} & \frac{1}{3} & \frac{1}{3} \\ \frac{1}{3} & \frac{1}{3} & \frac{1}{3} \end{bmatrix}.$$

This matrix has two properties that we will see as typical of projections:

(i) P is a symmetric matrix
(ii) Its square is itself: $P^2 = P$.

It is also a great example for understanding the four fundamental subspaces:

The rank is $r = 1$
The column space consists of the line through $a = (1, 1, 1)$
The nullspace consists of the plane perpendicular to a.

Every column is a multiple of a, so Pb lies on the line through a. The vectors that project to $p = 0$ are especially important. They are the vectors that satisfy $a^Tb = 0$—they are perpendicular to a and their component along the line is zero. They lie in the perpendicular plane, which is the nullspace of the projection matrix P.

Actually that example is too perfect. It has the nullspace orthogonal to the column space, which is haywire. The nullspace should be orthogonal to the *row space*. But because P is symmetric, its row and column spaces are the same.

Remark on scaling The projection matrix aa^T/a^Ta is the same if a is doubled:

$$a = \begin{bmatrix} 2 \\ 2 \\ 2 \end{bmatrix} \quad \text{gives} \quad P = \frac{1}{12} \begin{bmatrix} 2 \\ 2 \\ 2 \end{bmatrix} \begin{bmatrix} 2 & 2 & 2 \end{bmatrix} = \begin{bmatrix} \frac{1}{3} & \frac{1}{3} & \frac{1}{3} \\ \frac{1}{3} & \frac{1}{3} & \frac{1}{3} \\ \frac{1}{3} & \frac{1}{3} & \frac{1}{3} \end{bmatrix}.$$

The line through a is the same, and that's all the projection matrix cares about. If a has unit length then the denominator is $a^Ta = 1$ and the rank-one matrix is just $P = aa^T$.

EXAMPLE 2 Projection onto the "θ-direction" in the x-y plane. The line goes through $a = (\cos \theta, \sin \theta)$ and the matrix is

$$P = \frac{aa^T}{a^Ta} = \frac{\begin{bmatrix} c \\ s \end{bmatrix} \begin{bmatrix} c & s \end{bmatrix}}{\begin{bmatrix} c & s \end{bmatrix} \begin{bmatrix} c \\ s \end{bmatrix}} = \begin{bmatrix} c^2 & cs \\ cs & s^2 \end{bmatrix}.$$

Here c is $\cos \theta$, s is $\sin \theta$, and $c^2 + s^2 = 1$ in the denominator. This matrix P was discovered earlier, in Section 2.6 on linear transformations. Now we can go beyond the projection of the x-y plane onto a line, and compute P in any number of dimensions. We emphasize that it produces the projection p:

To project b onto a, multiply by P: $p = Pb$.

The Transpose of a Matrix

Finally we go back to transposes. Up to now, A^T has been defined simply by reflecting A across its main diagonal; the rows of A become the columns of A^T, and vice versa. In other words, the entry in row i and column j of A^T is the (j, i) entry of A:

$$(A^T)_{ij} = (A)_{ji}.$$

There is a deeper significance to the transpose, which comes from its close connection to inner products. In fact this connection can be used to give a new and much more "abstract" definition of the transpose:

3J   The transpose A^T can be defined by the following property: The inner product of Ax with y equals the inner product of x with $A^T y$. Formally, this simply means that

$$(Ax)^T y = x^T A^T y = x^T (A^T y). \tag{8}$$

This definition has two purposes:

(i)   It tells us how, when we measure the inner product in a different way, to make the proper change in the transpose. This becomes significant in the case of complex numbers; the new inner product is in Section 5.5.

(ii)   It gives us another way to verify the formula for the transpose of a product:

$$(AB)^T = B^T A^T. \tag{9}$$

This is confirmed by using equation (8) twice, first for A and then for B:

$$(ABx)^T y = (Bx)^T (A^T y) = x^T (B^T A^T y).$$

The transposes turn up in reverse order on the right side, just as the inverses do in the analogous formula $(AB)^{-1} = B^{-1} A^{-1}$. We mention again that these two formulas meet to give the remarkable combination $(A^{-1})^T = (A^T)^{-1}$.

EXERCISES

3.2.1   (a)   Given any two positive numbers x and y, choose the vector b equal to $(\sqrt{x}, \sqrt{y})$, and choose $a = (\sqrt{y}, \sqrt{x})$. Apply the Schwarz inequality to compare the arithmetic mean $\frac{1}{2}(x + y)$ with the geometric mean $\sqrt{xy}$.

(b) Suppose we start with a vector from the origin to the point x, and then add a vector of length $\|y\|$ connecting x to $x + y$. The third side of the triangle goes from the origin to $x + y$. *The triangle inequality asserts that this distance cannot be greater than the sum of the first two:*

$$\|x + y\| \le \|x\| + \|y\|.$$

After squaring both sides, and expanding $(x + y)^{\mathrm{T}}(x + y)$, reduce this to the Schwarz inequality.

3.2.2 Verify that the length of the projection is $\|p\| = \|b\| \, |\cos \theta|$, using formula (5).

3.2.3 What multiple of $a = (1, 1, 1)$ is closest to the point $b = (2, 4, 4)$? Find also the point closest to a on the line through b.

3.2.4 Explain why the Schwarz inequality becomes an equality in case a and b lie on the same line through the origin, and only in that case. What if they lie on opposite sides of the origin?

3.2.5 In n dimensions, what angle does the vector $(1, 1, \ldots, 1)$ make with the coordinate axes? What is the projection matrix P onto that vector?

3.2.6 The Schwarz inequality has a one-line proof if a and b are normalized ahead of time to be unit vectors:

$$|a^{\mathrm{T}}b| = \left| \sum a_j b_j \right| \le \sum |a_j| \, |b_j| \le \sum \frac{|a_j|^2 + |b_j|^2}{2} = \frac{1}{2} + \frac{1}{2} = \|a\| \, \|b\|.$$

Which previous exercise justifies the middle step?

3.2.7 By choosing the right vector b in the Schwarz inequality, prove that

$$(a_1 + \cdots + a_n)^2 \le n(a_1^2 + \cdots + a_n^2).$$

When does equality hold?

3.2.8 The methane molecule CH_4 is arranged as if the carbon atom were at the center of a regular tetrahedron with four hydrogen atoms at the vertices. If vertices are placed at $(0, 0, 0)$, $(1, 1, 0)$, $(1, 0, 1)$, and $(0, 1, 1)$—note that all six edges have length $\sqrt{2}$, so the tetrahedron is regular—what is the cosine of the angle between the rays going from the center $(\frac{1}{2}, \frac{1}{2}, \frac{1}{2})$ to the vertices? (The bond angle itself is about $109.5°$, an old friend of chemists.)

3.2.9 Square the matrix $P = aa^{\mathrm{T}}/a^{\mathrm{T}}a$, which projects onto a line, and show that $P^2 = P$. (Note the number $a^{\mathrm{T}}a$ in the middle of the matrix $aa^{\mathrm{T}}aa^{\mathrm{T}}$!)

3.2.10 Is the projection matrix P invertible? Why or why not?

3.2.11 (a) Find the projection matrix P_1 onto the line through $a = \left[\begin{smallmatrix} 1 \\ 3 \end{smallmatrix} \right]$ and also the matrix P_2 that projects onto the line perpendicular to a.
(b) Compute $P_1 + P_2$ and $P_1 P_2$ and explain.

3.2.12 Find the matrix that projects every point in the plane onto the line $x + 2y = 0$.

3.2.13 Prove that the *"trace"* of $P = aa^{\mathrm{T}}/a^{\mathrm{T}}a$—which is the sum of its diagonal entries—always equals one.

3.2.14 What matrix P projects every point in $\mathbf{R}^3$ onto the line of intersection of the planes $x + y + t = 0$ and $x - t = 0$?

3.2.15 Show that the length of Ax equals the length of $A^{\mathrm{T}}x$ if $AA^{\mathrm{T}} = A^{\mathrm{T}}A$.

3.2.16 Suppose P is the projection matrix onto the line through a.
(a)   Why is the inner product of x with Py equal to the inner product of Px with y?
(b)   Are the two angles the same? Find their cosines if $a = (1, 1, -1)$, $x = (2, 0, 1)$, $y = (2, 1, 2)$.
(c)   Why is the inner product of Px with Py again the same? What is the angle between those two?

PROJECTIONS AND LEAST SQUARES APPROXIMATIONS ■ 3.3

Up to this point, a system $Ax = b$ either has a solution or not. If b is not in the column space $\mathscr{R}(A)$, the system is inconsistent and Gaussian elimination fails. This is almost certain to be the case for a system of several equations in only one unknown. For example, the simultaneous equations

$$2x = b_1$$
$$3x = b_2$$
$$4x = b_3$$

will be solvable only if the right-hand sides are in the ratio 2:3:4. The solution x is certainly unique if it exists, but it will exist only if b is on the same line as the vector

$$a = \begin{bmatrix} 2 \\ 3 \\ 4 \end{bmatrix}.$$

In spite of their unsolvability, inconsistent equations arise in practice and have to be solved. One possibility is to determine x from part of the system, and ignore the rest; this is hard to justify if all m equations come from the same source. Rather than expecting no error in some equations and large errors in the others, it is much better *to choose the x that minimizes the average error in the m equations.* There are many ways to define such an average, but the most convenient is the *sum of squares:*

$$E^2 = (2x - b_1)^2 + (3x - b_2)^2 + (4x - b_3)^2.$$

If there is an exact solution to $ax = b$, the minimum error is $E = 0$. In the more likely case that b is not proportional to a, the function E^2 will be a parabola with its minimum at the point where

$$\frac{dE^2}{dx} = 2[(2x - b_1)2 + (3x - b_2)3 + (4x - b_3)4] = 0.$$

Solving for x, the least squares solution of the system $ax = b$ is

$$\bar{x} = \frac{2b_1 + 3b_2 + 4b_3}{2^2 + 3^2 + 4^2}.$$

You recognize a^Tb in the numerator and a^Ta in the denominator.
The general case is the same. We "solve" $ax = b$ by minimizing

$$E^2 = \|ax - b\|^2 = (a_1x - b_1)^2 + \cdots + (a_mx - b_m)^2.$$

The derivative of E^2 is zero at the point where

$$(a_1\bar{x} - b_1)a_1 + \cdots + (a_m\bar{x} - b_m)a_m = 0$$

We are minimizing the distance from b to the line through a, and calculus gives the same answer that geometry did earlier:

3K The least squares solution to a problem $ax = b$ in one unknown is $\bar{x} = \dfrac{a^{\mathrm{T}}b}{a^{\mathrm{T}}a}$.

You see that we keep coming back to the geometrical interpretation of a least squares problem—to minimize a distance. By differentiating E^2 and setting its derivative to zero, we have used calculus to confirm the geometry of the previous section; the line connecting b to p must be perpendicular to a:

$$a^{\mathrm{T}}(b - \bar{x}a) = a^{\mathrm{T}}b - \frac{a^{\mathrm{T}}b}{a^{\mathrm{T}}a}\, a^{\mathrm{T}}a = 0.$$

As a side remark, notice the degenerate case $a = 0$. All multiples of a are zero, and the line is only a point. Therefore $p = 0$ is the only candidate for the projection. But the formula for $\bar{x}$ becomes a meaningless $0/0$, and correctly reflects the fact that the multiple $\bar{x}$ is left completely undetermined. All values of x give the same error $E = \|0x - b\|$, so E^2 is a horizontal line instead of a parabola. One purpose of the pseudoinverse in the appendix is to assign some definite value to $\bar{x}$; in this case it would assign $\bar{x} = 0$, which at least seems a more "symmetric" choice than any other number.

Least Squares Problems with Several Variables

Now we are ready for the serious step, *to project b onto a subspace*—rather than just onto a line. This problem arises in the following way. Suppose we start from $Ax = b$, but this time let A be an m by n matrix. Instead of permitting only one unknown, with a single column vector a, the matrix now has n columns. We still imagine that the number m of observations is larger than the number n of unknowns, so it must be expected that $Ax = b$ will be inconsistent. *Probably there will not exist a choice of x that perfectly fits the data b.* In other words, probably the vector b will not be a combination of the columns of A; it will be outside the column space.

Again the problem is to choose $\bar{x}$ so as to minimize the error, and again this minimization will be done in the least squares sense. The error is $E = \|Ax - b\|$, and *this is exactly the distance from b to the point Ax in the column space*. (Remember that Ax is the combination of the columns with coefficients $x_1, \ldots, x_n$.) Therefore searching for the least squares solution $\bar{x}$, which minimizes E, is the same as locating the point $p = A\bar{x}$ that is closer to b than any other point in the column space.

We may use geometry or calculus to determine $\bar{x}$. In n dimensions, we prefer the appeal of geometry; p must be the "projection of b onto the column space." *The error vector $b - A\bar{x}$ must be perpendicular to that space* (Fig. 3.8).

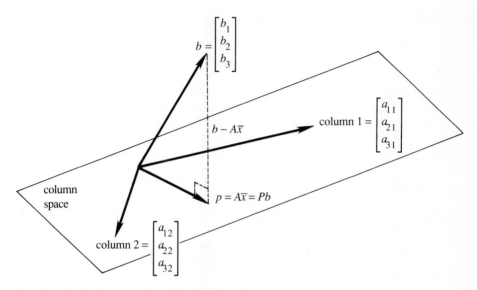

Fig. 3.8. Projection onto the column space of a 3 by 2 matrix.

The calculation of $\bar{x}$ and the projection $p = A\bar{x}$ is so fundamental that we do it in two ways:

1. The vectors perpendicular to the column space lie in the *left nullspace*. Thus the error vector $b - A\bar{x}$ must be in the nullspace of A^T:

$$A^T(b - A\bar{x}) = 0 \quad \text{or} \quad A^T A\bar{x} = A^T b.$$

2. The error vector must be perpendicular to *every column* of A:

$$
\begin{matrix}
a_1^T(b - A\bar{x}) = 0 \\
\vdots \\
a_n^T(b - A\bar{x}) = 0
\end{matrix}
\quad \text{or} \quad
\begin{bmatrix} a_1^T \\ \vdots \\ a_n^T \end{bmatrix}
\begin{bmatrix} b - A\bar{x} \end{bmatrix} = 0.
$$

This is again the equation $A^T(b - A\bar{x}) = 0$ or $A^T A\bar{x} = A^T b$. A third way is to take partial derivatives of $E^2 = (Ax - b)^T(Ax - b)$. That gives $2A^T Ax - 2A^T b = 0$. And the fastest way, given an unsolvable equation $Ax = b$, is just **to multiply through by A^T**. All these equivalent methods produce a square coefficient matrix

$A^T A$. It is symmetric (its transpose is not AA^T!) and it is the fundamental matrix of this chapter.

The equations $A^T A \bar{x} = A^T b$ are known in statistics as the "**normal equations**":

3L   The least squares solution to an inconsistent system $Ax = b$ of m equations in n unknowns satisfies

$$A^T A \bar{x} = A^T b. \tag{1}$$

If the columns of A are linearly independent, then $A^T A$ is invertible and

$$\bar{x} = (A^T A)^{-1} A^T b. \tag{2}$$

The projection of b onto the column space is therefore

$$p = A\bar{x} = A(A^T A)^{-1} A^T b. \tag{3}$$

We choose an example in which our intuition is as good as the formulas:

$$A = \begin{bmatrix} 1 & 2 \\ 1 & 3 \\ 0 & 0 \end{bmatrix}, \qquad b = \begin{bmatrix} 4 \\ 5 \\ 6 \end{bmatrix}$$

The column space of A is easy to visualize, since both columns end with a zero. It is the x-y plane within three-dimensional space. The projection of $b = (4, 5, 6)$ will be $p = (4, 5, 0)$—the x and y components stay the same but $z = 6$ will disappear. That is confirmed by the formulas:

$$A^T A = \begin{bmatrix} 1 & 1 & 0 \\ 2 & 3 & 0 \end{bmatrix} \begin{bmatrix} 1 & 2 \\ 1 & 3 \\ 0 & 0 \end{bmatrix} = \begin{bmatrix} 2 & 5 \\ 5 & 13 \end{bmatrix}$$

$$\bar{x} = (A^T A)^{-1} A^T b = \begin{bmatrix} 13 & -5 \\ -5 & 2 \end{bmatrix} \begin{bmatrix} 1 & 1 & 0 \\ 2 & 3 & 0 \end{bmatrix} \begin{bmatrix} 4 \\ 5 \\ 6 \end{bmatrix} = \begin{bmatrix} 2 \\ 1 \end{bmatrix}$$

$$p = A\bar{x} = \begin{bmatrix} 1 & 2 \\ 1 & 3 \\ 0 & 0 \end{bmatrix} \begin{bmatrix} 2 \\ 1 \end{bmatrix} = \begin{bmatrix} 4 \\ 5 \\ 0 \end{bmatrix}.$$

In this special case, the best we can do is to solve the first two equations of $Ax = b$:

$$x_1 + 2x_2 = 4$$
$$x_1 + 3x_2 = 5$$
$$0x_1 + 0x_2 = 6.$$

Thus $\bar{x}_1 = 2$ and $\bar{x}_2 = 1$. The error in the third equation is bound to be 6.

Remark 1 Suppose b is actually *in* the column space of A—it is a combination $b = Ax$ of the columns. Then the projection of b is still b:

$$p = A(A^T A)^{-1} A^T Ax = Ax = b.$$

The closest point p is just b itself—which is obvious.

Remark 2 At the other extreme, suppose b is *perpendicular* to the column space. It is perpendicular to every column, so $A^T b = 0$. In this case it projects to the zero vector:

$$p = A(A^T A)^{-1} A^T b = A(A^T A)^{-1} 0 = 0.$$

Remark 3 When A is square and invertible, the column space is the whole space. Every vector projects to itself, and p equal b:

$$p = A(A^T A)^{-1} A^T b = A A^{-1} (A^T)^{-1} A^T b = b.$$

This is the only case when we can take apart $(A^T A)^{-1}$, *and write it as* $A^{-1}(A^T)^{-1}$. When A is rectangular that is not possible.

Remark 4 (Projection onto a line) Suppose A has only one column, containing the vector a. Then the matrix $A^T A$ is the number $a^T a$ and $\bar{x}$ is $a^T b / a^T a$. This is the case where we can divide by $A^T A$, instead of stopping at $(A^T A)^{-1}$, and we get back to the earlier formula.

The Cross-Product Matrix $A^T A$

The one technical point that remains is to check the properties of $A^T A$. It is certainly symmetric; its transpose is $(A^T A)^T = A^T A^{TT}$, which is $A^T A$ again. The i, j entry is an inner product, column i of A with column j of A. That agrees with the j, i entry, which is column j times column i. The key question is the invertibility of $A^T A$, and fortunately

$$A^T A \text{ has the same nullspace as } A.$$

Certainly if $Ax = 0$ then $A^T Ax = 0$. Vectors x in the nullspace of A are also in the nullspace of $A^T A$. To go in the other direction, start by supposing that $A^T Ax = 0$ and take the inner product with x:

$$x^T A^T Ax = 0, \quad \text{or} \quad \|Ax\|^2 = 0, \quad \text{or} \quad Ax = 0.$$

Thus x is in the nullspace of A; the two nullspaces are identical. In particular, if A has independent columns (and only $x = 0$ is in its nullspace) then the same is true for $A^T A$:

3M If A has linearly independent columns, then $A^T A$ is *square, symmetric,* and *invertible.*

We show later that A^TA is also positive definite (all pivots and eigenvalues are positive).

In this case, which is by far the most common and most important, the normal equations can be solved for $\bar{x}$. As in the numerical example, A has "full column rank." Its n columns are independent—not so hard in m-dimensional space if $m > n$, but not automatic. We assume it in what follows.

Projection Matrices

Our computations have shown that the closest point to b is $p = A(A^TA)^{-1}A^Tb$. *This formula expresses in matrix terms the construction of a perpendicular line from b to the column space of A.* The matrix in that formula is a projection matrix, and it will be denoted by P:

$$P = A(A^TA)^{-1}A^T.$$

This matrix projects any vector b onto the column space of A.† In other words, $p = Pb$ is the component of b in the column space, and the error $b - Pb$ is the component in the orthogonal complement. (Or, as it seems natural to say, $I - P$ is also a projection matrix. It projects any vector b onto the orthogonal complement, and the projection is $(I - P)b = b - Pb$.) In short, we have a matrix formula for splitting a vector into two perpendicular components. Pb is in the column space $\mathscr{R}(A)$, and the other component $(I - P)b$ is in the left nullspace $\mathscr{N}(A^T)$—which is orthogonal to the column space.

These projection matrices can be understood geometrically and algebraically. They are a family of matrices with very special properties, to be used later as the fundamental building blocks for all symmetric matrices. Therefore we pause for a moment, before returning to least squares, to identify the properties of P.

3N The projection matrix $P = A(A^TA)^{-1}A^T$ has two basic properties:

  (i) It equals its square: $P^2 = P$.
  (ii) It equals its transpose: $P^T = P$.

Conversely, any symmetric matrix with $P^2 = P$ represents a projection.

Proof It is easy to see why $P^2 = P$. If we start with any vector b, then Pb lies in the subspace we are projecting onto. Therefore ***when we project again nothing is changed***. The vector Pb is already in the subspace, and $P(Pb)$ is still Pb. In other words $P^2 = P$. Two or three or fifty projections give the same point p as the first

---

† There may be a risk of confusion with permutation matrices, also denoted by P. But the risk should be small, and we try never to let both appear on the same page.

projection:

$$P^2 = A(A^TA)^{-1}A^TA(A^TA)^{-1}A^T = A(A^TA)^{-1}A^T = P.$$

To prove that P is also symmetric, take its transpose. Multiply the transposes in reverse order and use the identity $(B^{-1})^T = (B^T)^{-1}$ with $B = A^TA$, to come back to P:

$$P^T = (A^T)^T((A^TA)^{-1})^TA^T = A((A^TA)^T)^{-1}A^T = A(A^TA)^{-1}A^T = P.$$

For the converse, we have to deduce from $P^2 = P$ and $P^T = P$ that P is a projection. Like any other matrix, P will take every vector b into its column space. Pb is a combination of the columns. To prove it is the *projection* onto that space, what we have to show is that the remaining part of b—the error vector $b - Pb$—is *orthogonal to the space*. For any vector Pc in the space, the inner product is zero:

$$(b - Pb)^TPc = b^T(I - P)^TPc = b^T(P - P^2)c = 0. \qquad (4)$$

Thus $b - Pb$ is orthogonal to the space, and Pb is the projection onto the column space.

EXAMPLE Suppose A is actually invertible. If it is 4 by 4, then its four columns are independent and its column space is all of $\mathbf{R}^4$. What is the projection *onto the whole space*? It is the identity matrix.

$$P = A(A^TA)^{-1}A^T = AA^{-1}(A^T)^{-1}A^T = I. \qquad (5)$$

The identity matrix is symmetric, and $I^2 = I$, and the error $b - Ib$ is zero.

The point of all other examples is that what happened in (5) is *not allowed*. To repeat: We cannot invert the separate parts A^T and A, when those matrices are rectangular. It is the square matrix A^TA that is invertible.

Least Squares Fitting of Data

Suppose we do a series of experiments, and expect the output b to be a linear function of the input t. We look for a straight line $b = C + Dt$. For example:

(1) At different times we measure the distance to a satellite on its way to Mars. In this case t is the time and b is the distance. Unless the motor was left on or gravity is strong, the satellite should move with nearly constant velocity v: $b = b_0 + vt$.

(2) We vary the load on a structure, and measure the strain it produces. In this experiment t is the load and b is the reading from the strain gauge. Unless the load is so great that the material becomes plastic, a linear relation $b = C + Dt$ is normal in the theory of elasticity.

(3) The cost of producing t books like this one is nearly linear, $b = C + Dt$, with editing and typesetting in C and then printing and binding in D (the cost for each additional book).

The question is, How does one compute C and D from the results of experiments? If the relationship is truly linear, and there is no experimental error, then there is no problem; two measurements of b will completely determine the line $b = C + Dt$. All further measurements will lie on this line. But if there is error, and the additional points fail to land on the line, then we must be prepared to "average" the experiments and find an optimal line. That line is not to be confused with the line on which b was projected in the previous section! In fact, since there are two unknowns C and D to be determined, we shall be involved with projections onto a *two-dimensional* subspace. The least squares problem comes directly from the experimental results

$$
\begin{aligned}
C + Dt_1 &= b_1 \\
C + Dt_2 &= b_2 \\
&\vdots \\
C + Dt_m &= b_m.
\end{aligned}
$$
(6)

This is an overdetermined system, with m equations and two unknowns. If errors are present, it will have no solution. We emphasize that the matrix A has two columns, and the unknown x has two components C and D:

$$
\begin{bmatrix} 1 & t_1 \\ 1 & t_2 \\ \vdots & \vdots \\ 1 & t_m \end{bmatrix} \begin{bmatrix} C \\ D \end{bmatrix} = \begin{bmatrix} b_1 \\ b_2 \\ \vdots \\ b_m \end{bmatrix}, \quad \text{or} \quad Ax = b.
$$
(7)

The best solution $\bar{C}, \bar{D}$ is the one that minimizes

$$
E^2 = \|b - Ax\|^2 = (b_1 - C - Dt_1)^2 + \cdots + (b_m - C - Dt_m)^2.
$$

In matrix terminology, we choose $\bar{x}$ so that $p = A\bar{x}$ is as close as possible to b. Of all straight lines $b = C + Dt$, we are choosing the one that best fits the data (Fig. 3.9). On the graph, the errors are the **vertical distances** $b - C - Dt$ to the

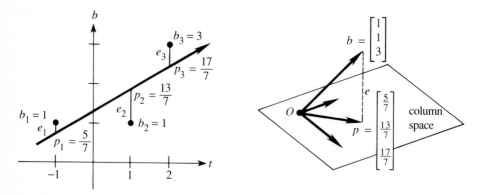

Fig. 3.9. Straight line approximation and the corresponding projection.

straight line (not the perpendicular distances!). It is the vertical distances that are squared, summed, and minimized.

EXAMPLE   Suppose we are given the three measurements marked on the figure at the left:

$$b = 1 \quad \text{at} \quad t = -1, \qquad b = 1 \quad \text{at} \quad t = 1, \qquad b = 3 \quad \text{at} \quad t = 2.$$

Note that the values of t are not required to be equally spaced. The experimenter may choose convenient values (even negative values) without any effect on the mathematics. The first step is *to write down the equations which* **would** *hold if a line could go through all three points*:

$$
\begin{matrix}
C - D = 1 \\
C + D = 1 \\
C + 2D = 3
\end{matrix}
\quad \text{or} \quad
\begin{bmatrix} 1 & -1 \\ 1 & 1 \\ 1 & 2 \end{bmatrix}
\begin{bmatrix} C \\ D \end{bmatrix}
=
\begin{bmatrix} 1 \\ 1 \\ 3 \end{bmatrix}.
$$

If those equations $Ax = b$ could be solved, there would be no errors. They can't be solved; the points are not on a line. Therefore they are solved by least squares:

$$
A^T A \bar{x} = A^T b \quad \text{is} \quad
\begin{bmatrix} 3 & 2 \\ 2 & 6 \end{bmatrix}
\begin{bmatrix} \bar{C} \\ \bar{D} \end{bmatrix}
=
\begin{bmatrix} 5 \\ 6 \end{bmatrix}.
$$

The solution is $\bar{C} = \frac{9}{7}$, $\bar{D} = \frac{4}{7}$ and the best line is $\frac{9}{7} + \frac{4}{7}t$.

Note the beautiful connections between the two figures. The problem is the same but the art shows it differently. In the right figure, b is not a combination of the columns; $(1, 1, 3)$ is not a combination of $(1, 1, 1)$ and $(-1, 1, 2)$. On the left, the three points are not on a line. Least squares replaces b by p, and it replaces points that are not on a line by points that are. Unable to solve $Ax = b$, we solve $A\bar{x} = p$.

Even better: The line $\frac{9}{7} + \frac{4}{7}t$ has heights $\frac{5}{7}, \frac{13}{7}, \frac{17}{7}$ at the measurement times $-1, 1, 2$. Those points *do* lie on a line. Therefore the vector $p = (\frac{5}{7}, \frac{13}{7}, \frac{17}{7})$ is in the column space. Furthermore, the line is the best line and this vector is the best vector. *It is the projection.* The right figure is in three dimensions (or m dimensions if there are m points) and the left figure is in two dimensions (or n dimensions if there are n parameters).

Subtracting p from b, the errors are $e = (\frac{2}{7}, -\frac{6}{7}, \frac{4}{7})$. Those are the vertical errors in the left figure, and they are the components of the dotted vector in the right figure. This error vector is orthogonal to the first column $(1, 1, 1)$, since $\frac{2}{7} - \frac{6}{7} + \frac{4}{7} = 0$. It is orthogonal to the second column $(-1, 1, 2)$, because $-\frac{2}{7} - \frac{6}{7} + \frac{8}{7} = 0$. It is orthogonal to the column space, and it is in the left nullspace.

Question: If the original measurements had been $b = (\frac{2}{7}, -\frac{6}{7}, \frac{4}{7})$, what would have been the best line and the best $\bar{x}$? Answer: The zero line—which is the horizontal axis—and $\bar{x} = 0$.

We can quickly summarize the equations for fitting by a straight line. The first column of A contains 1's, and the second column contains the times t_i. Therefore $A^T A$ contains the sum of m 1's, and the sum of the t_i, and the sum of the t_i^2:

30   Given the measurements $b_1, \ldots, b_m$ at distinct points $t_1, \ldots, t_m$, the straight line $\bar{C} + \bar{D}t$ which minimizes E^2 comes from least squares:

$$A^T A \begin{bmatrix} \bar{C} \\ \bar{D} \end{bmatrix} = A^T b \quad \text{or} \quad \begin{bmatrix} m & \Sigma\, t_i \\ \Sigma\, t_i & \Sigma\, t_i^2 \end{bmatrix} \begin{bmatrix} \bar{C} \\ \bar{D} \end{bmatrix} = \begin{bmatrix} \Sigma\, b_i \\ \Sigma\, t_i b_i \end{bmatrix}.$$

Remark   It makes no special difference to the mathematics of least squares that we are fitting the data by straight lines. In many experiments there is no reason to expect a linear relationship, and it would be crazy to look for one. Suppose we are handed some radioactive material. The output b will be the reading on a Geiger counter at various times t. We may know that we are holding a mixture of two radioactive chemicals, and we may know their half-lives (or rates of decay), but we do not know how much of each is in our hands. If these two unknown amounts are C and D, then the Geiger counter readings would behave like the sum of two exponentials (and not like a straight line):

$$b = Ce^{-\lambda t} + De^{-\mu t}. \tag{8}$$

In practice, this law is not reflected exactly by the counter. Instead, we make readings $b_1, \ldots, b_m$ at times $t_1, \ldots, t_m$, and (8) is approximately satisfied:

$$Ce^{-\lambda t_1} + De^{-\mu t_1} \approx b_1,$$
$$\vdots$$
$$Ce^{-\lambda t_m} + De^{-\mu t_m} \approx b_m.$$

If there are more than two readings, $m > 2$, then in all likelihood we cannot solve for C and D. But the least squares principle will give optimal values $\bar{C}$ and $\bar{D}$.

The situation would be completely different if we knew the amounts C and D, and were trying to discover the decay rates λ and μ. This is a problem in *nonlinear least squares*, and it is harder. We would still form E^2, the sum of the squares of the errors, and minimize it. But setting its derivatives to zero will not give linear equations for the optimal λ and μ. In the exercises, we stay with linear least squares.

EXERCISES

3.3.1   Find the best least squares solution $\bar{x}$ to $3x = 10, 4x = 5$. What error E^2 is minimized? Check that the error vector $(10 - 3\bar{x}, 5 - 4\bar{x})$ is perpendicular to the column $(3, 4)$.

3.3.2   Suppose the values $b_1 = 1$ and $b_2 = 7$ at times $t_1 = 1$ and $t_2 = 2$ are fitted by a line $b = Dt$ *through the origin*. Solve $D = 1$ and $2D = 7$ by least squares, and sketch the best line.

3.3.3 Solve $Ax = b$ by least squares and find $p = A\bar{x}$ if

$$A = \begin{bmatrix} 1 & 0 \\ 0 & 1 \\ 1 & 1 \end{bmatrix}, \qquad b = \begin{bmatrix} 1 \\ 1 \\ 0 \end{bmatrix}.$$

Verify that the error $b - p$ is perpendicular to the columns of A.

3.3.4 Write out $E^2 = \|Ax - b\|^2$ and set to zero its derivatives with respect to u and v, if

$$A = \begin{bmatrix} 1 & 0 \\ 0 & 1 \\ 1 & 1 \end{bmatrix}, \qquad x = \begin{bmatrix} u \\ v \end{bmatrix}, \qquad b = \begin{bmatrix} 1 \\ 3 \\ 4 \end{bmatrix}.$$

Compare the resulting equations with $A^T A\bar{x} = A^T b$, confirming that calculus as well as geometry gives the normal equations. Find the solution $\bar{x}$ and the projection $p = A\bar{x}$. Why is $p = b$?

3.3.5 The following system has no solution:

$$Ax = \begin{bmatrix} 1 & -1 \\ 1 & 0 \\ 1 & 1 \end{bmatrix} \begin{bmatrix} C \\ D \end{bmatrix} = \begin{bmatrix} 4 \\ 5 \\ 9 \end{bmatrix} = b.$$

Sketch and solve a straight line fit that leads to the minimization of the quadratic $(C - D - 4)^2 + (C - 5)^2 + (C + D - 9)^2$. What is the projection of b onto the column space of A?

3.3.6 Find the projection of b onto the column space of A:

$$A = \begin{bmatrix} 1 & 1 \\ 2 & -1 \\ -2 & 4 \end{bmatrix}, \qquad b = \begin{bmatrix} 1 \\ 2 \\ 7 \end{bmatrix}.$$

Split b into $p + q$, with p in the column space and q perpendicular to that space. Which of the four subspaces contains q?

3.3.7 Find the projection matrix P onto the space spanned by $a_1 = (1, 0, 1)$ and $a_2 = (1, 1, -1)$.

3.3.8 If P is the projection matrix onto a k-dimensional subspace S of the whole space $\mathbf{R}^n$, what is the column space of P and what is its rank?

3.3.9 (a) If $P = P^T P$ show that P is a projection matrix.
(b) What subspace does the matrix $P = 0$ project onto?

3.3.10 If the vectors a_1, a_2, and b are orthogonal, what are $A^T A$ and $A^T b$? What is the projection of b onto the plane of a_1 and a_2?

3.3.11 Suppose P is the projection matrix onto the subspace S and Q is the projection onto the orthogonal complement $S^\perp$. What are $P + Q$ and PQ? Show that $P - Q$ is its own inverse.

3.3.12 If V is the subspace spanned by $(1, 1, 0, 1)$ and $(0, 0, 1, 0)$, find
  (a) a basis for the orthogonal complement $V^\perp$
  (b) the projection matrix P onto V
  (c) the vector in V closest to the vector $b = (0, 1, 0, -1)$ in $V^\perp$.

3.3.13 Find the best straight line fit (least squares) to the measurements

$$b = 4 \quad \text{at} \quad t = -2, \qquad b = 3 \quad \text{at} \quad t = -1,$$
$$b = 1 \quad \text{at} \quad t = 0, \qquad b = 0 \quad \text{at} \quad t = 2.$$

Then find the projection of $b = (4, 3, 1, 0)$ onto the column space of

$$A = \begin{bmatrix} 1 & -2 \\ 1 & -1 \\ 1 & 0 \\ 1 & 2 \end{bmatrix}.$$

3.3.14 The vectors $a_1 = (1, 1, 0)$ and $a_2 = (1, 1, 1)$ span a plane in $\mathbf{R}^3$. Find the projection matrix P onto the plane, and find a nonzero vector b that is projected to zero.

3.3.15 If P is the projection matrix onto a line in the x-y plane, draw a figure to describe the effect of the "reflection matrix" $H = I - 2P$. Explain both geometrically and algebraically why $H^2 = I$.

3.3.16 Show that if u has unit length, then the rank one matrix $P = uu^T$ is a projection matrix: It has properties (i) and (ii). By choosing $u = a/\| a \|$, P becomes the projection onto the line through a, and Pb is the point $p = \bar{x}a$. Rank-one projections correspond exactly to least squares problems in one unknown.

3.3.17 What 2 by 2 matrix projects the x-y plane onto the $-45°$ line $x + y = 0$?

3.3.18 We want to fit a plane $y = C + Dt + Ez$ to the four points

$$y = 3 \quad \text{at} \quad t = 1, z = 1 \qquad y = 6 \quad \text{at} \quad t = 0, z = 3$$
$$y = 5 \quad \text{at} \quad t = 2, z = 1 \qquad y = 0 \quad \text{at} \quad t = 0, z = 0.$$

  (1) Find 4 equations in 3 unknowns to pass a plane through the points (if there is such a plane).
  (2) Find 3 equations in 3 unknowns for the best least squares solution.

3.3.19 If $P_C = A(A^T A)^{-1} A^T$ is the projection onto the column space of A, what is the projection P_R onto the row space? (It is not P_C^T!)

3.3.20 If P_R is the projection onto the row space of A, what is the projection P_N onto the nullspace? (The two subspaces are orthogonal.)

3.3.21 Suppose L_1 is the line through the origin in the direction of a_1, and L_2 is the line through b in the direction of a_2. To find the closest points $x_1 a_1$ and $b + x_2 a_2$ on the two lines, write down the two equations for the x_1 and x_2 that minimize $\| x_1 a_1 - x_2 a_2 - b \|$. Solve for x if $a_1 = (1, 1, 0)$, $a_2 = (0, 1, 0)$, $b = (2, 1, 4)$.

3.3.22 Find the best line $C + Dt$ to fit $b = 4, 2, -1, 0, 0$ at times $t = -2, -1, 0, 1, 2$.

3.3.23 Show that the best least squares fit to a set of measurements $y_1, \ldots, y_m$ by a horizontal line—in other words, by a constant function $y = C$—is their average

$$C = \frac{y_1 + \cdots + y_m}{m}.$$

In statistical terms, the choice $\bar{y}$ that minimizes the error $E^2 = (y_1 - y)^2 + \cdots + (y_m - y)^2$ is the *mean* of the sample, and the resulting E^2 is the *variance* σ^2.

3.3.24 Find the best straight line fit to the following measurements, and sketch your solution:

$$y = 2 \quad \text{at} \quad t = -1, \qquad y = 0 \quad \text{at} \quad t = 0,$$
$$y = -3 \quad \text{at} \quad t = 1, \qquad y = -5 \quad \text{at} \quad t = 2.$$

3.3.25 Suppose that instead of a straight line, we fit the data in the previous exercise by a parabola: $y = C + Dt + Et^2$. In the inconsistent system $Ax = b$ that comes from the four measurements, what are the coefficient matrix A, the unknown vector x, and the data vector b? You need not compute $\bar{x}$.

3.3.26 A middle-aged man was stretched on a rack to lengths $L = 5$, 6, and 7 feet under applied forces of $F = 1$, 2, and 4 tons. Assuming Hooke's law $L = a + bF$, find his normal length a by least squares.

3.4 ■ ORTHOGONAL BASES, ORTHOGONAL MATRICES, AND GRAM-SCHMIDT ORTHOGONALIZATION

In an orthogonal basis, every vector is perpendicular to every other vector. The coordinate axes are mutually orthogonal. That is just about optimal, and the one possible improvement is easy to make: We divide each vector by its length, to make it a *unit vector*. That step changes an ***orthogonal*** basis into an ***orthonormal*** basis. We will denote the basis vectors by q, introducing that letter to indicate orthonormality.

3P   The vectors $q_1, \ldots, q_k$ are *orthonormal* if

$$q_i^T q_j = \begin{cases} 0 & \text{whenever } i \neq j, \quad \text{giving the orthogonality} \\ 1 & \text{whenever } i = j, \quad \text{giving the normalization.} \end{cases}$$

When orthonormal vectors go into the columns of a matrix, that matrix will be called Q.

The most important example is the "*standard basis.*" For the x–y plane, the best-known axes are not only perpendicular but horizontal and vertical; the basis is $e_1 = (1, 0)$, $e_2 = (0, 1)$. Q is the identity matrix. In n dimensions the standard basis again consists of the columns of $Q = I$:

$$e_1 = \begin{bmatrix} 1 \\ 0 \\ 0 \\ \vdots \\ 0 \end{bmatrix}, \qquad e_2 = \begin{bmatrix} 0 \\ 1 \\ 0 \\ \vdots \\ 0 \end{bmatrix}, \qquad \cdots, \qquad e_n = \begin{bmatrix} 0 \\ 0 \\ 0 \\ \vdots \\ 1 \end{bmatrix}.$$

That is by no means the only orthonormal basis; we can rotate the axes without changing the right angles at which they meet. These rotation matrices will be introduced below as examples of Q. On the other hand, if we are thinking not about $\mathbf{R}^n$ but about one of its subspaces, the standard vectors e_i might not lie in the subspace. In that case, it is not so clear that even one orthonormal basis can be found. But we shall show that there does always exist such a basis, and that it can be constructed in a simple way out of any basis whatsoever. This construction, which converts a skew set of axes into a perpendicular set, is known as *Gram-Schmidt orthogonalization*.

To summarize, the three topics basic to this section are:

(1)   The definition and properties of orthogonal matrices Q.
(2)   The solution of $Qx = b$, ordinary or least squares.
(3)   The Gram-Schmidt process and its interpretation as a new factorization $A = QR$.

Orthogonal Matrices

An **orthogonal matrix** is simply **a square matrix with orthonormal columns.**†
We use the letter Q for the matrix, and $q_1, \ldots, q_n$ for its columns. The properties
of the columns were $q_i^T q_j = 0$ and $q_i^T q_i = 1$, and this should translate into a property
of the matrix Q. In that translation, which is so simple, you will see again
why matrix notation is worth using.

3Q If the columns of Q are orthonormal then

$$
Q^T Q = \begin{bmatrix} \underline{\quad} q_1^T \underline{\quad} \\ \underline{\quad} q_2^T \underline{\quad} \\ \vdots \\ \underline{\quad} q_n^T \underline{\quad} \end{bmatrix} \begin{bmatrix} \, \Big| & \Big| & & \Big| \\ q_1 & q_2 & & q_n \\ \Big| & \Big| & & \Big| \end{bmatrix} = \begin{bmatrix} 1 & 0 & \cdot & 0 \\ 0 & 1 & \cdot & 0 \\ \cdot & \cdot & \cdot & \cdot \\ 0 & 0 & \cdot & 1 \end{bmatrix}.
$$

Therefore $Q^T Q = I$ and $Q^T = Q^{-1}$. For orthogonal matrices, *the transpose is the
inverse.*

When row i of Q^T multiplies column j of Q, the result is $q_i^T q_j = 0$. Those are the
zeros off the diagonal. On the diagonal where $i = j$, we have $q_i^T q_i = 1$. That is the
normalization to unit vectors, of length one.

Note that $Q^T Q = I$ even if Q is rectangular. But then Q^T is only a left-inverse.

EXAMPLE 1

$$
Q = \begin{bmatrix} \cos\theta & -\sin\theta \\ \sin\theta & \cos\theta \end{bmatrix}, \qquad Q^T = Q^{-1} = \begin{bmatrix} \cos\theta & \sin\theta \\ -\sin\theta & \cos\theta \end{bmatrix}.
$$

Q rotates every vector through the angle θ, and Q^T rotates it back through $-\theta$.
The columns are clearly orthogonal, and they are orthonormal because $\sin^2\theta +
\cos^2\theta = 1$. The matrix Q^T is just as much an orthogonal matrix as Q.

EXAMPLE 2   Any permutation matrix P is an orthogonal matrix. The columns are
certainly unit vectors and certainly orthogonal—because the 1 appears in a different
place in each column:

$$
\text{if} \quad P = \begin{bmatrix} 0 & 1 & 0 \\ 0 & 0 & 1 \\ 1 & 0 & 0 \end{bmatrix} \quad \text{then} \quad P^{-1} = P^T = \begin{bmatrix} 0 & 0 & 1 \\ 1 & 0 & 0 \\ 0 & 1 & 0 \end{bmatrix}.
$$

---

† Orthonormal matrix would have been a better name, but it is too late to change. Also,
there is no accepted word for a rectangular matrix with orthonormal columns. We still
write Q, but we won't call it an "orthogonal matrix" unless it is square.

It is immediate that $P^{\mathrm{T}}P = I$; the transpose is the inverse. Another P, with $P_{13} = P_{22} = P_{31} = 1$, takes the x-y-z axes into the z-y-x axes—a "right-handed" system into a "left-handed" system. So we were wrong if we suggested that every orthogonal Q represents a rotation. *A reflection is also allowed.* The matrix $P = \begin{bmatrix} 0 & 1 \\ 1 & 0 \end{bmatrix}$ is not one of the rotations of Example 1; there is no value of θ that will produce P. Instead P reflects every point (x, y) into (y, x), its mirror image across the $45°$ line. Geometrically, an orthogonal Q is the product of a rotation and a reflection.

There does remain one property that is shared by rotations and reflections, and in fact by every orthogonal matrix. It is not shared by projections, which are not orthogonal or even invertible. Projections reduce the length of a vector, whereas orthogonal matrices have a property that is the most important and most characteristic of all:

> **3R** Multiplication by an orthogonal matrix Q preserves lengths:
>
> $$\|Qx\| = \|x\| \quad \text{for every vector } x.$$
>
> It also preserves inner products and angles, since $(Qx)^{\mathrm{T}}(Qy) = x^{\mathrm{T}}Q^{\mathrm{T}}Qy = x^{\mathrm{T}}y$.

The preservation of lengths comes directly from $Q^{\mathrm{T}}Q = I$:

$$\|Qx\|^2 = \|x\|^2 \quad \text{because} \quad (Qx)^{\mathrm{T}}(Qx) = x^{\mathrm{T}}Q^{\mathrm{T}}Qx = x^{\mathrm{T}}x.$$

All inner products and lengths are preserved, when the space is rotated or reflected. Of course the length comes from the inner product of a vector with itself. The angles come from inner products of x with y—because $\cos\theta$ is given by $x^{\mathrm{T}}y/\|x\|\,\|y\|$, and that fraction is not changed by Q.

We come now to the calculation that uses the special property $Q^{\mathrm{T}} = Q^{-1}$. In principle, this calculation can be done for any basis. In practice, it is exceptionally simple for an orthonormal basis—and we show later that this leads to the key idea behind Fourier series. If we have a basis, then any vector is a combination of the basis vectors, and the problem is *to find the coefficients in that combination*:

$$\text{Write } b \text{ as a combination } b = x_1 q_1 + x_2 q_2 + \cdots + x_n q_n.$$

To compute x_1 there is a neat trick. *Multiply both sides of the equation by q_1^{T}.* On the left side is $q_1^{\mathrm{T}}b$. On the right side all terms disappear (because $q_1^{\mathrm{T}}q_j = 0$) except the first term. We are left with

$$q_1^{\mathrm{T}}b = x_1 q_1^{\mathrm{T}}q_1.$$

Since $q_1^{\mathrm{T}}q_1 = 1$, we have found x_1. It is $q_1^{\mathrm{T}}b$. Similarly the second coefficient is $x_2 = q_2^{\mathrm{T}}b$; that is the only surviving term when we multiply by q_2^{T}. The other terms die of orthogonality. Each piece of b has a simple formula, and recombining the pieces gives back b:

$$\boxed{\textit{Any vector } b \textit{ is equal to } (q_1^{\mathrm{T}}b)q_1 + (q_2^{\mathrm{T}}b)q_2 + \cdots + (q_n^{\mathrm{T}}b)q_n.} \tag{1}$$

For that we need a basis, so Q is square.

I can't resist writing this calculation in matrix terms. We were looking for the coefficients in the vector equation $x_1q_1 + \cdots + x_nq_n = b$. That is identical to the matrix equation $Qx = b$. (The columns of Q multiply the components of x.) Its solution is $x = Q^{-1}b$. But since $Q^{-1} = Q^T$—this is where orthonormality enters—the solution is also $x = Q^Tb$:

$$x = Q^Tb = \begin{bmatrix} -q_1^T- \\ \\ -q_n^T- \end{bmatrix} \begin{bmatrix} \\ b \\ \\ \end{bmatrix} = \begin{bmatrix} q_1^Tb \\ \\ q_n^Tb \end{bmatrix} \tag{2}$$

All components of x are displayed, and they are the inner products q_i^Tb as before.

The matrix form also shows what happens when the columns are *not* orthonormal. Expressing b as a combination $x_1a_1 + \cdots + x_na_n$ is the same as solving $Ax = b$. The basis vectors go into the columns of the matrix. In that case we need A^{-1}, which takes work, but in the orthonormal case we only need Q^T.

Remark 1   The ratio $q_1^Tb/q_1^Tq_1$ appeared earlier, when we projected b onto a line. There it was a line through a, and the projection was $(a^Tb/a^Ta)a$. Here it is a line through q_1, and the denominator is 1, and the projection is $(q_1^Tb)q_1$. Thus we have a new interpretation for the formula $b = \Sigma(q_i^Tb)q_i$ in the box: *Every vector b is the sum of its one-dimensional projections onto the lines through the q's.*

One more thing. Since those projections are orthogonal, Pythagoras should still be correct. The square of the hypotenuse should still be the sum of squares of the components:

$$\|b\|^2 = (q_1^Tb)^2 + (q_2^Tb)^2 + \cdots + (q_n^Tb)^2.$$

That must be the same as $\|b\|^2 = \|Q^Tb\|^2$, proved earlier.

Remark 2   Since $Q^T = Q^{-1}$ we also have $QQ^T = I$. When Q comes before Q^T, multiplication takes the inner products of the *rows* of Q. (For Q^TQ it was the columns.) Since the result is again the identity matrix, we come to a surprising conclusion: *The rows of a square matrix are orthonormal whenever the columns are.* The rows point in completely different directions from the columns, as in the matrix below, and I don't see geometrically why they are forced to be orthonormal—but they are.

EXAMPLE:
$$Q = \begin{bmatrix} 1/\sqrt{3} & 1/\sqrt{2} & 1/\sqrt{6} \\ 1/\sqrt{3} & 0 & -2/\sqrt{6} \\ 1/\sqrt{3} & -1/\sqrt{2} & 1/\sqrt{6} \end{bmatrix}.$$

Rectangular Matrices with Orthonormal Columns

This chapter is about $Ax = b$, but A is not necessarily square. This section is about $Qx = b$, and we now admit the same possibility—there may be more rows than columns. We have n orthonormal vectors q_i, which are the columns of Q, but those vectors have $m > n$ components. In other words, Q is an m by n matrix and we cannot expect to solve $Qx = b$ exactly. Therefore *we solve it by least squares*.

If there is any justice, orthonormal columns should make the problem simple. It worked for square matrices, and now it will work for rectangular matrices. The key is to notice that *we still have $Q^TQ = I$*:

$$
\begin{bmatrix}
\text{---} q_1^T \text{---} \\
q_n^T
\end{bmatrix}
\begin{bmatrix}
| & & | \\
q_1 & & q_n \\
| & & |
\end{bmatrix}
=
\begin{bmatrix}
1 & . & 0 \\
. & . & . \\
0 & . & 1
\end{bmatrix}.
\tag{3}
$$

It is no longer true that Q^T is the inverse of Q, but it is still the **left-inverse**. For least squares that is all we need. The normal equations came from multiplying $Ax = b$ by the transpose matrix, to give $A^TA\bar{x} = A^Tb$. Here A is Q, and the normal equations are $Q^TQ\bar{x} = Q^Tb$. But Q^TQ is the identity matrix! Therefore *the solution is Q^Tb*, whether Q is square and Q^Tb is an exact solution, or Q is rectangular and we have a least squares solution.

3S If Q has orthonormal columns, then the least squares problem becomes easy:

$Qx = b$ (rectangular system with no solution for most b)

$Q^TQ\bar{x} = Q^Tb$ (normal equation for the best $\bar{x}$—in which $Q^TQ = I$)

$\bar{x} = Q^Tb$ ($\bar{x}_i$ is q_i^Tb)

$p = Q\bar{x}$ (projection of b onto columns is $(q_1^Tb)q_1 + \cdots + (q_n^Tb)q_n$)

$p = QQ^Tb$ (so the projection matrix is $P = QQ^T$).

The last formulas are like $p = A\bar{x}$ and $P = A(A^TA)^{-1}A^T$, which give the projection and the projection matrix for any A. When the columns are orthonormal and A

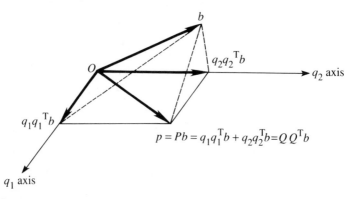

Fig. 3.10. Projection onto a plane = sum of projections onto orthonormal q_1 and q_2.

is Q, the "cross-product matrix" $A^T A$ becomes the identity matrix $Q^T Q = I$. The hard part of least squares disappears, when vectors are orthonormal. The projections onto the axes are uncoupled, and p is the sum of those one-dimensional projections: $p = (q_1^T b)q_1 + \cdots + (q_n^T b)q_n$.

We emphasize that those projections do not reconstruct b. In the square case $m = n$, they did. In the rectangular case $m > n$, they don't. They give the projection p and not the original vector b—which is all we can expect when there are more equations than unknowns, and the q's are no longer a basis. The projection matrix is usually $A(A^T A)^{-1}A^T$, and here it simplifies to

$$P = Q(Q^T Q)^{-1}Q^T \quad \text{or} \quad \boxed{P = QQ^T.} \tag{4}$$

Notice that $Q^T Q$ is the n by n identity matrix, while QQ^T is an m by m projection P. It is the identity matrix on the columns of Q (P leaves them alone), but it is the zero matrix on the orthogonal complement (the nullspace of Q^T).

EXAMPLE 1 The following case is simple but typical. Suppose we project a point $b = (x, y, z)$ onto the x-y plane. Its projection is $p = (x, y, 0)$, and this is the sum of the separate projections onto the x and y axes:

$$q_1 = \begin{bmatrix} 1 \\ 0 \\ 0 \end{bmatrix} \quad \text{and} \quad (q_1^T b)q_1 = \begin{bmatrix} x \\ 0 \\ 0 \end{bmatrix}; \qquad q_2 = \begin{bmatrix} 0 \\ 1 \\ 0 \end{bmatrix} \quad \text{and} \quad (q_2^T b)q_2 = \begin{bmatrix} 0 \\ y \\ 0 \end{bmatrix}.$$

The overall projection matrix is

$$P = q_1 q_1^T + q_2 q_2^T = \begin{bmatrix} 1 & 0 & 0 \\ 0 & 1 & 0 \\ 0 & 0 & 0 \end{bmatrix}, \quad \text{and} \quad P \begin{bmatrix} x \\ y \\ z \end{bmatrix} = \begin{bmatrix} x \\ y \\ 0 \end{bmatrix}.$$

EXAMPLE 2 There is one case in which fitting a straight line leads to orthogonal columns. If measurements y_1, y_2, and y_3 are taken at times which average to zero, say at $t_1 = -3, t_2 = 0$, and $t_3 = 3$, then the attempt to fit $y = C + Dt$ leads to three equations in two unknowns:

$$\begin{matrix} C + Dt_1 = y_1 \\ C + Dt_2 = y_2 \\ C + Dt_3 = y_3 \end{matrix} \quad \text{or} \quad \begin{bmatrix} 1 & -3 \\ 1 & 0 \\ 1 & 3 \end{bmatrix} \begin{bmatrix} C \\ D \end{bmatrix} = \begin{bmatrix} y_1 \\ y_2 \\ y_3 \end{bmatrix}.$$

The two columns are orthogonal. Therefore we can project y separately onto each column, and the best coefficients $\bar{C}$ and $\bar{D}$ can be found separately:

$$\bar{C} = \frac{[1 \ \ 1 \ \ 1][y_1 \ \ y_2 \ \ y_3]^T}{1^2 + 1^2 + 1^2}, \qquad \bar{D} = \frac{[-3 \ \ 0 \ \ 3][y_1 \ \ y_2 \ \ y_3]^T}{(-3)^2 + 0^2 + 3^2}.$$

Notice that $\bar{C} = (y_1 + y_2 + y_3)/3$ is especially simple; it is the *mean* of the data. It gives the best fit by a horizontal line, while $\bar{D}t$ is the best fit by a straight line through the origin. Because *the columns are orthogonal, the sum of these two separate pieces is the best fit by any straight line whatsoever.* But since the columns are not unit vectors—they are not ortho*normal*—the projection formula has the length squared in the denominator.

Orthogonal columns are so much better that it is worth changing to that case. If the average of the observation times is not zero—it is $\bar{t} = (t_1 + \cdots + t_m)/m$— then the time origin can be shifted by $\bar{t}$. Instead of $y = C + Dt$ we work with $y = c + d(t - \bar{t})$. The best line is the same! As in the example we find

$$\bar{c} = \frac{[1 \cdots 1][y_1 \cdots y_m]^T}{1^2 + 1^2 + \cdots + 1^2} = \frac{y_1 + \cdots + y_m}{m}$$

$$\bar{d} = \frac{[(t_1 - \bar{t}) \cdots (t_m - \bar{t})][y_1 \cdots y_m]^T}{(t_1 - \bar{t})^2 + \cdots + (t_m - \bar{t})^2} = \frac{\sum (t_i - \bar{t}) y_i}{\sum (t_i - \bar{t})^2}. \tag{5}$$

The best $\bar{c}$ is the mean, and we also get a convenient formula for $\bar{d}$. The earlier $A^T A$ had the off-diagonal entries $\sum t_i$, and shifting the time by $\bar{t}$ made these entries zero. This shift is an example of the Gram-Schmidt process, which *orthogonalizes the situation in advance*.

Orthogonal matrices are crucial to numerical linear algebra, because they introduce no instability. While lengths stay the same, roundoff is under control. Therefore the orthogonalization of vectors has become an essential technique. Probably it comes second only to elimination. And it leads to a factorization $A = QR$ that is nearly as famous as $A = LU$.

The Gram-Schmidt Process

Suppose you are given three independent vectors a, b, c. If they are orthonormal, life is easy. To project a vector v onto the first one, you compute $(a^T v)a$. To project the same vector v onto the plane of the first two, you just add $(a^T v)a + (b^T v)b$. To project onto the subspace in which the axes are a, b, c, you add up three projections. All calculations require only the inner products $a^T v$, $b^T v$, and $c^T v$. But to make this true, we are forced to start by saying "*If* they are orthonormal." Now we propose to find a way to make them orthonormal.

The method is simple. We are given a, b, c and we want q_1, q_2, q_3. There is no problem with q_1; it can go in the direction of a. We divide by the length, so that $q_1 = a/\|a\|$ is a unit vector. The real problem begins with q_2—which has to be orthogonal to q_1. The idea is to start with b, and if that vector has any component in the direction of q_1 (which is the direction of a), *that component has to be subtracted off*:

$$b' = b - (q_1^T b) q_1. \tag{6}$$

Now b' is orthogonal to q_1. It is the part of b that goes in a new direction, and not in the direction of a. In Fig. 3.11, b' is perpendicular to q_1. It sets the direction for the vector q_2. Since q_2 is required to be a unit vector we divide b' by its length: $q_2 = b'/\|b'\|$.

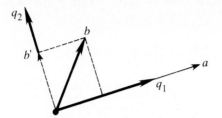

Fig. 3.11. q_1 component of b is removed; a and b' are normalized.

At this point q_1 and q_2 are set. The third orthogonal direction starts with c. It will not be in the plane of q_1 and q_2, which is the plane of a and b. However it may have a component in that plane, and that has to be subtracted off. (If the result is $c' = 0$, this signals that a, b, c were not independent in the first place.) What is left is the component c' we want, the part which is in a new direction perpendicular to the plane:

$$c' = c - (q_1^T c)q_1 - (q_2^T c)q_2. \qquad (7)$$

This is the one idea of the whole Gram-Schmidt process, ***to subtract from every new vector its components in the directions that are already settled***. That idea is used over and over again.† When there is a fourth vector, we take away its components in the directions of q_1, q_2, q_3. Of course the unit vector q_3 was formed from c' by dividing by its length: $q_3 = c'/\|c'\|$.

EXAMPLE   Suppose the independent vectors are

$$a = \begin{bmatrix} 1 \\ 0 \\ 1 \end{bmatrix}, \qquad b = \begin{bmatrix} 1 \\ 0 \\ 0 \end{bmatrix}, \qquad c = \begin{bmatrix} 2 \\ 1 \\ 0 \end{bmatrix}.$$

To find q_1, make the first vector into a unit vector: $q_1 = a/\sqrt{2}$. To find q_2, subtract from the second vector its component in the first direction:

$$b' = b - (q_1^T b)q_1 = \begin{bmatrix} 1 \\ 0 \\ 0 \end{bmatrix} - \frac{1}{\sqrt{2}} \begin{bmatrix} 1/\sqrt{2} \\ 0 \\ 1/\sqrt{2} \end{bmatrix} = \begin{bmatrix} \frac{1}{2} \\ 0 \\ -\frac{1}{2} \end{bmatrix}.$$

The normalized q_2 is b' divided by the length of b', which is $1/\sqrt{2}$:

$$q_2 = \begin{bmatrix} 1/\sqrt{2} \\ 0 \\ -1/\sqrt{2} \end{bmatrix}.$$

† If Gram thought of it first, what was left for Schmidt?

To find q_3, subtract from c its components along q_1 and q_2:

$$c' = c - (q_1^{\mathrm{T}}c)q_1 - (q_2^{\mathrm{T}}c)q_2$$

$$= \begin{bmatrix} 2 \\ 1 \\ 0 \end{bmatrix} - \sqrt{2}\begin{bmatrix} 1/\sqrt{2} \\ 0 \\ 1/\sqrt{2} \end{bmatrix} - \sqrt{2}\begin{bmatrix} 1/\sqrt{2} \\ 0 \\ -1/\sqrt{2} \end{bmatrix} = \begin{bmatrix} 0 \\ 1 \\ 0 \end{bmatrix}.$$

This is already a unit vector, so it is q_3. I went to desperate lengths to cut down the number of square roots (which are the painful part of Gram-Schmidt). The result is a set of orthonormal vectors q_1, q_2, q_3, which go into the columns of an orthogonal matrix Q:

$$Q = \begin{bmatrix} q_1 & q_2 & q_3 \end{bmatrix} = \begin{bmatrix} 1/\sqrt{2} & 1/\sqrt{2} & 0 \\ 0 & 0 & 1 \\ 1/\sqrt{2} & -1/\sqrt{2} & 0 \end{bmatrix}.$$

It is clear that those columns are orthonormal.

3T The Gram-Schmidt process starts with independent vectors $a_1, \ldots, a_n$ and ends with orthonormal vectors $q_1, \ldots, q_n$. At step j it subtracts from a_j its components in the directions that are already settled:

$$a_j' = a_j - (q_1^{\mathrm{T}}a_j)q_1 - \cdots - (q_{j-1}^{\mathrm{T}}a_j)q_{j-1}. \tag{8}$$

Then q_j is the unit vector $a_j'/\|a_j'\|$.

Remark on the calculations It is easier to orthogonalize the vectors, without forcing their lengths to equal one. Then square roots enter only at the end, when dividing by those lengths. The vectors a_j' are the same:

$$a_j' = a_j - (\text{projection of } a_j \text{ on } a_1') - \cdots - (\text{projection of } a_j \text{ on } a_{j-1}').$$

It is only the projection formula that looks different, by using the unnormalized a' instead of the unit vector q:

$$\text{projection of } a_j \text{ on } a_1' = \frac{(a_1')^{\mathrm{T}}a_j}{(a_1')^{\mathrm{T}}a_1'}\, a_1' = (q_1^{\mathrm{T}}a_j)q_1 = \text{projection of } a_j \text{ on } q_1.$$

The example above would have the same $b' = a_2'$ and $c' = a_3'$, without square roots:

$$b' = \begin{bmatrix} 1 \\ 0 \\ 0 \end{bmatrix} - \frac{1}{2}\begin{bmatrix} 1 \\ 0 \\ 1 \end{bmatrix} \quad \text{and then} \quad c' = \begin{bmatrix} 2 \\ 1 \\ 0 \end{bmatrix} - \begin{bmatrix} 1 \\ 0 \\ 1 \end{bmatrix} - 2\begin{bmatrix} \frac{1}{2} \\ 0 \\ -\frac{1}{2} \end{bmatrix}.$$

The Factorization $A = QR$

We started with a matrix A, whose columns were a, b, c. We ended with a matrix Q, whose columns are q_1, q_2, q_3. What is the relation between those matrices?

And what is the relation in higher dimensions, when we start with $a_1, \ldots, a_n$ and end with $q_1, \ldots, q_n$? The matrices A and Q are m by n, when the vectors are in m-dimensional space, and there has to be a third matrix that connects them.

The idea is to write the a's as combinations of the q's. For example, the vector b in Fig. 3.11 is a combination of the orthonormal vectors q_1 and q_2, and we know what combination it is:

$$b = (q_1^\mathsf{T} b)q_1 + (q_2^\mathsf{T} b)q_2.$$

This goes back to the boxed equation (1). Every vector in the plane is the sum of its q_1 and q_2 components. Similarly c is the sum of its q_1, q_2, q_3 components: $c = (q_1^\mathsf{T} c)q_1 + (q_2^\mathsf{T} c)q_2 + (q_3^\mathsf{T} c)q_3$. If we express that in matrix form we have **_the new factorization_** $A = QR$:

$$\begin{bmatrix} & & \\ a & b & c \\ & & \end{bmatrix} = \begin{bmatrix} & & \\ q_1 & q_2 & q_3 \\ & & \end{bmatrix} \begin{bmatrix} q_1^\mathsf{T} a & q_1^\mathsf{T} b & q_1^\mathsf{T} c \\ & q_2^\mathsf{T} b & q_2^\mathsf{T} c \\ & & q_3^\mathsf{T} c \end{bmatrix}$$

Notice the zeros in the last matrix! It is triangular, because of the way Gram-Schmidt was done. The first vectors a and q_1 fell on the same line. Then a, b and q_1, q_2 were in the same plane. The third vectors c and q_3 were not involved until step 3.

The factorization is like $A = LU$, except now the first factor Q has orthonormal columns. The second factor is called R, because the nonzeros are to the *right* of the diagonal (and the letter U is already taken). The off-diagonal entries of R are the numbers $q_1^\mathsf{T} b = 1/\sqrt{2}$ and $q_1^\mathsf{T} c = q_2^\mathsf{T} c = \sqrt{2}$ found above. The diagonal entries are the lengths $\sqrt{2}, 1/\sqrt{2}, 1$ that we divided by. The whole factorization is

$$A = \begin{bmatrix} 1 & 1 & 2 \\ 0 & 0 & 1 \\ 1 & 0 & 0 \end{bmatrix} = \begin{bmatrix} 1/\sqrt{2} & 1/\sqrt{2} & 0 \\ 0 & 0 & 1 \\ 1/\sqrt{2} & -1/\sqrt{2} & 0 \end{bmatrix} \begin{bmatrix} \sqrt{2} & 1/\sqrt{2} & \sqrt{2} \\ & 1/\sqrt{2} & \sqrt{2} \\ & & 1 \end{bmatrix} = QR.$$

You see the lengths of a', b', c' on the diagonal of R. Off the diagonal are the multiples of q_1 and q_2 that were subtracted by Gram-Schmidt. The orthonormal vectors themselves, which are q_1, q_2, q_3 and are the whole object of orthogonalization, are in the first factor Q.

Maybe QR is not as beautiful as LU (because of the square roots). Both factorizations are important to the theory of linear algebra, and absolutely central to the calculations. If LU is Hertz, then QR is Avis.

The general case is exactly the same. The matrix R is n by n, and its i,j entry is $q_i^\mathsf{T} a_j$. This is zero when i is greater than j (q_i is constructed perpendicular to a_j), so R is upper triangular. Its entries appear in formula (8), especially when $\|a_j'\| q_j$ is substituted for a_j':

$$a_j = (q_1^\mathsf{T} a_j)q_1 + \cdots + (q_{j-1}^\mathsf{T} a_j)q_{j-1} + \|a_j'\| q_j. \tag{9}$$

The right side is exactly Q times R, written out in full.

3U Every m by n matrix A with linearly independent columns can be factored into $A = QR$. The columns of Q are orthonormal, and R is upper triangular and invertible. When $m = n$ and all matrices are square, Q becomes an orthogonal matrix.

I must not forget the main point of orthogonalization. It simplifies the least squares problem $Ax = b$. The normal equations are still correct, but $A^{\mathrm{T}}A$ is easier to invert; it becomes

$$A^{\mathrm{T}}A = R^{\mathrm{T}}Q^{\mathrm{T}}QR = R^{\mathrm{T}}R.$$

Therefore the fundamental equation $A^{\mathrm{T}}A\bar{x} = A^{\mathrm{T}}b$ simplifies to

$$R^{\mathrm{T}}R\bar{x} = R^{\mathrm{T}}Q^{\mathrm{T}}b \qquad \text{or} \qquad R\bar{x} = Q^{\mathrm{T}}b. \tag{10}$$

Instead of solving $QRx = b$, which can't be done, we solve $R\bar{x} = Q^{\mathrm{T}}b$, which can be done very quickly—because R is triangular. The real cost is in the mn^2 operations of Gram-Schmidt, which are needed to find Q and R in the first place.

The same idea of orthogonality applies to functions. The sines and cosines are orthogonal; the powers $1, x, x^2, \ldots$ are not. When f is written as a combination of sines and cosines, that is a *Fourier series*. Each term is a projection onto a line—the line in function space containing multiples of $\cos nx$ or $\sin nx$. It is so completely parallel to the vector case, and so important, that it deserves to be seen. And finally we have a job for Schmidt: to orthogonalize the powers of x and produce the Legendre polynomials.

Function Spaces and Fourier Series

This is a brief and optional section, but it has a number of good intentions:

(1)   to introduce the most famous infinite-dimensional vector space;
(2)   to extend the ideas of length and inner product from vectors v to functions $f(x)$;
(3)   to recognize the Fourier series of f as a sum of one-dimensional projections; the orthogonal "columns" which span the space are the sines and cosines;
(4)   to apply Gram-Schmidt orthogonalization to the polynomials $1, x, x^2, \ldots$;
(5)   to find the best approximation to $f(x)$ by a straight line.

We will try to follow this outline, which opens up a range of new applications for linear algebra, in a systematic way.

1.   After studying all the finite-dimensional spaces $\mathbf{R}^n$, it is natural to think of the space $\mathbf{R}^\infty$. It contains all vectors $v = (v_1, v_2, v_3, \ldots)$ with an infinite sequence of components. This space is actually too big to be very useful, when there is no control on the components v_j. A much better idea is to keep the familiar definition of length, as the square root of Pythagoras' sum of squares, and *to include only those* *vectors that have a **finite length***: The infinite series

$$\|v\|^2 = v_1^2 + v_2^2 + v_3^2 + \cdots \tag{11}$$

must converge to a finite sum. This still leaves an infinite-dimensional set of vectors, including the vector $(1, \frac{1}{2}, \frac{1}{3}, \ldots)$ but excluding $(1, 1, 1, \ldots)$. The vectors with finite length can be added together ($\|v + w\| \le \|v\| + \|w\|$) and multiplied by scalars, so they form a vector space. It is the celebrated **Hilbert space**.

Hilbert space is the natural way to let the number of dimensions become infinite, and at the same time to keep the geometry of ordinary Euclidean space. Ellipses become infinite-dimensional ellipsoids, parabolas become paraboloids, and perpendicular lines are recognized in the same way as before: The vectors v and w are orthogonal when their inner product is zero,

$$v^T w = v_1 w_1 + v_2 w_2 + v_3 w_3 + \cdots = 0.$$

This sum is guaranteed to converge, and for any two vectors it still obeys the Schwarz inequality $|v^T w| \le \|v\| \, \|w\|$. The cosine, even in Hilbert space, is never larger than one.

There is another remarkable thing about this space: It is found under a great many different disguises. Its "vectors" can turn into functions, and that brings us to the second point.

2. Suppose we think of the function $f(x) = \sin x$, on the interval $0 \le x \le 2\pi$. This f is like a vector with a whole continuum of components, the values of $\sin x$ along the whole interval. To find the length of such a vector, the usual rule of adding the squares of the components becomes impossible. This summation is replaced, in a natural and inevitable way, by *integration*:

$$\|f\|^2 = \int_0^{2\pi} (f(x))^2 \, dx = \int_0^{2\pi} (\sin x)^2 \, dx = \pi. \tag{12}$$

Our Hilbert space has become a **function space**. The vectors are functions, we have a way to measure their length, and the space contains all those functions that have a finite length—just as in (11) above. It does not contain the function $F(x) = 1/x$, because the integral of $1/x^2$ is infinite.

The same idea of replacing summation by integration produces the **inner product of two functions**: If $f(x) = \sin x$ and $g(x) = \cos x$, then their inner product is

$$(f, g) = \int_0^{2\pi} f(x)g(x) \, dx = \int_0^{2\pi} \sin x \cos x \, dx = 0. \tag{13}$$

This is exactly like the vector inner product $f^T g$. It is still related to the length by $(f, f) = \|f\|^2$. The Schwarz inequality is still satisfied: $|(f, g)| \le \|f\| \, \|g\|$. Of course two functions like $\sin x$ and $\cos x$—whose inner product is zero—will be called orthogonal. They are even orthonormal, after division by their length $\sqrt{\pi}$.

3. The **Fourier series** of a function is an expansion into sines and cosines:

$$y(x) = a_0 + a_1 \cos x + b_1 \sin x + a_2 \cos 2x + b_2 \sin 2x + \cdots.$$

To compute a coefficient like b_1, *multiply* both sides by the corresponding function $\sin x$ and *integrate* from 0 to 2π. (The function y is given on that interval.) In other words, take the inner product of both sides with $\sin x$:

$$\int_0^{2\pi} y(x) \sin x \, dx = a_0 \int_0^{2\pi} \sin x \, dx + a_1 \int_0^{2\pi} \cos x \sin x \, dx + b_1 \int_0^{2\pi} (\sin x)^2 \, dx + \cdots.$$

On the right side, every integral is zero except one—the one in which $\sin x$ multiplies itself. *The sines and cosines are mutually orthogonal* as in (13). Therefore b_1 is the left side divided by that one nonzero integral:

$$b_1 = \frac{\displaystyle\int_0^{2\pi} y(x) \sin x \, dx}{\displaystyle\int_0^{2\pi} (\sin x)^2 \, dx} = \frac{(y, \sin x)}{(\sin x, \sin x)}.$$

The Fourier coefficient a_1 would have $\cos x$ in place of $\sin x$, and a_2 would use $\cos 2x$.

The point of this calculation is to see the analogy with projections. The component of the vector b along the line spanned by a was computed much earlier:

$$\bar{x} = \frac{b^{\mathsf{T}} a}{a^{\mathsf{T}} a}.$$

In a Fourier series, *we are projecting y onto* $\sin x$. Its component p in this direction is exactly $b_1 \sin x$. (For vectors it was $\bar{x}a$.) The coefficient b_1 is the least squares solution of the inconsistent equation $b_1 \sin x = y$; in other words, it brings $b_1 \sin x$ as close as possible to y. The same is true for all the terms in the series; every one is a projection of y onto a sine or cosine. Since the sines and cosines are orthogonal, *the Fourier series gives the coordinates of the "vector" y with respect to a set of (infinitely many) perpendicular axes.*

4. There are plenty of useful functions other than sines and cosines, and they are not always orthogonal. The simplest are the polynomials, and unfortunately there is no interval on which even the first three coordinate axes—the functions 1, x, and x^2—are perpendicular. (The inner product of 1 and x^2 is always positive, because it is the integral of x^2.) Therefore the closest parabola to $y(x)$ is not the sum of its projections onto 1, x, and x^2. There will be a coupling term, exactly like $(A^{\mathsf{T}}A)^{-1}$ in the matrix case, and in fact the coupling is given by the ill-conditioned Hilbert matrix. On the interval $0 \le x \le 1$,

$$A^{\mathsf{T}}A = \begin{bmatrix} (1, 1) & (1, x) & (1, x^2) \\ (x, 1) & (x, x) & (x, x^2) \\ (x^2, 1) & (x^2, x) & (x^2, x^2) \end{bmatrix} = \begin{bmatrix} \int 1 & \int x & \int x^2 \\ \int x & \int x^2 & \int x^3 \\ \int x^2 & \int x^3 & \int x^4 \end{bmatrix} = \begin{bmatrix} 1 & \frac{1}{2} & \frac{1}{3} \\ \frac{1}{2} & \frac{1}{3} & \frac{1}{4} \\ \frac{1}{3} & \frac{1}{4} & \frac{1}{5} \end{bmatrix}.$$

This matrix has a large inverse, because the axes 1, x, x^2 are far from perpendicular. Even for a computer, the situation becomes impossible if we add a few more axes. *It is virtually hopeless to solve $A^{\mathsf{T}}A\bar{x} = A^{\mathsf{T}}b$ for the closest polynomial of degree ten.*

More precisely, it is hopeless to solve this by Gaussian elimination; every roundoff error would be amplified by more than 10^{13}. On the other hand, we cannot just give up; approximation by polynomials has to be possible. The right idea is to switch to orthogonal axes, and this means a Gram-Schmidt orthogonalization: We look for combinations of 1, x, and x^2 that *are* orthogonal.

It is convenient to work with a symmetrically placed interval like $-1 \le x \le 1$, because this makes all the odd powers of x orthogonal to all the even powers:

$$(1, x) = \int_{-1}^{1} x \, dx = 0, \qquad (x, x^2) = \int_{-1}^{1} x^3 \, dx = 0.$$

Therefore the Gram-Schmidt process can begin by accepting $v_1 = 1$ and $v_2 = x$ as the first two perpendicular axes, and it only has to correct the angle between 1 and x^2. By formula (7) the third orthogonal polynomial is

$$v_3 = x^2 - \frac{(1, x^2)}{(1, 1)} 1 - \frac{(x, x^2)}{(x, x)} x = x^2 - \frac{\int_{-1}^{1} x^2 \, dx}{\int_{-1}^{1} 1 \, dx} = x^2 - \frac{1}{3}.$$

The polynomials constructed in this way are called the **Legendre polynomials** and they are orthogonal to each other over the interval $-1 \le x \le 1$.

Check

$$(1, x^2 - \tfrac{1}{3}) = \int_{-1}^{1} (x^2 - \tfrac{1}{3}) \, dx = \left[\frac{x^3}{3} - \frac{x}{3} \right]_{-1}^{1} = 0.$$

The closest polynomial of degree ten is now computable, without disaster, by projecting onto each of the first 10 (or 11) Legendre polynomials.

5. Suppose we want to approximate $y = x^5$ by a straight line $C + Dx$ between $x = 0$ and $x = 1$. There are at least three ways of finding that line, and if you compare them the whole chapter might become clear!

(1) Solve $\begin{bmatrix} 1 & x \end{bmatrix} \begin{bmatrix} C \\ D \end{bmatrix} = x^5$ by least squares. The equation $A^T A \bar{x} = A^T b$ is

$$\begin{bmatrix} (1, 1) & (1, x) \\ (x, 1) & (x, x) \end{bmatrix} \begin{bmatrix} C \\ D \end{bmatrix} = \begin{bmatrix} (1, x^5) \\ (x, x^5) \end{bmatrix} \qquad \text{or} \qquad \begin{bmatrix} 1 & \tfrac{1}{2} \\ \tfrac{1}{2} & \tfrac{1}{3} \end{bmatrix} \begin{bmatrix} C \\ D \end{bmatrix} = \begin{bmatrix} \tfrac{1}{6} \\ \tfrac{1}{7} \end{bmatrix}.$$

(2) Minimize $E^2 = \int_0^1 (x^5 - C - Dx)^2 \, dx = \tfrac{1}{11} - \tfrac{2}{6}C - \tfrac{2}{7}D + C^2 + CD + \tfrac{1}{3}D^2.$ The derivatives with respect to C and D, after dividing by 2, bring back the normal equations of method (1):

$$-\tfrac{1}{6} + C + \tfrac{1}{2}D = 0 \qquad \text{and} \qquad -\tfrac{1}{7} + \tfrac{1}{2}C + \tfrac{1}{3}D = 0.$$

(3) Apply Gram-Schmidt to replace x by $x - (1, x)/(1, 1)$. That is $x - \tfrac{1}{2}$, which is orthogonal to 1. Now the one-dimensional projections give the best line:

$$C + Dx = \frac{(x^5, 1)}{(1, 1)} 1 + \frac{(x^5, x - \tfrac{1}{2})}{(x - \tfrac{1}{2}, x - \tfrac{1}{2})} (x - \tfrac{1}{2}) = \tfrac{1}{6} + \tfrac{5}{7}(x - \tfrac{1}{2}).$$

EXERCISES

3.4.1 (a) Write down the four equations for fitting $y = C + Dt$ to the data

$$y = -4 \text{ at } t = -2, \qquad y = -3 \text{ at } t = -1.$$
$$y = -1 \text{ at } t = 1, \qquad y = 0 \quad \text{at } t = 2.$$

Show that the columns are orthogonal.
(b) Find the optimal straight line, draw a graph, and write down the error E^2.
(c) Interpret the fact that the error is zero in terms of the original system of four equations in two unknowns: Where is the right side b with relation to the column space, and what is its projection p?

3.4.2 Project $b = (0, 3, 0)$ onto each of the orthonormal vectors $a_1 = (\frac{2}{3}, \frac{2}{3}, -\frac{1}{3})$ and $a_2 = (-\frac{1}{3}, \frac{2}{3}, \frac{2}{3})$, and then find its projection p onto the plane of a_1 and a_2.

3.4.3 Find also the projection of $b = (0, 3, 0)$ onto $a_3 = (\frac{2}{3}, -\frac{1}{3}, \frac{2}{3})$, add up the three one-dimensional projections, and interpret the result. Why is $P = a_1 a_1^T + a_2 a_2^T + a_3 a_3^T$ equal to the identity?

3.4.4 If Q_1 and Q_2 are orthogonal matrices, and therefore satisfy $Q^T Q = I$, show that $Q_1 Q_2$ is also orthogonal. If Q_1 is rotation through θ, and Q_2 is rotation through ϕ, what is $Q_1 Q_2$? Can you find the trigonometric identities for $\sin(\theta + \phi)$ and $\cos(\theta + \phi)$ in the matrix multiplication $Q_1 Q_2$?

3.4.5 If u is a unit vector, show that $Q = I - 2uu^T$ is an orthogonal matrix. (It is a reflection, also known as a Householder transformation.) Compute Q when $u^T = [\frac{1}{2} \quad \frac{1}{2} \quad -\frac{1}{2} \quad -\frac{1}{2}]$.

3.4.6 Find a third column so that the matrix

$$Q = \begin{bmatrix} 1/\sqrt{3} & 1/\sqrt{14} \\ 1/\sqrt{3} & 2/\sqrt{14} \\ 1/\sqrt{3} & -3/\sqrt{14} \end{bmatrix}$$

is orthogonal. It must be a unit vector that is orthogonal to the other columns; how much freedom does this leave? Verify that the rows automatically become orthonormal at the same time.

3.4.7 Show, by forming $b^T b$ directly, that Pythagoras' law holds for any combination $b = x_1 q_1 + \cdots + x_n q_n$ of orthonormal vectors: $\|b\|^2 = x_1^2 + \cdots + x_n^2$. In matrix terms $b = Qx$, so this again proves that lengths are preserved: $\|Qx\|^2 = \|x\|^2$.

3.4.8 Project the vector $b = (1, 2)$ onto two vectors that are not orthogonal, $a_1 = (1, 0)$ and $a_2 = (1, 1)$. Show that unlike the orthogonal case, the sum of the two one-dimensional projections does not equal b.

3.4.9 If the vectors q_1, q_2, q_3 are orthonormal, what combination of q_1 and q_2 is closest to q_3?

3.4.10 If q_1 and q_2 are orthonormal what combination is closest to the vector b? Verify that the error vector is orthogonal to q_1 and q_2.

3.4.11 Show that an orthogonal matrix which is also upper triangular must be diagonal.

3.4.12 What multiple of $a_1 = \begin{bmatrix} 1 \\ 1 \end{bmatrix}$ should be subtracted from $a_2 = \begin{bmatrix} 4 \\ 0 \end{bmatrix}$ to make the result orthogonal to a_1? Factor $\begin{bmatrix} 1 & 4 \\ 1 & 0 \end{bmatrix}$ into QR with orthonormal vectors in Q.

3.4.13 Apply the Gram-Schmidt process to

$$a = \begin{bmatrix} 0 \\ 0 \\ 1 \end{bmatrix}, \qquad b = \begin{bmatrix} 0 \\ 1 \\ 1 \end{bmatrix}, \qquad c = \begin{bmatrix} 1 \\ 1 \\ 1 \end{bmatrix}$$

and write the result in the form $A = QR$.

3.4.14 Suppose the given vectors are

$$a = \begin{bmatrix} 1 \\ 1 \\ 0 \end{bmatrix}, \qquad b = \begin{bmatrix} 1 \\ 0 \\ 1 \end{bmatrix}, \qquad c = \begin{bmatrix} 0 \\ 1 \\ 1 \end{bmatrix}.$$

Find the orthonormal vectors q_1, q_2, q_3.

3.4.15 Find an orthonormal set q_1, q_2, q_3 for which q_1, q_2 span the column space of

$$A = \begin{bmatrix} 1 & 1 \\ 2 & -1 \\ -2 & 4 \end{bmatrix}.$$

Which fundamental subspace contains q_3? What is the least squares solution of $Ax = b$ if $b = \begin{bmatrix} 1 & 2 & 7 \end{bmatrix}^T$?

3.4.16 Express the Gram-Schmidt orthogonalization of

$$a_1 = \begin{bmatrix} 1 \\ 2 \\ 2 \end{bmatrix}, \qquad a_3 = \begin{bmatrix} 1 \\ 3 \\ 1 \end{bmatrix}$$

as $A = QR$. Given n vectors a_i, each with m components, what are the shapes of A, Q, and R?

3.4.17 With the same matrix A, and with $b = \begin{bmatrix} 1 & 1 & 1 \end{bmatrix}^T$, use $A = QR$ to solve the least squares problem $Ax = b$.

3.4.18 If $A = QR$, find a simple formula for the projection matrix P onto the column space of A.

3.4.19 Show that the *modified Gram-Schmidt* steps

$$c'' = c - (q_1^T c)q_1 \qquad \text{and} \qquad c' = c'' - (q_2^T c'')q_2$$

produce the same vector c' as in (7). This is much more stable, to subtract off the projections one at a time.

3.4.20 Find the length of the vector $v = (1/\sqrt{2}, 1/\sqrt{4}, 1/\sqrt{8}, \ldots)$ and of the function $f(x) = e^x$ (over the interval $0 \le x \le 1$). What is the inner product over this interval of e^x and e^{-x}?

3.4.21 What is the closest function $a \cos x + b \sin x$ to the function $f = \sin 2x$ on the interval from $-\pi$ to π? What is the closest straight line $c + dx$?

3.4.22 By setting the derivative to zero, find the value of b_1 that minimizes

$$\|b_1 \sin x - y\|^2 = \int_0^{2\pi} (b_1 \sin x - y(x))^2 \, dx.$$

Compare with the Fourier coefficient b_1. If $y(x) = \cos x$, what is b_1?

3.4.23 Find the Fourier coefficients a_0, a_1, b_1 of the step function $y(x)$, which equals 1 on the interval $0 \le x \le \pi$ and 0 on the remaining interval $\pi < x < 2\pi$:

$$a_0 = \frac{(y, 1)}{(1, 1)}, \qquad a_1 = \frac{(y, \cos x)}{(\cos x, \cos x)}, \qquad b_1 = \frac{(y, \sin x)}{(\sin x, \sin x)}.$$

3.4.24 Find the next Legendre polynomial—a cubic orthogonal to 1, x, and $x^2 - \frac{1}{3}$ over the interval $-1 \le x \le 1$.

3.4.25 What is the closest straight line to the parabola $y = x^2$ over $-1 \le x \le 1$?

3.4.26 In the Gram-Schmidt formula (7), verify that c' is orthogonal to q_1 and q_2.

3.4.27 Find an orthonormal basis for the subspace spanned by $a_1 = (1, -1, 0, 0)$, $a_2 = (0, 1, -1, 0)$, $a_3 = (0, 0, 1, -1)$.

3.4.28 Apply Gram-Schmidt to $(1, -1, 0)$, $(0, 1, -1)$, and $(1, 0, -1)$, to find an orthonormal basis on the plane $x_1 + x_2 + x_3 = 0$. What is the dimension of this subspace, and how many nonzero vectors come out of Gram-Schmidt?

At the end of the last section we mentioned Fourier series. That was linear algebra in infinite dimensions. The "vectors" were functions $f(x)$; they were projected onto the sines and cosines; that produced the Fourier coefficients. From this infinite sequence of sines and cosines, multiplied by their Fourier coefficients, we can reconstruct $f(x)$. That is the classical case, which Fourier dreamt about, but in actual calculations it is the *discrete Fourier transform* that we compute. Fourier still lives, but in finite dimensions.

The theory is pure linear algebra, based on orthogonality. The input is a sequence of numbers $y_0, \ldots, y_{n-1}$, instead of a function f. The output is another sequence of the same length—a set of n Fourier coefficients $c_0, \ldots, c_{n-1}$ instead of an infinite sequence. The relation between y and c is linear, so it must be given by a matrix. This is the *Fourier matrix* F, and the whole technology of digital signal processing depends on it. The signal is digitized, whether it comes from speech or images or sonar or telecommunications (or even oil exploration). It can be transformed by the matrix F, and later it can be transformed back—to reconstruct the original image. What is crucially important is that both transforms can be done quickly:

(1) the inverse matrix F^{-1} must be simple
(2) the multiplications by F and F^{-1} must be fast.

Both of those are now true. The matrix F^{-1} has been known for years, and it looks just like F. In fact F is symmetric and orthogonal (apart from a factor $\sqrt{n}$), and it has only one drawback: its entries are *complex numbers*. That is a small price to pay, and we pay it below. The difficulties are minimized by the fact that *all entries of F and F^{-1} are powers of a single number w.* Instead of a full introduction to complex numbers, we go only far enough to use the remarkable equation $w^n = 1$—which involves sines and cosines and exponentials, and lies at the heart of discrete Fourier analysis.

It is remarkable that F is so easy to invert. If that were all (and up to 1965 it was all) the discrete transform would have an important place. Now there is more. The multiplications by F and F^{-1} can be done in an extremely fast and ingenious way. Instead of n^2 separate multiplications, coming from the n^2 entries in the matrix, a matrix-vector product like $F^{-1}y$ requires only $\frac{1}{2}n \log n$ steps. It is the same multiplication, but arranged in a good way. This rearrangement is called the *Fast Fourier Transform*.

The section begins with w and its properties, moves on to F^{-1}, and ends with the FFT—the fast transform and its applications. The great application is called *convolution*, and the key to its success is the *convolution rule*.

Complex Roots of Unity

Real equations can have complex solutions. The equation $x^2 + 1 = 0$ led to the invention of i (and also to $-i$!). That was declared to be a solution, and the case was closed. If someone asked about $x^2 - i = 0$, there was an answer: The square roots of a complex number are again complex numbers. You must allow combinations $x + iy$, with a real part x and an imaginary part y, but no further inventions are necessary. Every polynomial of degree n has a full set of n roots (possibly complex and possibly repeated). That is the fundamental theorem of algebra, and the word "complex" allows the possibility that the imaginary part is $y = 0$ and the number is actually real.

We are interested in equations like $x^4 = 1$. That should have four solutions—there should be four "**fourth roots of unity**". The two square roots of unity are 1 and -1. The fourth roots are the square roots of the square roots, 1 and -1, i and $-i$. The number i will satisfy $i^4 = 1$ because it satisfies $i^2 = -1$. For the eighth roots of unity we need the square roots of i, and that brings us to complex numbers like $w = (1 + i)/\sqrt{2}$. Squaring w produces $(1 + 2i + i^2)/2$, which is the same as i—because $1 + i^2$ is zero. Since the square is $w^2 = i$, the eighth power is $w^8 = i^4 = 1$. There has to be a system here.

The complex numbers in the Fourier matrix are extremely special. Their real parts are cosines and their imaginary parts are sines:

$$\boxed{w = \cos\theta + i\sin\theta.} \tag{1}$$

Suppose the real part is plotted on the x-axis and the imaginary part on the y-axis (Fig. 3.12). Then the number w lies on the **unit circle**; its distance from the origin is $\cos^2\theta + \sin^2\theta = 1$. It makes an angle θ with the horizontal. The plane in that figure is called the **complex plane**, and every complex number $z = x + iy$ has a place. The whole plane enters in Chapter 5, where complex numbers will appear as eigenvalues (even of real matrices). Here we need only special points w, all of them on the unit circle, in order to solve $w^n = 1$.

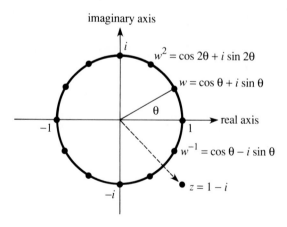

Fig. 3.12. The complex plane and the unit circle.

How do we compute the nth power of w? Certainly its square can be found directly:

$$w^2 = (\cos\theta + i\sin\theta)^2 = \cos^2\theta - \sin^2\theta + 2i\sin\theta\cos\theta.$$

That looks bad, until it is illuminated by the double-angle formulas of trigonometry. The real part $\cos^2\theta - \sin^2\theta$ is equal to $\cos 2\theta$, and the imaginary part $2\sin\theta\cos\theta$ is $\sin 2\theta$. (Note that i is not included; the imaginary part is a real number.) Thus $w^2 = \cos 2\theta + i\sin 2\theta$. The square of w is still on the unit circle, but *at the double angle* 2θ. That makes us suspect that w^n lies at the angle $n\theta$, and we are right.

There is a better way to take powers of w. The combination of cosine and sine is a complex exponential, with amplitude one and phase angle θ:

$$\boxed{\cos\theta + i\sin\theta = e^{i\theta}.} \tag{2}$$

The rules for multiplying and dividing powers, like $(e^2)(e^3) = e^5$ or $e^2/e^3 = e^{-1}$, continue to hold when the exponents are imaginary. Therefore the powers of $w = e^{i\theta}$ are exactly as anticipated:

$$\boxed{w^2 = e^{i2\theta}, \quad w^n = e^{in\theta}, \quad \frac{1}{w} = e^{-i\theta}.} \tag{3}$$

The nth power is at the angle $n\theta$. When $n = -1$, *the reciprocal* $1/w$ *is at the angle* $-\theta$. If we multiply $\cos\theta + i\sin\theta$ by $\cos(-\theta) + i\sin(-\theta)$, we should get the answer 1. Since $\cos(-\theta)$ is equal to $\cos\theta$ (the cosine is even), and $\sin(-\theta)$ is equal to $-\sin\theta$ (the sine is odd), that multiplication does give

$$e^{i\theta}e^{-i\theta} = (\cos\theta + i\sin\theta)(\cos\theta - i\sin\theta) = \cos^2\theta + \sin^2\theta = 1.$$

Note I remember the day when a letter came to MIT from a prisoner in New York, asking if Euler's formula (2) was true. It is really astonishing, when you think of it, that three of the key functions of mathematics should come together in such a graceful way. Our best answer was to look at the power series

$$e^{i\theta} = 1 + i\theta + \frac{(i\theta)^2}{2!} + \frac{(i\theta)^3}{3!} + \cdots.$$

The real part $1 - \theta^2/2 + \cdots$ is the cosine. The imaginary part $\theta - \theta^3/6 + \cdots$ is the sine. The formula is correct, and I wish we had sent a more beautiful proof.

With this formula, we can solve the equation $w^n = 1$. It becomes $e^{in\theta} = 1$, so that $n\theta$ must carry us around the unit circle and back to the start. The solution is to choose $\theta = 2\pi/n$: The "*primitive*" nth root of unity is

$$\boxed{w_n = e^{2\pi i/n} = \cos\frac{2\pi}{n} + i\sin\frac{2\pi}{n}.} \tag{4}$$

Its nth power is $e^{2\pi i}$, which equals 1. In the cases $n = 4$ and $n = 8$, this primitive root is

$$w_4 = \cos\frac{\pi}{2} + i\sin\frac{\pi}{2} = i$$

$$w_8 = \cos\frac{\pi}{4} + i\sin\frac{\pi}{4} = \frac{1+i}{\sqrt{2}}$$

The fourth root is at $\theta = 90°$, and the eighth root is at $\theta = 45°$. Those are $\frac{1}{4}(360°)$ and $\frac{1}{8}(360°)$, but they are not the only fourth and eighth roots of unity! The other fourth roots are the powers $i^2 = -1$, $i^3 = -i$, and $i^4 = 1$. The other eighth roots are the powers $w_8^2, w_8^3, \ldots, w_8^8$. The roots are equally spaced around the unit circle, at intervals of $2\pi/n$. Note again that the square of w_8 is w_4, which will be essential in the Fast Fourier Transform. Note also that *the roots add up to zero*. For $w_4 = i$ this is just $1 + i - 1 - i = 0$, and for w_8 it is

$$1 + w_8 + w_8^2 + \cdots + w_8^7 = 0. \tag{5}$$

One proof is to multiply the left side by w_8, which leaves it unchanged. (It yields $w_8 + w_8^2 + \cdots + w_8^8$, and w_8^8 equals 1.) The eight points each move through $45°$, but they remain the same eight points. Since zero is the only number that is unchanged when multiplied by w_8, the sum must be zero.†

The Fourier Matrix and Its Inverse

In the continuous case, the Fourier series can reproduce $f(x)$ over a whole interval. It uses infinitely many sines and cosines (or exponentials). In the discrete case, with only n coefficients to choose, that is too much to expect. We only ask for *equality at n points*, which gives n equations. In a typical problem, with $n = 4$, the equations to reproduce the four values 2, 4, 6, 8 are

$$\begin{aligned}
c_0 + c_1 + c_2 + c_3 &= 2 \\
c_0 + ic_1 - c_2 - ic_3 &= 4 \\
c_0 - c_1 + c_2 - c_3 &= 6 \\
c_0 - ic_1 - c_2 + ic_3 &= 8.
\end{aligned} \tag{6}$$

The input sequence is $y = 2, 4, 6, 8$. The output sequence is c_0, c_1, c_2, c_3. The equations look for a 4-term Fourier series that matches the inputs at four equally spaced points on the interval from 0 to 2π:

$$c_0 + c_1 e^{ix} + c_2 e^{2ix} + c_3 e^{3ix} = \begin{array}{ll} 2 & \text{at} \quad x = 0 \\ 4 & \text{at} \quad x = \pi/2 \\ 6 & \text{at} \quad x = \pi \\ 8 & \text{at} \quad x = 3\pi/2. \end{array}$$

---

† In this case $w^5 = -w$, $w^6 = -w^2$, $w^7 = -w^3$, $w^8 = -w^4$. When n is even the roots can be paired off. But the three cube roots of 1 also add to zero, without cancelling in pairs.

Those are the four equations in (6). At $x = 2\pi$ the series returns to the first value $y_0 = 2$ and continues periodically.

Note that the Fourier series is written in its complex form, as a combination of exponentials e^{ikx} rather than sines and cosines. The two are equivalent when all harmonics are included, because of the matching

$$c_k e^{ikx} + c_{-k} e^{-ikx} = a_k \cos kx + b_k \sin kx. \tag{7}$$

It is possible to stay with real series, but when they stop after four terms it requires special care and the formulas are not elegant. A better decision is to use the complex form, and to solve equation (6) for the coefficients c_0, c_1, c_2, c_3.

That is not hard to do. If we add the four equations, there is tremendous cancellation on the left side. The result is $4c_0 = 20$, so that c_0 is the average value 5 of the signals 2, 4, 6, 8.

There is also a way to find c_1. Multiply the equations by $1, -i, -1, i$, and add. Everything cancels on the left except $4c_1$, which equals $2 - 4i - 6 + 8i$. Therefore $c_1 = -1 + i$ (note that it is complex). There has to be a similar method for c_2 and c_3.

The pattern is clearer with the equations in matrix form. The four equations are $Fc = y$, and the matrix is

$$F = \begin{bmatrix} 1 & 1 & 1 & 1 \\ 1 & i & i^2 & i^3 \\ 1 & i^2 & i^4 & i^6 \\ 1 & i^3 & i^6 & i^9 \end{bmatrix}.$$

Those 16 entries are the same as the coefficients in (6). Instead of i^9 we previously wrote i, and i^6 agrees with -1, and i^3 is $-i$. In the present form it will be easier to recognize F^{-1}. Apart from a factor $\frac{1}{4}$, the inverse matrix has the same form!

$$F^{-1} = \frac{1}{4} \begin{bmatrix} 1 & 1 & 1 & 1 \\ 1 & (-i) & (-i)^2 & (-i)^3 \\ 1 & (-i)^2 & (-i)^4 & (-i)^6 \\ 1 & (-i)^3 & (-i)^6 & (-i)^9 \end{bmatrix}.$$

A direct multiplication gives $FF^{-1} = I$. For example, the second row of F times the second column of F^{-1} is $\frac{1}{4}(1 + 1 + 1 + 1)$. The other diagonal entries equal 1 in the same way. Off the diagonal a typical product is $1 + i + i^2 + i^3 = 0$. (Remember that the fourth roots add to zero.) From the top row of $c = F^{-1}y$ we see again that c_0 is the average of the four signals, $c_0 = \frac{1}{4}(y_0 + y_1 + y_2 + y_3)$.

Note For Fourier matrices it is natural to number the rows and columns from 0 to $n - 1$, instead of 1 to n.

The pattern found above is not limited to $n = 4$. For every n the matrix connecting y to c can be written down and inverted. It represents n equations, each one requiring the finite series $c_0 + c_1 e^{ix} + \cdots$ (**n terms**) to agree with y (**at n points**).

The first agreement is at $x = 0$:

$$c_0 + c_1 + \cdots + c_{n-1} = y_0.$$

The next point is at $x = 2\pi/n$, which introduces the crucial number $w = e^{2\pi i/n}$:

$$c_0 + c_1 w + \cdots + c_{n-1} w^{n-1} = y_1.$$

The third point $x = 4\pi/n$ involves $e^{4\pi i/n}$, which is w^2:

$$c_0 + c_1 w^2 + \cdots + c_{n-1} w^{2(n-1)} = y_2.$$

The remaining points bring higher powers of w, and the full problem is

$$\begin{bmatrix} 1 & 1 & 1 & \cdot & 1 \\ 1 & w & w^2 & \cdot & w^{n-1} \\ 1 & w^2 & w^4 & \cdot & w^{2(n-1)} \\ \cdot & & \cdot & & \cdot \\ 1 & w^{n-1} & w^{2(n-1)} & \cdot & w^{(n-1)^2} \end{bmatrix} \begin{bmatrix} c_0 \\ c_1 \\ c_2 \\ \cdot \\ c_{n-1} \end{bmatrix} = \begin{bmatrix} y_0 \\ y_1 \\ y_2 \\ \cdot \\ y_{n-1} \end{bmatrix}. \tag{8}$$

There stands the Fourier matrix F.

For $n = 4$ the number w was i, the fourth root of unity. In general w is $e^{2\pi i/n}$, the primitive nth root of unity. The entry in row j and column k of F is a power of w:

$$\boxed{F_{jk} = w^{jk}.}$$

The first row has $j = 0$, the first column has $k = 0$, and all their entries are $w^0 = 1$.

To find the c's we have to invert F. In the 4 by 4 case the inverse matrix contained powers of $-i$; in other words F^{-1} was built from $1/i$. That is the general rule, that F^{-1} comes from the complex number w^{-1}. It lies at the angle $-2\pi/n$, where w was at the angle $+2\pi/n$:

3V The n by n inverse matrix is built from the powers of $w^{-1} = 1/w$:

$$F^{-1} = \frac{1}{n} \begin{bmatrix} 1 & 1 & 1 & \cdot & 1 \\ 1 & w^{-1} & w^{-2} & \cdot & w^{-(n-1)} \\ 1 & w^{-2} & w^{-4} & \cdot & \cdot \\ \cdot & \cdot & \cdot & \cdot & \cdot \\ 1 & w^{-(n-1)} & w^{-2(n-1)} & \cdot & w^{-(n-1)^2} \end{bmatrix}. \tag{9}$$

In the cases $n = 2$ and $n = 3$ this means that

$$F = \begin{bmatrix} 1 & 1 \\ 1 & -1 \end{bmatrix} \quad \text{has} \quad F^{-1} = \frac{1}{2}\begin{bmatrix} 1 & 1 \\ 1 & -1 \end{bmatrix}$$

$$F = \begin{bmatrix} 1 & 1 & 1 \\ 1 & e^{2\pi i/3} & e^{4\pi i/3} \\ 1 & e^{4\pi i/3} & e^{8\pi i/3} \end{bmatrix} \quad \text{has} \quad F^{-1} = \frac{1}{3}\begin{bmatrix} 1 & 1 & 1 \\ 1 & e^{-2\pi i/3} & e^{-4\pi i/3} \\ 1 & e^{-4\pi i/3} & e^{-8\pi i/3} \end{bmatrix}$$

We need to confirm that FF^{-1} equals the identity matrix.

On the main diagonal that is clear. Row j of F times column j of F^{-1} is $(1/n)(1 + 1 + \cdots + 1)$, which is 1. The harder part is off the diagonal, to show that row j of F times column k of F^{-1} gives zero:

$$1 \cdot 1 + w^j w^{-k} + w^{2j} w^{-2k} + \cdots + w^{(n-1)j} w^{-(n-1)k} = 0 \quad \text{if} \quad j \neq k. \tag{10}$$

The key is to notice that those terms are the powers of $W = w^j w^{-k}$:

$$\boxed{1 + W + W^2 + \cdots + W^{n-1} = 0.} \tag{11}$$

This number W is still one of the roots of unity: $W^n = w^{nj} w^{-nk}$ is equal to $1^j 1^{-k} = 1$. Since j is different from k, W is different from 1. It is one of the *other* roots, equally spaced around the unit circle. *Those roots all satisfy* $1 + W + \cdots + W^{n-1} = 0$. The argument is exactly as it was in (5) above: Multiplying by W, the left side does not change (since W^n at the right end equals 1 at the left end). The sum is unchanged when multiplied by W, so it must be zero.

Remark   Another proof is immediate from the identity

$$1 - W^n = (1 - W)(1 + W + W^2 + \cdots + W^{n-1}). \tag{12}$$

Since $W^n = 1$, the left side is zero. But W is not 1, so the last factor must be zero. *The columns of F are orthogonal.* Except for taking complex conjugates ($w^{jk} \rightarrow \bar{w}^{jk}$), and dividing by n, the transpose of F is F^{-1}.

The Fast Fourier Transform

Fourier analysis is a beautiful theory, but what makes it important is that it is also so practical. To analyze a waveform into its frequencies is the best way to take it apart. The reverse process brings it back. For physical and mathematical reasons the exponentials are special, and we can pinpoint one central cause: *If you differentiate e^{ikx}, or integrate it, or translate x to $x + h$, the result is still a multiple of e^{ikx}.* Exponentials are exactly suited to differential equations and integral equations and difference equations. Each frequency component goes its own way, and then they are recombined into the solution. Therefore the analysis and synthesis of signals—the computation of c from y and y from c—is an absolutely central part of scientific computing.

We want to show that it can be done quickly. The key is in the relation of F_4 to F_2—or rather to *two copies* of F_2, which go into a matrix F_2^*:

$$F_4 = \begin{bmatrix} 1 & 1 & 1 & 1 \\ 1 & i & i^2 & i^3 \\ 1 & i^2 & i^4 & i^6 \\ 1 & i^3 & i^6 & i^9 \end{bmatrix} \quad \text{is close to} \quad F_2^* = \begin{bmatrix} 1 & 1 & & \\ 1 & -1 & & \\ & & 1 & 1 \\ & & 1 & -1 \end{bmatrix}.$$

F_4 contains the powers of $w_4 = i$, the *fourth root* of 1. F_2^* contains the powers of $w_2 = -1$, the *square root* of 1. Note especially that half the entries in F_2^* are zero. The 2 by 2 transform, even when done twice, requires only half as much work as a direct 4 by 4 transform. Similarly, if a 64 by 64 transform could be replaced by two 32 by 32 transforms, the work would be cut in half (plus the cost of reassembling the results). What makes this true, and possible in practice, is the simple connection between w_{64} and w_{32}:

$$(w_{64})^2 = w_{32}, \quad \text{or} \quad (e^{2\pi i/64})^2 = e^{2\pi i/32}.$$

The 32nd root is twice as far around the circle as the 64th root. If $w^{64} = 1$, then $(w^2)^{32} = 1$. In general the mth root is the square of the nth root, if m is half of n:

$$\boxed{w_n^2 = w_m \quad \text{if} \quad m = \tfrac{1}{2}n.} \tag{13}$$

The speed of the FFT, in the standard form presented here, depends on working with highly composite numbers like $2^{12} = 4096$. There will be n^2 entries in the Fourier matrix, so without the fast transform it takes $(4096)^2 = 2^{24}$ multiplications to produce F times x. Repeated multiplications by F become expensive. By contrast, a fast transform can do each multiplication in only $6 \cdot 2^{12}$ steps. *It is more than* 680 *times faster*, because it replaces one factor of 4096 by 6. In general it replaces n^2 multiplications by $\tfrac{1}{2}nl$, when n is 2^l. By connecting the matrix F_n to two copies of $F_{n/2}$, and then to four copies of $F_{n/4}$, and eventually to n copies of F_1 (which is trivial), the usual n^2 steps are reduced to $\tfrac{1}{2}n \log_2 n$.

We need to see how $y = F_n c$ (a vector with n components) can be recovered from two vectors that are only half as long. The first step is to divide up c itself. The vector $(c_0, c_1, \ldots, c_{n-1})$ is split into two shorter pieces, by separating its even-numbered components from its odd components:

$$c' = (c_0, c_2, \ldots, c_{n-2}) \quad \text{and} \quad c'' = (c_1, c_3, \ldots, c_{n-1}).$$

The coefficients just go alternately into c' and c''. Then from those vectors we form $y' = F_m c'$ and $y'' = F_m c''$. *Those are the two multiplications by the smaller matrix* F_m (remember that $m = \tfrac{1}{2}n$). As in the replacement of F_4 by F_2^*, the work has been cut in half. The central problem is to recover y from the half-size vectors y' and y'', and Cooley and Tukey noticed how it could be done:

3W The first m and the last m components of the vector $y = F_n c$ are

$$y_j = y_j' + w_n^j y_j'', \quad j = 0, \ldots, m-1$$
$$y_{j+m} = y_j' - w_n^j y_j'', \quad j = 0, \ldots, m-1. \tag{14}$$

Thus the three steps are: split c into c' and c'', transform them by F_m into y' and y'', and reconstruct y from equation (14).

We verify in a moment that this gives the correct y. (You may prefer the flow graph to the algebra.) It means that the calculation of F_n needs only twice as many steps as F_m, plus the m extra multiplications by the m numbers w_n^j in (14). Furthermore, this idea can be repeated. We go from F_{1024} to F_{512} to F_{256}. The final count is $\frac{1}{2}nl$, when starting with the power $n = 2^l$ and going all the way to $n = 1$—where no multiplication is needed. This number $\frac{1}{2}nl$ satisfies the rule given above: *twice the count for m, plus m extra multiplications, produces the count for n:*

$$2(\tfrac{1}{2}m(l - 1)) + m = \tfrac{1}{2}nl.$$

The cost is only slightly more than linear, and the logarithm l reflects the multiplications $w_n^j y_j''$ in (14). However $\frac{1}{2}nl$ is so far below n^2 that discrete Fourier analysis has been completely transformed by the FFT.

Verification of formula (14) *for y:* Separate each component of $y = Fc$ into a part from c' (the even components c_{2k}) and a part from c'' (the odd components c_{2k+1}). That splits the formula for y_j into two parts:

$$y_j = \sum_{k=0}^{n-1} w_n^{jk} c_k \quad \text{is identical to} \quad \sum_{k=0}^{m-1} w_n^{2kj} c_{2k} + \sum_{k=0}^{m-1} w_n^{(2k+1)j} c_{2k+1}.$$

Each sum on the right has $m = \frac{1}{2}n$ terms. Since w_n^2 is w_m, the two sums are

$$y_j = \sum_{k=0}^{m-1} w_m^{kj} c_k' + w_n^j \sum_{k=0}^{m-1} w_m^{kj} c_k''. \tag{15}$$

These sums are exactly y_j' and y_j'', coming from the half-size transforms $F_m c'$ and $F_m c''$. The first part of (14) is the same as (15). For the second part we need $j + m$ in place of j, and this produces a sign change:

inside the sums, $w_m^{k(j+m)}$ remains w_m^{kj} since $w_m^{km} = 1^k = 1$

outside, $w_n^{j+m} = -w_n^j$ because $w_n^m = e^{2\pi i m/n} = e^{\pi i} = -1.$

The sign change yields the second part of (14), and the FFT formula is verified. The idea is easily modified to allow other prime factors of n (not only powers of 2). If n itself is a prime, a completely different algorithm is necessary.

EXAMPLE The steps from $n = 4$ to $m = 2$ are

$$\begin{bmatrix} c_0 \\ c_1 \\ c_2 \\ c_3 \end{bmatrix} \rightarrow \begin{bmatrix} c_0 \\ c_2 \\ c_1 \\ c_3 \end{bmatrix} \rightarrow \begin{bmatrix} F_2 c' \\ \\ F_2 c'' \end{bmatrix} \rightarrow \begin{bmatrix} y \end{bmatrix}$$

Combined, the three steps multiply c by F_4 to give y. Since each step is linear it must come from a matrix, and the product of those matrices must be F_4:

$$\begin{bmatrix} 1 & 1 & 1 & 1 \\ 1 & i & i^2 & i^3 \\ 1 & i^2 & i^4 & i^6 \\ 1 & i^3 & i^6 & i^9 \end{bmatrix} = \begin{bmatrix} 1 & & 1 & \\ & 1 & & i \\ 1 & & -1 & \\ & 1 & & -i \end{bmatrix} \begin{bmatrix} 1 & 1 & & \\ 1 & -1 & & \\ & & 1 & 1 \\ & & 1 & -1 \end{bmatrix} \begin{bmatrix} 1 & & & \\ & & 1 & \\ & 1 & & \\ & & & 1 \end{bmatrix}. \tag{16}$$

You recognize the two copies of F_2 in the center. At the right is the permutation matrix that separates c into c' and c''. At the left is the matrix that multiplies by

w_n^j. If we started with F_8 the middle matrix would contain two copies of F_4. **Each of those would be split as above.** Thus the FFT amounts to a giant factorization of the Fourier matrix! The single matrix F with n^2 nonzeros is a product of approximately $l = \log_2 n$ matrices, with a total of only nl nonzeros.

The Complete FFT and the Butterfly

The first step of the FFT changes multiplication by F_n to two multiplications by $F_m = F_{n/2}$. The even-numbered components $(c_0, c_2, \ldots)$ are transformed separately from $(c_1, c_3, \ldots)$. We give a flow graph for $n = 8$:

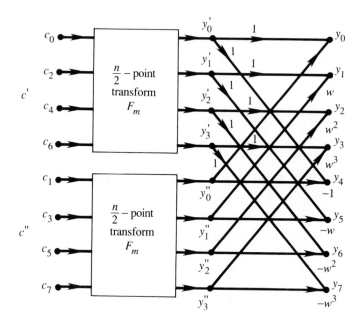

The key idea is **to replace each F_4 box by a similar picture involving two F_2 boxes**. The new factor $w_4 = i$ is the square of the old factor $w = w_8 = e^{2\pi i/8} = (1 + i)/\sqrt{2}$. The top half of the graph changes from F_4 to

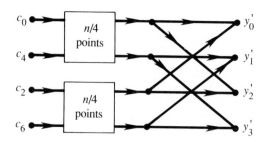

Then each of those boxes for $F_2 = \begin{bmatrix} 1 & 1 \\ 1 & -1 \end{bmatrix}$ is a single butterfly:

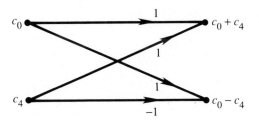

By combining the graphs you can see the whole picture. It shows the order that the nc's enter the FFT and the $\log n$ stages that take them through it—and it also shows the simplicity of the logic:

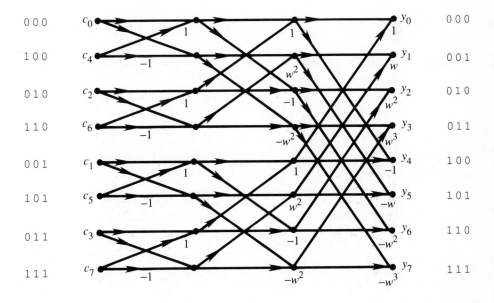

Every stage needs $\frac{1}{2}n$ multiplications so the final count is $\frac{1}{2}n \log n$. There is an amazing rule for the permutation of c's before entering the FFT: Write the subscripts $0, \ldots, 7$ in binary and *reverse the order of their bits*. The subscripts appear in "bit-reversed order" on the left side of the graph. Even numbers come before odd (numbers ending in 0 come before numbers ending in 1) and this is repeated at every stage.

A code is available at no cost through the electronic mail service *netlib* (instructions in Appendix C).

EXERCISES

3.5.1   What are F^2 and F^4 for the 4 by 4 Fourier matrix F?

3.5.2   What are the real and imaginary parts of the three cube roots of unity?

3.5.3   If you form a 3 by 3 submatrix of the 6 by 6 matrix F_6, keeping only the entries in its first, third, and fifth rows and columns, what is that submatrix?

3.5.4   Mark all the sixth roots of 1 in the complex plane. What is the primitive root w_6 (find its real and imaginary part)? Which power of w_6 is equal to $1/w_6$? What is $1 + w + w^2 + w^3 + w^4 + w^5$?

3.5.5   Find all solutions to the equation $e^{ix} = -1$, and all solutions to $e^{i\theta} = i$.

3.5.6   What are the square and the square root of w_{128}, the primitive 128th root of 1?

3.5.7   Solve the 4 by 4 system (6) if the right sides are $y_0 = 2$, $y_1 = 0$, $y_2 = 2$, $y_3 = 0$. In other words, solve $F_4 c = y$.

3.5.8   Solve the same system with $y = (2, 0, -2, 0)$, by knowing F_4^{-1} and computing $c = F_4^{-1} y$. Verify that $c_0 + c_1 e^{ix} + c_2 e^{2ix} + c_3 e^{3ix}$ takes the values 2, 0, -2, 0 at the points $x = 0$, $\pi/2$, π, $3\pi/2$.

3.5.9   (a)   If $y = (1, 1, 1, 1)$ show that $c = (1, 0, 0, 0)$ satisfies $F_4 c = y$.
              (b)   Now suppose $y = (1, 0, 0, 0)$ and find c.

3.5.10   For $n = 2$ write down y_0 from the first line of (14) and y_1 from the second line. For $n = 4$ use the first line to find y_0 and y_1, and the second to find y_2 and y_3, all in terms of y' and y''.

3.5.11   Compute $y = F_4 c$ by the three steps of the Fast Fourier Transform if $c = (1, 0, 1, 0)$.

3.5.12   Compute $y = F_8 c$ by the three steps of the Fast Fourier Transform if $c = (1, 0, 1, 0, 1, 0, 1, 0)$. Repeat the computation with $c = (0, 1, 0, 1, 0, 1, 0, 1)$.

3.5.13   For the 4 by 4 matrix write out the formulas for c_0, c_1, c_2, c_3 and verify that *if f is odd then c is odd*. The vector f is odd if $f_{n-j} = -f_j$; for $n = 4$ that means $f_0 = 0$, $f_3 = -f_1$, $f_2 = 0$. This is copied by c and by $(\sin 0, \sin \pi/2, \sin \pi, \sin 3\pi/2)$, and it leads to a fast sine transform.

The "first half" of linear algebra is now concluded. It concentrated on $Ax = b$ and it took more than its half of the semester (at least in the author's class). It started with square matrices, pushed ahead to rectangular matrices, and went back to the square matrix $A^{\mathrm{T}}A$:

Chapter 1. *Solution of $Ax = b$ for invertible A* (by elimination)
Chapter 2. *Solution of $Ax = b$ for any A* (by the four subspaces)
Chapter 3. *Least squares solution* (by projection and $A^{\mathrm{T}}A\bar{x} = A^{\mathrm{T}}b$).

I am unhappy about boiling down this subject to such a small list! You recognize that there is a lot of mathematics in those three lines—and more to be said about each of them. In looking back at what was done, we have the chance to make new connections that are clearer with hindsight. They are optional and they vary in importance—but subspaces are absolutely fundamental, and the ideas behind them will come through when we take their sum and intersection.

The review starts as the book did, with computations. The key point is that every solution came from a factorization of the matrix:

$$\text{Chapters 1 and 2:} \quad A = LU \qquad \text{Chapter 3:} \quad A = QR$$

Whether square or rectangular, $Ax = b$ is reduced to two easy steps:

$$\text{first} \quad Lc = b \quad \text{and then} \quad Ux = c: \quad x = U^{-1}L^{-1}b$$
$$\text{first} \quad Q\bar{c} = b \quad \text{and then} \quad R\bar{x} = \bar{c}: \quad \bar{x} = R^{-1}Q^{\mathrm{T}}b$$

Three of those matrices—L, U, and R—are triangular. The odd one is Q. It has m rows and only n columns, with the special feature that the columns are orthonormal. Therefore $Q^{\mathrm{T}}Q$ is the identity matrix. The least squares solution to $Qc = b$ is $\bar{c} = Q^{\mathrm{T}}b$. Geometrically $\bar{c}$ comes from one-dimensional projections onto each separate column. The final $\bar{x}$ is the same as the solution to $A^{\mathrm{T}}A\bar{x} = A^{\mathrm{T}}b$.

Thus the system $Ax = b$ leads us (almost involuntarily) to LU and QR. The factorizations come out of the steps of the solution. We have also been close to three other factorizations, and slipped by them in order to keep the text clear. The first was mentioned earlier and is important in practice; it is the special form of $A = LU$, or rather $A = LDU$, when A is symmetric.

Chapter 1: *Cholesky factorization* for symmetric A:

$$A = LDL^{\mathrm{T}} = (LD^{1/2})(LD^{1/2})^{\mathrm{T}}$$

Chapter 2: *Reduced factorization*—m by r times r by n, for any A: $A = \underline{L}\underline{U}$
Chapter 3: *Singular value decomposition* for any A:

$$A = Q_1 \Sigma Q_2^{\mathrm{T}}$$

We take those one at a time.
1. When A is symmetric, U is the same as L^{T}. In that way the LDU factorization is $A = LDL^{\mathrm{T}}$, and the right side is perfectly symmetric. Cholesky took it one step further, and split the diagonal D by taking square roots of the pivots. (He assumed

they were positive; that is the "positive definite" case of Chapter 6.) Then A is down to two triangular factors $(LD^{1/2})(LD^{1/2})^T$.

Here is a curious fact: If Cholesky had started with $A^T A$ (which has positive pivots), his two factors would have been exactly the R^T and R of Gram-Schmidt. In other words

$$A^T A = R^T R \qquad (\text{because} \quad A^T A = R^T Q^T Q R \quad \text{and} \quad Q^T Q = I).$$

3X The multipliers l_{ij} from Gaussian elimination on $A^T A$ are identical to the multipliers that orthogonalize the columns of A.

EXAMPLE (our favorite matrix for Gram-Schmidt)

$$A = \begin{bmatrix} 1 & 0 & 0 \\ -1 & 1 & 0 \\ 0 & -1 & 1 \\ 0 & 0 & -1 \end{bmatrix} \quad \text{and} \quad A^T A = \begin{bmatrix} 2 & -1 & 0 \\ -1 & 2 & -1 \\ 0 & -1 & 2 \end{bmatrix}.$$

Look first at $A^T A$. Elimination adds $\frac{1}{2}$ the first row to the second row, and then $\frac{2}{3}$ of the second row to the third row. The pivots are $2, \frac{3}{2}, \frac{4}{3}$. What does Gram-Schmidt do to A?

If the shaded statement is correct, $\frac{1}{2}$ the first column of A is added to the second column. The result is $a_2' = (\frac{1}{2}, \frac{1}{2}, -1, 0)$, and *it is orthogonal to the first column.* Then add $\frac{2}{3}$ of the new second column to the third column. The result is $a_3' = (\frac{1}{3}, \frac{1}{3}, \frac{1}{3}, -1)$, and *it is orthogonal to the first two columns.* Those are the steps of Gram-Schmidt, and they lead to a matrix with orthogonal columns:

$$A' = \begin{bmatrix} 1 & \frac{1}{2} & \frac{1}{3} \\ -1 & \frac{1}{2} & \frac{1}{3} \\ 0 & -1 & \frac{1}{3} \\ 0 & 0 & -1 \end{bmatrix}.$$

In a larger example the next column would begin with $\frac{1}{4}, \frac{1}{4}, \frac{1}{4}, \frac{1}{4}, -1$. That is a pattern I had never seen before, with orthogonal columns that look good.

It remains to divide the columns by their lengths. What is special is that *the lengths squared are identical to the pivots of $A^T A$*:

$$1^2 + (-1)^2 = 2, \qquad (\tfrac{1}{2})^2 + (\tfrac{1}{2})^2 + (-1)^2 = \tfrac{3}{2}, \qquad (\tfrac{1}{3})^2 + (\tfrac{1}{3})^2 + (\tfrac{1}{3})^2 + (-1)^2 = \tfrac{4}{3}.$$

This cannot be an accident! The columns of $A' = A(L^T)^{-1}$ are orthogonal, with squared lengths appearing in D:

$$\text{if} \quad A^T A = LDL^T \quad \text{then} \quad (A')^T A' = L^{-1} A^T A (L^T)^{-1} = D.$$

The final step divides by the square roots of those pivots, to normalize columns:

$$Q = A(L^T)^{-1} D^{-1/2} \qquad \text{or} \qquad A = Q D^{1/2} L^T = QR.$$

That brings us back to Cholesky's $D^{1/2} L^T$, identical to the Gram-Schmidt R.

The triangularity of R reflects the order: Multiples of earlier columns are subtracted from later columns. That is like elimination: Multiples of earlier rows are subtracted from later rows. L records the elimination steps on $A^T A$, and R records the orthogonalization steps on A.

2. The "reduced factorization" is $A = \underline{L}\,\underline{U}$. Like the symmetric factorization LDL^T, it is more balanced than the old $A = LU$, but in a different way. When A has rank r, we really need only r columns in L and r rows in U. The last $m - r$ rows of U can be thrown away (since they were all zero). The last $n - r$ columns of L can also be thrown away (since they multiplied those zero rows of U, and had no effect). We are left with $\underline{L}$ and $\underline{U}$, and their product is still the same matrix A:

A matrix of rank r can be factored into an m by r matrix times an r by n matrix.

In case A needed row exchanges, so that $PA = LU$, there is a slight change: $\underline{L}$ is formed from the first r columns of $P^{-1}L$ instead of L. In every case the r columns of $\underline{L}$ are a basis for the column space of A, and the r rows of $\underline{U}$ are a basis for the row space of A.

3. The *singular value decomposition* is far too important to be summarized in a paragraph, and it is the subject of Appendix A. It puts a diagonal matrix Σ between two orthogonal matrices. They give orthogonal bases for the four subspaces, and they lead to the pseudoinverse A^+ and the "condition number." The SVD has become fundamental in scientific computing.

Vector Spaces and Subspaces

We turn from computations to algebra. That began with the idea of a *vector space*, in which two operations are possible:

vector addition $x + y$ and scalar multiplication cx.

Those operations are easy in the n-dimensional space $\mathbf{R}^n$—which is the outstanding example of a vector space. They are also possible within smaller sets, like lines and planes through the origin in $\mathbf{R}^3$. Subsets which are vector spaces on their own are *subspaces*, and the four fundamental subspaces are the key to $Ax = b$:

existence of x : b must be in the column space $\mathscr{R}(A)$
               : b must be perpendicular to the left nullspace $\mathscr{N}(A^T)$
uniqueness of x : the nullspace $\mathscr{N}(A)$ must contain only the zero vector
               : the row space $\mathscr{R}(A^T)$ must be all of $\mathbf{R}^n$

When there is a solution for every b, the rows are independent and the rank is $r = m$. When the solution is unique, the columns are independent and the rank is $r = n$. For every matrix, the rank is the dimension of both the column space and the row space. If $r = m = n$, then A is square and invertible.

It is fair to say that these chapters pursued one fixed goal, the understanding of $Ax = b$. Each new idea and definition—including vector spaces and linear independence, basis and dimension, rank and nullspace, inner product and orthogonality—was introduced as it was needed for this one purpose. Now we look again at those same ideas, to find some of the relationships that were missed.

1. The intersection of two vector spaces   The key idea goes back to the definition of a vector space and a subspace. New questions arise from considering not just a single subspace or a single matrix A, but the interconnections between two subspaces or two matrices. The first point is the most important:

> **3Y**   If V and W are both subspaces of a given vector space, then so is their *intersection* $V \cap W$. The vectors belonging to both spaces form another subspace.

The proof is immediate. Suppose x and y belong to $V \cap W$, in other words they are vectors *in V and also in W*. Then, because V and W are vector spaces in their own right, $x + y$ and cx are in V and in W. *The results of addition and scalar multiplication stay within the intersection.* Geometrically, the intersection of two planes through the origin (or "hyperplanes" in $\mathbf{R}^n$) is again a subspace. The same will be true of the intersection of several subspaces, or even of infinitely many.

EXAMPLE 1   The intersection of two orthogonal subspaces V and W is the one-point subspace $\{0\}$. Only the zero vector is orthogonal to itself.

EXAMPLE 2   If the sets of n by n upper and lower triangular matrices are the subspaces V and W, their intersection is the set of diagonal matrices. This is certainly a subspace. Adding two diagonal matrices, or multiplying by a scalar, leaves us with a diagonal matrix.

EXAMPLE 3   Suppose V is the nullspace of A and W is the nullspace of B. Then $V \cap W$ is the smaller nullspace of the larger matrix

$$C = \begin{bmatrix} A \\ B \end{bmatrix}.$$

$Cx = 0$ requires both $Ax = 0$ and $Bx = 0$, so x has to be in both nullspaces.

2. The sum of two vector spaces   Usually, after discussing and illustrating the intersection of two sets, it is natural to look at their union. With vector spaces, however, it is not natural. *The union $V \cup W$ of two subspaces will not in general be a subspace.* Consider the x axis and the y axis in the plane. Each axis by itself is a subspace, but taken together they are not. The sum of $(1, 0)$ and $(0, 1)$ is not on either axis. This will always happen unless one of the subspaces is contained in the other; only then is their union (which coincides with the larger one) again a subspace.

Nevertheless, we do want to combine two subspaces, and therefore in place of their union we turn to their sum.

> **DEFINITION** If V and W are both subspaces of a given space, then so is their *sum* $V + W$. It is made up of all possible combinations $x = v + w$, where v is an arbitrary vector in V and w is an arbitrary vector in W.

This is nothing but the space spanned by $V \cup W$. It is the smallest vector space that contains both V and W. The sum of the x axis and the y axis is the whole x-y plane; so is the sum of any two different lines, *perpendicular or not*. If V is the x axis and W is the $45°$ line $x = y$, then any vector like $(5, 3)$ can be split into $v + w = (2, 0) + (3, 3)$. Thus $V + W$ is all of $\mathbf{R}^2$.

EXAMPLE 4 Suppose V and W are orthogonal complements of one another in $\mathbf{R}^n$. Then their sum is $V + W = \mathbf{R}^n$. Every x is the sum of its projection v in V and its projection w in W.

EXAMPLE 5 If V is the space of upper triangular matrices, and W is the space of lower triangular matrices, then $V + W$ is the space of all matrices. Every matrix can be written as the sum of an upper and a lower triangular matrix—in many ways, because the diagonals are not uniquely determined.

EXAMPLE 6 If V is the column space of a matrix A, and W is the column space of B, then $V + W$ is the column space of the larger matrix $D = [A \quad B]$. The dimension of $V + W$ may be less than the combined dimensions of V and W (because the two spaces may overlap), but it is easy to find:

$$\dim(V + W) = \text{rank of } D. \tag{1}$$

Surprisingly, *the computation of $V \cap W$ is much more subtle.* Suppose we are given the two bases $v_1, \ldots, v_k$ and $w_1, \ldots, w_l$; this time we want a basis for the intersection of the two subspaces. Certainly it is not enough just to check whether any of the v's equal any of the w's. The two spaces could even be identical, $V = W$, and still the bases might be completely different.

The most efficient method is this. Form the same matrix D whose columns are $v_1, \ldots, v_k, w_1, \ldots, w_l$, and compute its nullspace $\mathcal{N}(D)$. We shall show that a basis for this nullspace leads to a basis for $V \cap W$, and that *the two spaces have the same dimension.* The dimension of the nullspace is called the "nullity," so

$$\dim(V \cap W) = \text{nullity of } D. \tag{2}$$

This leads to a formula which is important in its own right. Adding (1) and (2),

$$\dim(V + W) + \dim(V \cap W) = \text{rank of } D + \text{nullity of } D.$$

From our computations with the four fundamental subspaces, we know that the rank plus the nullity equals the number of columns. In this case D has $k + l$ columns, and since $k = \dim V$ and $l = \dim W$, we are led to the following conclusion:

$$\dim(V + W) + \dim(V \cap W) = \dim V + \dim W. \tag{3}$$

Not a bad formula.

EXAMPLE 7   The spaces V and W of upper and lower triangular matrices both have dimension $n(n + 1)/2$. The space $V + W$ of all matrices has dimension n^2, and the space $V \cap W$ of diagonal matrices has dimension n. As predicted by (3), $n^2 + n = n(n + 1)/2 + n(n + 1)/2$.

We now look at the proof of (3). For once in this book, the interest is less in the actual computation than in the technique of proof. It is the only time we will use the trick of understanding one space by matching it with another. Note first that the nullspace of D is a subspace of $\mathbf{R}^{k+l}$, whereas $V \cap W$ is a subspace of $\mathbf{R}^m$. We have to prove that these two spaces have the same dimension. The trick is to show that these two subspaces are perfectly matched by the following correspondence.

Given any vector x in the nullspace of D, write the equation $Dx = 0$ in terms of the columns as follows:

$$x_1 v_1 + \cdots + x_k v_k + x_{k+1} w_1 + \cdots + x_{k+l} w_l = 0, \tag{4}$$

or

$$x_1 v_1 + \cdots + x_k v_k = -x_{k+1} w_1 - \cdots - x_{k+l} w_l. \tag{5}$$

The left side of this last equation is in V, being a combination of the v_k, and the right side is in W. Since the two are equal, they represent a vector y in $V \cap W$. This provides the correspondence between the vector x in $\mathcal{N}(D)$ and the vector y in $V \cap W$. It is easy to check that the correspondence preserves addition and scalar multiplication: If x corresponds to y and x' to y', then $x + y$ corresponds to $x' + y'$ and cx corresponds to cx'. Furthermore, every y in $V \cap W$ comes from one and only one x in $\mathcal{N}(D)$ (Exercise 3.6.18).

This is a perfect illustration of an *isomorphism* between two vector spaces. The spaces are different, but *for all algebraic purposes they are exactly the same*. They match completely: Linearly independent sets correspond to linearly independent sets, and a basis in one corresponds to a basis in the other. So their dimensions are equal, which completes the proof of (2) and (3). This is the kind of result an algebraist is after, to identify two different mathematical objects as being fundamentally the same.† It is a fact that any two spaces with the same scalars and the same (finite) dimension are always isomorphic, but this is too general to be very

---

† Another isomorphism is between the row space and column space, both of dimension r.

exciting. The interest comes in matching two superficially dissimilar spaces, like $\mathcal{N}(D)$ and $V \cap W$.

EXAMPLE 8 V is the x-y plane and W is the x-z plane:

$$D = \begin{bmatrix} 1 & 0 & 1 & 0 \\ 0 & 1 & 0 & 0 \\ 0 & 0 & 0 & 1 \end{bmatrix} \quad \begin{array}{l} \text{first 2 columns:} \quad \text{basis for } V \\ \text{last 2 columns:} \quad \text{basis for } W \end{array}$$

The rank of D is 3, and $V + W$ is all of $\mathbf{R}^3$. The nullspace contains $x = (1, 0, -1, 0)$, and has dimension 1. The corresponding vector y is 1(column 1) + 0(column 2), pointing along the x-axis—which is the intersection $V \cap W$. Formula (3) for the dimensions of $V + W$ and $V \cap W$ becomes $3 + 1 = 2 + 2$.

The Fundamental Spaces for Products AB

We turn from pairs of subspaces to products of matrices. As always, it is not the individual entries of AB that are particularly interesting; they probably have no similarity to the entries of A and B. Instead, it is at the level of vectors—the rows or columns of a matrix, rather than its entries—that properties of A and B may be inherited by AB. And it is not even so much the individual rows or columns, as the subspaces they span; these subspaces describe the whole matrix at once.

Our basic question is this: What are the relationships between the four fundamental subspaces associated with A, the four associated with B, and the four associated with the product AB? All these matrices may be rectangular, and there are four principal relationships:

   (i)   *The nullspace of* AB *contains the nullspace of B*
   (ii) *The column space of* AB *is contained in the column space of A*
   (iii) *The left nullspace of* AB *contains the left nullspace of A*
   (iv) *The row space of* AB *is contained in the row space of B.*

The proofs are extremely simple.

   (i)   If $Bx = 0$, then $ABx = 0$. Every x in the nullspace of B is also in $\mathcal{N}(AB)$.
   (ii) Each column of AB is a combination of the columns of A.
   (iii) If $y^T A = 0$ then $y^T AB = 0$.
   (iv) Each row of AB is a combination of the rows of B.

By knowing this much about the subspaces associated with AB, we also know bounds on the dimensions of three of them:

COROLLARY   The rank and nullity of AB satisfy

$$r(AB) \le r(A) \qquad r(AB) \le r(B) \qquad \dim \mathcal{N}(AB) \ge \dim \mathcal{N}(B).$$

There is no attempt to prove that AB has a larger nullspace than A, which cannot be guaranteed. In fact A and B could be 2 by 3 and 3 by 2 matrices of zeros—giving AB the smaller nullspace $\mathbf{R}^2$ when A has the larger nullspace $\mathbf{R}^3$.

Finally, recall that a *submatrix* C is formed by striking out some (or none) of the rows of A, and some (or none) of its columns. It is not hard to guess a limitation on the rank of C.

3Z   Suppose A is an m by n matrix of rank r. Then:

(i)   Every submatrix C is of rank $\leq r$.

(ii)   At least one r by r submatrix is of rank exactly r.

Proof   We shall reduce A to C in two stages. The first keeps the number of columns intact, and removes only the rows that are not wanted in C. The row space of this intermediate matrix B is obviously contained in the row space of A, so that $\text{rank}(B) \leq \text{rank}(A) = r$. At the second stage B is reduced to C by excluding the unwanted columns. Therefore the column space of C is contained in the column space of B, and $\text{rank}(C) \leq \text{rank}(B) \leq r$. This establishes (i).

To prove (ii), suppose that B is formed from r independent rows of A. Then the row space of B is of dimension r; $\text{rank}(B) = r$, and the column space of B must also have dimension r. Suppose next that C is formed from r independent columns of B. Then the column space of C has dimension r, and $\text{rank}(C) = r$. This completes the proof of (ii): Every matrix of rank r contains a nonsingular r by r submatrix.

EXAMPLE   Consider once more that 3 by 4 matrix

$$A = \begin{bmatrix} 1 & 3 & 3 & 2 \\ 2 & 6 & 9 & 5 \\ -1 & -3 & 3 & 0 \end{bmatrix} \quad \text{with submatrix} \quad C = \begin{bmatrix} 1 & 3 \\ 2 & 9 \end{bmatrix}.$$

Every 3 by 3 submatrix of A is singular, but C is not. Therefore the rank is 2.

This theorem does not deserve to be overemphasized. Superficially, it resembles a theorem that *is* important—the one next to Fig. 3.4. There we proved that every A is an invertible transformation from its r-dimensional row space to its r-dimensional column space. Those spaces, and that transformation, give total information about A; the whole matrix can be reassembled once the transformation is known. Here it is only a question of finding an invertible submatrix C, and there is nothing special about the one that is chosen. There may be, and in the example there are, many other invertible submatrices of order r. The only thing we do get is a new and equivalent definition of **rank**: *It is the order of the largest nonsingular submatrix.*

Weighted Least Squares

Suppose we return for a moment to the simplest kind of least squares problem, the estimate $\bar{x}$ of a patient's weight on the basis of two observations $x = b_1$ and $x = b_2$. Unless these measurements are identical, we are faced with an inconsistent system of two equations in one unknown:

$$\begin{bmatrix} 1 \\ 1 \end{bmatrix} [x] = \begin{bmatrix} b_1 \\ b_2 \end{bmatrix}$$

Up to now, we have regarded the two observations as equally reliable, and looked for the value $\bar{x}$ that minimized $E^2 = (x - b_1)^2 + (x - b_2)^2$:

$$\frac{dE^2}{dx} = 0 \quad \text{at} \quad \bar{x} = \frac{b_1 + b_2}{2}.$$

The optimal $\bar{x}$ is the average of the measurements, and the same conclusion comes from the equation $A^T A \bar{x} = A^T b$. In fact $A^T A$ is a 1 by 1 matrix, and the equation is $2\bar{x} = b_1 + b_2$.

As further review, note the projection of b onto the line through $a = \begin{bmatrix} 1 \\ 1 \end{bmatrix}$. The projection is $p = \bar{x}a$, with $\bar{x} = a^T b / a^T a$—and that ratio is again the average $\frac{1}{2}(b_1 + b_2)$. This is the natural answer, except for the following variation which is important in practice.

Suppose that the two observations are not trusted to the same degree. The value $x = b_1$ may be obtained from a more accurate scale—or, in a statistical problem, from a larger sample—than the value $x = b_2$. Nevertheless, if the second observation contains some information, we are not willing to rely totally on $x = b_1$. The simplest compromise is to attach different weights w_i^2 to the two observations, and to choose the $\bar{x}$ that minimizes the *weighted sum of squares*

$$E^2 = w_1^2 (x - b_1)^2 + w_2^2 (x - b_2)^2.$$

If $w_1 > w_2$, then more importance is attached to the first observation, and the minimizing process tries harder to make $(x - b_1)^2$ small. We can easily compute

$$\frac{dE^2}{dx} = 2[w_1^2 (x - b_1) + w_2^2 (x - b_2)],$$

and setting this to zero gives the new solution $\bar{x}_W$:

$$\boxed{\bar{x}_W = \frac{w_1^2 b_1 + w_2^2 b_2}{w_1^2 + w_2^2}} \tag{6}$$

Instead of the average of b_1 and b_2, as we had when $w_1 = w_2 = 1$, $\bar{x}_W$ is a *weighted average* of the data. This average is closer to b_1 than to b_2.

It is easy to find the ordinary least squares problem that leads to $\bar{x}_W$. It comes from changing $Ax = b$ to the new system $WAx = Wb$. **This changes the solution** *from $\bar{x}$ to $\bar{x}_W$.* In our example

$$A = \begin{bmatrix} 1 \\ 1 \end{bmatrix} \quad \text{and} \quad W = \begin{bmatrix} w_1 & 0 \\ 0 & w_2 \end{bmatrix}.$$

The normal equations for the new problem have WA in place of A and Wb in place of b. Otherwise nothing is new, and the matrix W^TW turns up on both sides of the weighted normal equations:

> **The least squares solution to $WAx = Wb$ is determined from**
>
> $$(A^TW^TWA)\bar{x}_W = A^TW^TWb.$$

What happens to the geometric picture, in which b was projected to $A\bar{x}$? That must also change. The projection is again the point in the column space that is closest to b. But the word "closest" has a new meaning when the length involves W. We are working with a weighted length of x, equal to the ordinary length of Wx. Perpendicularity no longer means $y^Tx = 0$; in the new system the test is $(Wy)^T(Wx) = 0$. The matrix W^TW appears in the middle. In this new sense, the projection $A\bar{x}_W$ and the error $b - A\bar{x}_W$ are again perpendicular.

That last paragraph describes *all inner products*: They come from invertible matrices W. They actually involve only the symmetric combination $C = W^TW$; the inner product of x and y is y^TCx. Note that for an orthogonal matrix $W = Q$, when this combination is $C = Q^TQ = I$, the inner product is not new or different. An orthogonal weighting matrix has no effect, since rotating the space leaves the inner product unchanged. Every other W changes the length and inner product.

For any invertible matrix W, the following rules define a new inner product and length:

$$(x, y)_W = (Wy)^T(Wx) \quad \text{and} \quad \|x\|_W = \|Wx\|. \tag{7}$$

Since W is invertible, no vector is assigned length zero (except the zero vector). All possible inner products—which depend linearly on x and y and are positive when $x = y \neq 0$—are found in this way, from some matrix W.

In practice, the important question is the choice of W (or C). The best answer comes from statisticians, and originally from Gauss. We may know that the average error is zero. That is the "expected value" of the error in b—although the error is not really expected to be zero! We may also know the average of the *square* of the error; that is the *variance*. If the errors in the measurements b_i are independent of each other, and their variances are σ_i^2, then the right weights are $w_i = 1/\sigma_i$. The more we know about the measurement, which is reflected in a smaller variance, the more heavily it is weighted.

In addition to unequal reliability, *the observations may not be independent.* If the errors are coupled—the polls for President are not independent of those for Senator, and certainly not of those for Vice-President—then W has off-diagonal terms. The best unbiased matrix $C = W^TW$ is the inverse of the *covariance matrix*—

whose i, j entry is the expected value of (error in b_i) times (error in b_j). Its main diagonal contains the variances σ_i^2, which are the average of (error in b_i)^2.

EXAMPLE Suppose two bridge partners both guess (after the bidding) the total number of spades they hold. For each guess, the errors $-1, 0, 1$ might have equal probability $\frac{1}{3}$. Then the expected error is zero and the variance is $\frac{2}{3}$:

$$E(e) = \tfrac{1}{3}(-1) + \tfrac{1}{3}(0) + \tfrac{1}{3}(1) = 0$$
$$E(e^2) = \tfrac{1}{3}(-1)^2 + \tfrac{1}{3}(0)^2 + \tfrac{1}{3}(1)^2 = \tfrac{2}{3}.$$

The two guesses are dependent, because they are based on the same bidding—but not identical, because they are looking at different hands. Say the chance that they are both too high or both too low is zero, but the chance of opposite errors is $\frac{1}{3}$. Then $E(e_1 e_2) = \frac{1}{3}(-1)$, and the inverse of the covariance matrix is

$$\begin{bmatrix} \frac{2}{3} & -\frac{1}{3} \\ -\frac{1}{3} & \frac{2}{3} \end{bmatrix}^{-1} = \begin{bmatrix} 2 & 1 \\ 1 & 2 \end{bmatrix} = C = W^{\mathsf{T}} W.$$

This matrix goes into the middle of the weighted normal equations.

EXERCISES

3.6.1 Suppose S and T are subspaces of $\mathbf{R}^{13}$, with dim $S = 7$ and dim $T = 8$.
(a) What is the largest possible dimension of $S \cap T$?
(b) What is the smallest possible dimension of $S \cap T$?
(c) What is the smallest possible dimension of $S + T$?
(d) What is the largest possible dimension of $S + T$?

3.6.2 What are the intersections of the following pairs of subspaces?
(a) The x-y plane and the y-z plane in $\mathbf{R}^3$.
(b) The line through $(1, 1, 1)$ and the plane through $(1, 0, 0)$ and $(0, 1, 1)$.
(c) The zero vector and the whole space $\mathbf{R}^3$.
(d) The plane perpendicular to $(1, 1, 0)$ and the plane perpendicular to $(0, 1, 1)$ in $\mathbf{R}^3$.
What are the *sums* of those pairs of subspaces?

3.6.3 Within the space of all 4 by 4 matrices, let V be the subspace of *tridiagonal* matrices and W the subspace of *upper triangular* matrices. Describe the subspace $V + W$, whose members are the upper Hessenberg matrices, and the subspace $V \cap W$. Verify formula (3).

3.6.4 If $V \cap W$ contains only the zero vector then (3) becomes dim$(V + W) =$ dim $V +$ dim W. Check this when V is the row space of A, W is the nullspace, and A is m by n of rank r. What are the dimensions?

3.6.5 Give an example in $\mathbf{R}^3$ for which $V \cap W$ contains only the zero vector but V is not orthogonal to W.

3.6.6   If $V \cap W = \{0\}$ then $V + W$ is called the *direct sum* of V and W, with the special notation $V \oplus W$. If V is spanned by $(1, 1, 1)$ and $(1, 0, 1)$, choose a subspace W so that $V \oplus W = \mathbf{R}^3$.

3.6.7   Explain why any vector x in the direct sum $V \oplus W$ can be written in one *and only one* way as $x = v + w$ (with v in V and w in W).

3.6.8   Find a basis for the sum $V + W$ of the space V spanned by $v_1 = (1, 1, 0, 0)$, $v_2 = (1, 0, 1, 0)$ and the space W spanned by $w_1 = (0, 1, 0, 1)$, $w_2 = (0, 0, 1, 1)$. Find also the dimension of $V \cap W$ and a basis for it.

3.6.9   Show by example that the nullspace of AB need not contain the nullspace of A, and the column space of AB is not necessarily contained in the column space of B.

3.6.10   Find the largest invertible submatrix and the rank of

$$A_1 = \begin{bmatrix} 1 & 0 & 1 \\ 2 & 0 & 2 \\ 3 & 0 & 4 \end{bmatrix} \quad \text{and} \quad A_2 = \begin{bmatrix} 1 & 1 & 0 & 0 \\ 1 & 1 & 0 & 0 \\ 0 & 0 & 1 & 1 \\ 0 & 0 & 1 & 1 \end{bmatrix}$$

3.6.11   Suppose A is m by n and B is n by m, with $n < m$. Prove that their product AB is singular.

3.6.12   Prove from (3) that $\text{rank}(A + B) \le \text{rank}(A) + \text{rank}(B)$.

3.6.13   If A is square and invertible prove that AB has the same nullspace (and the same row space and the same rank) as B itself. *Hint*: Apply relationship (i) also to the product of A^{-1} and AB.

3.6.14   Factor A into an m by r matrix L times an r by n matrix U:

$$A = \begin{bmatrix} 0 & 1 & 4 & 0 \\ 0 & 2 & 8 & 0 \end{bmatrix} \quad \text{and also} \quad A = \begin{bmatrix} 1 & 0 & 0 \\ 0 & 1 & 0 \\ 0 & 0 & 0 \end{bmatrix}.$$

3.6.15   Multiplying each column of L by the corresponding row of U, and adding, gives the product $A = LU$ as the sum of r matrices of rank one. Construct L and U and the two matrices of rank one that add to

$$A = \begin{bmatrix} 1 & -1 & 0 \\ 0 & 1 & -1 \\ 1 & 0 & -1 \end{bmatrix}.$$

3.6.16   Prove that the intersection of three 6-dimensional subspaces of $\mathbf{R}^8$ is not the single point $\{0\}$. *Hint*: How small can the intersection of the first two subspaces be?

3.6.17   Find the factorization $A = LDL^{\mathrm{T}}$, and then the two Cholesky factors in $(LD^{1/2})(LD^{1/2})^{\mathrm{T}}$, for

$$A = \begin{bmatrix} 4 & 12 \\ 12 & 45 \end{bmatrix}.$$

3.6.18 Verify the statement that "every y in $V \cap W$ comes from one and only one x in $\mathcal{N}(D)$"—by describing, for a given y, how to go back to equation (5) and find x.

3.6.19 What happens to the weighted average $\bar{x}_W = (w_1^2 b_1 + w_2^2 b_2)/(w_1^2 + w_2^2)$ if the first weight w_1 approaches zero? The measurement b_1 is totally unreliable.

3.6.20 From m independent measurements $b_1, \ldots, b_m$ of your pulse rate, weighted by $w_1, \ldots, w_m$, what is the weighted average that replaces (6)? It is the best estimate when the statistical variances are $\sigma_i^2 = 1/w_i^2$.

3.6.21 If $W = \begin{bmatrix} 2 & 0 \\ 0 & 1 \end{bmatrix}$, find the W-inner product of $x = (2, 3)$ and $y = (1, 1)$ and the W-length of x. What line of vectors is W-perpendicular to y?

3.6.22 Find the weighted least squares solution $\bar{x}_W$ to $Ax = b$:

$$A = \begin{bmatrix} 1 & 0 \\ 1 & 1 \\ 1 & 2 \end{bmatrix} \quad b = \begin{bmatrix} 0 \\ 1 \\ 1 \end{bmatrix} \quad W = \begin{bmatrix} 2 & 0 & 0 \\ 0 & 1 & 0 \\ 0 & 0 & 1 \end{bmatrix}.$$

Check that the projection $A\bar{x}_W$ is still perpendicular (in the W-inner product!) to the error $b - A\bar{x}_W$.

3.6.23 (a) Suppose you guess your professor's age, making errors $e = -2, -1, 5$ with probabilities $\frac{1}{2}, \frac{1}{4}, \frac{1}{4}$. Check that the expected error $E(e)$ is zero and find the variance $E(e^2)$.
(b) If the professor guesses too (or tries to remember), making errors $-1, 0, 1$ with probabilities $\frac{1}{8}, \frac{6}{8}, \frac{1}{8}$, what weights w_1 and w_2 give the reliability of your guess and the professor's guess?

3.6.24 Suppose p rows and q columns, taken together, contain all the nonzero entries of A. Show that the rank is not greater than $p + q$. How large does a square block of zeros have to be, in a corner of a 9 by 9 matrix, to guarantee that the matrix is singular?

REVIEW EXERCISES: Chapter 3

3.1 Find the length of $a = (2, -2, 1)$ and write down two independent vectors that are perpendicular to a.

3.2 Find all vectors that are perpendicular to $(1, 3, 1)$ and $(2, 7, 2)$, by making those the rows of A and solving $Ax = 0$.

3.3 What is the angle between $a = (2, -2, 1)$ and $b = (1, 2, 2)$?

3.4 What is the projection p of $b = (1, 2, 2)$ onto $a = (2, -2, 1)$?

3.5 Find the cosine of the angle between the vectors $(3, 4)$ and $(4, 3)$.

3.6 Where is the projection of $b = (1, 1, 1)$ onto the plane spanned by $(1, 0, 0)$ and $(1, 1, 0)$?

3.7 The system $Ax = b$ has a solution if and only if b is orthogonal to which of the four fundamental subspaces?

3.8 Which straight line gives the best fit to the following data: $b = 0$ at $t = 0$, $b = 0$ at $t = 1$, $b = 12$ at $t = 3$?

3.9 Construct the projection matrix P onto the space spanned by $(1, 1, 1)$ and $(0, 1, 3)$.

3.10 Which constant function is closest to $y = x^4$ in the least squares sense over the interval $0 \le x \le 1$?

3.11 If Q is orthogonal, is the same true of Q^3?

3.12 Find all 3 by 3 orthogonal matrices whose entries are zeros and ones.

3.13 What multiple of a_1 should be subtracted from a_2, to make the result orthogonal to a_1? Sketch a figure.

3.14 Factor

$$\begin{bmatrix} \cos\theta & \sin\theta \\ \sin\theta & 0 \end{bmatrix}$$

into QR, recognizing that the first column is already a unit vector.

3.15 If every entry in an orthogonal matrix is either $\frac{1}{4}$ or $-\frac{1}{4}$, how big is the matrix?

3.16 Suppose the vectors $q_1, \ldots, q_n$ are orthonormal. If $b = c_1 q_1 + \cdots + c_n q_n$, give a formula for the first coefficient c_1 in terms of b and the q's.

3.17 What words describe the equation $A^T A\bar{x} = A^T b$ and the vector $p = A\bar{x} = Pb$ and the matrix $P = A(A^T A)^{-1} A^T$?

3.18 If the orthonormal vectors $q_1 = (\frac{2}{3}, \frac{2}{3}, -\frac{1}{3})$ and $q_2 = (-\frac{1}{3}, \frac{2}{3}, \frac{2}{3})$ are the columns of Q, what are the matrices $Q^T Q$ and QQ^T? Show that QQ^T is a projection matrix (onto the plane of q_1 and q_2).

3.19 If $v_1, \ldots, v_n$ is an orthonormal basis for $\mathbf{R}^n$, show that $v_1 v_1^T + \cdots + v_n v_n^T = I$.

3.20 *True or false*: If the vectors x and y are orthogonal, and P is a projection, then Px and Py are orthogonal.

3.21   Try to fit a line $b = C + Dt$ through the points $b = 0$, $t = 2$, and $b = 6$, $t = 2$, and show that the normal equations break down. Sketch all the optimal lines, minimizing the sum of squares of the two errors.

3.22   What point on the plane $x + y - z = 0$ is closest to $b = (2, 1, 0)$?

3.23   Find an orthonormal basis for $\mathbf{R}^3$ starting with the vector $(1, 1, -1)$.

3.24   The new X-ray machines (CT scanners) examine the patient from different directions and produce a matrix giving the densities of bone and tissue at each point. Mathematically, the problem is to recover a matrix from its projections. In the 2 by 2 case, can you recover the matrix A if you know the sum along each row and down each column?

3.25   Can you recover a 3 by 3 matrix if you know its row sums and column sums, and also the sums down the main diagonal and the four other parallel diagonals?

3.26   Find an orthonormal basis for the plane $x - y + z = 0$, and find the matrix P which projects onto the plane. What is the nullspace of P?

3.27   Let $A = \begin{bmatrix} 3 & 1 & -1 \end{bmatrix}$ and let V be the nullspace of A.
   (a)   Find a basis for V and a basis for $V^\perp$.
   (b)   Write down an orthonormal basis for $V^\perp$, and find the projection matrix P_1 which projects vectors in $\mathbf{R}^3$ onto $V^\perp$.
   (c)   Find the projection matrix P_2 which projects vectors in $\mathbf{R}^3$ onto V.

3.28   Use Gram-Schmidt to construct an orthonormal pair q_1, q_2 from $a_1 = (4, 5, 2, 2)$ and $a_2 = (1, 2, 0, 0)$. Express a_1 and a_2 as combinations of q_1 and q_2 and write down the triangular R in $A = QR$.

3.29   For any A, b, x, and y, show that

   (i)   if $Ax = b$ and $y^T A = 0$, then $y^T b = 0$;
   (ii)   if $Ax = 0$ and $A^T y = b$, then $x^T b = 0$.

   What theorem does this prove about the fundamental subspaces?

3.30   Is there a matrix whose row space contains $(1, 1, 0)$ and whose nullspace contains $(0, 1, 1)$?

3.31   The distance from a plane $a^T x = c$ in m-dimensional space to the origin is $|c|/\|a\|$. How far is the plane $x_1 + x_2 - x_3 - x_4 = 8$ from the origin and what point on it is nearest?

3.32   In the parallelogram with corners at 0, v, w, and $v + w$, show that the sum of the lengths squared of the four sides equals the sum of the lengths squared of the two diagonals.

3.33   (a)   Find an orthonormal basis for the column space of

$$A = \begin{bmatrix} 1 & -6 \\ 3 & 6 \\ 4 & 8 \\ 5 & 0 \\ 7 & 8 \end{bmatrix}.$$

(b)   Write A as QR, where Q has orthonormal columns and R is upper triangular.
(c)   Find the least squares solution to $Ax = b$, if $b = (-3, 7, 1, 0, 4)$.

3.34   Find the intersection $V \cap W$ and the sum $V + W$ if
(a) $V =$ nullspace of a matrix A and $W =$ row space of A.
(b) $V =$ symmetric 3 by 3 matrices and $W =$ upper triangular 3 by 3 matrices.

3.35   With weighting matrix $W = \begin{bmatrix} 2 & 1 \\ 1 & 0 \end{bmatrix}$, what is the W-inner product of $(1, 0)$ with $(0, 1)$?

3.36   To solve a rectangular system $Ax = b$ we replace A^{-1} (which doesn't exist) by $(A^T A)^{-1} A^T$ (which exists if A has independent columns). Show that this is a left-inverse of A but not a right-inverse. On the left of A it gives the identity; on the right it gives the projection P.

3.37   Find the straight line $C + Dt$ that best fits the measurements $b = 0, 1, 2, 5$ at times $t = 0, 1, 3, 4$.

3.38   Find the curve $y = C + D2^t$ which gives the best least squares fit to the measurements $y = 6$ at $t = 0$, $y = 4$ at $t = 1$, $y = 0$ at $t = 2$. Write down the three equations that are solved if the curve goes through the three points, and find the best C and D.

3.39   If the columns of A are orthogonal to each other, what can you say about the form of $A^T A$? If the columns are orthonormal, what can you say then?

3.40   Under what condition on the columns of A (which may be rectangular) is $A^T A$ invertible?

4

DETERMINANTS

It is hard to know what to say about determinants. Seventy years ago they seemed more interesting and more important than the matrices they came from, and Muir's *History of Determinants* filled four volumes. Mathematics keeps changing direction, however, and determinants are now far from the center of linear algebra. After all, a single number can tell only so much about a matrix. Still it is amazing how much this number can do.

One viewpoint is this: The determinant provides an explicit "formula," a concise and definite expression in closed form, for quantities such as A^{-1}. This formula will not change the way we compute A^{-1}, or $A^{-1}b$; even the determinant itself is found by elimination. In fact, elimination can be regarded as the most efficient way to substitute the entries of an n by n matrix into the formula. What the formula does is to show how A^{-1} depends on the n^2 entries of the matrix, and how it varies when those entries vary.

We can list some of the main uses of determinants:

(1) It gives a test for invertibility. *If the determinant of A is zero, then A is singular. If* det $A \neq 0$, *then A is invertible.* The most important application, and the reason this chapter is essential to the book, is to the family of matrices $A - \lambda I$. The parameter λ is subtracted all along the main diagonal, and the problem is to find those values of λ (the *eigenvalues*) for which $A - \lambda I$ is singular. The test is to see if the determinant of this matrix is zero. We shall see that $\det(A - \lambda I)$ is a polynomial of degree n in λ, and therefore, counting multiplicities, it has exactly n roots. The matrix has n eigenvalues. This is a fact which follows from the determinant formula, and not from a computer.

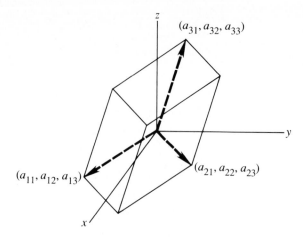

Fig. 4.1. The parallelepiped formed from the rows of A.

(2)   The determinant of A equals the **volume** of a parallelepiped P in n-dimensional space, provided the edges of P come from the rows of A (Fig. 4.1.)†
This volume may seem an odd thing to want to compute. In practice, P is often the infinitesimal volume element in a multiple integral. The simplest element is a little cube $dV = dx\,dy\,dz$, as in $\iiint f(x, y, z)\,dV$. Suppose, in order to simplify the integral, we decide to change to cylindrical coordinates r, θ, z. The relation is $x = r\cos\theta$, $y = r\sin\theta$, $z = z$. Then, just as we have to remember that a small integral dx is stretched to $(dx/du)\,du$—when u replaces x in a single integral—so the volume element $dV = dx\,dy\,dz$ is stretched in three dimensions. It becomes $J\,dr\,d\theta\,dz$, where the *Jacobian determinant* is the three-dimensional analogue of the stretching factor dx/du:

$$J = \begin{bmatrix} \partial x/\partial r & \partial x/\partial\theta & \partial x/\partial z \\ \partial y/\partial r & \partial y/\partial\theta & \partial y/\partial z \\ \partial z/\partial r & \partial z/\partial\theta & \partial z/\partial z \end{bmatrix} = \begin{bmatrix} \cos\theta & -r\sin\theta & 0 \\ \sin\theta & r\cos\theta & 0 \\ 0 & 0 & 1 \end{bmatrix}.$$

The value of this determinant is $J = r$. It is the r in the cylindrical volume element $r\,dr\,d\theta\,dz$; this element is our little parallelepiped. (It looks curved if we try to draw it, but probably it gets straighter as the edges become infinitesimal.)

(3)   The determinant gives formulas for the pivots. Theoretically, we could use it to predict when a pivot entry will be zero, and a row exchange will be necessary. More importantly, from the formula **determinant** $= \pm$(**product of the pivots**), it follows that *regardless of the order of elimination, the product of the pivots remains the same apart from sign*. Years ago, this led to the belief that it was useless to escape

---

† Or the edges could come from the columns of A, giving an entirely different parallelepiped with the same volume.

a very small pivot by exchanging rows, since eventually the small pivot would catch up with us. But what usually happens in practice, if an abnormally small pivot is not avoided, is that it is very soon followed by an abnormally large one. This brings the product back to normal but it leaves the numerical solution in ruins.

(4)   The determinant measures the dependence of $A^{-1}b$ on each element of b. If one parameter is changed in an experiment, or one observation is corrected, the "influence coefficient" on $x = A^{-1}b$ is a ratio of determinants.

There is one more problem about the determinant. It is difficult not only to decide on its importance, and its proper place in the theory of linear algebra, but also to decide on its definition. Obviously, det A will not be some extremely simple function of n^2 variables; otherwise A^{-1} would be much easier to find than it actually is. The explicit formula given in Section 4.3 will require a good deal of explanation, and its connection with the inverse is far from evident.

The simple things about the determinant are not the explicit formulas, but the properties it possesses. This suggests the natural place to begin. The determinant can be (and will be) defined by its three most basic properties. The problem is then to show how, by systematically using these properties, the determinant can be computed. This will bring us back to Gaussian elimination, and to the product of the pivots. And the more difficult theoretical problem is to show that whatever the order in which the properties are used, the result is always the same—the defining properties are self-consistent.

The next section lists the defining properties of the determinant, and their most important consequences. Then Section 4.3 gives several formulas for the determinant—one is an explicit formula with $n!$ terms, another is a formula "by induction," and the third is the one involving pivots (from which the determinant of a large matrix is actually computed). In Section 4.4 the determinant is applied to find A^{-1} and then to solve for $x = A^{-1}b$; the latter is *Cramer's rule*. And finally, in an optional remark on permutations, we prove that the properties are self-consistent—so there is no ambiguity in the definition.

4.2 ■ THE PROPERTIES OF THE DETERMINANT

This will be a pretty long list. Fortunately each rule is easy to understand, and even easier to illustrate, for a 2 by 2 example. Therefore we shall verify that the familiar definition in the 2 by 2 case,

$$\det \begin{bmatrix} a & b \\ c & d \end{bmatrix} = \begin{vmatrix} a & b \\ c & d \end{vmatrix} = ad - bc,$$

possesses every property in the list. (Notice the two accepted notations for the determinant of A: either det A or $|A|$.) From the fourth property on, we also indicate how it can be deduced from the previous ones; then every property is a consequence of the first three. We emphasize that the rules apply to square matrices *of any size*, and the first rule is tricky but crucial.

1. *The determinant depends linearly on the first row.* Suppose three matrices A, B, C are identical from the second row onward—and the first row of A is a combination of the first rows of B and C. Then the rule says: det A is the same combination of det B and det C.

Linear combinations involve two operations—adding vectors and multiplying by scalars. Therefore this rule can be split into two parts:

$$\begin{vmatrix} a + a' & b + b' \\ c & d \end{vmatrix} = \begin{vmatrix} a & b \\ c & d \end{vmatrix} + \begin{vmatrix} a' & b' \\ c & d \end{vmatrix}$$

$$\begin{vmatrix} ta & tb \\ c & d \end{vmatrix} = t \begin{vmatrix} a & b \\ c & d \end{vmatrix}.$$

Notice that the first part is *not* the false statement $\det(B + C) = \det B + \det C$. You cannot add all the rows: only one row is allowed to change. Both sides give the answer $ad + a'd - bc - b'c$.

Similarly the second part is not the false statement $\det(tA) = t \det A$. The matrix tA has a factor t in *every* row (and eventually each one multiplies the determinant by t). It is like the volume of a box, when all sides are stretched by 4. In n dimensions the volume and determinant go up by 4^n. If only the first side is stretched, the volume and determinant go up by 4; that is rule 1.

The next rule shows that there is nothing special about the first row.

2. *The determinant changes sign when two rows are exchanged.*

$$\begin{vmatrix} c & d \\ a & b \end{vmatrix} = cb - ad = - \begin{vmatrix} a & b \\ c & d \end{vmatrix}.$$

It follows that the determinant depends linearly **on each row separately**. If the factor 4 multiplies row 2, we can exchange with row 1 (changing the sign of the determinant), then factor out the 4 (by rule 1) and exchange rows again (changing the sign back to where it started). So rule 1 applies to every row.

Rule 2 is important in itself; it leads to the determinant of a *permutation matrix*. By a series of row exchanges, we can turn the permutation matrix into the identity matrix. Each row exchange switches the sign of the determinant—and then we need the determinant of the identity matrix. That is the simplest rule of all.

3. *The determinant of the identity matrix is 1.*

$$\begin{vmatrix} 1 & 0 \\ 0 & 1 \end{vmatrix} = 1 \quad \text{and} \quad \begin{vmatrix} 1 & 0 & 0 \\ 0 & 1 & 0 \\ 0 & 0 & 1 \end{vmatrix} = 1 \quad \text{and} \ldots$$

Rules 1 and 2 left a scaling constant undecided; rule 3 decides it. There was a "one-dimensional space" of possible determinants, and this rule picks out one of them—by normalizing det I to 1.

The determinant is now settled, but that fact is not at all obvious. Therefore we gradually use these rules to find the determinant of any matrix.

4. *If two rows of A are equal, then* det $A = 0$.

$$\begin{vmatrix} a & b \\ a & b \end{vmatrix} = ab - ba = 0.$$

This follows from rule 2, since if the equal rows are exchanged, the determinant is supposed to change sign. But it also has to stay the same, because the matrix stays the same. The only number which can do that is zero, so det $A = 0$. (The reasoning fails if $1 = -1$, which is the case in Boolean algebra. Then rule 4 should replace rule 2 as one of the defining properties.)

5. *The elementary operation of subtracting a multiple of one row from another row leaves the determinant unchanged.*

$$\begin{vmatrix} a - lc & b - ld \\ c & d \end{vmatrix} = \begin{vmatrix} a & b \\ c & d \end{vmatrix}.$$

Rule 1 would say that there is further term $-l \begin{vmatrix} c & d \\ c & d \end{vmatrix}$, but that term is zero by rule 4. The usual elimination steps do not affect the determinant.

6. *If A has a zero row, then* det $A = 0$.

$$\begin{vmatrix} 0 & 0 \\ c & d \end{vmatrix} = 0.$$

One proof is to add some other row to the zero row. The determinant is unchanged, by rule 5, and because the matrix will now have two identical rows, det $A = 0$ by rule 4.

7. *If A is triangular, then* det A *is the product* $a_{11}a_{22} \cdots a_{nn}$ *of the entries on the main diagonal. In particular, if A has 1's along the diagonal,* det $A = 1$.

$$\begin{vmatrix} a & b \\ 0 & d \end{vmatrix} = ad, \qquad \begin{vmatrix} a & 0 \\ c & d \end{vmatrix} = ad.$$

Proof Suppose the diagonal entries are nonzero. Then elimination steps can remove all the off-diagonal entries, without changing the determinant (by rule 5). If A is lower triangular the steps are in the usual order, starting with the first pivot. If A is upper triangular, then the *last* column is cleared out first—using multiples of a_{nn} from the bottom corner. Either way we reach the diagonal matrix

$$D = \begin{bmatrix} a_{11} & & \\ & \ddots & \\ & & a_{nn} \end{bmatrix}.$$

To find its determinant we patiently apply rule 1. Factoring out a_{11} and then a_{22} and finally a_{nn} leaves the identity matrix. Finally we have a use for rule 3!

$$\det D = a_{11}a_{22} \cdots a_{nn} \det I = a_{11}a_{22} \cdots a_{nn}.$$

In contrast, *if a diagonal entry is zero then elimination will produce a zero row.* By rule 5 these elimination steps do not change the determinant. By rule 6 the zero row means a zero determinant. Rule 7 is proved.

When a triangular matrix is *singular* (because of a zero on the main diagonal) its determinant is *zero*. The next rule shows that this is the determinant of all singular matrices.

8. *If A is singular, then* det $A = 0$. *If A is invertible, then* det $A \neq 0$.

$$\begin{bmatrix} a & b \\ c & d \end{bmatrix} \quad \text{is singular if and only if} \quad ad - bc = 0.$$

If A is singular, elimination leads to a matrix U with a zero row. By rules 5 and 6, det $A = \det U = 0$. If A is not singular, elimination leads to an upper triangular U with nonzeros along the main diagonal. These are the pivots $d_1, \ldots, d_n$, and by rule 7, det $A = \pm \det U = \pm d_1 d_2 \cdots d_n$. **Here we have our first formula for the determinant.** The plus or minus sign depends on whether the number of row exchanges is even or odd.

The next property is the most surprising.

9. *For any two n by n matrices, the determinant of the product AB is the product of the determinants:* $\det AB = (\det A)(\det B)$.

$$\begin{vmatrix} a & b \\ c & d \end{vmatrix} \begin{vmatrix} e & f \\ g & h \end{vmatrix} = \begin{vmatrix} ae + bg & af + bh \\ ce + dg & cf + dh \end{vmatrix}.$$

A particular case of this rule gives the determinant of A^{-1} as $1/\det A$:

$$(\det A)(\det A^{-1}) = \det AA^{-1} = \det I = 1, \quad \text{so } \det A^{-1} = \frac{1}{\det A}.$$

In the 2 by 2 case the product rule is the same as

$$(ad - bc)(eh - fg) = (ae + bg)(cf + dh) - (af + bh)(ce + dg).$$

In the n by n case we suggest two possible proofs—since this is the least obvious rule. Both proofs assume that A and B are nonsingular; otherwise AB is singular, and the equation $\det AB = (\det A)(\det B)$ is easily verified. By rule 8, it becomes $0 = 0$.

(i) We consider the ratio $d(A) = \det AB/\det B$, and prove that it has properties 1–3. Then, because these properties define the determinant, $d(A)$ must equal $\det A$. For example, if A is the identity, then certainly $d(I) = \det B/\det B = 1$; thus rule 3 is satisfied by $d(A)$. If two rows of A are exchanged, so are the same two rows of AB, and the sign of d changes as required by rule 2. And a linear combination appearing in the first row of A gives the same linear combination in the first row of AB. Therefore rule 1 for the determinant of AB, divided by the fixed quantity $\det B$, leads to rule 1 for the ratio $d(A)$. Thus $d(A)$ *coincides with the determinant.* That means $\det AB/\det B = \det A$, which is our product formula.

(ii) This second proof is less elegant. It starts by supposing that A is a diagonal matrix D. Then $\det DB = (\det D)(\det B)$ follows from rule 1, by factoring out each diagonal element d_i from its row. For a general matrix A, we reduce to D by "Gauss-Jordan" elimination steps—from A to U by the usual sequence, and then from U to D by using each pivot to produce zeros above it. The determinant does not change, except for a sign reversal when rows are exchanged. The same steps reduce AB to DB, with precisely the same effect on the determinant. But for DB it is already confirmed that rule 9 is correct.

10. *The transpose of A has the same determinant as A itself:* $\det A^T = \det A$.

$$\begin{vmatrix} a & b \\ c & d \end{vmatrix} = \begin{vmatrix} a & c \\ b & d \end{vmatrix}.$$

Again the singular case is separate; A is singular if and only if A^T is singular, and we have $0 = 0$. If A is not singular, then it allows the factorization $PA = LDU$, and we apply the previous rule 9 for the determinant of a product:

$$\det P \det A = \det L \det D \det U. \tag{1}$$

Transposing $PA = LDU$ gives $A^\mathrm{T} P^\mathrm{T} = U^\mathrm{T} D^\mathrm{T} L^\mathrm{T}$, and again by rule 9,

$$\det A^\mathrm{T} \det P^\mathrm{T} = \det U^\mathrm{T} \det D^\mathrm{T} \det L^\mathrm{T}. \tag{2}$$

This is simpler than it looks, because L, U, L^T, and U^T are triangular with unit diagonal. By rule 7, their determinants all equal one. Also, any diagonal matrix is the same as its transpose: $D = D^\mathrm{T}$. This leaves only the permutation matrices.

Certainly the determinant of P is either 1 or -1, because it comes from the identity matrix by a sequence of row exchanges. Observe also that $PP^\mathrm{T} = I$. (Multiplying P and P^T, the 1 in the first row of P matches the 1 in the first column of P^T, and misses the 1's in the other columns.) Therefore $\det P \det P^\mathrm{T} = \det I = 1$, and P and P^T must have the same determinant; both equal 1 or both equal -1.

We conclude that the products (1) and (2) are the same, and $\det A = \det A^\mathrm{T}$. This fact practically doubles our list of properties, because every rule which applied to the rows can now be applied to the columns: *The determinant changes sign when two columns are exchanged, two equal columns* (or a column of zeros) *produce a zero determinant, and the determinant depends linearly on each individual column.* The proof is just to transpose the matrix and work with the rows.

I think it is time to stop and call the list complete. It only remains to find a definite formula for the determinant, and to put that formula to use.

EXERCISES

4.2.1   How are $\det(2A)$, $\det(-A)$, and $\det(A^2)$ related to $\det A$, when A is n by n?

4.2.2   Show—by carrying out each step on a 2 by 2 example—that an exchange of rows i and j can be produced by adding row i to row j, then subtracting the new row j from row i, then adding the new row i to row j, and finally multiplying row i by -1. Which rules could we then use to deduce rule 2?

4.2.3   By applying row operations to produce an upper triangular U, compute

$$\det \begin{bmatrix} 1 & 2 & -2 & 0 \\ 2 & 3 & -4 & 1 \\ -1 & -2 & 0 & 2 \\ 0 & 2 & 5 & 3 \end{bmatrix} \quad \text{and} \quad \det \begin{bmatrix} 2 & -1 & 0 & 0 \\ -1 & 2 & -1 & 0 \\ 0 & -1 & 2 & -1 \\ 0 & 0 & -1 & 2 \end{bmatrix}.$$

Exchange rows 3 and 4 of the second matrix and recompute the pivots and determinant.

Note   Some readers will already know a formula for 3 by 3 determinants. It has six terms (equation (2) of the next section), three going parallel to the main diagonal and three others going the opposite way with minus signs. It is natural to hope for a similar formula for 4 by 4 determinants. There is a formula, **but it contains** 4! = 24 **terms** (*not just eight*). You cannot even be sure that a minus sign goes with the opposite diagonal, as the next exercises show.

4.2.4 Explain why

$$\det \begin{bmatrix} 0 & 0 & 0 & 1 \\ 0 & 0 & 1 & 0 \\ 0 & 1 & 0 & 0 \\ 1 & 0 & 0 & 0 \end{bmatrix} = +1 \quad \text{and} \quad \det \begin{bmatrix} 0 & 1 & 0 & 0 \\ 0 & 0 & 1 & 0 \\ 0 & 0 & 0 & 1 \\ 1 & 0 & 0 & 0 \end{bmatrix} = -1.$$

4.2.5 How many exchanges does it take to get (row n, row $n-1$, ..., row 1) into the normal order (row 1, ..., row $n-1$, row n)? When is det $P = 1$ and when is det $P = -1$, for the n by n permutation with 1's on the opposite diagonal? The previous exercise had $n = 4$.

4.2.6 Find the determinants of:
(a) a rank one matrix

$$A = \begin{bmatrix} 1 \\ 4 \\ 2 \end{bmatrix} \begin{bmatrix} 2 & -1 & 2 \end{bmatrix}$$

(b) the upper triangular matrix

$$U = \begin{bmatrix} 4 & 4 & 8 & 8 \\ 0 & 1 & 2 & 2 \\ 0 & 0 & 2 & 6 \\ 0 & 0 & 0 & 2 \end{bmatrix};$$

(c) the lower triangular matrix U^{T};
(d) the inverse matrix U^{-1};
(e) the "reverse-triangular" matrix that results from row exchanges,

$$M = \begin{bmatrix} 0 & 0 & 0 & 2 \\ 0 & 0 & 2 & 6 \\ 0 & 1 & 2 & 2 \\ 4 & 4 & 8 & 8 \end{bmatrix}.$$

4.2.7 Show how rule 6 (det $= 0$ if a row is zero) comes directly from rules 1 and 2.

4.2.8 Suppose you do two row operations *at once*, going from

$$\begin{bmatrix} a & b \\ c & d \end{bmatrix} \quad \text{to} \quad \begin{bmatrix} a - mc & b - md \\ c - la & d - lb \end{bmatrix}.$$

Find the determinant of the new matrix, by rule 1 or by direct calculation (and simplification).

4.2.9 If Q is an orthogonal matrix, so that $Q^TQ = I$, prove that det Q equals $+1$ or -1. What kind of parallelepiped is formed from the rows (or columns) of an orthogonal matrix Q?

4.2.10 Use row operations to verify that the 3 by 3 "Vandermonde determinant" is

$$\det \begin{bmatrix} 1 & a & a^2 \\ 1 & b & b^2 \\ 1 & c & c^2 \end{bmatrix} = (b - a)(c - a)(c - b).$$

The 4 by 4 case is among the review exercises.

4.2.11 (a) A skew-symmetric matrix satisfies $K^T = -K$, as in

$$K = \begin{bmatrix} 0 & a & b \\ -a & 0 & c \\ -b & -c & 0 \end{bmatrix}.$$

In the 3 by 3 case why is $\det(-K) = (-1)^3 \det K$? On the other hand $\det K^T = \det K$ (always). Deduce that $-\det K = \det K$ and the determinant must be zero.
(b) Write down a 4 by 4 skew-symmetric matrix with det K *not* zero.

4.2.12 True or false, with reason if true and counterexample if false:
(a) If A and B are identical except in the upper left corner, where $b_{11} = 2a_{11}$, then det $B = 2$ det A.
(b) The determinant is the product of the pivots.
(c) If A is invertible and B is singular, then $A + B$ is invertible.
(d) If A is invertible and B is singular, then AB is singular.

4.2.13 If every row of A adds to zero prove that det $A = 0$. If every row adds to 1 prove that $\det(A - I) = 0$. Show by example that this does not imply det $A = 1$.

4.2.14 Find these 4 by 4 determinants by Gaussian elimination:

$$\det \begin{bmatrix} 11 & 12 & 13 & 14 \\ 21 & 22 & 23 & 24 \\ 31 & 32 & 33 & 34 \\ 41 & 42 & 43 & 44 \end{bmatrix} \quad \text{and} \quad \det \begin{bmatrix} 1 & t & t^2 & t^3 \\ t & 1 & t & t^2 \\ t^2 & t & 1 & t \\ t^3 & t^2 & t & 1 \end{bmatrix}.$$

4.2.15 Find the determinants of

$$A = \begin{bmatrix} 4 & 2 \\ 1 & 3 \end{bmatrix}, \quad A^{-1} = \frac{1}{10}\begin{bmatrix} 3 & -2 \\ -1 & 4 \end{bmatrix}, \quad A - \lambda I = \begin{bmatrix} 4 - \lambda & 2 \\ 1 & 3 - \lambda \end{bmatrix}.$$

For which values of λ is $A - \lambda I$ a singular matrix?

4.2.16 Evaluate det A by reducing the matrix to triangular form (rules 5 and 7).

$$A = \begin{bmatrix} 1 & 1 & 3 \\ 0 & 4 & 6 \\ 1 & 5 & 8 \end{bmatrix}, \quad B = \begin{bmatrix} 1 & 1 & 3 \\ 0 & 4 & 6 \\ 0 & 0 & 1 \end{bmatrix}, \quad C = \begin{bmatrix} 1 & 1 & 3 \\ 0 & 4 & 6 \\ 1 & 5 & 9 \end{bmatrix}.$$

What are the determinants of B, C, AB, $A^T A$, and C^T?

4.2.17 Suppose that $CD = -DC$, and find the flaw in the following argument: Taking determinants gives $(\det C)(\det D) = -(\det D)(\det C)$, so either C or D must have zero determinant. Thus $CD = -DC$ is only possible if C or D is singular.

4.3 ■ FORMULAS FOR THE DETERMINANT

The first formula has already appeared:

4A If A is nonsingular, then $A = P^{-1}LDU$, and

$$\det A = \det P^{-1} \det L \det D \det U$$
$$= \pm(\text{product of the pivots}). \tag{1}$$

The sign ± 1 is the determinant of P^{-1} (or of P), and depends on whether the number of row exchanges is even or odd. The triangular factors have $\det L = \det U = 1$ and $\det D = d_1 \cdots d_n$.

In the 2 by 2 case, the standard LDU factorization is

$$\begin{bmatrix} a & b \\ c & d \end{bmatrix} = \begin{bmatrix} 1 & 0 \\ c/a & 1 \end{bmatrix} \begin{bmatrix} a & 0 \\ 0 & (ad - bc)/a \end{bmatrix} \begin{bmatrix} 1 & b/a \\ 0 & 1 \end{bmatrix}.$$

The product of the pivots is $ad - bc$. If the first step is a row exchange, then

$$PA = \begin{bmatrix} c & d \\ a & b \end{bmatrix} = \begin{bmatrix} 1 & 0 \\ a/c & 1 \end{bmatrix} \begin{bmatrix} c & 0 \\ 0 & (cb - da)/c \end{bmatrix} \begin{bmatrix} 1 & d/c \\ 0 & 1 \end{bmatrix}.$$

The product of the pivots is now $-\det A$.

EXAMPLE The finite difference matrix in Section 1.7 had the $A = LDU$ factorization

$$\begin{bmatrix} 2 & -1 & & & \\ -1 & 2 & -1 & & \\ & -1 & 2 & \cdot & \\ & & \cdot & \cdot & -1 \\ & & & -1 & 2 \end{bmatrix} = L \begin{bmatrix} 2 & & & & \\ & 3/2 & & & \\ & & 4/3 & & \\ & & & \cdot & \\ & & & & (n+1)/n \end{bmatrix} U.$$

Its determinant is the product of its pivots:

$$\det A = 2 \left(\frac{3}{2}\right)\left(\frac{4}{3}\right) \cdots \left(\frac{n+1}{n}\right) = n + 1.$$

This is the way determinants are calculated, except for very special matrices. The code in Appendix C finds $\det A$ from the pivots. In fact, the pivots are the result of condensing the information that was originally spread over all n^2 entries of the matrix. From a theoretical point of view, however, concentrating all information into the pivots has a disadvantage: It is impossible to figure out how a change in one entry would affect the determinant. Therefore we now propose to find an explicit expression for the determinant in terms of the n^2 entries.

For $n = 2$, we will be proving that $ad - bc$ is correct. For $n = 3$, the corresponding formula is again pretty well known:

$$\begin{vmatrix} a_{11} & a_{12} & a_{13} \\ a_{21} & a_{22} & a_{23} \\ a_{31} & a_{32} & a_{33} \end{vmatrix} = \begin{array}{l} +a_{11}a_{22}a_{33} + a_{12}a_{23}a_{31} + a_{13}a_{21}a_{32} \\ -a_{11}a_{23}a_{32} - a_{12}a_{21}a_{33} - a_{13}a_{22}a_{31}. \end{array} \qquad (2)$$

Our goal is to derive these formulas directly from the defining properties 1–3 of the previous section. If we can handle $n = 2$ and $n = 3$ in a sufficiently organized way, you will see the pattern for all matrices.

To start, each row can be broken down into vectors that go in the coordinate directions:

$$[a \quad b] = [a \quad 0] + [0 \quad b] \quad \text{and} \quad [c \quad d] = [c \quad 0] + [0 \quad d].$$

Then we apply the key property of linearity in each row separately—first in row 1 and then in row 2:

$$\begin{vmatrix} a & b \\ c & d \end{vmatrix} = \begin{vmatrix} a & 0 \\ c & d \end{vmatrix} + \begin{vmatrix} 0 & b \\ c & d \end{vmatrix}$$

$$= \begin{vmatrix} a & 0 \\ c & 0 \end{vmatrix} + \begin{vmatrix} a & 0 \\ 0 & d \end{vmatrix} + \begin{vmatrix} 0 & b \\ c & 0 \end{vmatrix} + \begin{vmatrix} 0 & b \\ 0 & d \end{vmatrix}. \qquad (3)$$

For an n by n matrix, every row will be split into n coordinate directions. This expansion has n^n terms: In our case $2^2 = 4$. Fortunately, most of them (like the first and last terms above) will be automatically zero. Whenever two rows are in the same coordinate direction, one will be a multiple of the other, and

$$\begin{vmatrix} a & 0 \\ c & 0 \end{vmatrix} = 0, \qquad \begin{vmatrix} 0 & b \\ 0 & d \end{vmatrix} = 0.$$

There is a column of zeros, and a zero determinant. Therefore, we pay attention *only when the rows point in different directions; **the nonzero terms have to come in different columns.*** Suppose the first row has a nonzero entry in column α, the second row is nonzero in column β, and finally the nth row is nonzero in column ν. The column numbers $\alpha, \beta, \ldots, \nu$ are all different; they are a reordering, or *permutation*, of the numbers $1, 2, \ldots, n$. For the 3 by 3 case there are six terms:

$$\begin{vmatrix} a_{11} & a_{12} & a_{13} \\ a_{21} & a_{22} & a_{23} \\ a_{31} & a_{32} & a_{33} \end{vmatrix} = \begin{vmatrix} a_{11} & & \\ & a_{22} & \\ & & a_{33} \end{vmatrix} + \begin{vmatrix} & a_{12} & \\ & & a_{23} \\ a_{31} & & \end{vmatrix} + \begin{vmatrix} & & a_{13} \\ a_{21} & & \\ & a_{32} & \end{vmatrix}$$

$$+ \begin{vmatrix} a_{11} & & \\ & & a_{23} \\ & a_{32} & \end{vmatrix} + \begin{vmatrix} & a_{12} & \\ a_{21} & & \\ & & a_{33} \end{vmatrix} + \begin{vmatrix} & & a_{13} \\ & a_{22} & \\ a_{31} & & \end{vmatrix}. \quad (4)$$

To repeat, the expansion could have $3^3 = 27$ terms; all but these $3! = 6$ are zero, because a column is repeated. In general, $n!$ terms are left. (There are n choices for the first column α, $n - 1$ remaining choices for β, and finally only one choice for the last column ν—all but one column will be used by that time, when we "snake" down the rows of the matrix). In other words, **there are $n!$ ways to permute the numbers** $1, 2, \ldots, n$. We look at the sequence of column numbers to find the associated permutations; the six terms in (4) come from the following columns:

$$(\alpha, \beta, \nu) = (1, 2, 3), (2, 3, 1), (3, 1, 2), (1, 3, 2), (2, 1, 3), (3, 2, 1).$$

Those are the $3! = 6$ permutations of $(1, 2, 3)$; the first one is the identity.

The determinant of A is now reduced to six separate and much simpler determinants. Factoring out the a_{ij}, there is a term for every one of the six permutations:

$$\det A = a_{11}a_{22}a_{33} \begin{vmatrix} 1 & & \\ & 1 & \\ & & 1 \end{vmatrix} + a_{12}a_{23}a_{31} \begin{vmatrix} & 1 & \\ & & 1 \\ 1 & & \end{vmatrix} + a_{13}a_{21}a_{32} \begin{vmatrix} & & 1 \\ 1 & & \\ & 1 & \end{vmatrix}$$

$$+ a_{11}a_{23}a_{32} \begin{vmatrix} 1 & & \\ & & 1 \\ & 1 & \end{vmatrix} + a_{12}a_{21}a_{33} \begin{vmatrix} & 1 & \\ 1 & & \\ & & 1 \end{vmatrix} + a_{13}a_{22}a_{31} \begin{vmatrix} & & 1 \\ & 1 & \\ 1 & & \end{vmatrix}. \quad (5)$$

Every term is a product of $n = 3$ entries a_{ij}, with each row and column represented once. In other words, there is a term corresponding to every path that goes down through the matrix and uses each column once. If the columns are used in the order $(\alpha, \ldots, \nu)$, then that term is the product $a_{1\alpha} \cdots a_{n\nu}$ times the determinant of a permutation matrix P_σ. The determinant of the whole matrix is the sum of these terms, and **that sum is the explicit formula we are after**:

$$\det A = \sum_\sigma (a_{1\alpha}a_{2\beta} \cdots a_{n\nu}) \det P_\sigma. \quad (6)$$

For an n by n matrix, this sum is taken over all $n!$ permutations $\sigma = (\alpha, \ldots, \nu)$ of the numbers $(1, \ldots, n)$. The permutation gives the sequence of column numbers as we go down the rows of the matrix, and it also specifies the permutation matrix P_σ: The 1's appear in P_σ at the same places where the a's appeared in A.

It remains to find the determinant of P_σ. Row exchanges transform it to the identity matrix, and each exchange reverses the sign of the determinant:

$\det P_\sigma = +1$ or -1 **depending on whether the number of exchanges is even or odd.**

$$P_\sigma = \begin{bmatrix} 1 & & \\ & & 1 \\ & 1 & \end{bmatrix} \text{ has column sequence } (\alpha, \beta, v) = (1, 3, 2);$$

$$P_\sigma = \begin{bmatrix} & & 1 \\ 1 & & \\ & 1 & \end{bmatrix} \text{ has column sequence } \sigma = (3, 1, 2).$$

The first requires one exchange (the second and third rows), so that $\det P_\sigma = -1$. The second requires two exchanges to recover the identity (the first and second rows, followed by the second and third), so that $\det P_\sigma = (-1)^2 = 1$. These are two of the six $\pm$ signs that appear in (2).

Equation (6) is an explicit expression for the determinant, and it is easy enough to check the 2 by 2 case: The $2! = 2$ permutations are $\sigma = (1, 2)$ and $\sigma = (2, 1)$, and therefore

$$\det A = a_{11}a_{22} \det \begin{bmatrix} 1 & 0 \\ 0 & 1 \end{bmatrix} + a_{12}a_{21} \det \begin{bmatrix} 0 & 1 \\ 1 & 0 \end{bmatrix} = a_{11}a_{22} - a_{12}a_{21} \text{ (or } ad - bc\text{)}.$$

No one can claim that the explicit formula (6) is particularly simple. Nevertheless, it is possible to see why it has properties 1–3. Property 3, the fact that $\det I = 1$, is of course the simplest; the products of the a_{ij} will always be zero, except for the special column sequence $\sigma = (1, 2, \dots, n)$, in other words the identity permutation. This term gives $\det I = 1$. Property 2 will be checked in the next section, because here we are most interested in property 1: The determinant should depend linearly on the row $a_{11}, a_{12}, \dots, a_{1n}$. To see this dependence, look at the terms in formula (6) involving a_{11}. They occur when the choice of the first column is $\alpha = 1$, leaving some permutation $\sigma' = (\beta, \dots, v)$ of the remaining column numbers $(2, \dots, n)$. We collect all these terms together as $a_{11}A_{11}$, where the coefficient of a_{11} is

$$A_{11} = \sum_{\sigma'} (a_{2\beta} \cdots a_{nv}) \det P_{\sigma'}. \tag{7}$$

Similarly, the entry a_{12} is multiplied by some messy expression A_{12}. Grouping all the terms which start with the same a_{1j}, the formula (6) becomes

$$\det A = a_{11}A_{11} + a_{12}A_{12} + \cdots + a_{1n}A_{1n}. \tag{8}$$

Thus $\det A$ depends linearly on the entries $a_{11}, \dots, a_{1n}$ of the first row.

EXAMPLE   For a 3 by 3 matrix, this way of collecting terms gives

$$\det A = a_{11}(a_{22}a_{33} - a_{23}a_{32}) + a_{12}(a_{23}a_{31} - a_{21}a_{33}) + a_{13}(a_{21}a_{32} - a_{22}a_{31}). \quad (9)$$

The "*cofactors*" A_{11}, A_{12}, A_{13} are written out in the three parentheses.

Expansion of det A in Cofactors

We want one more formula for the determinant. If this meant starting again from scratch, it would be too much. But *the formula is already discovered—it is* (8), *and the only point is to identify the cofactors* A_{1j}.

We know that this number A_{1j} depends on rows $2, \ldots, n$; row 1 is already accounted for by the factor a_{1j}. Furthermore, a_{1j} also accounts for the jth column, so its cofactor A_{1j} must depend entirely on *the other columns*. No row or column can be used twice in the same term. What we are really doing is splitting the determinant into the following sum:

$$\begin{vmatrix} a_{11} & a_{12} & a_{13} \\ a_{21} & a_{22} & a_{23} \\ a_{31} & a_{32} & a_{33} \end{vmatrix} = \begin{vmatrix} a_{11} & & \\ & a_{22} & a_{23} \\ & a_{32} & a_{33} \end{vmatrix} + \begin{vmatrix} & a_{12} & \\ a_{21} & & a_{23} \\ a_{31} & & a_{33} \end{vmatrix} + \begin{vmatrix} & & a_{13} \\ a_{21} & a_{22} & \\ a_{31} & a_{32} & \end{vmatrix}.$$

For a determinant of order n, this splitting gives n smaller determinants (called **minors**) of order $n - 1$; you can see the 2 by 2 submatrices that appear on the right-hand side. The submatrix M_{1j} is formed by *throwing away row 1 and column j*. Each term on the right is a product of a_{1j} and the determinant of M_{1j}—with the correct plus or minus sign. These signs alternate as we go along the row, and the cofactors are finally identified as

$$A_{1j} = (-1)^{1+j} \det M_{1j}.$$

For example, the second cofactor A_{12} is $a_{23}a_{31} - a_{21}a_{33}$, which is $\det M_{12}$ times -1. This same technique works on square matrices of any size. A close look at equation (7) confirms that A_{11} is the determinant of the lower right corner M_{11}.

There is a similar expansion on any other row, say row i, which could be proved by exchanging row i with row 1:

4B   The determinant of A is a combination of row i and the cofactors of row i:

$$\det A = a_{i1}A_{i1} + a_{i2}A_{i2} + \cdots + a_{in}A_{in}. \quad (10)$$

The **cofactor** A_{ij} is the determinant of M_{ij} with the correct sign:

$$A_{ij} = (-1)^{i+j} \det M_{ij}. \quad (11)$$

M_{ij} is formed by deleting row i and column j of A.

These formulas express det A as a combination of determinants of order $n - 1$. Therefore *we could have defined the determinant by induction on n.* For 1 by 1 matrices, we would set det $A = a_{11}$—and then use (10) to define successively the determinants of 2 by 2 matrices, 3 by 3 matrices, and so on indefinitely. We preferred to define the determinant by its properties, which are much simpler to explain, and then to deduce the explicit formula (6) and the cofactor formula (10) from these properties.

There is one more consequence of the fact that det $A =$ det A^T. This property allows us to expand in cofactors of a *column* instead of a row. Down column j,

$$\det A = a_{1j}A_{1j} + a_{2j}A_{2j} + \cdots + a_{nj}A_{nj}. \tag{12}$$

The proof is simply to expand det A^T in the cofactors of its jth row, which is the jth column of A.

EXAMPLE Consider the 4 by 4 finite difference matrix

$$A_4 = \begin{bmatrix} 2 & -1 & 0 & 0 \\ -1 & 2 & -1 & 0 \\ 0 & -1 & 2 & -1 \\ 0 & 0 & -1 & 2 \end{bmatrix}.$$

The cofactor method is most useful for a row with a lot of zeros. Here row 1 produces only two terms. The cofactor of a_{11} comes from erasing row 1 and column 1, which leaves a 3 by 3 matrix with the same pattern:

$$A_3 = \begin{bmatrix} 2 & -1 & 0 \\ -1 & 2 & -1 \\ 0 & -1 & 2 \end{bmatrix}.$$

For a_{12} it is column 2 that gets removed, and we need

$$(-1)^{1+2} \det \begin{bmatrix} -1 & -1 & 0 \\ 0 & 2 & -1 \\ 0 & -1 & 2 \end{bmatrix} = +\det \begin{bmatrix} 2 & -1 \\ -1 & 2 \end{bmatrix}.$$

Notice how cofactors got used again at the last step! This time the good choice was column 1 (because it only had one nonzero—which was the -1 that produced the plus sign). It left us with the 2 by 2 determinant of the same form; this is the cofactor of the original entry $a_{12} = -1$. Altogether row 1 has produced

$$\det A_4 = 2(\det A_3) - \det A_2.$$

The same idea applies to the 5 by 5 case, and 6 by 6, and n by n:

$$\det A_n = 2(\det A_{n-1}) - \det A_{n-2}. \tag{13}$$

Since we know det $A_1 = 2$ and det $A_2 = 3$, this *recursion formula* gives the determinant of increasingly bigger matrices. At every step the determinant of A_n is $n + 1$, because that satisfies (13):

$$n + 1 = 2(n) - (n - 1).$$

The answer $n + 1$ agrees with the product of pivots at the start of this section.

EXERCISES

4.3.1   For the matrix

$$A = \begin{bmatrix} 0 & 1 & 0 & 0 \\ 1 & 0 & 1 & 0 \\ 0 & 1 & 0 & 1 \\ 0 & 0 & 1 & 0 \end{bmatrix}$$

find the only nonzero term to appear in formula (6)—the only way of choosing four entries which come from different rows and different columns, without choosing any zeros. By deciding whether this permutation is even or odd, compute det A.

4.3.2   Carry out the expansion in cofactors for the first row of the preceding matrix A, and reduce det A to a 3 by 3 determinant. Do the same for that determinant (still watching the sign $(-1)^{i+j}$) and again for the resulting 2 by 2 determinant. Finally compute det A.

4.3.3   *True or false*: (1)   The determinant of $S^{-1}AS$ equals the determinant of A.
(2)   If det $A = 0$ then at least one of the cofactors must be zero.
(3)   A matrix whose entries are 0's and 1's has determinant 1, 0, or -1.

4.3.4   (a)   Find the LU factorization and the pivots and the determinant of the 4 by 4 matrix whose entries are $a_{ij} = $ smaller of i and j. (Write out the matrix.)
(b)   Find the determinant if $a_{ij} = $ smaller of n_i and n_j, where $n_1 = 2$, $n_2 = 6$, $n_3 = 8$, $n_4 = 10$. Can you give a general rule for any $n_1 \le n_2 \le n_3 \le n_4$?

4.3.5   Let D_n be the determinant of the 1, 1, -1 tridiagonal matrix (n by n)

$$D_n = \det \begin{bmatrix} 1 & -1 & & & \\ 1 & 1 & -1 & & \\ & 1 & 1 & -1 & \\ & & 1 & 1 & -1 \\ & & & \ddots & \ddots & \ddots \\ & & & & 1 & 1 \end{bmatrix}.$$

By expanding in cofactors along row 1 show that $D_n = D_{n-1} + D_{n-2}$. This yields the *Fibonacci sequence* 1, 2, 3, 5, 8, 13, ... for the determinants.

4.3.6 Suppose A_n is the n by n tridiagonal matrix with 1's everywhere on the three diagonals:

$$A_1 = [1], \quad A_2 = \begin{bmatrix} 1 & 1 \\ 1 & 1 \end{bmatrix}, \quad A_3 = \begin{bmatrix} 1 & 1 & 0 \\ 1 & 1 & 1 \\ 0 & 1 & 1 \end{bmatrix}, \ldots$$

Let D_n be the determinant of A_n; we want to find it.
(a) Expand in cofactors along the first row of A_n to show that $D_n = D_{n-1} - D_{n-2}$.
(b) Starting from $D_1 = 1$ and $D_2 = 0$ find $D_3, D_4, \ldots, D_8$. By noticing how these numbers cycle around (with what period?) find D_{1000}.

4.3.7 (a) Evaluate by cofactors of row 1:
$$\begin{vmatrix} 7 & 7 & 7 & 7 \\ 1 & 2 & 0 & 1 \\ 2 & 0 & 1 & 2 \\ 1 & 1 & 0 & 2 \end{vmatrix}.$$

(b) Check by subtracting column 1 from the other columns and recomputing.

4.3.8 Explain why a 5 by 5 matrix with a 3 by 3 zero submatrix is sure to be singular (regardless of the 16 nonzeros marked by x's):

$$\text{the determinant of } A = \begin{bmatrix} x & x & x & x & x \\ x & x & x & x & x \\ 0 & 0 & 0 & x & x \\ 0 & 0 & 0 & x & x \\ 0 & 0 & 0 & x & x \end{bmatrix} \text{ is zero.}$$

4.3.9 With 4 by 4 matrices show that in general

$$\det \begin{bmatrix} A & B \\ 0 & D \end{bmatrix} = \det A \det D \quad \text{but} \quad \det \begin{bmatrix} A & B \\ C & D \end{bmatrix} \neq \det A \det D - \det B \det C.$$

Here A, B, C, D are 2 by 2; give an example to establish the inequality part.

4.3.10 Compute the determinant of

$$A_4 = \begin{bmatrix} 0 & 1 & 1 & 1 \\ 1 & 0 & 1 & 1 \\ 1 & 1 & 0 & 1 \\ 1 & 1 & 1 & 0 \end{bmatrix}$$

either by using row operations to produce zeros or by expanding in cofactors of the first row. Find also the determinants of the smaller matrices A_3 and A_2, with the same pattern of zeros on the diagonal and ones elsewhere. Can you predict $\det A_n$?

4.3.11 How many multiplications to find an n by n determinant from
(a)   the explicit formula (6)?
(b)   the cofactor formula (10)?
(c)   the pivot formula (1)?
In (b) relate the count for n to the count for $n - 1$; in (c) remember elimination.

4.3.12 In a 5 by 5 matrix, does a $+$ sign or $-$ sign go with the product $a_{15}a_{24}a_{33}a_{42}a_{51}$ down the counterdiagonal? In other words, is $\sigma = (5, 4, 3, 2, 1)$ even or odd?
   Note: The checkerboard pattern of $\pm$ signs for cofactors does *not* give the sign of σ! It would give a $+$ sign for the 3 by 3 counterdiagonal $a_{13}a_{22}a_{31}$—which is wrong. You must decide whether the permutation is even or odd.

4.3.13 If A is m by n and B is n by m, show that

$$\det \begin{bmatrix} 0 & A \\ -B & I \end{bmatrix} = \det AB. \quad \left(Hint\text{: Postmultiply by } \begin{bmatrix} I & 0 \\ B & I \end{bmatrix}. \right)$$

Do an example with $m < n$ and an example with $m > n$. Why does the second example have $\det AB = 0$?

4.3.14 Suppose the matrix A is fixed, except that a_{11} varies from $-\infty$ to $+\infty$. Give examples in which $\det A$ is always zero or never zero. Then show (from the cofactor expansion (8)) that otherwise $\det A = 0$ for exactly one value of a_{11}.

APPLICATIONS OF DETERMINANTS ■ 4.4

This section follows through on the applications described in the introduction: *inverses*, *solution of* $Ax = b$, *volumes*, and *pivots*. They are among the key computations in linear algebra (done by elimination), and now we have formulas for the answers.

1. *The computation of* A^{-1}. This starts with the determinant from the cofactors along row i. Remember that each entry is multiplied by its cofactor:

$$\det A = a_{i1}A_{i1} + \cdots + a_{in}A_{in}. \tag{1}$$

We can create a matrix multiplication in which *that equation gives the answer along the main diagonal*:

$$
\begin{bmatrix}
a_{11} & a_{12} & \cdots & a_{1n} \\
a_{21} & a_{22} & \cdots & a_{2n} \\
\vdots & \vdots & & \vdots \\
a_{n1} & a_{n2} & \cdots & a_{nn}
\end{bmatrix}
\begin{bmatrix}
A_{11} & A_{21} & \cdots & A_{n1} \\
A_{12} & A_{22} & \cdots & A_{n2} \\
\vdots & \vdots & & \vdots \\
A_{1n} & A_{2n} & \cdots & A_{nn}
\end{bmatrix}
$$

$$
= \begin{bmatrix}
\det A & 0 & \cdots & 0 \\
0 & \det A & \cdots & 0 \\
\vdots & \vdots & & \vdots \\
0 & 0 & \cdots & \det A
\end{bmatrix}. \tag{2}
$$

Notice that the second matrix—the "*cofactor matrix*"—is **transposed**. We had to put $A_{11}, \ldots, A_{1n}$ into the first *column* and not the first row, so they would multiply $a_{11}, \ldots, a_{1n}$ and give the diagonal entry $\det A$. The other rows of A and columns of their cofactors give the same answer, $\det A$, on the diagonal. The critical question is: *Why do we get zeros everywhere off the diagonal?* If we combine the entries from row 1 with the cofactors associated with row 2, why is

$$a_{11}A_{21} + a_{12}A_{22} + \cdots + a_{1n}A_{2n} = 0? \tag{3}$$

The answer is: We are really computing the determinant of a new matrix B, which is the same as A except in row 2. The first row of A is copied into the second row of B. Therefore B has two equal rows, and $\det B = 0$. Equation (3) is the expansion of $\det B$ along row 2, where B has exactly the same cofactors as A (because the second row is thrown away to find the cofactors, and only that row is different from A). Thus the remarkable matrix multiplication (2) is correct.

That multiplication immediately gives A^{-1}. On the right side of (2) we have a multiple of the identity, $\det A$ times I:

$$(A)(A_{\text{cof}}) = (\det A)I. \tag{4}$$

A_{cof} is the **cofactor matrix**, or "adjugate matrix," that appears in equation (2). Remember that the cofactor from deleting row i and column j of A goes into *row j and column i* of the cofactor matrix. Dividing by the number det A (if it is not zero!) gives the formula for A^{-1}:

4C The entries of A^{-1} are the cofactors of A, transposed as in (2), divided by the determinant to give $A_{ji}/\det A$:

$$A^{-1} = \frac{1}{\det A} A_{\text{cof}}. \tag{5}$$

If det $A = 0$ then A is not invertible.

EXAMPLE 1   The cofactors of $\begin{bmatrix} a & b \\ c & d \end{bmatrix}$ are $A_{11} = d, A_{12} = -c, A_{21} = -b, A_{22} = a$:

$$(A)(A_{\text{cof}}) = \begin{bmatrix} a & b \\ c & d \end{bmatrix} \begin{bmatrix} d & -b \\ -c & a \end{bmatrix} = \begin{bmatrix} ad - bc & 0 \\ 0 & ad - bc \end{bmatrix}.$$

Dividing by $ad - bc$, which is det A, this is A times A^{-1}:

$$\begin{bmatrix} a & b \\ c & d \end{bmatrix}^{-1} = \frac{1}{ad - bc} \begin{bmatrix} d & -b \\ -c & a \end{bmatrix}. \tag{6}$$

Notice again the transposing in A_{cof} on the right.

EXAMPLE 2   The inverse of

$$A = \begin{bmatrix} 1 & 1 & 1 \\ 0 & 1 & 1 \\ 0 & 0 & 1 \end{bmatrix} \quad \text{is} \quad \frac{A_{\text{cof}}}{\det A} = \begin{bmatrix} 1 & -1 & 0 \\ 0 & 1 & -1 \\ 0 & 0 & 1 \end{bmatrix}.$$

The minus signs enter because cofactors include $(-1)^{i+j}$.

2. **The solution of** $Ax = b$. This second application is just multiplication of b by the matrix A^{-1} in (5):

$$x = A^{-1}b = \frac{1}{\det A} A_{\text{cof}} b.$$

This is simply the product of a matrix and a vector, divided by the number det A, but there is a famous way in which to write the answer:

4D   *Cramer's rule*: The jth component of $x = A^{-1}b$ is

$$x_j = \frac{\det B_j}{\det A}, \quad \text{where} \quad B_j = \begin{bmatrix} a_{11} & a_{12} & b_1 & a_{1n} \\ \vdots & \vdots & \vdots & \vdots \\ a_{n1} & a_{n2} & b_n & a_{nn} \end{bmatrix}. \tag{7}$$

In B_j, the vector b replaces the jth column of A.

Proof   Expand $\det B_j$ in cofactors of the jth column (which is b). Since the cofactors ignore that column, the result is

$$\det B_j = b_1 A_{1j} + b_2 A_{2j} + \cdots + b_n A_{nj}.$$

This is exactly the jth component in the matrix-vector product $A_{\text{cof}}b$. Dividing by $\det A$, the result is the jth component of x.

Thus each component of x is a ratio of two determinants, a polynomial of degree n divided by another polynomial of degree n. This fact might have been recognized from Gaussian elimination, but it never was.

EXAMPLE   The solution of

$$x_1 + 3x_2 = 0$$
$$2x_1 + 4x_2 = 6$$

is

$$x_1 = \frac{\begin{vmatrix} 0 & 3 \\ 6 & 4 \end{vmatrix}}{\begin{vmatrix} 1 & 3 \\ 2 & 4 \end{vmatrix}} = \frac{-18}{-2} = 9, \quad x_2 = \frac{\begin{vmatrix} 1 & 0 \\ 2 & 6 \end{vmatrix}}{\begin{vmatrix} 1 & 3 \\ 2 & 4 \end{vmatrix}} = \frac{6}{-2} = -3.$$

The denominators are the same, always $\det A$, while the right sides 0 and 6 appear in the first column of x_1 and the second column of x_2. For 1000 equations there would be 1001 determinants. To my dismay I found in an old book that this was actually recommended (and elimination was thrown aside):

"To deal with a set involving the four variables u, v, w, z, we first have to eliminate one of them in each of three pairs to derive three equations in three variables and then proceed as for the three-fold left-hand set to derive values for two of them. The reader who does so as an exercise will begin to realize how formidably laborious the method of elimination becomes, when we have to deal with more than three variables. This consideration invites us to explore the possibility of a *speedier method*" (which is Cramer's rule!!).†

For special matrices, when Cramer's rule can be carried through, it gives full information about the solution.

3.   *The volume of a parallelepiped.* The connection between the determinant and the volume is not at all obvious, but suppose first that all angles are *right angles*— the edges are mutually perpendicular, and we have a rectangular box. Then the volume is just the product of the lengths of the edges: volume $= l_1 l_2 \cdots l_n$.

We want to obtain the same formula from the determinant. Recall that the edges of the box were the rows of A. In our right-angled case, these rows are mutually orthogonal, and so

$$AA^T = \begin{bmatrix} \text{row } 1 \\ \vdots \\ \text{row } n \end{bmatrix} \begin{bmatrix} r & & r \\ o & & o \\ w & \cdots & w \\ 1 & & n \end{bmatrix} = \begin{bmatrix} l_1^2 & & 0 \\ & \ddots & \\ 0 & & l_n^2 \end{bmatrix}.$$

The l_i are the lengths of the rows (the edges), and the zeros off the diagonal come because the rows are orthogonal. Taking determinants, and using rules 9 and 10,

$$l_1^2 l_2^2 \cdots l_n^2 = \det(AA^T) = (\det A)(\det A^T) = (\det A)^2.$$

The square root of this equation is the required result: ***The determinant equals the volume.*** The *sign* of det A will indicate whether the edges form a "right-handed" set of coordinates, as in the usual *x-y-z* system, or a left-handed system like *y-x-z*.

If the region is not rectangular, then the volume is no longer the product of the edge lengths. In the plane (Fig. 4.2), the volume of a parallelogram equals the base l_1 times the height h. The vector pb of length h is the second row $b = (a_{21}, a_{22})$, minus its projection p onto the first row. The key point is this: By rule 5, det A is unchanged if a multiple of the first row is subtracted from the second row. At the same time, the volume is correct if we switch to a rectangle of base l_1 and height h. *We can return the problem to the rectangular case*, where it is already proved that volume = determinant.

In n dimensions, it takes longer to make each parallelepiped into a rectangular box, but the idea is the same. Neither the volume nor the determinant will be changed if, systematically for rows 2, 3, . . . , n, we subtract from each row its projection onto the space spanned by the preceding rows—leaving a "height" vector like pb which is perpendicular to the base. The result of this Gram-Schmidt process is a set of mutually orthogonal rows—with the same determinant and the same

---

† This quotation is from *Mathematics for the Millions* by Lancelot Hogben (1937). If he plans to use Cramer's rule, I call it *Mathematics for the Millionaire.*

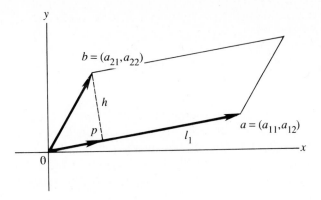

Fig. 4.2. Volume of a parallelogram = det A.

volume as the original set. Since volume = determinant in the rectangular case, the same equality must have held for the original rows.

This completes the link between volumes and determinants, but it is worth coming back one more time to the simplest case. We know that

$$\det \begin{bmatrix} 1 & 0 \\ 0 & 1 \end{bmatrix} = 1, \qquad \det \begin{bmatrix} 1 & 0 \\ c & 1 \end{bmatrix} = 1.$$

These determinants give the volumes—or areas, since we are in two dimensions— of the "parallelepipeds" drawn in Fig. 4.3. The first is the unit square, whose area is certainly 1. The second is a parallelogram with unit base and unit height; independent of the "shearing" produced by the coefficient c, its area is also equal to 1.

4.   *A formula for the pivots.* The last application is to the question of zero pivots; we can finally discover when Gaussian elimination is possible without row exchanges. The key observation is that the first k pivots are completely determined by the submatrix A_k in the upper left corner of A. *The remaining rows and columns of A have no effect on this corner of the problem.*

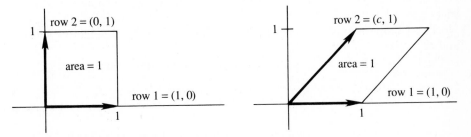

Fig. 4.3. The areas of a square and of a parallelogram.

EXAMPLE

$$A = \begin{bmatrix} a & b & e \\ c & d & f \\ g & h & i \end{bmatrix} \rightarrow \begin{bmatrix} a & b & e \\ 0 & (ad - bc)/a & (af - ec)/a \\ g & h & i \end{bmatrix}.$$

Certainly the first pivot depended only on the first row and column; it was $d_1 = a$. And the second pivot has become visible after a single elimination step; it is $(ad - bc)/a$, and it depends only on the entries a, b, c, and d. The rest of A does not enter until the third pivot. Actually it is not just the pivots, but the entire upper left corners of L, D, and U, which are determined by the upper left corner of A:

$$A = LDU = \begin{bmatrix} 1 & & \\ c/a & 1 & \\ * & * & 1 \end{bmatrix} \begin{bmatrix} a & & \\ & (ad - bc)/a & \\ & & * \end{bmatrix} \begin{bmatrix} 1 & b/a & * \\ & 1 & * \\ & & 1 \end{bmatrix}.$$

What we see in the first two rows and columns is exactly the factorization of the submatrix $A_2 = \begin{bmatrix} a & b \\ c & d \end{bmatrix}$. This is a general rule if there are no row exchanges:

4E If A is factored into LDU, then the upper left corners satisfy

$$A_k = L_k D_k U_k.$$

For every k, the submatrix A_k is going through a Gaussian elimination of its own.

The proof is either just to see that this corner can be settled first, before even looking at the eliminations elsewhere, or to use the laws for **block multiplication of matrices**. These laws are the same as the ordinary element by element rule: $LDU = A$ becomes

$$\begin{bmatrix} L_k & 0 \\ B & C \end{bmatrix} \begin{bmatrix} D_k & 0 \\ 0 & E \end{bmatrix} \begin{bmatrix} U_k & F \\ 0 & G \end{bmatrix} = \begin{bmatrix} L_k D_k U_k & L_k D_k F \\ B D_k U_k & B D_k F + CEG \end{bmatrix}.$$

As long as matrices are properly partitioned—the square or rectangular submatrices are the right sizes for multiplication—they can be multiplied this way in blocks.† Comparing the last matrix with A, the corner $L_k D_k U_k$ coincides with A_k and **4E** is correct.

The formulas for the pivots follow immediately by taking determinants:

$$\det A_k = \det L_k \det D_k \det U_k = \det D_k = d_1 d_2 \cdots d_k. \tag{8}$$

---

† This is a very useful rule, even though we have not met it since Chapter 1.

The product of the first k pivots is the determinant of A_k. This is the same rule for A_k that we know already for the whole matrix $A = A_n$. Since the determinant of A_{k-1} will be given similarly by $d_1 d_2 \cdots d_{k-1}$, we can isolate the pivot d_k as a *ratio of determinants*:

$$\frac{\det A_k}{\det A_{k-1}} = \frac{d_1 d_2 \cdots d_k}{d_1 d_2 \cdots d_{k-1}} = d_k. \tag{9}$$

In our example above, the second pivot was exactly this ratio $(ad - bc)/a$. It is the determinant of A_2 divided by the determinant of A_1. (By convention $\det A_0 = 1$, so that the first pivot is $a/1 = a$.) Multiplying together all the individual pivots, we recover

$$d_1 d_2 \cdots d_n = \frac{\det A_1}{\det A_0} \frac{\det A_2}{\det A_1} \cdots \frac{\det A_n}{\det A_{n-1}} = \frac{\det A_n}{\det A_0} = \det A.$$

From (9) we can finally read off the answer to our original question: ***The pivot entries are all nonzero whenever the numbers*** $\det A_k$ ***are all nonzero***:

4F Gaussian elimination can be carried out on A, without row exchanges or a permutation matrix, if and only if the leading submatrices $A_1, A_2, \ldots, A_n$ are all nonsingular.

That does it for determinants, except for the optional remark promised at the beginning of the chapter. That remark concerns the self-consistency of the defining properties 1-3. The key is property 2, the sign reversal on row exchanges, which led to the ***determinant of a permutation matrix***. This was the only questionable point in the explicit formula (6): Is it true that, independent of the particular sequence of row exchanges linking P_σ to the identity, the number of exchanges is either always even or always odd? If so, we are justified in calling the permutation "even" or "odd," and its determinant is well defined by rule 2 as either $+1$ or -1.

Starting from the permutation $(3, 2, 1)$, a single exchange of 3 and 1 would achieve the natural order $(1, 2, 3)$. So would an exchange of 3 and 2, then 3 and 1, and then 2 and 1. In both sequences, the number of exchanges is odd. The assertion is that *an even number of exchanges can never produce the natural order, beginning with* $(3, 2, 1)$.

Here is a proof. Look at each pair of numbers in the permutation, and let N count the pairs in which the larger number comes first. Certainly $N = 0$ for the natural order $(1, 2, 3)$, the identity permutation; and $N = 3$ for the order $(3, 2, 1)$, since all the pairs $(3, 2)$, $(3, 1)$, and $(2, 1)$ are wrong. The point is to show that the permutation is odd or even according as N is odd or even. In other words, starting with any permutation, every exchange will alter N by an odd number. Then to arrive at $N = 0$ (the natural order) takes a number of exchanges having the same parity—evenness or oddness—as the initial N.

If the pair being exchanged lie next to one another, obviously N changes by $+1$ or -1, both of which are odd numbers. Once it is observed that *any exchange can be achieved by an odd number of exchanges of neighbors*, the proof is complete; an odd number of odd numbers is odd. This observation is easy to confirm by an example; to exchange the first and fourth entries below, which happen to be 2 and 3, we use five exchanges (an odd number) of neighbors:

$$(2, 1, 4, 3) \to (1, 2, 4, 3) \to (1, 4, 2, 3) \to (1, 4, 3, 2) \to (1, 3, 4, 2) \to (3, 1, 4, 2).$$

In general we need $l - k$ exchanges of neighbors to move the entry in place k to place l. Then $l - k - 1$ exchanges move the one originally in place l (and now found in place $l - 1$) back down to place k. Since $(l - k) + (l - k - 1)$ is odd, the proof is complete. The determinant not only has all the properties found earlier, it even exists.

EXERCISES

4.4.1   Find the determinant and all nine cofactors of

$$A = \begin{bmatrix} 1 & 2 & 3 \\ 0 & 4 & 0 \\ 0 & 0 & 5 \end{bmatrix}.$$

Then form the cofactor matrix whose i, j entry is the cofactor A_{ji}. Verify that A times A_{cof} is the identity matrix times the determinant. What is A^{-1}?

4.4.2   From the formula $A_{cof}/\det A$ for the inverse, explain why A^{-1} is upper triangular if A is upper triangular (and invertible).

4.4.3   Use the cofactor matrix to invert

$$A = \begin{bmatrix} 2 & -1 & 0 \\ -1 & 2 & -1 \\ 0 & -1 & 2 \end{bmatrix} \quad \text{and} \quad B = \begin{bmatrix} 1 & 1 & 1 \\ 1 & 2 & 2 \\ 1 & 2 & 3 \end{bmatrix}.$$

4.4.4   From the formula $A_{cof}/\det A$ for the inverse, explain why A^{-1} is symmetric if A is symmetric (and invertible).

4.4.5   Find x, y, and z by Cramer's rule:

$$\begin{aligned} ax + by &= 1 \\ cx + dy &= 0 \end{aligned} \quad \text{and} \quad \begin{aligned} x + 4y - z &= 1 \\ x + y + z &= 0 \\ 2x \quad\quad + 3z &= 0. \end{aligned}$$

4.4.6 (a) Find the determinant when a vector x replaces column j of the identity:

$$\text{if} \quad M = \begin{bmatrix} 1 & & x_1 & & \\ & 1 & \cdot & & \\ & & x_j & & \\ & & \cdot & 1 & \\ & & x_n & & 1 \end{bmatrix} \quad \text{then} \quad \det M = ?$$

(b) If $Ax = b$ show that AM is the matrix B_j in equation (7); the right side b enters column j.

(c) Derive *Cramer's rule* by taking determinants in $AM = B_j$.

4.4.7 Find the Jacobian determinant J for the change from rectangular coordinates x, y, z to spherical coordinates r, θ, ϕ: $x = r \cos \theta \cos \phi$, $y = r \sin \theta \cos \phi$, $z = r \sin \phi$.

4.4.8 (a) Draw the triangle whose vertices are $A = (2, 2)$, $B = (-1, 3)$, and $C = (0, 0)$. By regarding it as half of a parallelogram, explain why its area equals

$$\text{area}(ABC) = \tfrac{1}{2} \det \begin{bmatrix} 2 & 2 \\ -1 & 3 \end{bmatrix}.$$

(b) Suppose the third vertex is $C = (1, -4)$ instead of $(0, 0)$. Justify the formula

$$\text{area}(ABC) = \tfrac{1}{2} \det \begin{bmatrix} x_1 & y_1 & 1 \\ x_2 & y_2 & 1 \\ x_3 & y_3 & 1 \end{bmatrix} = \tfrac{1}{2} \det \begin{bmatrix} 2 & 2 & 1 \\ -1 & 3 & 1 \\ 1 & -4 & 1 \end{bmatrix}.$$

Hint: Subtracting the last row from each of the others leaves

$$\det \begin{bmatrix} 2 & 2 & 1 \\ -1 & 3 & 1 \\ 1 & -4 & 1 \end{bmatrix} = \det \begin{bmatrix} 1 & 6 & 0 \\ -2 & 7 & 0 \\ 1 & -4 & 1 \end{bmatrix} = \det \begin{bmatrix} 1 & 6 \\ -2 & 7 \end{bmatrix}.$$

Sketch the vertices $A' = (1, 6)$, $B' = (-2, 7)$, $C' = (0, 0)$ and their relation to A, B, C.

4.4.9 Explain in terms of volumes why $\det 3A = 3^n \det A$ for an n by n matrix A.

4.4.10 Block elimination gives, if the pivot block A is invertible,

$$\begin{bmatrix} I & 0 \\ -CA^{-1} & I \end{bmatrix} \begin{bmatrix} A & B \\ C & D \end{bmatrix} = \begin{bmatrix} A & B \\ 0 & D - CA^{-1}B \end{bmatrix}.$$

The matrix $D - CA^{-1}B$ is called a *Schur complement*. Show that its determinant times $\det A$ equals the determinant of the original block matrix on the left. If $AC = CA$ show that this becomes $\det(AD - CB)$.

4.4.11 What rules for the product AB come from block multiplication with long thin blocks, which are columns b and c and rows r?

$$AB = \begin{bmatrix} & A & \end{bmatrix} \begin{bmatrix} b_1 \cdots b_n \end{bmatrix} \quad \text{and} \quad AB = \begin{bmatrix} c_1 \cdots c_n \end{bmatrix} \begin{bmatrix} r_1 \\ \vdots \\ r_n \end{bmatrix}.$$

4.4.12 Predict in advance, and confirm by elimination, the pivot entries of

$$A = \begin{bmatrix} 2 & 1 & 2 \\ 4 & 5 & 0 \\ 2 & 7 & 0 \end{bmatrix} \quad \text{and} \quad B = \begin{bmatrix} 2 & 1 & 2 \\ 4 & 5 & 3 \\ 2 & 7 & 0 \end{bmatrix}.$$

4.4.13 Find all the odd permutations of the numbers $\{1, 2, 3, 4\}$. These are the permutations coming from an odd number of exchanges and leading to $\det P_\sigma = -1$.

4.4.14 Suppose the permutation σ takes $(1, 2, 3, 4, 5)$ to $(5, 4, 1, 2, 3)$.
(a) What does σ^2 do to $(1, 2, 3, 4, 5)$?
(b) What does σ^{-1} do to $(1, 2, 3, 4, 5)$?

4.4.15 If σ is an odd permutation, explain why σ^2 is even but σ^{-1} is odd. Give an example with $n = 3$.

4.4.16 Prove that if you keep multiplying A by the same permutation matrix P, the first row eventually comes back to its original place.

4.4.17 If A is a 5 by 5 matrix with all $|a_{ij}| \leq 1$, then $\det A \leq$?. (I do not know the best bound, but volumes or formula (6) or pivots should give some upper bound on the determinant.)

REVIEW EXERCISES: Chapter 4

4.1 Where can you put zeros into a 4 by 4 matrix, using as few as possible but enough to guarantee that the determinant is zero?

4.2 Where can you put zeros and ones into a 4 by 4 matrix, using as few as possible but enough to guarantee that the determinant is one?

4.3 Find the determinants of

$$\begin{bmatrix} 1 & 1 & 1 & 1 \\ 1 & 1 & 1 & 2 \\ 1 & 1 & 3 & 1 \\ 1 & 4 & 1 & 1 \end{bmatrix} \quad \text{and} \quad \begin{bmatrix} 2 & -1 & 0 & -1 \\ -1 & 2 & -1 & 0 \\ 0 & -1 & 2 & -1 \\ -1 & 0 & -1 & 2 \end{bmatrix}.$$

4.4 If $B = M^{-1} AM$, why is $\det B = \det A$? Show also that $\det A^{-1}B = 1$.

4.5 Give a counterexample to $\det(A + B) = \det A + \det B$. For what size matrices is that statement true?

4.6 Starting with a matrix A, multiply its first row by 3 to produce B, and then subtract the first row of B from the second to produce C. How is $\det C$ related to $\det A$?

4.7 Solve $3u + 2v = 7, 4u + 3v = 11$ by Cramer's rule.

4.8 If the entries of A are integers, and $\det A$ is 1 or -1, how do you know that the entries of A^{-1} are integers? Give a 2 by 2 example (not diagonal).

4.9 If the entries of A and A^{-1} are all integers, how do you know that both determinants are 1 or -1? *Hint*: What is $\det A$ times $\det A^{-1}$?

4.10 Find all the cofactors, and the inverse, of

$$\begin{bmatrix} 3 & 5 \\ 6 & 9 \end{bmatrix} \quad \text{and} \quad \begin{bmatrix} a & b \\ b & a \end{bmatrix} \quad \text{and} \quad \begin{bmatrix} a & b \\ c & d \end{bmatrix}^{-1}.$$

4.11 What is the volume of the parallelepiped with four of its vertices at $(0, 0, 0), (-1, 2, 2)$, $(2, -1, 2)$, and $(2, 2, -1)$? Where are the other four vertices?

4.12 How many terms are in the expansion of a 5 by 5 determinant, and how many are sure to be zero if $a_{21} = 0$?

4.13 If every row of A adds up to zero, and x is a column vector of ones, what is Ax? How do you know that $\det A = 0$?

4.14 Why are there an even number of permutations of $(1, 2, \ldots, 9)$, and why are exactly half of them odd permutations?

4.15 If P_1 is an even permutation matrix and P_2 is odd, deduce from $P_1 + P_2 = P_1(P_1^T + P_2^T)P_2$ that $\det(P_1 + P_2) = 0$.

4.16 Find the determinant of A, if $a_{ij} = i + j$.

4.17 If A has a positive determinant, show that it can be connected to the identity matrix by a continuous chain of matrices all having positive determinants. Then A is changed into the identity without becoming singular on the way. (The straight path from A to I, with a chain $A(t) = A + t(I - A)$, does go from $A(0) = A$ to $A(1) = I$, but in between $A(t)$ might be singular. The problem is not so easy, and solutions are welcomed by the author.)

4.18 If A is nonsingular, show that there must be some permutation P for which PA has no zeros on its main diagonal. It is *not* the P from elimination.

4.19 Explain why the point (x, y) is on the line through $(2, 8)$ and $(4, 7)$ if

$$\det \begin{bmatrix} x & y & 1 \\ 2 & 8 & 1 \\ 4 & 7 & 1 \end{bmatrix} = 0, \quad \text{or} \quad x + 2y - 18 = 0.$$

4.20 In analogy with the previous exercise, what is the equation for (x, y, z) to be on the plane through $(2, 0, 0)$, $(0, 2, 0)$, and $(0, 0, 4)$? It involves a 4 by 4 determinant.

4.21 If the points (x, y, z), $(2, 1, 0)$, and $(1, 1, 1)$ lie on a plane through the origin, what determinant is zero? Are the vectors $(1, 0, -1)$, $(2, 1, 0)$, $(1, 1, 1)$ dependent or independent?

4.22 If every row of A has either a single $+1$, or a single -1, or one of each (and is otherwise zero) show that $\det A = 1$ or -1 or 0.

4.23 If $C = \begin{bmatrix} a & b \\ c & d \end{bmatrix}$ and $D = \begin{bmatrix} u & v \\ w & z \end{bmatrix}$ then $CD = -DC$ in Exercise 1.4.14 becomes

$$CD + DC = 0 \quad \text{or} \quad \begin{bmatrix} 2a & c & b & 0 \\ b & a+d & 0 & b \\ c & 0 & a+d & c \\ 0 & c & b & 2d \end{bmatrix} \begin{bmatrix} u \\ v \\ w \\ z \end{bmatrix} = \begin{bmatrix} 0 \\ 0 \\ 0 \\ 0 \end{bmatrix}.$$

(a) Find the determinant of this 4 by 4 coefficient matrix A.
(b) Show that $\det A = 0$ in two cases: $a + d = 0$ or $ad - bc = 0$.

In all other cases, $CD = -DC$ is only possible with $D = 0$.

4.24 Explain why the 4 by 4 Vandermonde determinant

$$V_4 = \det \begin{bmatrix} 1 & a & a^2 & a^3 \\ 1 & b & b^2 & b^3 \\ 1 & c & c^2 & c^3 \\ 1 & x & x^2 & x^3 \end{bmatrix}$$

must be a cubic polynomial in x, and why that polynomial is zero at $x = a$, $x = b$, and $x = c$. Using the cofactor of x^3 from Exercise 4.2.10 leads to

$$V_4 = (b - a)(c - a)(c - b)(x - a)(x - b)(x - c).$$

4.25 The circular shift permutes $(1, 2, \ldots, n)$ into $(2, 3, \ldots, 1)$. What is the corresponding permutation matrix P, and (depending on n) what is its determinant?

5

EIGENVALUES AND EIGENVECTORS

This chapter begins the "second half" of matrix theory. The first part was almost completely involved with linear systems $Ax = b$, and the fundamental technique was elimination. From now on row operations will play only a minor role. The new problems will still be solved by simplifying a matrix—making it diagonal or upper triangular—but *the basic step is no longer to subtract a multiple of one row from another*. We are not interested any more in preserving the row space of a matrix, but in preserving its eigenvalues. Those are changed by elimination.

The chapter on determinants was really a transition from the old problem $Ax = b$ to the new problem of eigenvalues. In both cases the determinant leads to a "formal solution": to Cramer's rule for $x = A^{-1}b$ and to the polynomial $\det(A - \lambda I)$ whose roots will be the eigenvalues. (We emphasize that all matrices are now square; the eigenvalues of a rectangular matrix make no more sense than its determinant.) As always, the determinant can actually be used to solve the problem, if $n = 2$ or 3. For large n the computation of eigenvalues is a longer and more difficult task than solving $Ax = b$, and even Gauss himself did not help much. But that can wait.

The first step is to understand what eigenvalues are and how they can be useful. One of their applications, the one by which we want to introduce them, is to the solution of ordinary differential equations. We shall not assume that the reader is an expert on differential equations! If you can differentiate the usual functions like x^n, $\sin x$, and e^x, you know enough. As a specific example, consider the coupled pair of equations

$$\frac{dv}{dt} = 4v - 5w, \quad v = 8 \text{ at } t = 0,$$

$$\frac{dw}{dt} = 2v - 3w, \quad w = 5 \text{ at } t = 0. \tag{1}$$

This is an initial-value problem. The unknown is specified at time $t = 0$, and not at both endpoints of an interval; we are interested in a transient rather than a steady state. The system evolves in time from the given initial values 8 and 5, and the problem is to follow this evolution.

It is easy to write the system in matrix form. Let the unknown vector be u, its initial value be u_0, and the coefficient matrix be A:

$$u(t) = \begin{bmatrix} v(t) \\ w(t) \end{bmatrix}, \quad u_0 = \begin{bmatrix} 8 \\ 5 \end{bmatrix}, \quad A = \begin{bmatrix} 4 & -5 \\ 2 & -3 \end{bmatrix}.$$

In this notation, the system becomes a vector equation

$$\boxed{\frac{du}{dt} = Au, \quad u = u_0 \text{ at } t = 0.} \tag{2}$$

This is the basic statement of the problem. Note that it is a first-order equation—no higher derivatives appear—and it is *linear* in the unknowns. It also has *constant coefficients*; the matrix A is independent of time.

How do we find the solution? If there were only one unknown instead of two, that question would be easy to answer. We would have a scalar instead of a vector differential equation. If it is again homogeneous with constant coefficients, it can only be

$$\frac{du}{dt} = au, \quad u = u_0 \text{ at } t = 0. \tag{3}$$

The solution is the one thing you need to know:

$$u(t) = e^{at}u_0. \tag{4}$$

At the initial time $t = 0$, u equals u_0 because $e^0 = 1$. The derivative of e^{at} has the required factor a, so that $du/dt = au$. Thus the initial condition and the equation are both satisfied.

Notice the behavior of u for large times. The equation is unstable if $a > 0$, neutrally stable if $a = 0$, or stable if $a < 0$; the solution approaches infinity, remains bounded, or goes to zero. If a were a complex number, $a = \alpha + i\beta$, then the same tests would be applied to the real part α. The complex part produces oscillations $e^{i\beta t} = \cos \beta t + i \sin \beta t$; but stability is governed by the factor $e^{\alpha t}$.

So much for a single equation. We shall take a direct approach to systems, and look for solutions with the *same exponential dependence on t* just found in the scalar case. In other words, we look for solutions of the form

$$v(t) = e^{\lambda t}y$$
$$w(t) = e^{\lambda t}z, \tag{5}$$

or in vector notation

$$u(t) = e^{\lambda t}x. \tag{6}$$

This is the whole key to differential equations $du/dt = Au$: **Look for pure exponential solutions.** Substituting $v = e^{\lambda t}y$ and $w = e^{\lambda t}z$ into the equation we find

$$\lambda e^{\lambda t}y = 4e^{\lambda t}y - 5e^{\lambda t}z$$
$$\lambda e^{\lambda t}z = 2e^{\lambda t}y - 3e^{\lambda t}z.$$

The factor $e^{\lambda t}$ is common to every term, and can be removed. This cancellation is the reason for assuming the same exponent λ for both unknowns; it leaves

$$4y - 5z = \lambda y$$
$$2y - 3z = \lambda z. \tag{7}$$

That is the basic equation; in matrix form it is $Ax = \lambda x$. You can see it again if we use the vector solution $u = e^{\lambda t}x$—a number $e^{\lambda t}$ that grows or decays times a fixed vector x. **The substitution of $u = e^{\lambda t}x$ into $du/dt = Au$ gives $\lambda e^{\lambda t}x = Ae^{\lambda t}x$, and the cancellation produces**

$$\boxed{Ax = \lambda x.} \tag{8}$$

Now we have the fundamental equation of this chapter. It involves two unknowns λ and x, and it is an algebra problem. The differential equations can be forgotten! The number λ (lambda) is called an *eigenvalue* of the matrix A, and the vector x is the associated *eigenvector*. Our goal is to find the eigenvalues and eigenvectors, and to use them.

The Solutions of $Ax = \lambda$

Notice that $Ax = \lambda x$ is a nonlinear equation; λ multiplies x. If we could discover λ, then the equation for x would be linear. In fact we could write λIx in place of λx,† and bring this term over to the left side:

$$\boxed{(A - \lambda I)x = 0.} \tag{9}$$

---

† The identity matrix is needed to keep matrices and vectors and scalars straight; the equation $(A - \lambda)x = 0$ is shorter, but mixed up.

This is the key to the problem:

> **The vector x is in the nullspace of $A - \lambda I$**
>
> **The number λ is chosen so that $A - \lambda I$ has a nullspace.**

Of course every matrix has a nullspace. It was ridiculous to suggest otherwise, but you see the point. We want a *nonzero* eigenvector x. The vector $x = 0$ always satisfies $Ax = \lambda x$, and it is always in the nullspace, but it is useless in solving differential equations. The goal is to build $u(t)$ out of exponentials $e^{\lambda t}x$, and *we are interested only in those particular values λ for which there is a nonzero eigenvector x.* To be of any use, the nullspace of $A - \lambda I$ must contain vectors other than zero. In short, $A - \lambda I$ **must be singular.**

For this, the determinant gives a conclusive test.

5A   The number λ is an eigenvalue of A if and only if

$$\det(A - \lambda I) = 0. \tag{10}$$

This is the characteristic equation, and each solution λ has a corresponding eigenvector x:

$$(A - \lambda I)x = 0 \quad \text{or} \quad Ax = \lambda x. \tag{11}$$

In our example, shifting A by λI gives

$$A - \lambda I = \begin{bmatrix} 4 - \lambda & -5 \\ 2 & -3 - \lambda \end{bmatrix}.$$

Note that λ is subtracted only from the main diagonal (because it multiplies I). The determinant of $A - \lambda I$ is

$$(4 - \lambda)(-3 - \lambda) + 10 \quad \text{or} \quad \lambda^2 - \lambda - 2.$$

This is the "*characteristic polynomial.*" Its roots, where the determinant is zero, are the eigenvalues. They come from the general formula for the roots of a quadratic, or from factoring into $\lambda^2 - \lambda - 2 = (\lambda + 1)(\lambda - 2)$. That is zero if $\lambda = -1$ or $\lambda = 2$, as the general formula confirms:

$$\lambda = \frac{-b \pm \sqrt{b^2 - 4ac}}{2a} = \frac{1 \pm \sqrt{9}}{2} = -1 \text{ or } 2.$$

There are two eigenvalues, because a quadratic has two roots. Every 2 by 2 matrix $A - \lambda I$ has λ^2 (and no higher power) in its determinant.

Each of these special values, $\lambda = -1$ and $\lambda = 2$, leads to a solution of $Ax = \lambda x$ or $(A - \lambda I)x = 0$. A matrix with zero determinant is singular, so there must be a nonzero vector x in its nullspace.† In fact the nullspace contains a whole *line* of eigenvectors; it is a subspace!

$$\lambda_1 = -1: \qquad (A - \lambda_1 I)x = \begin{bmatrix} 5 & -5 \\ 2 & -2 \end{bmatrix} \begin{bmatrix} y \\ z \end{bmatrix} = \begin{bmatrix} 0 \\ 0 \end{bmatrix}.$$

The solution (the first eigenvector) is any multiple of

$$x_1 = \begin{bmatrix} 1 \\ 1 \end{bmatrix}.$$

The computation for λ_2 is done separately:

$$\lambda_2 = 2: \qquad (A - \lambda_2 I)x = \begin{bmatrix} 2 & -5 \\ 2 & -5 \end{bmatrix} \begin{bmatrix} y \\ z \end{bmatrix} = \begin{bmatrix} 0 \\ 0 \end{bmatrix}.$$

The second eigenvector is any multiple of

$$x_2 = \begin{bmatrix} 5 \\ 2 \end{bmatrix}.$$

Note on computing eigenvectors: In the 2 by 2 case, both rows of $A - \lambda I$ will be multiples of the same vector (a, b). Then the eigenvector is any multiple of $(-b, a)$. The rows of $A - \lambda_2 I$ were $(2, -5)$ and the eigenvector was $(5, 2)$. In the 3 by 3 case, I often set a component of x equal to one and solve $(A - \lambda I)x = 0$ for the other components. Of course if x is an eigenvector then so is $7x$ and so is $-x$. All vectors in the nullspace of $A - \lambda I$ (which we call the *eigenspace*) will satisfy $Ax = \lambda x$. In this case the eigenspaces are the lines through $x_1 = (1, 1)$ and $x_2 = (5, 2)$.

Before going back to the application (the differential equation), we emphasize the steps in solving the eigenvalue problem:

1. *Compute the determinant of $A - \lambda I$.* With λ subtracted along the diagonal, this determinant is a polynomial of degree n.
2. *Find the roots of this polynomial.* The n roots are the eigenvalues.
3. *For each eigenvalue solve the equation $(A - \lambda I)x = 0$.* Since the determinant is zero, there are solutions other than $x = 0$. Those are the eigenvectors.

---

† If solving $(A - \lambda I)x = 0$ leads you to $x = 0$, then λ is not an eigenvalue.

In the differential equation, this produces the special solutions $u = e^{\lambda t}x$. They are the *pure exponential solutions*

$$u = e^{\lambda_1 t}x_1 = e^{-t}\begin{bmatrix}1\\1\end{bmatrix} \quad \text{and} \quad u = e^{\lambda_2 t}x_2 = e^{2t}\begin{bmatrix}5\\2\end{bmatrix}.$$

More than that, these two special solutions give the complete solution. They can be multiplied by any numbers c_1 and c_2, and they can be added together. When two functions u_1 and u_2 satisfy the linear equation $du/dt = Au$, so does their sum $u_1 + u_2$. Thus any combination

$$u = c_1 e^{\lambda_1 t}x_1 + c_2 e^{\lambda_2 t}x_2 \tag{12}$$

is again a solution. This is **superposition**, and it applies to differential equations (homogeneous and linear) just as it applied to algebraic equations $Ax = 0$. The nullspace is always a subspace, and combinations of solutions are still solutions.

Now we have two free parameters c_1 and c_2, and it is reasonable to hope that they can be chosen to satisfy the initial condition $u = u_0$ at $t = 0$:

$$c_1 x_1 + c_2 x_2 = u_0 \quad \text{or} \quad \begin{bmatrix}1 & 5\\1 & 2\end{bmatrix}\begin{bmatrix}c_1\\c_2\end{bmatrix} = \begin{bmatrix}8\\5\end{bmatrix}. \tag{13}$$

The constants are $c_1 = 3$ and $c_2 = 1$, and **the solution to the original equation is**

$$u(t) = 3e^{-t}\begin{bmatrix}1\\1\end{bmatrix} + e^{2t}\begin{bmatrix}5\\2\end{bmatrix}. \tag{14}$$

Writing the two components separately, this means that

$$v(t) = 3e^{-t} + 5e^{2t}, \quad w(t) = 3e^{-t} + 2e^{2t}.$$

The initial conditions $v_0 = 8$ and $w_0 = 5$ are easily checked.

The message seems to be that the key to an equation is in its eigenvalues and eigenvectors. But what the example does not show is their physical significance; they are important in themselves, and not just part of a trick for finding u. Probably the homeliest example† is that of soldiers going over a bridge. Traditionally, they stop marching and just walk across. The reason is that they might happen to march at a frequency equal to one of the eigenvalues of the bridge, and it would begin to oscillate. (Just as a child's swing does; you soon notice the natural frequency of a swing, and by matching it you make the swing go higher.) An engineer tries to keep the natural frequencies of his bridge or rocket away from those of the wind or the sloshing of fuel. And at the other extreme, a stockbroker spends

---

† One which I never really believed. But a bridge did crash this way in 1831.

his life trying to get in line with the natural frequencies of the market. The eigenvalues are the most important feature of practically any dynamical system.

We stop now to summarize what has been done, and what there remains to do. This introduction has shown how the eigenvalues and eigenvectors of A appear naturally and automatically when solving $du/dt = Au$. Such an equation has *pure exponential solutions* $u = e^{\lambda t}x$; the eigenvalue gives the rate of growth or decay, and the eigenvector x develops at this rate. The other solutions will be mixtures of these pure solutions, and the mixture is adjusted to fit the initial conditions.

The key equation was $Ax = \lambda x$. Most vectors x will not satisfy such an equation. A typical x changes direction when multiplied by A, so that Ax is not a multiple of x. This means that *only certain special numbers λ are eigenvalues, and only certain special vectors x are eigenvectors*. Of course, if A were a multiple of the identity matrix, then no vector would change direction, and all vectors would be eigenvectors. But in the usual case, eigenvectors are few and far between. They are the "normal modes" of the system, and they act independently. We can watch the behavior of each eigenvector, and then combine these normal modes to find the solution. To say the same thing in another way, *the underlying matrix can be diagonalized*.

We plan to devote Section 5.2 to the theory of diagonalization, and the following sections to its applications: first to difference equations and Fibonacci numbers and Markov processes, and afterward to differential equations. In every example, we start by computing the eigenvalues and eigenvectors; there is no shortcut to avoid that. But then the examples go in so many directions that a quick summary is impossible, except to emphasize that symmetric matrices are especially easy and certain other "defective matrices" are especially hard. They lack a full set of eigenvectors, they are not diagonalizable, and they produce a breakdown in the technique of normal modes. Certainly they have to be discussed, but we do not intend to allow them to take over the book.

We start with examples of particularly good matrices.

EXAMPLE 1 Everything is clear when A is *diagonal*:

$$A = \begin{bmatrix} 3 & 0 \\ 0 & 2 \end{bmatrix} \text{ has } \lambda_1 = 3 \text{ with } x_1 = \begin{bmatrix} 1 \\ 0 \end{bmatrix}, \quad \lambda_2 = 2 \text{ with } x_2 = \begin{bmatrix} 0 \\ 1 \end{bmatrix}.$$

On each eigenvector A acts like a multiple of the identity: $Ax_1 = 3x_1$ and $Ax_2 = 2x_2$. Other vectors like $x = (1, 5)$ are mixtures $x_1 + 5x_2$ of the two eigenvectors, and when A multiplies x it gives

$$Ax = \lambda_1 x_1 + 5\lambda_2 x_2 = \begin{bmatrix} 3 \\ 10 \end{bmatrix}.$$

This was a typical vector x—not an eigenvector—but the action of A was still determined by its eigenvectors and eigenvalues.

EXAMPLE 2   The situation is also good for a **projection**:

$$P = \begin{bmatrix} \frac{1}{2} & \frac{1}{2} \\ \frac{1}{2} & \frac{1}{2} \end{bmatrix} \text{ has } \lambda_1 = 1 \text{ with } x_1 = \begin{bmatrix} 1 \\ 1 \end{bmatrix}, \quad \lambda_2 = 0 \text{ with } x_2 = \begin{bmatrix} 1 \\ -1 \end{bmatrix}.$$

The eigenvalues of a projection are one or zero! We have $\lambda = 1$ when the vector projects to itself, and $\lambda = 0$ when it projects to the zero vector. The column space of P is filled with eigenvectors and so is the nullspace. If those spaces have dimension r and $n - r$, then $\lambda = 1$ is repeated r times and $\lambda = 0$ is repeated $n - r$ times:

$$P = \begin{bmatrix} 1 & 0 & 0 & 0 \\ 0 & 0 & 0 & 0 \\ 0 & 0 & 0 & 0 \\ 0 & 0 & 0 & 1 \end{bmatrix} \quad \text{has} \quad \lambda = 1, 1, 0, 0.$$

There are still four eigenvalues, even if not distinct, when P is 4 by 4.

Notice that **there is nothing exceptional about** $\lambda = 0$. Like every other number, zero might be an eigenvalue and it might not. If it is, then its eigenvectors satisfy $Ax = 0x$. Thus x is in the nullspace of A. A zero eigenvalue signals that A has linearly dependent columns and rows; its determinant is zero. Invertible matrices have all $\lambda \neq 0$, whereas singular matrices include zero among their eigenvalues.

EXAMPLE 3   The eigenvalues are still obvious when A is **triangular**:

$$\det(A - \lambda I) = \begin{vmatrix} 1 - \lambda & 4 & 5 \\ 0 & \frac{3}{4} - \lambda & 6 \\ 0 & 0 & \frac{1}{2} - \lambda \end{vmatrix} = (1 - \lambda)(\tfrac{3}{4} - \lambda)(\tfrac{1}{2} - \lambda).$$

The determinant is just the product of the diagonal entries. It is zero if $\lambda = 1$, or $\lambda = \frac{3}{4}$, or $\lambda = \frac{1}{2}$; the eigenvalues were already sitting along the main diagonal.

This example, in which the eigenvalues can be found by inspection, points to one main theme of the whole chapter: To transform A into a diagonal or triangular matrix *without changing its eigenvalues*. We emphasize once more that the Gaussian factorization $A = LU$ is not suited to this purpose. The eigenvalues of U may be visible on the diagonal, but they are **not** the eigenvalues of A.

For most matrices, there is no doubt that the eigenvalue problem is computationally more difficult than $Ax = b$. With linear systems, a finite number of elimination steps produced the exact answer in a finite time. (Or equivalently, Cramer's rule gave an exact formula for the solution.) In the case of eigenvalues, no such steps and no such formula can exist, or Galois would turn in his grave. The characteristic polynomial of a 5 by 5 matrix is a quintic, and he proved that there can be no algebraic formula for the roots of a fifth degree polynomial. All he will allow is a few simple checks on the eigenvalues, *after* they have been computed, and we mention two of them:

5B The *sum* of the n eigenvalues equals the sum of the n diagonal entries:

$$\lambda_1 + \cdots + \lambda_n = a_{11} + \cdots + a_{nn}. \tag{15}$$

This sum is known as the *trace* of A. Furthermore, the *product* of the n eigenvalues equals the *determinant* of A.

The projection matrix P had diagonal entries $\frac{1}{2}, \frac{1}{2}$ and eigenvalues $1, 0$—and $\frac{1}{2} + \frac{1}{2}$ agrees with $1 + 0$ as it should. So does the determinant, which is $0 \cdot 1 = 0$. We see again that a singular matrix, with zero determinant, has one or more of its eigenvalues equal to zero.

There should be no confusion between the diagonal entries and the eigenvalues. For a triangular matrix they are the same—but that is exceptional. Normally the pivots and diagonal entries and eigenvalues are completely different. And for a 2 by 2 matrix, we know everything:

$$\begin{bmatrix} a & b \\ c & d \end{bmatrix} \text{ has trace } a + d, \text{ determinant } ad - bc$$

$$\det \begin{bmatrix} a - \lambda & b \\ c & d - \lambda \end{bmatrix} = \lambda^2 - (\text{trace})\lambda + \text{determinant}$$

$$\lambda = \frac{\text{trace} \pm [(\text{trace})^2 - 4 \det]^{1/2}}{2}.$$

Those two λ's add up to the trace; Exercise 5.1.9 gives $\sum \lambda_i = $ trace for all matrices.

EXERCISES

5.1.1 Find the eigenvalues and eigenvectors of the matrix $A = \begin{bmatrix} 1 & -1 \\ 2 & 4 \end{bmatrix}$. Verify that the trace equals the sum of the eigenvalues, and the determinant equals their product.

5.1.2 With the same matrix A, solve the differential equation $du/dt = Au$, $u_0 = \begin{bmatrix} 0 \\ 6 \end{bmatrix}$. What are the two pure exponential solutions?

5.1.3   Suppose we shift the preceding A by subtracting $7I$:

$$B = A - 7I = \begin{bmatrix} -6 & -1 \\ 2 & -3 \end{bmatrix}.$$

What are the eigenvalues and eigenvectors of B, and how are they related to those of A?

5.1.4   Solve $du/dt = Pu$ when P is a projection:

$$\frac{du}{dt} = \begin{bmatrix} \frac{1}{2} & \frac{1}{2} \\ \frac{1}{2} & \frac{1}{2} \end{bmatrix} u \qquad \text{with} \qquad u_0 = \begin{bmatrix} 5 \\ 3 \end{bmatrix}.$$

The column space component of u_0 increases exponentially while the nullspace component stays fixed.

5.1.5   Find the eigenvalues and eigenvectors of

$$A = \begin{bmatrix} 3 & 4 & 2 \\ 0 & 1 & 2 \\ 0 & 0 & 0 \end{bmatrix} \qquad \text{and} \qquad B = \begin{bmatrix} 0 & 0 & 2 \\ 0 & 2 & 0 \\ 2 & 0 & 0 \end{bmatrix}.$$

Check that $\lambda_1 + \lambda_2 + \lambda_3$ equals the trace and $\lambda_1\lambda_2\lambda_3$ equals the determinant.

5.1.6   Give an example to show that the eigenvalues can be changed when a multiple of one row is subtracted from another.

5.1.7   Suppose that λ is an eigenvalue of A, and x is its eigenvector: $Ax = \lambda x$.
(a)   Show that this same x is an eigenvector of $B = A - 7I$, and find the eigenvalue. This should confirm Exercise 5.1.3.
(b)   Assuming $\lambda \neq 0$, show that x is also an eigenvector of A^{-1}—and find the eigenvalue.

5.1.8   Show that the determinant equals the product of the eigenvalues by imagining that the characteristic polynomial is factored into

$$\det(A - \lambda I) = (\lambda_1 - \lambda)(\lambda_2 - \lambda) \cdots (\lambda_n - \lambda), \tag{15}$$

and making a clever choice of λ.

5.1.9   Show that the trace equals the sum of the eigenvalues, in two steps. First, find the coefficient of $(-\lambda)^{n-1}$ on the right side of (15). Next, look for all the terms in

$$\det(A - \lambda I) = \det \begin{bmatrix} a_{11} - \lambda & a_{12} & \cdots & a_{1n} \\ a_{21} & a_{22} - \lambda & \cdots & a_{2n} \\ \vdots & \vdots & & \vdots \\ a_{n1} & a_{n2} & \cdots & a_{nn} - \lambda \end{bmatrix}$$

which involve $(-\lambda)^{n-1}$. Explain why they all come from the product down the main diagonal, and find the coefficient of $(-\lambda)^{n-1}$ on the left side of (15). Compare.

5.1.10 (a) Construct 2 by 2 matrices such that the eigenvalues of AB are not the products of the eigenvalues of A and B, and the eigenvalues of $A + B$ are not the sums of the individual eigenvalues.

(b) Verify however that the sum of the eigenvalues of $A + B$ equals the sum of all the individual eigenvalues of A and B, and similarly for products. Why is this true?

5.1.11 Prove that A and A^T have the same eigenvalues, by comparing their characteristic polynomials.

5.1.12 Find the eigenvalues and eigenvectors of $A = \begin{bmatrix} 3 & 4 \\ 4 & -3 \end{bmatrix}$ and $A = \begin{bmatrix} a & b \\ b & a \end{bmatrix}$.

5.1.13 If B has eigenvalues 1, 2, 3 and C has eigenvalues 4, 5, 6, and D has eigenvalues 7, 8, 9, what are the eigenvalues of the 6 by 6 matrix $A = \begin{bmatrix} B & C \\ 0 & D \end{bmatrix}$?

5.1.14 Find the rank and all four eigenvalues for both the matrix of ones and the checkerboard matrix:

$$A = \begin{bmatrix} 1 & 1 & 1 & 1 \\ 1 & 1 & 1 & 1 \\ 1 & 1 & 1 & 1 \\ 1 & 1 & 1 & 1 \end{bmatrix} \quad \text{and} \quad C = \begin{bmatrix} 0 & 1 & 0 & 1 \\ 1 & 0 & 1 & 0 \\ 0 & 1 & 0 & 1 \\ 1 & 0 & 1 & 0 \end{bmatrix}.$$

Which eigenvectors correspond to nonzero eigenvalues?

5.1.15 What are the rank and eigenvalues when A and C in the previous exercise are n by n? Remember that the eigenvalue $\lambda = 0$ is repeated $n - r$ times.

5.1.16 If A is the 4 by 4 matrix of ones, find the eigenvalues and the determinant of $A - I$ (compare Ex. 4.3.10).

5.1.17 Choose the third row of the "companion matrix"

$$A = \begin{bmatrix} 0 & 1 & 0 \\ 0 & 0 & 1 \\ . & . & . \end{bmatrix}$$

so that its characteristic polynomial $|A - \lambda I|$ is $-\lambda^3 + 4\lambda^2 + 5\lambda + 6$.

5.1.18 Suppose the matrix A has eigenvalues 0, 1, 2 with eigenvectors v_0, v_1, v_2. Describe the nullspace and the column space. Solve the equation $Ax = v_1 + v_2$. Show that $Ax = v_0$ has no solution.

5.2 ■ THE DIAGONAL FORM OF A MATRIX

We start right off with the one essential computation. It is perfectly simple and will be used in every section of this chapter.

5C Suppose the n by n matrix A has n linearly independent eigenvectors. Then if these vectors are chosen to be the columns of a matrix S, it follows that $S^{-1}AS$ is a diagonal matrix Λ, with the eigenvalues of A along its diagonal:

$$S^{-1}AS = \Lambda = \begin{bmatrix} \lambda_1 & & & \\ & \lambda_2 & & \\ & & \ddots & \\ & & & \lambda_n \end{bmatrix}. \tag{1}$$

We call S the "eigenvector matrix" and Λ the "eigenvalue matrix"—using a capital lambda because of the small lambdas for the eigenvalues on its diagonal.

Proof Put the eigenvectors x_i in the columns of S, and compute the product AS one column at a time:

$$AS = A \begin{bmatrix} | & | & & | \\ x_1 & x_2 & \cdots & x_n \\ | & | & & | \end{bmatrix} = \begin{bmatrix} | & | & & | \\ \lambda_1 x_1 & \lambda_2 x_2 & \cdots & \lambda_n x_n \\ | & | & & | \end{bmatrix}.$$

Then the trick is to split this last matrix into a quite different product:

$$\begin{bmatrix} \lambda_1 x_1 & \lambda_2 x_2 & \cdots & \lambda_n x_n \end{bmatrix} = \begin{bmatrix} x_1 & x_2 & \cdots & x_n \end{bmatrix} \begin{bmatrix} \lambda_1 & & & \\ & \lambda_2 & & \\ & & \ddots & \\ & & & \lambda_n \end{bmatrix}.$$

Regarded simply as an exercise in matrix multiplication, it is crucial to keep these matrices in the right order. If Λ came before S instead of after, then λ_1 would multiply the entries in the first row, whereas we want it to appear in the first column. As it is, we have the correct product $S\Lambda$. Therefore

$$AS = S\Lambda, \quad \text{or} \quad S^{-1}AS = \Lambda, \quad \text{or} \quad A = S\Lambda S^{-1}. \tag{2}$$

The matrix S is invertible, because its columns (the eigenvectors) were assumed to be linearly independent.

We add four remarks before giving any examples or applications.

Remark 1 If the matrix A has no repeated eigenvalues—the numbers $\lambda_1, \ldots, \lambda_n$ are distinct—then the n eigenvectors are automatically independent (see 5D below). Therefore ***any matrix with distinct eigenvalues can be diagonalized***.

Remark 2 The diagonalizing matrix S is *not unique*. In the first place, an eigenvector x can be multiplied by a constant, and will remain an eigenvector. Therefore we can multiply the columns of S by any nonzero constants, and produce a new diagonalizing S. Repeated eigenvalues leave even more freedom, and for the trivial example $A = I$, any invertible S will do: $S^{-1}IS$ is always diagonal (and the diagonal matrix Λ is just I). This reflects the fact that all vectors are eigenvectors of the identity.

Remark 3 The equation $AS = S\Lambda$ holds if the columns of S are the eigenvectors of A, and not otherwise. *Other matrices S will not produce a diagonal Λ.* The reason lies in the rules for matrix multiplication. Suppose the first column of S is y. Then the first column of $S\Lambda$ is $\lambda_1 y$. If this is to agree with the first column of AS, which by matrix multiplication is Ay, then y must be an eigenvector: $Ay = \lambda_1 y$. In fact, the order of the eigenvectors in S and the eigenvalues in Λ is automatically the same.

Remark 4 Not all matrices possess n linearly independent eigenvectors, and therefore ***not all matrices are diagonalizable***. The standard example of a "defective matrix" is

$$A = \begin{bmatrix} 0 & 1 \\ 0 & 0 \end{bmatrix}.$$

Its eigenvalues are $\lambda_1 = \lambda_2 = 0$, since it is triangular with zeros on the diagonal:

$$\det(A - \lambda I) = \det \begin{bmatrix} -\lambda & 1 \\ 0 & -\lambda \end{bmatrix} = \lambda^2.$$

If x is an eigenvector, it must satisfy

$$\begin{bmatrix} 0 & 1 \\ 0 & 0 \end{bmatrix} x = \begin{bmatrix} 0 \\ 0 \end{bmatrix}, \quad \text{or} \quad x = \begin{bmatrix} x_1 \\ 0 \end{bmatrix}.$$

Although $\lambda = 0$ is a double eigenvalue—its *algebraic multiplicity* is 2—it has only a one-dimensional space of eigenvectors. The *geometric multiplicity* is 1—there is only one independent eigenvector—and we cannot construct S.

Here is a more direct proof that A is not diagonalizable. Since $\lambda_1 = \lambda_2 = 0$, Λ would have to be the zero matrix. But if $S^{-1}AS = 0$, then we premultiply by S

and postmultiply by S^{-1}, to deduce that $A = 0$. Since A is not 0, the contradiction proves that no S can achieve $S^{-1}AS = \Lambda$.

I hope that example is not misleading. The failure of diagonalization was *not* a result of zero eigenvalues. The matrices

$$A = \begin{bmatrix} 3 & 1 \\ 0 & 3 \end{bmatrix} \quad \text{and} \quad A = \begin{bmatrix} 2 & -1 \\ 1 & 0 \end{bmatrix}$$

also fail to be diagonalizable, but their eigenvalues are 3, 3 and 1, 1. The problem is the shortage of eigenvectors—which are needed for S. That needs to be emphasized:

> *Diagonalizability is concerned with the eigenvectors.*
> *Invertibility is concerned with the eigenvalues.*

There is no connection between diagonalizability (n independent eigenvectors) and invertibility (no zero eigenvalues). The only indication that comes from the eigenvalues is this: *Diagonalization can fail only if there are repeated eigenvalues.* Even then, it does not always fail. The matrix $A = I$ has repeated eigenvalues $1, 1, \ldots, 1$ but it is already diagonal! There is no shortage of eigenvectors in that case. The test is to check, for an eigenvalue that is repeated p times, whether there are p independent eigenvectors—in other words, whether $A - \lambda I$ has rank $n - p$.

To complete that circle of ideas, we have to show that *distinct* eigenvalues present no problem.

5D   If the eigenvectors $x_1, \ldots, x_k$ correspond to *different eigenvalues* $\lambda_1, \ldots, \lambda_k$, then those eigenvectors are linearly independent.

Suppose first that $k = 2$, and that some combination of x_1 and x_2 produces zero: $c_1 x_1 + c_2 x_2 = 0$. Multiplying by A, we find $c_1 \lambda_1 x_1 + c_2 \lambda_2 x_2 = 0$. Subtracting λ_2 times the previous equation, the vector x_2 disappears:

$$c_1(\lambda_1 - \lambda_2)x_1 = 0.$$

Since $\lambda_1 \neq \lambda_2$ and $x_1 \neq 0$, we are forced into $c_1 = 0$. Similarly $c_2 = 0$, and the two vectors are independent; only the trivial combination gives zero.

This same argument extends to any number of eigenvectors: We assume some combination produces zero, multiply by A, subtract λ_k times the original combination, and the vector x_k disappears—leaving a combination of $x_1, \ldots, x_{k-1}$ which produces zero. By repeating the same steps (or by saying the words *mathematical induction*) we end up with a multiple of x_1 that produces zero. This forces $c_1 = 0$, and ultimately every $c_i = 0$. Therefore eigenvectors that come from distinct eigenvalues are automatically independent.

A matrix with n distinct eigenvalues can be diagonalized. This is the typical case.

Examples of Diagonalization

We go back to the main point of this section, which was $S^{-1}AS = \Lambda$. The eigenvector matrix S converts the original A into the eigenvalue matrix Λ—which is diagonal. That can be seen for projections and rotations.

EXAMPLE 1   The projection $A = \begin{bmatrix} \frac{1}{2} & \frac{1}{2} \\ \frac{1}{2} & \frac{1}{2} \end{bmatrix}$ has eigenvalue matrix $\Lambda = \begin{bmatrix} 1 & 0 \\ 0 & 0 \end{bmatrix}$. The eigenvectors computed earlier go into the columns of S:

$$S = \begin{bmatrix} 1 & 1 \\ 1 & -1 \end{bmatrix} \quad \text{and} \quad AS = S\Lambda = \begin{bmatrix} 1 & 0 \\ 1 & 0 \end{bmatrix}.$$

That last equation can be verified at a glance. Therefore $S^{-1}AS = \Lambda$.

EXAMPLE 2   The eigenvalues themselves are not so clear for a *rotation*:

$$K = \begin{bmatrix} 0 & -1 \\ 1 & 0 \end{bmatrix} \quad \text{has} \quad \det(K - \lambda I) = \lambda^2 + 1.$$

That matrix rotates the plane through 90°, and **how can a vector be rotated and still have its direction unchanged?** Apparently it can't—except for the zero vector which is useless. But there must be eigenvalues, and we must be able to solve $du/dt = Ku$. The characteristic polynomial $\lambda^2 + 1$ should still have two roots—but those roots are not real.

You see the way out. The eigenvalues of K are *imaginary numbers*, $\lambda_1 = i$ and $\lambda_2 = -i$. The eigenvectors are also not real. Somehow, in turning through 90°, they are multiplied by i or $-i$:

$$(K - \lambda_1 I)x_1 = \begin{bmatrix} -i & -1 \\ 1 & -i \end{bmatrix} \begin{bmatrix} y \\ z \end{bmatrix} = \begin{bmatrix} 0 \\ 0 \end{bmatrix} \quad \text{and} \quad x_1 = \begin{bmatrix} 1 \\ -i \end{bmatrix}$$

$$(K - \lambda_2 I)x_2 = \begin{bmatrix} i & -1 \\ 1 & i \end{bmatrix} \begin{bmatrix} y \\ z \end{bmatrix} = \begin{bmatrix} 0 \\ 0 \end{bmatrix} \quad \text{and} \quad x_2 = \begin{bmatrix} 1 \\ i \end{bmatrix}.$$

The eigenvalues are distinct, even if imaginary, and the eigenvectors are independent. They go into the columns of S:

$$S = \begin{bmatrix} 1 & 1 \\ -i & i \end{bmatrix} \quad \text{and} \quad S^{-1}KS = \begin{bmatrix} i & 0 \\ 0 & -i \end{bmatrix}.$$

Remark   We are faced with an inescapable fact, that **complex numbers are needed even for real matrices**. If there are too few real eigenvalues, there are always

n complex eigenvalues. (Complex includes real, when the imaginary part is zero.) If there are too few eigenvectors in the real world $\mathbf{R}^3$, or in $\mathbf{R}^n$, we look in $\mathbf{C}^3$ or $\mathbf{C}^n$. The space $\mathbf{C}^n$ contains all column vectors with complex components, and it has new definitions of length and inner product and orthogonality. But it is not more difficult than $\mathbf{R}^n$, and in Section 5.5 we make an easy conversion to the complex case.

Powers and Products: A^k and AB

There is one more situation in which the calculations are easy. Suppose we have already found the eigenvalues and eigenvectors of a matrix A. Then *the eigenvalues of A^2 are exactly* $\lambda_1^2, \ldots, \lambda_n^2$, *and every eigenvector of A is also an eigenvector of A^2.* We start from $Ax = \lambda x$, and multiply again by A:

$$A^2 x = A\lambda x = \lambda Ax = \lambda^2 x. \tag{3}$$

Thus λ^2 is an eigenvalue of A^2, with the same eigenvector x. If the first multiplication by A leaves the direction of x unchanged, then so does the second.

The same result comes from diagonalization. If $S^{-1}AS = \Lambda$, then squaring both sides gives

$$(S^{-1}AS)(S^{-1}AS) = \Lambda^2 \qquad \text{or} \qquad S^{-1}A^2 S = \Lambda^2.$$

The matrix A^2 is diagonalized by the same S, so the eigenvectors are unchanged. The eigenvalues are squared.

This continues to hold for any power of A:

5E   The eigenvalues of A^k are $\lambda_1^k, \ldots, \lambda_n^k$, the kth powers of the eigenvalues of A. Each eigenvector of A is still an eigenvector of A^k, and if S diagonalizes A it also diagonalizes A^k:

$$\Lambda^k = (S^{-1}AS)(S^{-1}AS) \cdots (S^{-1}AS) = S^{-1}A^k S. \tag{4}$$

Each S^{-1} cancels an S, except for the first S^{-1} and the last S.

If A is invertible this rule also applies to its inverse (the power $k = -1$). *The eigenvalues of A^{-1} are* $1/\lambda_i$. That can be seen even without diagonalizing:

$$\text{if} \quad Ax = \lambda x \quad \text{then} \quad x = \lambda A^{-1}x \quad \text{and} \quad \frac{1}{\lambda}x = A^{-1}x.$$

EXAMPLE   If K is rotation through $90°$, then K^2 is rotation through $180°$ and K^{-1} is rotation through $-90°$:

$$K = \begin{bmatrix} 0 & -1 \\ 1 & 0 \end{bmatrix} \quad \text{and} \quad K^2 = \begin{bmatrix} -1 & 0 \\ 0 & -1 \end{bmatrix} \quad \text{and} \quad K^{-1} = \begin{bmatrix} 0 & 1 \\ -1 & 0 \end{bmatrix}.$$

The eigenvalues are i and $-i$; their squares are -1 and -1; their reciprocals are $1/i = -i$ and $1/(-i) = i$. We can go on to K^4, which is a complete rotation through $360°$:

$$K^4 = \begin{bmatrix} 1 & 0 \\ 0 & 1 \end{bmatrix} \quad \text{and also} \quad \Lambda^4 = \begin{bmatrix} i^4 & 0 \\ 0 & (-i)^4 \end{bmatrix} = \begin{bmatrix} 1 & 0 \\ 0 & 1 \end{bmatrix}.$$

The power i^4 is 1, and a $360°$ rotation matrix is the identity.

Now we turn to a **product of two matrices**, and ask about the eigenvalues of AB. It is very tempting to try the same reasoning, in an attempt to prove what is *not in general true*. If λ is an eigenvalue of A and μ is an eigenvalue of B, here is the false proof that AB has the eigenvalue $\mu\lambda$:

$$ABx = A\mu x = \mu Ax = \mu\lambda x.$$

The fallacy lies in assuming that A and B share the *same* eigenvector x. In general, they do not. We could have two matrices with zero eigenvalues, while their product has an eigenvalue $\lambda = 1$:

$$AB = \begin{bmatrix} 0 & 1 \\ 0 & 0 \end{bmatrix} \begin{bmatrix} 0 & 0 \\ 1 & 0 \end{bmatrix} = \begin{bmatrix} 1 & 0 \\ 0 & 0 \end{bmatrix}.$$

The eigenvectors of this A and B are completely different, which is typical. For the same reason, the eigenvalues of $A + B$ have nothing to do with $\lambda + \mu$.

This false proof does suggest what is true. If the eigenvector *is* the same for A and B, then the eigenvalues multiply and AB has the eigenvalue $\mu\lambda$. But there is something more important. There is an easy way to recognize when A and B share a full set of eigenvectors, and that is a key question in quantum mechanics:

5F If A and B are diagonalizable, they share the same eigenvector matrix S if and only if $AB = BA$.

Proof If the same S diagonalizes both $A = S\Lambda_1 S^{-1}$ and $B = S\Lambda_2 S^{-1}$, we can multiply in either order:

$$AB = S\Lambda_1 S^{-1} S\Lambda_2 S^{-1} = S\Lambda_1\Lambda_2 S^{-1} \text{ and } BA = S\Lambda_2 S^{-1} S\Lambda_1 S^{-1} = S\Lambda_2\Lambda_1 S^{-1}.$$

Since $\Lambda_1\Lambda_2 = \Lambda_2\Lambda_1$ (diagonal matrices always commute) we have $AB = BA$.

In the opposite direction, suppose $AB = BA$. Starting from $Ax = \lambda x$, we have

$$ABx = BAx = B\lambda x = \lambda Bx.$$

Thus x and Bx are both eigenvectors of A, sharing the same λ (or else $Bx = 0$). If we assume for convenience that the eigenvalues of A are distinct—the eigenspaces are all one-dimensional—then Bx *must be a multiple of* x. In other words x is an eigenvector of B as well as A. The proof with repeated eigenvalues is a little longer.

Remark   In quantum mechanics it is matrices that don't commute—like position P and momentum Q—which suffer from Heisenberg's *uncertainty principle*. Position is symmetric, momentum is skew-symmetric, and together they satisfy $QP - PQ = I$. The uncertainty principle comes directly from the Schwarz inequality $(Qx)^T(Px) \leq \|Qx\| \|Px\|$ of Section 3.2:

$$\|x\|^2 = x^Tx = x^T(QP - PQ)x \leq 2\|Qx\| \|Px\|.$$

The product of $\|Qx\|/\|x\|$ and $\|Px\|/\|x\|$—which can represent momentum and position errors, when the wave function is x—is at least $\frac{1}{2}$. It is impossible to get both errors small, because when you try to measure the position of a particle you change its momentum.

At the end we come back to $A = S\Lambda S^{-1}$. That is the factorization produced by the eigenvalues. It is particularly suited to take powers of A, and the simplest case A^2 makes the point. The LU factorization is hopeless when squared, but $S\Lambda S^{-1}$ is perfect. The square is $S\Lambda^2 S^{-1}$, the eigenvectors are unchanged, and by following those eigenvectors we will solve difference equations and differential equations.

EXERCISES

5.2.1   Factor the following matrices into $S\Lambda S^{-1}$:

$$A = \begin{bmatrix} 1 & 1 \\ 1 & 1 \end{bmatrix} \quad \text{and} \quad A = \begin{bmatrix} 2 & 1 \\ 0 & 0 \end{bmatrix}.$$

5.2.2   Find the matrix A whose eigenvalues are 1 and 4, and whose eigenvectors are $\begin{bmatrix} 3 \\ 1 \end{bmatrix}$ and $\begin{bmatrix} 2 \\ 1 \end{bmatrix}$, respectively. (*Hint:* $A = S\Lambda S^{-1}$.)

5.2.3   Find *all* the eigenvalues and eigenvectors of

$$A = \begin{bmatrix} 1 & 1 & 1 \\ 1 & 1 & 1 \\ 1 & 1 & 1 \end{bmatrix}$$

and write down two different diagonalizing matrices S.

5.2.4   If a 3 by 3 upper triangular matrix has diagonal entries 1, 2, 7, how do you know it can be diagonalized? What is Λ?

5.2.5   Which of these matrices cannot be diagonalized?

$$A_1 = \begin{bmatrix} 2 & -2 \\ 2 & -2 \end{bmatrix} \quad A_2 = \begin{bmatrix} 2 & 0 \\ 2 & -2 \end{bmatrix} \quad A_3 = \begin{bmatrix} 2 & 0 \\ 2 & 2 \end{bmatrix}.$$

5.2.6   (a)   If $A^2 = I$ what are the possible eigenvalues of A?
(b)   If this A is 2 by 2, and not I or $-I$, find its trace and determinant.
(c)   If the first row is $(3, -1)$ what is the second row?

5.2.7 If $A = \begin{bmatrix} 4 & 3 \\ 1 & 2 \end{bmatrix}$ find A^{100} by diagonalizing A.

5.2.8 Suppose $A = uv^T$ is a column times a row (a rank-one matrix).
(a) By multiplying A times u show that u is an eigenvector. What is λ?
(b) What are the other eigenvalues (and why)?
(c) Compute trace$(A) = v^Tu$ in two ways, from the sum on the diagonal and the sum of λ's.

5.2.9 Show by direct calculation that AB and BA have the same trace when

$$A = \begin{bmatrix} a & b \\ c & d \end{bmatrix} \quad \text{and} \quad B = \begin{bmatrix} q & r \\ s & t \end{bmatrix}.$$

Deduce that $AB - BA = I$ is impossible. (It only happens in infinite dimensions.)

5.2.10 Suppose A has eigenvalues 1, 2, 4. What is the trace of A^2? What is the determinant of $(A^{-1})^T$?

5.2.11 If the eigenvalues of A are 1, 1, 2, which of the following are certain to be true? Give a reason if true or a counterexample if false:
(1) A is invertible
(2) A is diagonalizable
(3) A is not diagonalizable

5.2.12 Suppose the only eigenvectors of A are multiples of $x = (1, 0, 0)$:

T F A is not invertible
T F A has a repeated eigenvalue
T F A is not diagonalizable

5.2.13 Diagonalize the matrix $A = \begin{bmatrix} 5 & 4 \\ 4 & 5 \end{bmatrix}$ and find one of its square roots—a matrix such that $R^2 = A$. How many square roots will there be?

5.2.14 If A is diagonalizable, show that the determinant of $A = S\Lambda S^{-1}$ is the product of the eigenvalues.

5.2.15 Show that every matrix is the sum of two nonsingular matrices.

5.3 ■ DIFFERENCE EQUATIONS AND THE POWERS A^k

Difference equations are not as well known as differential equations, but they should be. They move forward in a finite number of finite steps, while a differential equation takes an infinite number of infinitesimal steps—but the two theories stay absolutely in parallel. It is the same analogy between the discrete and the continuous that appears over and over in mathematics. Perhaps the best illustration is one which really does not involve n-dimensional linear algebra, because money in a bank is only a scalar.

Suppose you invest $1000 for five years at 6% interest. If it is compounded once a year, then the principal is multiplied by 1.06, and $P_{k+1} = 1.06 \, P_k$. *This is a difference equation with a time step of one year.* It relates the principal after $k + 1$ years to the principal the year before, and it is easy to solve: After 5 years, the original principal $P_0 = 1000$ has been multiplied 5 times, and

$$P_5 = (1.06)^5 P_0 = (1.06)^5 1000 = \$1338.$$

Now suppose the time step is reduced to a month. The new difference equation is $p_{k+1} = (1 + .06/12)p_k$. After 5 years, or 60 months,

$$p_{60} = \left(1 + \frac{.06}{12}\right)^{60} p_0 = (1.005)^{60} 1000 = \$1349.$$

The next step is to compound the interest daily:

$$\left(1 + \frac{.06}{365}\right)^{5 \cdot 365} 1000 = \$1349.83.$$

Finally, to keep their employees really moving, banks offer *continuous compounding*. The interest is added on at every instant, and the difference equation breaks down. In fact you can hope that the treasurer does not know calculus, and cannot figure out what he owes you. But he has two different possibilities: Either he can compound the interest more and more frequently, and see that the limit is

$$\left(1 + \frac{.06}{N}\right)^{5N} 1000 \to e^{.30} 1000 = \$1349.87.$$

Or he can switch to a differential equation—the limit of the difference equation $p_{k+1} = (1 + .06 \, \Delta t)p_k$. Moving p_k to the left side and dividing by the time step Δt,

$$\frac{p_{k+1} - p_k}{\Delta t} = .06 p_k \qquad \text{approaches} \qquad \frac{dp}{dt} = .06p.$$

The solution is $p(t) = e^{.06t} p_0$, and after 5 years this again amounts to $1349.87. The principal stays finite, even when it is compounded every instant—and the difference is only four cents.

This example included both difference equations and differential equations, with one approaching the other as the time step disappeared. But there are plenty of difference equations that stand by themselves, and our second example comes from the famous **Fibonacci sequence**:

$$0, 1, 1, 2, 3, 5, 8, 13, \ldots.$$

Probably you see the pattern: Every Fibonacci number is the sum of its two predecessors,

$$\boxed{F_{k+2} = F_{k+1} + F_k.} \tag{1}$$

That is the difference equation. It turns up in a most fantastic variety of applications, and deserves a book of its own. Thorns and leaves grow in a spiral pattern, and on the hawthorn or apple or oak you find five growths for every two turns around the stem. The pear tree has eight for every three turns, and the willow is even more complicated, 13 growths for every five spirals. The champion seems to be a sunflower of Daniel T. O'Connell (*Scientific American*, November, 1951) whose seeds chose an almost unbelievable ratio of $F_{12}/F_{13} = 144/233$.†

How could we find the 1000th Fibonacci number, other than by starting with $F_0 = 0$, $F_1 = 1$, and working all the way out to F_{1000}? The goal is to solve the difference equation $F_{k+2} = F_{k+1} + F_k$, and as a first step it can be reduced to a one-step equation $u_{k+1} = Au_k$. This is just like compound interest, $P_{k+1} = 1.06P_k$, except that now the unknown has to be a vector and the multiplier A has to be a matrix: if

$$u_k = \begin{bmatrix} F_{k+1} \\ F_k \end{bmatrix},$$

then

$$\begin{matrix} F_{k+2} = F_{k+1} + F_k \\ F_{k+1} = F_{k+1} \end{matrix} \qquad \text{becomes} \qquad u_{k+1} = \begin{bmatrix} 1 & 1 \\ 1 & 0 \end{bmatrix} u_k.$$

This is a standard trick for an equation of order s; $s - 1$ trivial equations like $F_{k+1} = F_{k+1}$ combine with the given equation in a one-step system. For Fibonacci, $s = 2$.

The equation $u_{k+1} = Au_k$ is easy to solve. It starts from u_0. After one step it produces $u_1 = Au_0$. At the second step u_2 is Au_1, which is A^2u_0. *Every step brings a multiplication by A*, and after k steps there are k multiplications:

The solution to $u_{k+1} = Au_k$ is $u_k = A^ku_0$.

---

† For these botanical applications, see D'Arcy Thompson's book *On Growth and Form* (Cambridge Univ. Press, 1942) or Peter Stevens' beautiful *Patterns in Nature* (Little, Brown, 1974). Hundreds of other properties of the F_n have been published in the *Fibonacci Quarterly*. Apparently Fibonacci brought Arabic numerals into Europe, about 1200 A.D.

The real problem is to find some quick way to compute the powers A^k, and thereby find the 1000th Fibonacci number. The key lies in the eigenvalues and eigenvectors:

5G If A can be diagonalized, $A = S\Lambda S^{-1}$, then

$$u_k = A^k u_0 = (S\Lambda S^{-1})(S\Lambda S^{-1}) \cdots (S\Lambda S^{-1})u_0 = S\Lambda^k S^{-1}u_0. \tag{2}$$

The columns of S are the eigenvectors of A, and by setting $S^{-1}u_0 = c$ the solution becomes

$$u_k = S\Lambda^k c = \begin{bmatrix} x_1 & \cdots & x_n \end{bmatrix} \begin{bmatrix} \lambda_1^k & & \\ & \ddots & \\ & & \lambda_n^k \end{bmatrix} \begin{bmatrix} c_1 \\ \vdots \\ c_n \end{bmatrix} = c_1\lambda_1^k x_1 + \cdots + c_n\lambda_n^k x_n. \tag{3}$$

The solution is a combination of the "pure solutions" $\lambda_i^k x_i$.

These formulas give two different approaches to the same solution $u_k = S\Lambda^k S^{-1}u_0$. The first formula recognized that A^k is identical with $S\Lambda^k S^{-1}$, and we could have stopped there. But the second approach brings out more clearly the analogy with solving a differential equation: *Instead of the pure exponential solutions $e^{\lambda_i t}x_i$, we now have the pure powers $\lambda_i^k x_i$.* The normal modes are again the eigenvectors x_i, and at each step they are amplified by the eigenvalues λ_i. By combining these special solutions in such a way as to match u_0—that is where c came from—we recover the correct solution $u_k = S\Lambda^k S^{-1}u_0$.

In any specific example like Fibonacci's, the first step is to find the eigenvalues:

$$A = \begin{bmatrix} 1 & 1 \\ 1 & 0 \end{bmatrix}, \qquad \det(A - \lambda I) = \lambda^2 - \lambda - 1,$$

$$\lambda_1 = \frac{1 + \sqrt{5}}{2}, \qquad \lambda_2 = \frac{1 - \sqrt{5}}{2}.$$

The second row of $A - \lambda I$ is $(1, -\lambda)$, so the eigenvector is $(\lambda, 1)$. The first Fibonacci numbers $F_0 = 0$ and $F_1 = 1$ go into u_0, and

$$c = S^{-1}u_0 = \begin{bmatrix} \lambda_1 & \lambda_2 \\ 1 & 1 \end{bmatrix}^{-1} \begin{bmatrix} 1 \\ 0 \end{bmatrix} = \begin{bmatrix} 1/(\lambda_1 - \lambda_2) \\ -1/(\lambda_1 - \lambda_2) \end{bmatrix}.$$

Those are the constants in $u_k = c_1\lambda_1^k x_1 + c_2\lambda_2^k x_2$. Since the second component of both eigenvectors is 1, that leaves $F_k = c_1\lambda_1^k + c_2\lambda_2^k$ in the second component of u_k:

$$F_k = \frac{\lambda_1^k}{\lambda_1 - \lambda_2} - \frac{\lambda_2^k}{\lambda_1 - \lambda_2} = \frac{1}{\sqrt{5}}\left[\left(\frac{1 + \sqrt{5}}{2}\right)^k - \left(\frac{1 - \sqrt{5}}{2}\right)^k\right].$$

This is the answer we wanted. In one way it is rather surprising, because Fibonacci's rule $F_{k+2} = F_{k+1} + F_k$ must always produce whole numbers, and we have ended up with fractions and square roots. Somehow these must cancel out, and leave an integer. In fact, since the second term $[(1 - \sqrt{5})/2]^k/\sqrt{5}$ is always less than $\frac{1}{2}$, it must just move the first term to the nearest integer. Subtraction leaves only the integer part, and

$$F_{1000} = \text{nearest integer to } \frac{1}{\sqrt{5}} \left(\frac{1 + \sqrt{5}}{2} \right)^{1000}.$$

Of course this is an enormous number, and F_{1001} will be even bigger. It is pretty clear that the fractions are becoming completely insignificant compared to the integers; the ratio F_{1001}/F_{1000} must be very close to the quantity $(1 + \sqrt{5})/2 \approx 1.618$, which the Greeks called the "golden mean."† In other words λ_2^k is becoming insignificant compared to λ_1^k, and the ratio F_{k+1}/F_k approaches $\lambda_1^{k+1}/\lambda_1^k = \lambda_1$.

That is a typical example, leading to the powers of $A = \begin{bmatrix} 1 & 1 \\ 1 & 0 \end{bmatrix}$. It involved $\sqrt{5}$ because the eigenvalues did. If we choose a different matrix whose eigenvalues are whole numbers, we can focus on the simplicity of the computation—*after the matrix has been diagonalized:*

$$A = \begin{bmatrix} -4 & -5 \\ 10 & 11 \end{bmatrix} \text{ has } \lambda_1 = 1, \quad x_1 = \begin{bmatrix} 1 \\ -1 \end{bmatrix}, \quad \lambda_2 = 6, \quad x_2 = \begin{bmatrix} -1 \\ 2 \end{bmatrix}$$

$$A^k = S\Lambda^k S^{-1} = \begin{bmatrix} 1 & -1 \\ -1 & 2 \end{bmatrix} \begin{bmatrix} 1^k & 0 \\ 0 & 6^k \end{bmatrix} \begin{bmatrix} 2 & 1 \\ 1 & 1 \end{bmatrix} = \begin{bmatrix} 2 - 6^k & 1 - 6^k \\ -2 + 2 \cdot 6^k & -1 + 2 \cdot 6^k \end{bmatrix}.$$

The powers 6^k and 1^k are visible in that last matrix A^k, mixed in by the eigenvectors.

For the difference equation $u_{k+1} = Au_k$, we emphasize the main point. If x is an eigenvector then

> ***one possible solution is*** $u_0 = x, u_1 = \lambda x, u_2 = \lambda^2 x, \ldots$

When the initial u_0 happens to equal an eigenvector, this is *the* solution: $u_k = \lambda^k x$. In general u_0 is not an eigenvector. But if u_0 is a *combination* of eigenvectors, the solution u_k is the same combination of these special solutions.

5H If $u_0 = c_1 x_1 + \cdots + c_n x_n$ then $u_k = c_1 \lambda_1^k x_1 + \cdots + c_n \lambda_n^k x_n$. The role of the c's is to match the initial conditions:

$$u_0 = \begin{bmatrix} x_1 & \cdots & x_n \end{bmatrix} \begin{bmatrix} c_1 \\ \vdots \\ c_n \end{bmatrix} = Sc \quad \text{and} \quad c = S^{-1} u_0. \tag{4}$$

We turn to important applications of difference equations—or powers of matrices.

---

† The most elegant rectangles have their sides in the ratio of 1.618 to 1.

A Markov Process

There was an exercise in Chapter 1, about moving in and out of California, which is worth another look. These were the rules:

Each year $\frac{1}{10}$ of the people outside California move in, and $\frac{2}{10}$ of the people inside California move out.

This suggests a difference equation. We start with y_0 people outside and z_0 inside, and at the end of the first year the numbers outside and inside are

$$
\begin{aligned}
y_1 &= .9y_0 + .2z_0 \\
z_1 &= .1y_0 + .8z_0
\end{aligned}
\quad \text{or} \quad
\begin{bmatrix} y_1 \\ z_1 \end{bmatrix} = \begin{bmatrix} .9 & .2 \\ .1 & .8 \end{bmatrix} \begin{bmatrix} y_0 \\ z_0 \end{bmatrix}.
$$

Of course this problem was produced out of thin air, but it has the two essential properties of a *Markov process*: The total number of people stays fixed, and the numbers outside and inside can never become negative.† The first property is reflected in the fact that *each column of the matrix adds up to* 1. Everybody is accounted for, and nobody is gained or lost. The second property is reflected in the fact that *the matrix has no negative entries*. As long as the initial y_0 and z_0 are nonnegative, the same will be true of y_1 and z_1, y_2 and z_2, and so on forever. The powers A^k are all nonnegative.

We propose first to solve this difference equation (using the formula $S\Lambda^k S^{-1} u_0$), then to see whether the population eventually approaches a "steady state," and finally to discuss Markov processes in general. To start the computations, A has to be diagonalized:

$$
A = \begin{bmatrix} .9 & .2 \\ .1 & .8 \end{bmatrix}, \qquad \det(A - \lambda I) = \lambda^2 - 1.7\lambda + .7,
$$

$$
\lambda_1 = 1 \quad \text{and} \quad \lambda_2 = .7
$$

$$
A = S\Lambda S^{-1} = \begin{bmatrix} \frac{2}{3} & \frac{1}{3} \\ \frac{1}{3} & -\frac{1}{3} \end{bmatrix} \begin{bmatrix} 1 & \\ & .7 \end{bmatrix} \begin{bmatrix} 1 & 1 \\ 1 & -2 \end{bmatrix}.
$$

To find A^k, and the distribution after k years, we change Λ to Λ^k:

$$
\begin{bmatrix} y_k \\ z_k \end{bmatrix} = A^k \begin{bmatrix} y_0 \\ z_0 \end{bmatrix} = \begin{bmatrix} \frac{2}{3} & \frac{1}{3} \\ \frac{1}{3} & -\frac{1}{3} \end{bmatrix} \begin{bmatrix} 1^k & \\ & .7^k \end{bmatrix} \begin{bmatrix} 1 & 1 \\ 1 & -2 \end{bmatrix} \begin{bmatrix} y_0 \\ z_0 \end{bmatrix}
$$

$$
= (y_0 + z_0) \begin{bmatrix} \frac{2}{3} \\ \frac{1}{3} \end{bmatrix} + (y_0 - 2z_0)(.7)^k \begin{bmatrix} \frac{1}{3} \\ -\frac{1}{3} \end{bmatrix}.
$$

---

† Furthermore, history is completely disregarded; each new situation u_{k+1} depends only on the current u_k, and the record of $u_0, \ldots, u_{k-1}$ can be thrown away. Perhaps even our lives are examples of Markov processes, but I hope not.

These are the terms $c_1 \lambda_1^k x_1 + c_2 \lambda_2^k x_2$. The factor $\lambda_1^k = 1$ is hidden in the first term, and it is easy to see what happens in the long run: The other factor $(.7)^k$ becomes extremely small, and *the solution approaches a limiting state*

$$\begin{bmatrix} y_\infty \\ z_\infty \end{bmatrix} = (y_0 + z_0) \begin{bmatrix} \frac{2}{3} \\ \frac{1}{3} \end{bmatrix}.$$

The total population is still $y_0 + z_0$, just as it was initially, but in the limit $\frac{2}{3}$ of this population is outside California and $\frac{1}{3}$ is inside. This is true no matter what the initial distribution may have been! You might recognize that this steady state is exactly the distribution that was asked for in Exercise 1.3.13. If the year starts with $\frac{2}{3}$ outside and $\frac{1}{3}$ inside, then it ends the same way:

$$\begin{bmatrix} .9 & .2 \\ .1 & .8 \end{bmatrix} \begin{bmatrix} \frac{2}{3} \\ \frac{1}{3} \end{bmatrix} = \begin{bmatrix} \frac{2}{3} \\ \frac{1}{3} \end{bmatrix}, \qquad \text{or} \qquad A u_\infty = u_\infty.$$

The steady state is the eigenvector of A corresponding to $\lambda = 1$. Multiplication by A, which takes us from one time step to the next, leaves u_∞ unchanged.

We summarize the theory of Markov processes:

5I   A Markov matrix is nonnegative, with each column adding to 1.

(a) $\lambda_1 = 1$ is an eigenvalue
(b)   its eigenvector x_1 is nonnegative—and it is a steady state since $A x_1 = x_1$
(c)   the other eigenvalues satisfy $|\lambda_i| \leq 1$
(d)   if any power of A has all *positive* entries, these other $|\lambda_i|$ are below 1. The solution $A^k u_0$ approaches a multiple of x_1—which is the steady state u_∞.

To find the right multiple of x_1, use the fact that the total population stays the same. If California started with all 90 million people out, it ended with 60 million out and 30 million in. It also ended that way if all 90 million were originally in.

We note that some authors work with A^T, and *rows* that add to 1.

Remark   Our description of a Markov process was completely deterministic; populations moved in fixed proportions. But if we look at a single individual, the rules for moving can be given a probabilistic interpretation. If the individual is outside California, then with probability $\frac{1}{10}$ he moves in; if he is inside, then with probability $\frac{2}{10}$ he moves out. His movement becomes a *random process*, and the matrix A that governs it is called a *transition matrix*. We no longer know exactly where he is, but every year the components of $u_k = A^k u_0$ specify the probability that he is outside the state, and the probability that he is inside. These probabilities add to 1—he has to be somewhere—and they are never negative. That brings us back to the two fundamental properties of a transition matrix: Each column adds to 1, and no entry is negative.

The key step in the theory is to understand why $\lambda = 1$ is always an eigenvalue, and why its eigenvector is the steady state. The first point is easy to explain: Each column of $A - I$ adds up to $1 - 1 = 0$. Therefore the rows of $A - I$ add up to the zero row, they are linearly dependent, $A - I$ is singular, and $\lambda_1 = 1$ is an eigenvalue. Except for very special cases,† u_k will eventually approach the corresponding eigenvector. This is suggested by the formula $u_k = c_1 \lambda_1^k x_1 + \cdots + c_n \lambda_n^k x_n$, in which no eigenvalue can be larger than 1. (Otherwise the probabilities u_k would blow up like Fibonacci numbers.) If all other eigenvalues are strictly smaller than $\lambda_1 = 1$, then the first term in the formula will be completely dominant; the other λ_i^k go to zero, and $u_k \to c_1 x_1 = u_\infty$.

This is an example of one of the central themes of this chapter: Given information about A, find information about its eigenvalues. Here we found $\lambda_{\max} = 1$.

There is an obvious difference between Fibonacci numbers and Markov processes. The numbers F_k become larger and larger, while by definition any "probability" is between 0 and 1. The Fibonacci equation is *unstable*, and so is the compound interest equation $P_{k+1} = 1.06 P_k$; the principal keeps growing forever. If the Markov probabilities decreased to zero, that equation would be stable; but they do not, since at every stage they must add to 1. Therefore a Markov process is *neutrally stable*.

Now suppose we are given any difference equation $u_{k+1} = Au_k$, and we want to study its behavior as $k \to \infty$. Assuming A can be diagonalized, the solution u_k will be a combination of pure solutions,

$$u_k = S\Lambda^k S^{-1} u_0 = c_1 \lambda_1^k x_1 + \cdots + c_n \lambda_n^k x_n.$$

The growth of u_k is governed by the factors λ_i^k, and **stability depends on the eigenvalues**.

5J   The difference equation $u_{k+1} = Au_k$ is

>   *stable* if all eigenvalues satisfy $|\lambda_i| < 1$
>   *neutrally stable* if some $|\lambda_i| = 1$ and the other $|\lambda_i| < 1$
>   *unstable* if at least one eigenvalue has $|\lambda_i| > 1$.

In the stable case, the powers A^k approach zero and so does the solution $u_k = A^k u_0$.

EXAMPLE   The matrix

$$A = \begin{bmatrix} 0 & 4 \\ 0 & \frac{1}{2} \end{bmatrix}$$

---

† If everybody outside moves in and everybody inside moves out, then the populations are reversed every year and there is no steady state. The transition matrix is $A = \begin{bmatrix} 0 & 1 \\ 1 & 0 \end{bmatrix}$ and -1 is an eigenvalue as well as $+1$—which cannot happen if all $a_{ij} > 0$.

is certainly stable; its eigenvalues are 0 and $\frac{1}{2}$, lying on the main diagonal because A is triangular. Starting from any initial vector u_0, and following the rule $u_{k+1} = Au_k$, the solution must eventually approach zero:

$$u_0 = \begin{bmatrix} 0 \\ 1 \end{bmatrix}, \quad u_1 = \begin{bmatrix} 4 \\ \frac{1}{2} \end{bmatrix}, \quad u_2 = \begin{bmatrix} 2 \\ \frac{1}{4} \end{bmatrix}, \quad u_3 = \begin{bmatrix} 1 \\ \frac{1}{8} \end{bmatrix}, \quad u_4 = \begin{bmatrix} \frac{1}{2} \\ \frac{1}{16} \end{bmatrix}, \ldots$$

You can see how the larger eigenvalue $\lambda = \frac{1}{2}$ governs the decay; after the first step every vector u_k is half of the preceding one. The real effect of the first step is to split u_0 into the two eigenvectors of A,

$$\begin{bmatrix} 0 \\ 1 \end{bmatrix} = \begin{bmatrix} 8 \\ 1 \end{bmatrix} + \begin{bmatrix} -8 \\ 0 \end{bmatrix},$$

and to annihilate the second eigenvector (corresponding to $\lambda = 0$). The first eigenvector is multiplied by $\lambda = \frac{1}{2}$ at every step.

Positive Matrices and Applications

By developing the Markov ideas we can find a small gold mine (*entirely optional*) of matrix applications in economics.

EXAMPLE 1 *Leontief's input-output matrix*
This is one of the first great successes of mathematical economics. To illustrate it, we construct a *consumption matrix*—in which a_{ij} gives the amount of product j that is needed to create one unit of product i:

$$A = \begin{bmatrix} .4 & 0 & .1 \\ 0 & .1 & .8 \\ .5 & .7 & .1 \end{bmatrix} \quad \begin{matrix} \text{(steel)} \\ \text{(food)} \\ \text{(labor)} \end{matrix}$$

The first question is: Can we produce y_1 units of steel, y_2 units of food, and y_3 units of labor? To do so, we must start with larger amounts p_1, p_2, p_3, because some part is consumed by the production itself. The amount consumed is Ap, and it leaves a net production of $p - Ap$.

Problem *To find a vector p such that $p - Ap = y$,   or $p = (I - A)^{-1}y$.*

On the surface, we are only asking if $I - A$ is invertible. But there is a nonnegative twist to the problem. Demand and production, y and p, are nonnegative. Since p is $(I - A)^{-1}y$, the real question is about the matrix that multiplies y:

When is $(I - A)^{-1}$ a nonnegative matrix?

Roughly speaking, A cannot be too large. If production consumes too much, nothing is left as output. The key is in the largest eigenvalue λ_1 of A, which must be below 1:

If $\lambda_1 > 1$, $(I - A)^{-1}$ fails to be nonnegative

If $\lambda_1 = 1$, $(I - A)^{-1}$ fails to exist

If $\lambda_1 < 1$, $(I - A)^{-1}$ is a sum of nonnegative matrices:

$$(I - A)^{-1} = I + A + A^2 + A^3 + \cdots. \tag{5}$$

The 3 by 3 example has $\lambda_1 = .9$ and output exceeds input. Production can go on.

Those are easy to prove, once we know the main fact about a nonnegative matrix like A: Not only is the largest eigenvalue positive, but so is the eigenvector x_1. Then $(I - A)^{-1}$ has the same eigenvector, with eigenvalue $1/(1 - \lambda_1)$.

If λ_1 exceeds 1, that last number is negative. The matrix $(I - A)^{-1}$ will take the positive vector x_1 to a negative vector $x_1/(1 - \lambda_1)$. In that case $(I - A)^{-1}$ is definitely not nonnegative. If $\lambda_1 = 1$, then $I - A$ is singular. The productive case of a healthy economy is $\lambda_1 < 1$, when the powers of A go to zero (stability) and the infinite series $I + A + A^2 + \cdots$ converges. Multiplying this series by $I - A$ leaves the identity matrix—all higher powers cancel—so $(I - A)^{-1}$ is a sum of nonnegative matrices. We give two examples:

$$A = \begin{bmatrix} 0 & 2 \\ 1 & 0 \end{bmatrix} \text{ has } \lambda_1 = 2 \text{ and the economy is lost}$$

$$A = \begin{bmatrix} .5 & 2 \\ 0 & .5 \end{bmatrix} \text{ has } \lambda_1 = \frac{1}{2} \text{ and we can produce anything.}$$

The matrices $(I - A)^{-1}$ in those two cases are $\begin{bmatrix} -1 & -2 \\ -1 & -1 \end{bmatrix}$ and $\begin{bmatrix} 2 & 8 \\ 0 & 2 \end{bmatrix}$.

Leontief's inspiration was to find a model which uses genuine data from the real economy; the U.S. table for 1958 contained 83 industries, each of whom sent in a complete "transactions table" of consumption and production. The theory also reaches beyond $(I - A)^{-1}$, to decide natural prices and questions of optimization. Normally labor is in limited supply and ought to be minimized. And, of course, the economy is not always linear.

EXAMPLE 2 *The prices in a closed input-output model*
The model is called "closed" when everything produced is also consumed. Nothing goes outside the system. In that case the matrix A goes back to a Markov matrix. *The columns add up to* 1. We might be talking about the *value* of steel and food and labor, instead of the number of units. The vector p represents prices instead of production levels.

Suppose p_0 is a vector of prices. Then Ap_0 multiplies prices by amounts to give the value of each product. That is a new set of prices, which the system uses for the next set of values $A^2 p_0$. The question is whether the prices approach equilibrium. Are there prices such that $p = Ap$, and does the system take us there?

You recognize p as the (nonnegative) eigenvector of the Markov matrix A—corresponding to $\lambda = 1$. It is the steady state p_∞, and it is approached from any starting point p_0. The economy depends on this, that by repeating a transaction over and over the price tends to equilibrium.

For completeness we give a quick explanation of the key properties of a **positive matrix**—not to be confused with a *positive definite* matrix, which is symmetric and has all its eigenvalues positive. Here all the entries a_{ij} are positive.

5K The largest eigenvalue λ_1 of a positive matrix is real and positive, and so are the components of the eigenvector x_1.

Proof Suppose $A > 0$. The key idea is to look at all numbers t such that $Ax \geq tx$ for some nonnegative vector x (other than $x = 0$). We are allowing inequality in $Ax \geq tx$ in order to have many positive candidates t. For the largest value t_{max} (which is attained), we will show that *equality holds*: $Ax = t_{max}x$.

Otherwise, if $Ax \geq t_{max}x$ is not an equality, multiply by A. Because A is positive that produces a strict inequality $A^2 x > t_{max}Ax$. Therefore the positive vector $y = Ax$ satisfies $Ay > t_{max}y$, and t_{max} could be increased. This contradiction forces the equality $Ax = t_{max}x$, and we have an eigenvalue. Its eigenvector x is positive because on the left side of that equality Ax is sure to be positive.

To see that no eigenvalue can be larger than t_{max}, suppose $Az = \lambda z$. Since λ and z may involve negative or complex numbers, we take absolute values: $|\lambda||z| = |Az| \leq A|z|$ by the "triangle inequality." This $|z|$ is a nonnegative vector, so $|\lambda|$ is one of the possible candidates t. Therefore $|\lambda|$ cannot exceed t_{max}—which must be the largest eigenvalue.

This is the *Perron-Frobenius theorem* for positive matrices, and it has one more application in mathematical economics.

EXAMPLE 3 *Von Neumann's model of an expanding economy*
We go back to the 3 by 3 matrix A that gave the consumption of steel, food, and labor. If the outputs are s_1, f_1, l_1, then the required inputs are

$$u_0 = \begin{bmatrix} .4 & 0 & .1 \\ 0 & .1 & .8 \\ .5 & .7 & .1 \end{bmatrix} \begin{bmatrix} s_1 \\ f_1 \\ l_1 \end{bmatrix} = Au_1.$$

In economics the difference equation is backward! Instead of $u_1 = Au_0$ we have $u_0 = Au_1$. If A is small (as it is), then production does not consume everything—and the economy can grow. The eigenvalues of A^{-1} will govern this growth. But again there is a nonnegative twist, since steel, food, and labor cannot come in negative amounts. Von Neumann asked for the maximum rate t at which the economy can expand and *still stay nonnegative*, meaning that $u_1 \geq tu_0 \geq 0$.

Thus the problem requires $u_1 \geq tAu_1$. It is like the Perron-Frobenius theorem, with A on the other side. As before, equality holds when t reaches t_{max}—which is the eigenvalue associated with the positive eigenvector of A^{-1}. In this example the expansion factor is $\frac{10}{9}$:

$$x = \begin{bmatrix} 1 \\ 5 \\ 5 \end{bmatrix} \quad \text{and} \quad Ax = \begin{bmatrix} .4 & 0 & .1 \\ 0 & .1 & .8 \\ .5 & .7 & .1 \end{bmatrix} \begin{bmatrix} 1 \\ 5 \\ 5 \end{bmatrix} = \begin{bmatrix} 0.9 \\ 4.5 \\ 4.5 \end{bmatrix} = \frac{9}{10} x.$$

With steel-food-labor in the ratio 1-5-5, the economy grows as quickly as possible: *The maximum growth rate is* $1/\lambda_1$.

EXERCISES

5.3.1 (a) For the Fibonacci matrix $A = \begin{bmatrix} 1 & 1 \\ 1 & 0 \end{bmatrix}$, write down A^2, A^3, A^4, and then (using the text) A^{100}.
   (b) Find B^{-101} if $B = \begin{bmatrix} 7 & 12 \\ -4 & -7 \end{bmatrix}$.

5.3.2 Suppose Fibonacci had started his sequence with $F_0 = 1$ and $F_1 = 3$, and then followed the same rule $F_{k+2} = F_{k+1} + F_k$. Find the new initial vector u_0, the new coefficients $c = S^{-1}u_0$, and the new Fibonacci numbers. Show that the ratios F_{k+1}/F_k still approach the golden mean.

5.3.3 If each number is the *average of the two previous numbers*, $G_{k+2} = \frac{1}{2}(G_{k+1} + G_k)$, set up the matrix A and diagonalize it. Starting from $G_0 = 0$ and $G_1 = \frac{1}{2}$, find a formula for G_k and compute its limit as $k \to \infty$.

5.3.4 Bernadelli studied a beetle "which lives three years only, and propagates in its third year." If the first age group survives with probability $\frac{1}{2}$, and then the second with probability $\frac{1}{3}$, and then the third produces six females on the way out, the matrix is

$$A = \begin{bmatrix} 0 & 0 & 6 \\ \frac{1}{2} & 0 & 0 \\ 0 & \frac{1}{3} & 0 \end{bmatrix}.$$

Show that $A^3 = I$, and follow the distribution of beetles for six years starting with 3000 beetles in each age group.

5.3.5 Suppose there is an epidemic in which every month half of those who are well become sick, and a quarter of those who are sick become dead. Find the steady state for the corresponding Markov process

$$\begin{bmatrix} d_{k+1} \\ s_{k+1} \\ w_{k+1} \end{bmatrix} = \begin{bmatrix} 1 & \frac{1}{4} & 0 \\ 0 & \frac{3}{4} & \frac{1}{2} \\ 0 & 0 & \frac{1}{2} \end{bmatrix} \begin{bmatrix} d_k \\ s_k \\ w_k \end{bmatrix}.$$

5.3.6 Write down the 3 by 3 transition matrix for a chemistry course that is taught in two sections, if every week $\frac{1}{4}$ of those in Section A and $\frac{1}{3}$ of those in Section B drop the course, and $\frac{1}{6}$ of each section transfer to the other section.

5.3.7 Find the limiting values of y_k and z_k $(k \to \infty)$ if

$$\begin{aligned} y_{k+1} &= .8y_k + .3z_k & y_0 &= 0 \\ z_{k+1} &= .2y_k + .7z_k & z_0 &= 5. \end{aligned}$$

Also find formulas for y_k and z_k from $A^k = S\Lambda^k S^{-1}$.

5.3.8 (a) From the fact that column 1 + column 2 = 2 (column 3), so the columns are linearly dependent, what is one of the eigenvalues of A and what is its corresponding eigenvector?

$$A = \begin{bmatrix} .2 & .4 & .3 \\ .4 & .2 & .3 \\ .4 & .4 & .4 \end{bmatrix}.$$

(b) Find the other eigenvalues of A.
(c) If $u_0 = (0, 10, 0)$ find the limit of $A^k u_0$ as $k \to \infty$.

5.3.9 Suppose there are three major centers for Move-It-Yourself trucks. Every month half of those in Boston and in Los Angeles go to Chicago, the other half stay where they are, and the trucks in Chicago are split equally between Boston and Los Angeles. Set up the 3 by 3 transition matrix A, and find the steady state u_∞ corresponding to the eigenvalue $\lambda = 1$.

5.3.10 (a) In what range of a and b is the following equation a Markov process?

$$u_{k+1} = Au_k = \begin{bmatrix} a & b \\ 1-a & 1-b \end{bmatrix} u_k, \quad u_0 = \begin{bmatrix} 1 \\ 1 \end{bmatrix}.$$

(b) Compute $u_k = S\Lambda^k S^{-1} u_0$ for any a and b.
(c) Under what condition on a and b does u_k approach a finite limit as $k \to \infty$, and what is the limit? Does A have to be a Markov matrix?

5.3.11 Multinational companies in the U.S., Japan, and Europe have assets of $4 trillion. At the start, $2 trillion are in the U.S. and $2 trillion in Europe. Each year $\frac{1}{2}$ the U.S. money stays home, $\frac{1}{4}$ goes to both Japan and Europe. For Japan and Europe, $\frac{1}{2}$ stays home and $\frac{1}{2}$ is sent to the U.S.
(a) Find the matrix that gives

$$\begin{bmatrix} US \\ J \\ E \end{bmatrix}_{\text{year } k+1} = A \begin{bmatrix} US \\ J \\ E \end{bmatrix}_{\text{year } k}$$

(b) Find the eigenvalues and eigenvectors of A.
(c) Find the limiting distribution of the $4 trillion as the world ends.
(d) Find the distribution at year k.

5.3.12 If A is a Markov matrix, show that the sum of the components of Ax equals the sum of the components of x. Deduce that if $Ax = \lambda x$ with $\lambda \neq 1$, the components of the eigenvector add to zero.

5.3.13 The solution to $du/dt = Au = \begin{bmatrix} 0 & -1 \\ 1 & 0 \end{bmatrix} u$ (eigenvalues i and $-i$) goes around in a circle: $u = (\cos t, \sin t)$. Suppose we approximate by forward, backward, and centered difference equations:

(a) $u_{n+1} - u_n = Au_n$ or $u_{n+1} = (I + A)u_n$

(b) $u_{n+1} - u_n = Au_{n+1}$ or $u_{n+1} = (I - A)^{-1}u_n$

(c) $u_{n+1} - u_n = \frac{1}{2}A(u_{n+1} + u_n)$ or $u_{n+1} = (I - \frac{1}{2}A)^{-1}(I + \frac{1}{2}A)u_n$.

Find the eigenvalues of $I + A$ and $(I - A)^{-1}$ and $(I - \frac{1}{2}A)^{-1}(I + \frac{1}{2}A)$. For which difference equation does the solution stay on a circle?

5.3.14 For the system $v_{n+1} = \alpha(v_n + w_n)$, $w_{n+1} = \alpha(v_n + w_n)$, what values of α produce instability?

5.3.15 Find the largest values a, b, c for which these matrices are stable or neutrally stable:

$$\begin{bmatrix} a & -.8 \\ .8 & .2 \end{bmatrix}, \quad \begin{bmatrix} b & .8 \\ 0 & .2 \end{bmatrix}, \quad \begin{bmatrix} c & .8 \\ .2 & c \end{bmatrix}.$$

5.3.16 Multiplying term by term, check that $(I - A)(I + A + A^2 + \cdots) = I$. This infinite series represents $(I - A)^{-1}$, and is nonnegative whenever A is nonnegative, provided it has a finite sum; the condition for that is $\lambda_1 < 1$. Add up the infinite series, and confirm that it equals $(I - A)^{-1}$, for the consumption matrix

$$A = \begin{bmatrix} 0 & 1 & 1 \\ 0 & 0 & 1 \\ 0 & 0 & 0 \end{bmatrix}.$$

5.3.17 For $A = \begin{bmatrix} 0 & 2 \\ 0 & 5 \end{bmatrix}$ find the powers A^k (including A^0) and show explicitly that their sum agrees with $(I - A)^{-1}$.

5.3.18 Explain by mathematics or economics why increasing any entry of the "consumption matrix" A must increase $t_{max} = \lambda_1$ (and slow down the expansion).

5.3.19 What are the limits as $k \to \infty$ (the steady states) of

$$\begin{bmatrix} .4 & .2 \\ .6 & .8 \end{bmatrix}^k \begin{bmatrix} 1 \\ 0 \end{bmatrix} \quad \text{and} \quad \begin{bmatrix} .4 & .2 \\ .6 & .8 \end{bmatrix}^k \begin{bmatrix} 0 \\ 1 \end{bmatrix} \quad \text{and} \quad \begin{bmatrix} .4 & .2 \\ .6 & .8 \end{bmatrix}^k ?$$

5.3.20 Prove that every third Fibonacci number is even.

<h2 style="text-align:center">DIFFERENTIAL EQUATIONS AND THE EXPONENTIAL e^{At} ■ 5.4</h2>

Wherever you find a system of equations, rather than a single equation, matrix theory has a part to play. This was true for difference equations, where the solution $u_k = A^k u_0$ depended on the powers of A. It is equally true for differential equations, where the solution $u(t) = e^{At} u_0$ depends on the **exponential** of A. To define this exponential, and to understand it, we turn right away to an example:

$$\frac{du}{dt} = Au = \begin{bmatrix} -2 & 1 \\ 1 & -2 \end{bmatrix} u. \tag{1}$$

The first step is always to find the eigenvalues and eigenvectors:

$$A \begin{bmatrix} 1 \\ 1 \end{bmatrix} = (-1) \begin{bmatrix} 1 \\ 1 \end{bmatrix}, \qquad A \begin{bmatrix} 1 \\ -1 \end{bmatrix} = (-3) \begin{bmatrix} 1 \\ -1 \end{bmatrix}.$$

Then there are several possibilities, all leading to the same answer. Probably the best way is to write down the general solution, and match it to the initial vector u_0 at $t = 0$. The general solution is a combination of pure exponential solutions. These are solutions of the special form $ce^{\lambda t} x$, where λ is an eigenvalue of A and x is its eigenvector; they satisfy the differential equation, since $d/dt(ce^{\lambda t}x) = A(ce^{\lambda t}x)$. (They were our introduction to eigenvalues, at the start of the chapter.) In this 2 by 2 example, there are two pure exponentials to be combined:

$$u(t) = c_1 e^{\lambda_1 t} x_1 + c_2 e^{\lambda_2 t} x_2 \qquad \text{or} \qquad u = \begin{bmatrix} 1 & 1 \\ 1 & -1 \end{bmatrix} \begin{bmatrix} e^{-t} & \\ & e^{-3t} \end{bmatrix} \begin{bmatrix} c_1 \\ c_2 \end{bmatrix}. \tag{2}$$

At time zero, when the exponentials are $e^0 = 1$, u_0 determines c_1 and c_2:

$$u_0 = c_1 x_1 + c_2 x_2 \qquad \text{or} \qquad u_0 = \begin{bmatrix} 1 & 1 \\ 1 & -1 \end{bmatrix} \begin{bmatrix} c_1 \\ c_2 \end{bmatrix} = Sc.$$

You recognize S, the matrix of eigenvectors. And the constants $c = S^{-1} u_0$ are the same as they were for difference equations. Substituting them back into (2), the problem is solved. In matrix form, the solution is

$$u(t) = \begin{bmatrix} 1 & 1 \\ 1 & -1 \end{bmatrix} \begin{bmatrix} e^{-t} & \\ & e^{-3t} \end{bmatrix} \begin{bmatrix} c_1 \\ c_2 \end{bmatrix} = S \begin{bmatrix} e^{-t} & \\ & e^{-3t} \end{bmatrix} S^{-1} u_0.$$

Here is the fundamental formula of this section: $u = Se^{\Lambda t} S^{-1} u_0$ solves the differential equation, just as $S\Lambda^k S^{-1} u_0$ solved the difference equation. The key matrices are

$$\Lambda = \begin{bmatrix} -1 & \\ & -3 \end{bmatrix} \qquad \text{and} \qquad e^{\Lambda t} = \begin{bmatrix} e^{-t} & \\ & e^{-3t} \end{bmatrix}.$$

There are two more things to be done with this example. One is to complete the mathematics, by giving a direct definition of the ***exponential of a matrix***. The other is to give a physical interpretation of the equation and its solution. It is the kind of differential equation that has useful applications.

First, we take up the exponential. For a diagonal matrix Λ it is easy; $e^{\Lambda t}$ just has the n numbers $e^{\lambda t}$ on the diagonal (as in the display above). For a general matrix A, the natural idea is to imitate the power series definition

$$e^x = 1 + x + \frac{x^2}{2!} + \frac{x^3}{3!} + \cdots.$$

If we replace x by At and 1 by I, this sum is an n by n matrix:

$$e^{At} = I + At + \frac{(At)^2}{2!} + \frac{(At)^3}{3!} + \cdots. \tag{3}$$

That is the exponential of At. The series always converges, and its sum has the right properties:

$$(e^{As})(e^{At}) = e^{A(s+t)} \tag{4}$$

$$(e^{At})(e^{-At}) = I \tag{5}$$

$$\frac{d}{dt}(e^{At}) = Ae^{At} \tag{6}$$

From the last one, $u = e^{At}u_0$ solves the differential equation. This solution must be the same as the form we used for computation, which was $u = Se^{\Lambda t}S^{-1}u_0$. To prove directly that those solutions agree, remember that each power $(S\Lambda S^{-1})^k$ telescopes into $A^k = S\Lambda^k S^{-1}$ (because S^{-1} cancels S). Therefore the whole exponential is diagonalized at the same time:

$$e^{At} = I + S\Lambda S^{-1}t + \frac{S\Lambda^2 S^{-1}t^2}{2!} + \frac{S\Lambda^3 S^{-1}t^3}{3!} + \cdots$$

$$= S\left(I + \Lambda t + \frac{(\Lambda t)^2}{2!} + \frac{(\Lambda t)^3}{3!} + \cdots\right)S^{-1} = Se^{\Lambda t}S^{-1}.$$

EXAMPLE   The exponential of $A = \begin{bmatrix} -2 & 1 \\ 1 & -2 \end{bmatrix}$ is

$$e^{At} = Se^{\Lambda t}S^{-1} = \begin{bmatrix} 1 & 1 \\ 1 & -1 \end{bmatrix}\begin{bmatrix} e^{-t} & \\ & e^{-3t} \end{bmatrix}\begin{bmatrix} 1 & 1 \\ 1 & -1 \end{bmatrix}^{-1}$$

$$= \frac{1}{2}\begin{bmatrix} e^{-t} + e^{-3t} & e^{-t} - e^{-3t} \\ e^{-t} - e^{-3t} & e^{-t} + e^{-3t} \end{bmatrix}.$$

At $t = 0$ we get the identity matrix: $e^0 = I$. The infinite series e^{At} gives the answer for all t, but it can be hard to compute. The form $Se^{\Lambda t}S^{-1}$ gives the same answer when A can be diagonalized; there must be n independent eigenvectors in S. But this form is much simpler, and it leads to a combination of n pure exponentials

$e^{\lambda t}x$—which is the best solution of all:

5L If A can be diagonalized, $A = S\Lambda S^{-1}$, then the differential equation $du/dt = Au$ has the solution

$$u(t) = e^{At}u_0 = Se^{\Lambda t}S^{-1}u_0.\tag{7}$$

The columns of S are the eigenvectors of A, so that

$$u(t) = \begin{bmatrix} x_1 & \cdots & x_n \end{bmatrix}\begin{bmatrix} e^{\lambda_1 t} & & \\ & \ddots & \\ & & e^{\lambda_n t} \end{bmatrix}S^{-1}u_0$$

$$= c_1 e^{\lambda_1 t}x_1 + \cdots + c_n e^{\lambda_n t}x_n.\tag{8}$$

The general solution is a combination of pure exponentials, and the constants c_i that match the initial condition u_0 are $c = S^{-1}u_0$.

This gives a complete analogy with difference equations—you could compare it with 5G. In both cases we assumed that A could be diagonalized, since otherwise it has fewer than n eigenvectors and we have not found enough special solutions. The missing solutions do exist, but they are more complicated than pure exponentials; they involve "generalized eigenvectors" and factors like $te^{\lambda t}$. Nevertheless the formula $u(t) = e^{At}u_0$ remains completely correct.†

The matrix e^{At} is *never singular*. One proof is to look at its eigenvalues; if λ is an eigenvalue of A, then $e^{\lambda t}$ is the corresponding eigenvalue of e^{At}—and a number like $e^{\lambda t}$ can never be zero. Another approach is to compute the determinant, which is the product of the eigenvalues:

$$\det e^{At} = e^{\lambda_1 t}e^{\lambda_2 t}\cdots e^{\lambda_n t} = e^{\text{trace}(At)}.$$

Again, this cannot be zero. And the best way is *to recognize e^{-At} as the inverse*, from (5).

This invertibility is fundamental in ordinary differential equations, since it has the following consequence: If n solutions are linearly independent at $t = 0$, then *they remain linearly independent forever*. If the initial vectors are $v_1, \ldots, v_n$, we can put the solutions $e^{At}v$ into a matrix:

$$\begin{bmatrix} e^{At}v_1 & \cdots & e^{At}v_n \end{bmatrix} = e^{At}\begin{bmatrix} v_1 & \cdots & v_n \end{bmatrix}.$$

The determinant of the left side is called the "*Wronskian*." It never becomes zero, because it is the product of two nonzero determinants. Both matrices on the right are invertible.

---

† To compute this defective case we can use the Jordan form in Appendix B, and find e^{Jt}.

Remark   Not all differential equations come to us as a first-order system $du/dt = Au$. In fact, we may start from a single equation of higher order, like

$$y''' - 3y'' + 2y' = 0.$$

To convert to a 3 by 3 system, introduce $v = y'$ and $w = v'$ as additional unknowns along with y itself. Then these two equations combine with the original one to give

$$
\begin{aligned}
y' &= v \\
v' &= w \\
w' &= 3w - 2v
\end{aligned}
\qquad \text{or} \qquad
u' =
\begin{bmatrix}
0 & 1 & 0 \\
0 & 0 & 1 \\
0 & -2 & 3
\end{bmatrix}
\begin{bmatrix}
y \\
v \\
w
\end{bmatrix}
= Au.
$$

We are back to a first-order system. The problem can therefore be solved in two ways. In a course on differential equations, you would substitute $y = e^{\lambda t}$ into the third-order equation:

$$(\lambda^3 - 3\lambda^2 + 2\lambda)e^{\lambda t} = 0 \qquad \text{or} \qquad \lambda(\lambda - 1)(\lambda - 2)e^{\lambda t} = 0.$$

The three pure exponential solutions are $y = e^{0t}$, $y = e^t$, and $y = e^{2t}$. No eigenvectors are involved. In a linear algebra course, we proceed as usual for a first-order system, and find the eigenvalues of A:

$$
\det(A - \lambda I) =
\begin{bmatrix}
-\lambda & 1 & 0 \\
0 & -\lambda & 1 \\
0 & -2 & 3 - \lambda
\end{bmatrix}
= -\lambda^3 + 3\lambda^2 - 2\lambda = 0. \tag{9}
$$

Again, the same three exponents appear: $\lambda = 0$, $\lambda = 1$, and $\lambda = 2$. This is a general rule, which makes the two methods consistent; the growth rates of the solutions are intrinsic to the problem, and they stay fixed when the equations change form. It seems to us that solving the third-order equation is quicker.

Now we turn to the physical significance of $du/dt = Au$, with $A = \begin{bmatrix} -2 & 1 \\ 1 & -2 \end{bmatrix}$. It is easy to explain and at the same time genuinely important. The differential equation describes a process of *diffusion*, which can be visualized by dividing an infinite pipe into four segments. The two in the middle are finite, and the two at the ends are semi-infinite (Fig. 5.1). At time $t = 0$, the finite segments contain concentrations v_0 and w_0 of some chemical solution. At the same time, and for all times, the concentration in the two infinite segments is zero; with an infinite volume, this will be a correct picture of the average concentration in these infinite

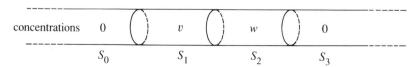

Fig. 5.1. A model of diffusion.

segments even after the chemical has started to diffuse. Diffusion starts at time $t = 0$, and is governed by the following law: *At each time t, the diffusion rate between two adjacent segments equals the difference in concentrations.* We are imagining that, within each segment, the concentration remains uniform. The process is continuous in time but discrete in space; the only unknowns are $v(t)$ and $w(t)$ in the two inner segments S_1 and S_2.

The concentration v is changing in two ways, by diffusion into the far left segment S_0 and by diffusion into or out of S_2. The net rate of change is therefore

$$\frac{dv}{dt} = (w - v) + (0 - v),$$

because the concentration in S_0 is identically zero. Similarly,

$$\frac{dw}{dt} = (0 - w) + (v - w).$$

Therefore the system exactly matches our example $du/dt = Au$ in equation (1):

$$u = \begin{bmatrix} v \\ w \end{bmatrix}, \qquad \frac{du}{dt} = \begin{bmatrix} -2v + w \\ v - 2w \end{bmatrix} = \begin{bmatrix} -2 & 1 \\ 1 & -2 \end{bmatrix} u.$$

The eigenvalues -1 and -3 will govern the behavior of the solution. They give the rate at which the concentrations decay, and λ_1 is the more important because only an exceptional set of starting conditions can lead to "superdecay" at the rate e^{-3t}. In fact, those conditions must come from the eigenvector $(1, -1)$. If the experiment admits only nonnegative concentrations, superdecay is impossible and the limiting rate must be e^{-t}. The solution that decays at this rate corresponds to the eigenvector $(1, 1)$, and therefore the two concentrations will become nearly equal as $t \to \infty$.

One more comment on this example: It is a discrete approximation, with only two unknowns, to the continuous diffusion described by the partial differential equation

$$\frac{\partial u}{\partial t} = \frac{\partial^2 u}{\partial x^2}, \quad u(0) = u(1) = 0.$$

That equation is approached by keeping the two infinite segments at zero concentration, and dividing the middle of the pipe into smaller and smaller segments, of length $1/N$. The discrete system with N unknowns is governed by

$$\frac{d}{dt} \begin{bmatrix} u_1 \\ \cdot \\ \cdot \\ u_N \end{bmatrix} = \begin{bmatrix} -2 & 1 & & \\ 1 & -2 & \cdot & \\ & \cdot & \cdot & 1 \\ & & 1 & -2 \end{bmatrix} \begin{bmatrix} u_1 \\ \cdot \\ \cdot \\ u_N \end{bmatrix} = Au. \tag{10}$$

This is the finite difference matrix with the $1, -2, 1$ pattern. The right side Au approaches the second derivative d^2u/dx^2, after a scaling factor N^2 comes from the flow problem. In the limit as $N \to \infty$, we reach the **heat equation** $\partial u/\partial t = \partial^2 u/\partial x^2$. Its solutions are still combinations of pure exponentials, but now there are infinitely many. Instead of an eigenvector in $Ax = \lambda x$, we have an *eigenfunction* in $d^2u/dx^2 = \lambda u$; it is $u = \sin n\pi x$. Then the solution to the heat equation is

$$u(t) = \sum_{n=1}^{\infty} c_n e^{-n^2\pi^2 t} \sin n\pi x.$$

The constants c_n are determined as always by the initial conditions. The rates of decay are the eigenvalues: $\lambda_n = -n^2\pi^2$. The normal modes are the eigenfunctions. The only novelty is that they are functions and not vectors, because the problem is continuous and not discrete.

Stability of Differential Equations

Just as for difference equations, it is the eigenvalues that decide how $u(t)$ behaves as $t \to \infty$. As long as A can be diagonalized, there will be n pure exponential solutions to the differential equation, and any specific solution $u(t)$ is some combination

$$u(t) = Se^{\Lambda t}S^{-1}u_0 = c_1 e^{\lambda_1 t}x_1 + \cdots + c_n e^{\lambda_n t}x_n.$$

Stability is governed by the factors $e^{\lambda_i t}$. If they all approach zero, then $u(t)$ approaches zero; if they all stay bounded, then $u(t)$ stays bounded; if one of them blows up, then except for very special starting conditions the solution will blow up. Furthermore, the size of $e^{\lambda t}$ depends only on the real part of λ. **It is only the real parts that govern stability**: If $\lambda = a + ib$, then

$$e^{\lambda t} = e^{at}e^{ibt} = e^{at}(\cos bt + i \sin bt) \qquad \text{and} \qquad |e^{\lambda t}| = e^{at}.$$

This decays for $a < 0$, it is constant for $a = 0$, and it explodes for $a > 0$. The imaginary part is producing oscillations, but the amplitude comes from the real part.

5M   The differential equation $du/dt = Au$ is

stable and $e^{At} \to 0$ whenever all Re $\lambda_i < 0$
neutrally stable when all Re $\lambda_i \le 0$ and some Re $\lambda_i = 0$
unstable and e^{At} is unbounded if any eigenvalue has Re $\lambda_i > 0$.

In some texts the condition Re $\lambda < 0$ is called *asymptotic* stability, because it guarantees decay for large times t. Our argument depended on having n pure exponential solutions, but even if A is not diagonalizable (and there are terms like $te^{\lambda t}$) the result is still true: **All solutions approach zero if and only if all eigenvalues have a negative real part**.

Stability is especially easy to decide for a 2 by 2 system (which is very common in applications). The equation is

$$\frac{du}{dt} = \begin{bmatrix} a & b \\ c & d \end{bmatrix} u,$$

and we need to know when both eigenvalues of that matrix have negative real parts. (Note again that the eigenvalues can be complex numbers.) The test for stability is very direct:

> **The trace $a + d$ must be negative.**
> **The determinant $ad - bc$ must be positive.**

When the eigenvalues are real, those tests guarantee them to be negative. Their product is the determinant; if it is positive, then the eigenvalues are both positive or both negative. Their sum is the trace; if it is negative then the eigenvalues had to be negative.

When the eigenvalues are a complex pair $x \pm iy$, the tests still succeed: the trace is their sum $2x$ (which is <0) and the determinant is $(x + iy)(x - iy) = x^2 + y^2$ (which is >0). The figure shows the one stable quadrant, and it also shows the parabolic boundary line between real and complex eigenvalues. The reason for the parabola is in the equation for the eigenvalues:

$$\det \begin{bmatrix} a - \lambda & b \\ c & d - \lambda \end{bmatrix} = \lambda^2 - (\text{trace})\lambda + (\det) = 0.$$

This is a quadratic equation, so the eigenvalues are

$$\lambda = \tfrac{1}{2}[\text{trace} \pm \sqrt{(\text{trace})^2 - 4(\det)}]. \tag{11}$$

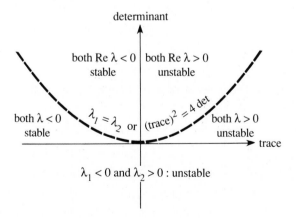

Fig. 5.2. Stability regions for a 2 by 2 matrix.

Above the parabola, the number under the square root is negative—so λ is not real. On the parabola, the square root is zero and λ is repeated. Below the parabola it is real. Every symmetric matrix is on or below, since if $b = c$ then

$$(\text{trace})^2 - 4(\det) = (a + d)^2 - 4(ad - b^2) = (a - d)^2 + 4b^2 \geq 0.$$

Complex eigenvalues are possible only when b and c have opposite signs and are sufficiently large.

AN EXAMPLE FROM EACH QUADRANT:

$$\begin{bmatrix} 1 & 0 \\ 0 & 2 \end{bmatrix} \quad \begin{bmatrix} -1 & 0 \\ 0 & -2 \end{bmatrix} \quad \begin{bmatrix} 1 & 0 \\ 0 & -2 \end{bmatrix} \quad \begin{bmatrix} -1 & 0 \\ 0 & 2 \end{bmatrix}$$

On the boundaries of the second quadrant the equation is neutrally stable. Crossing those boundaries gives instability. On the horizontal axis one eigenvalue is zero (because the determinant is $\lambda_1\lambda_2 = 0$). On the vertical axis above the origin, both eigenvalues are pure imaginary (because the trace is zero). Those crossings are the two fundamental ways in which stability can be lost.

The n by n case is more difficult. To find a condition that is both necessary and sufficient for stability—a full test for $\text{Re }\lambda_i < 0$—there are two possibilities. One is to go back to Routh and Hurwitz, who found a series of inequalities on the entries a_{ij}. I do not think this approach is much good for a large matrix; the computer can probably find the eigenvalues with more certainty than it can test these inequalities. The other possibility was discovered by Lyapunov in 1897. *It is to find a weighting matrix W so that the weighted length $\| Wu(t) \|$ is always decreasing.* If there exists such a W, then $u' = Au$ must be stable; $\| Wu \|$ will decrease steadily to zero, and after a few ups and downs u must get there too. The real value of Lyapunov's method is in the case of a nonlinearity, which may make the equation impossible to solve but still leave a decreasing $\| Wu(t) \|$—so that stability can be proved without knowing a formula for $u(t)$.

EXAMPLE 1 $du/dt = \begin{bmatrix} 0 & -1 \\ 1 & 0 \end{bmatrix} u$ sends u around a circle.

Since trace $= 0$ and det $= 1$, we have pure imaginary eigenvalues:

$$\begin{vmatrix} -\lambda & -1 \\ 1 & -\lambda \end{vmatrix} = \lambda^2 + 1 = 0 \quad \text{so} \quad \lambda = +i \text{ and } -i.$$

The eigenvectors are $(1, -i)$ and $(1, i)$, and the solution is

$$u(t) = c_1 e^{it} \begin{bmatrix} 1 \\ -i \end{bmatrix} + c_2 e^{-it} \begin{bmatrix} 1 \\ i \end{bmatrix}.$$

That is correct but not beautiful. By substituting $\cos t + i \sin t$ for e^{it} (and $\cos t - i \sin t$ for e^{-it}) *real numbers will reappear*:

$$u(t) = \begin{bmatrix} (c_1 + c_2) \cos t + i(c_1 - c_2) \sin t \\ -i(c_1 - c_2) \cos t + (c_1 + c_2) \sin t \end{bmatrix}. \tag{12}$$

At $t = 0$, where $\cos t = 1$, this should agree with $u_0 = \begin{bmatrix} a \\ b \end{bmatrix}$. Therefore the numbers a and b must multiply $\cos t$, and $u(t)$ ends up as

$$u(t) = \begin{bmatrix} a \cos t - b \sin t \\ b \cos t + a \sin t \end{bmatrix} = \begin{bmatrix} \cos t & -\sin t \\ \sin t & \cos t \end{bmatrix} \begin{bmatrix} a \\ b \end{bmatrix}. \tag{13}$$

There we have something important! The last matrix is multiplying u_0, so it must be the exponential e^{At}. (Remember that the solution is $u = e^{At} u_0$.) That matrix of cosines and sines is our leading example of an *orthogonal matrix*. The columns have length one, their inner product is zero, and we have a confirmation of the following wonderful fact:

If A is skew-symmetric then e^{At} is an orthogonal matrix.

These are exactly the matrices for conservative systems, when no energy is lost by damping or diffusion:

$$A^T = -A \quad \text{and} \quad (e^{At})^T = e^{-At} \quad \text{and} \quad \|e^{At} u_0\| = \|u_0\|.$$

That last equation expresses an essential property of orthogonal matrices. When they multiply a vector, the length is not changed. The vector u_0 is just rotated, and that describes the solution to $du/dt = Au$: *It goes around in a circle.*

In this very unusual case, e^{At} can also be recognized directly from the infinite series. Note that $A = \begin{bmatrix} 0 & -1 \\ 1 & 0 \end{bmatrix}$ has $A^2 = -I$, and use this over and over:

$$e^{At} = I + At + \frac{(At)^2}{2} + \frac{(At)^3}{6} + \cdots$$

$$= \begin{bmatrix} \left(1 - \frac{t^2}{2} + \cdots\right) & \left(-t + \frac{t^3}{6} - \cdots\right) \\ \left(t - \frac{t^3}{6} + \cdots\right) & \left(1 - \frac{t^2}{2} + \cdots\right) \end{bmatrix} = \begin{bmatrix} \cos t & -\sin t \\ \sin t & \cos t \end{bmatrix}$$

EXAMPLE 2 The diffusion equation $du/dt = \begin{bmatrix} -2 & 1 \\ 1 & -2 \end{bmatrix} u$ was stable, with $\lambda = -1$ and $\lambda = -3$.

EXAMPLE 3 If we close off the infinite segments, nothing leaves the system:

$$\frac{du}{dt} = \begin{bmatrix} -1 & 1 \\ 1 & -1 \end{bmatrix} u \quad \text{or} \quad \begin{array}{l} dv/dt = w - v \\ dw/dt = v - w \end{array}$$

This is a *continuous Markov process.* Instead of moving every year, the particles move every instant. Their total number $v + w$ is constant. That comes from adding the two equations on the right: $d/dt\,(v + w) = 0$.

A Markov matrix has its column sums equal to $\lambda_{max} = 1$. A *continuous* Markov matrix, for differential equations, has its column sums equal to $\lambda_{max} = 0$. A is a Markov matrix if and only if $B = A - I$ is a continuous Markov matrix. The steady state for both is the eigenvector for λ_{max}. It is multiplied by $1^k = 1$ in difference equations and by $e^{0t} = 1$ in differential equations, and it doesn't move.

In the example the steady state has $v = w$.

EXAMPLE 4   In nuclear engineering a reactor is called *critical* when it is neutrally stable; the fission balances the decay. Slower fission makes it stable, or *subcritical,* and eventually it runs down. Unstable fission is a bomb.

Second-Order Equations

The laws of diffusion led to a first-order system $du/dt = Au$. So do a lot of other applications, in chemistry, in biology, and elsewhere, but the most important law of physics does not. It is *Newton's law* $F = ma$, and the acceleration a is a second derivative. Inertial terms produce second-order equations (we have to solve $d^2u/dt^2 = Au$ instead of $du/dt = Au$) and the goal is to understand how this change to second derivatives alters the solution.† It is optional in linear algebra, but not in physics.

The comparison will be perfect if we keep the same A:

$$\frac{d^2u}{dt^2} = Au = \begin{bmatrix} -2 & 1 \\ 1 & -2 \end{bmatrix} u. \tag{14}$$

Two initial conditions have to be specified at $t = 0$ in order to get the system started—the "displacement" $u = u_0$ and the "velocity" $du/dt = u_0'$. To match these conditions, there will be $2n$ pure exponential solutions instead of n.

Suppose we use ω rather than λ, and write these special solutions as $u = e^{i\omega t}x$. Substituting this exponential into the differential equation, it must satisfy

$$\frac{d^2}{dt^2}(e^{i\omega t}x) = A(e^{i\omega t}x), \quad \text{or} \quad -\omega^2 x = Ax. \tag{15}$$

The vector x must be an eigenvector of A, exactly as before. The corresponding eigenvalue is now $-\omega^2$, so the frequency ω is connected to the decay rate λ by the law $-\omega^2 = \lambda$. Every special solution $e^{\lambda t}x$ of the first-order equation leads to *two* special solutions $e^{i\omega t}x$ of the second-order equation, and the two exponents are

---

† Fourth derivatives are also possible, in the bending of beams, but nature seems to resist going higher than four.

$\omega = \pm\sqrt{-\lambda}$. This breaks down only when $\lambda = 0$, which has just one square root; if the eigenvector is x, the two special solutions are x and tx.

For a genuine diffusion matrix, the eigenvalues λ are all negative and therefore the frequencies ω are all real: *Pure diffusion is converted into pure oscillation*. The factors $e^{i\omega t}$ produce neutral stability, the solution neither grows or decays, and the total energy stays precisely constant. It just keeps passing around the system. The general solution to $d^2u/dt^2 = Au$, if A has negative eigenvalues $\lambda_1, \ldots, \lambda_n$ and if $\omega_j = \sqrt{-\lambda_j}$, is

$$u(t) = (c_1 e^{i\omega_1 t} + d_1 e^{-i\omega_1 t})x_1 + \cdots + (c_n e^{i\omega_n t} + d_n e^{-i\omega_n t})x_n. \tag{16}$$

As always, the constants are found from the initial conditions. This is easier to do (at the expense of one extra formula) by switching from oscillating exponentials to the more familiar sine and cosine:

$$u(t) = (a_1 \cos \omega_1 t + b_1 \sin \omega_1 t)x_1 + \cdots + (a_n \cos \omega_n t + b_n \sin \omega_n t)x_n. \tag{17}$$

The initial displacement is easy to keep separate: $t = 0$ means that $\sin \omega t = 0$ and $\cos \omega t = 1$, leaving only

$$u_0 = a_1 x_1 + \cdots + a_n x_n, \quad \text{or} \quad u_0 = Sa, \quad \text{or} \quad a = S^{-1}u_0.$$

The displacement is matched by the a's; it leads to $S^{-1}u_0$ as before. Then differentiating $u(t)$ and setting $t = 0$, the b's are determined by the initial velocity:

$$u_0' = b_1\omega_1 x_1 + \cdots + b_n\omega_n x_n.$$

Substituting the a's and b's into the formula for $u(t)$, the equation is solved.

We want to apply these formulas to the example. Its eigenvalues were $\lambda_1 = -1$ and $\lambda_2 = -3$, so the frequencies are $\omega_1 = 1$ and $\omega_2 = \sqrt{3}$. If the system starts from rest (the initial velocity u_0' is zero), then the terms in $b \sin \omega t$ will disappear. And if the first oscillator is given a unit displacement, $u_0 = a_1 x_1 + a_2 x_2$ becomes

$$\begin{bmatrix} 1 \\ 0 \end{bmatrix} = a_1 \begin{bmatrix} 1 \\ 1 \end{bmatrix} + a_2 \begin{bmatrix} 1 \\ -1 \end{bmatrix}, \quad \text{or} \quad a_1 = a_2 = \tfrac{1}{2}.$$

Therefore the solution is

$$u(t) = \tfrac{1}{2} \cos t \begin{bmatrix} 1 \\ 1 \end{bmatrix} + \tfrac{1}{2} \cos \sqrt{3}t \begin{bmatrix} 1 \\ -1 \end{bmatrix}.$$

We can interpret this solution physically. There are two masses, connected to each other and to stationary walls by three identical springs (Fig. 5.3). The first mass is pushed to $v_0 = 1$, the second mass is held in place, and at $t = 0$ we let go. Their motion $u(t)$ becomes an average of two pure oscillations, corresponding to the two eigenvectors. In the first mode, the masses move exactly in unison and

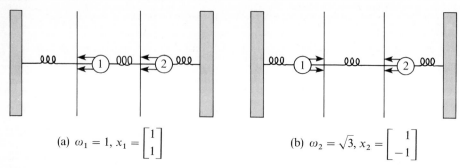

(a) $\omega_1 = 1$, $x_1 = \begin{bmatrix} 1 \\ 1 \end{bmatrix}$ (b) $\omega_2 = \sqrt{3}$, $x_2 = \begin{bmatrix} 1 \\ -1 \end{bmatrix}$

Fig. 5.3. The slow and fast modes of oscillation.

the spring in the middle is never stretched (Fig. 5.3a). The frequency $\omega_1 = 1$ is the same as for a single spring and a single mass. In the faster mode $x_2 = (1, -1)$, with components of opposite sign and with frequency $\sqrt{3}$, the masses move in opposite directions but with equal speeds (Fig. 5.3b). The general solution is a combination of these two normal modes, and our particular solution is half of each.

As time goes on, the motion is what we call "almost periodic." If the ratio ω_1/ω_2 had been a fraction, the two masses would eventually return to $v = 1$ and $w = 0$, and then begin the whole pattern again. A combination of $\sin 2t$ and $\sin 3t$ would have a period of 2π. But since $\sqrt{3}$ is irrational, the best we can say is that the masses will come *arbitrarily close* to reproducing the initial situation. They also come close, if we can wait long enough, to the opposite situation $v = 0$ and $w = 1$. Like a billiard ball bouncing forever on a perfectly smooth table, the total energy is fixed. Sooner or later the masses come near any state with this energy.

Again we cannot leave the problem without drawing a parallel to the continuous case: Instead of two masses, or N masses, there is a continuum. As the discrete masses and springs merge into a solid rod, the "second differences" given by the matrix coefficients 1, -2, 1 turn into second derivatives. This limit is described by the celebrated **wave equation** $\partial^2 u/\partial t^2 = \partial^2 u/\partial x^2$.

EXERCISES

5.4.1   Following the first example in this section, find the eigenvalues and eigenvectors and the exponential e^{At} for

$$A = \begin{bmatrix} -1 & 1 \\ 1 & -1 \end{bmatrix}.$$

5.4.2   For the previous matrix write down the general solution to $du/dt = Au$, and the specific solution that matches $u_0 = (3, 1)$. What is the *steady state* as $t \to \infty$? (This is a continuous Markov process; $\lambda = 0$ in a differential equation corresponds to to $\lambda = 1$ in a difference equation, since $e^{0t} = 1$.)

5.4.3 Suppose the time direction is reversed to give the matrix $-A$:

$$\frac{du}{dt} = \begin{bmatrix} 1 & -1 \\ -1 & 1 \end{bmatrix} u \quad \text{with} \quad u_0 = \begin{bmatrix} 3 \\ 1 \end{bmatrix}.$$

Find $u(t)$ and show that *it blows up* instead of decaying as $t \to \infty$. (Diffusion is irreversible, and the heat equation cannot run backward.)

5.4.4 If P is a projection matrix show from the infinite series that

$$e^P \approx I + 1.718P.$$

5.4.5 A diagonal matrix like $\Lambda = \begin{bmatrix} 1 & 0 \\ 0 & 2 \end{bmatrix}$ satisfies the usual rule $e^{\Lambda(t+T)} = e^{\Lambda t} e^{\Lambda T}$, because the rule holds for each diagonal entry.
(a) Explain why $e^{A(t+T)} = e^{At} e^{AT}$, using the formula $e^{At} = S e^{\Lambda t} S^{-1}$.
(b) Show that $e^{A+B} = e^A e^B$ is *not true* for matrices, from the example

$$A = \begin{bmatrix} 0 & 0 \\ 1 & 0 \end{bmatrix} \qquad B = \begin{bmatrix} 0 & -1 \\ 0 & 0 \end{bmatrix} \qquad \text{(use series for } e^A \text{ and } e^B\text{)}$$

5.4.6 The higher order equation $y'' + y = 0$ can be written as a first-order system by introducing the velocity y' as another unknown:

$$\frac{d}{dt} \begin{bmatrix} y \\ y' \end{bmatrix} = \begin{bmatrix} y' \\ y'' \end{bmatrix} = \begin{bmatrix} y' \\ -y \end{bmatrix}.$$

If this is $du/dt = Au$ what is the 2 by 2 matrix A? Find its eigenvalues and eigenvectors and compute the solution which starts from $y_0 = 2$, $y_0' = 0$.

5.4.7 Convert $y'' = 0$ to a first-order system $du/dt = Au$:

$$\frac{d}{dt} \begin{bmatrix} y \\ y' \end{bmatrix} = \begin{bmatrix} y' \\ 0 \end{bmatrix} = \begin{bmatrix} 0 & 1 \\ 0 & 0 \end{bmatrix} \begin{bmatrix} y \\ y' \end{bmatrix}.$$

This 2 by 2 matrix A is defective (it has only one eigenvector and cannot be diagonalized). Compute e^{At} from the series $I + At + \cdots$ and write down the solution $e^{At} u_0$ starting from $y_0 = 3$, $y_0' = 4$. Check that your $u = \begin{bmatrix} y \\ y' \end{bmatrix}$ satisfies $y'' = 0$.

5.4.8 Suppose the rabbit population r and the wolf population w are governed by

$$\frac{dr}{dt} = 4r - 2w,$$

$$\frac{dw}{dt} = r + w.$$

(a) Is this system stable, neutrally stable, or unstable?
(b) If initially $r = 300$ and $w = 200$, what are the populations at time t?
(c) After a long time, what is the proportion of rabbits to wolves?

5.4.9   Decide the stability of $u' = Au$ for the following matrices:

(a) $A = \begin{bmatrix} 2 & 3 \\ 4 & 5 \end{bmatrix}$ (b) $A = \begin{bmatrix} 1 & 2 \\ 3 & -1 \end{bmatrix}$

(c) $A = \begin{bmatrix} 1 & 1 \\ 1 & -2 \end{bmatrix}$ (d) $A = \begin{bmatrix} -1 & -1 \\ -1 & -1 \end{bmatrix}$

5.4.10   Decide on the stability or instability of $dv/dt = w$, $dw/dt = v$. Is there a solution that decays?

5.4.11   From their trace and determinant, at what time t do the following matrices change between stable with real eigenvalues, stable with complex eigenvalues, and unstable?

$$A_1 = \begin{bmatrix} 1 & -1 \\ t & -1 \end{bmatrix}, \quad A_2 = \begin{bmatrix} 0 & 4-t \\ 1 & -2 \end{bmatrix}, \quad A_3 = \begin{bmatrix} t & -1 \\ 1 & t \end{bmatrix}.$$

5.4.12   Find the eigenvalues and eigenvectors for

$$\frac{du}{dt} = Au = \begin{bmatrix} 0 & 3 & 0 \\ -3 & 0 & 4 \\ 0 & -4 & 0 \end{bmatrix} u.$$

Why do you know, without computing, that e^{At} will be an orthogonal matrix and $\|u(t)\|^2 = u_1^2 + u_2^2 + u_3^2$ will be constant?

5.4.13   For the skew-symmetric equation

$$\frac{du}{dt} = Au = \begin{bmatrix} 0 & c & -b \\ -c & 0 & a \\ b & -a & 0 \end{bmatrix} \begin{bmatrix} u_1 \\ u_2 \\ u_3 \end{bmatrix}$$

(a)   write out u_1', u_2', u_3' and confirm that $u_1'u_1 + u_2'u_2 + u_3'u_3 = 0$
(b)   deduce that the length $u_1^2 + u_2^2 + u_3^2$ is a constant
(c)   find the eigenvalues of A.
The solution will rotate around the axis $w = (a, b, c)$, because Au is the "cross product" $u \times w$—which is perpendicular to u and w.

5.4.14   What are the eigenvalues λ and frequencies ω for

$$\frac{d^2u}{dt^2} = \begin{bmatrix} -5 & 4 \\ 4 & -5 \end{bmatrix} u?$$

Write down the general solution as in equation (17).

5.4.15   Solve the second-order equation

$$\frac{d^2u}{dt^2} = \begin{bmatrix} -5 & -1 \\ -1 & -5 \end{bmatrix} u \text{ with } u_0 = \begin{bmatrix} 1 \\ 0 \end{bmatrix} \text{ and } u_0' = \begin{bmatrix} 0 \\ 0 \end{bmatrix}.$$

5.4.16 In most applications the second-order equation looks like $Mu'' + Ku = 0$, with a *mass matrix* multiplying the second derivatives. Substitute the pure exponential $u = e^{i\omega t}x$ and find the "generalized eigenvalue problem" that must be solved for the frequency ω and the vector x.

5.4.17 With a friction matrix F in the equation $u'' + Fu' - Au = 0$, substitute a pure exponential $u = e^{\lambda t}x$ and find a quadratic eigenvalue problem for λ.

5.4.18 For equation (14) in the text, with frequencies 1 and $\sqrt{3}$, find the motion of the second mass if the first one is hit at $t = 0$; $u_0 = \begin{bmatrix} 0 \\ 0 \end{bmatrix}$ and $u'_0 = \begin{bmatrix} 1 \\ 0 \end{bmatrix}$.

5.4.19 A matrix with trace zero can be written as

$$A = \begin{bmatrix} a & b + c \\ b - c & -a \end{bmatrix}.$$

Show that its eigenvalues are real exactly when $a^2 + b^2 \geq c^2$.

5.4.20 By back-substitution or by computing eigenvectors, solve

$$\frac{du}{dt} = \begin{bmatrix} 1 & 2 & 1 \\ 0 & 3 & 6 \\ 0 & 0 & 4 \end{bmatrix} u \quad \text{with} \quad u_0 = \begin{bmatrix} 1 \\ 0 \\ 1 \end{bmatrix}.$$

5.5 ■ COMPLEX MATRICES: SYMMETRIC VS. HERMITIAN
AND ORTHOGONAL VS. UNITARY

It is no longer possible to work only with real vectors and real matrices. In the first half of this book, when the basic problem was $Ax = b$, it was certain that x would be real whenever A and b were. Therefore there was no need for complex numbers; they could have been permitted, but would have contributed nothing new. Now we cannot avoid them. A real matrix has real coefficients in its characteristic polynomial, but the eigenvalues (as in rotations) may be complex.

We shall therefore introduce the space $\mathbf{C}^n$ of vectors with n complex components. Addition and matrix multiplication follow the same rules as before. But *the length of a vector has to be changed*. If $\|x\|^2 = x_1^2 + \cdots + x_n^2$, the vector with components $(1, i)$ will have zero length: $1^2 + i^2 = 0$. Instead its length should be $\sqrt{2}$; the length squared is $1^2 + |i|^2$. This change in the computation of length forces a whole series of other changes. The inner product of two vectors, the transpose of a matrix, the definitions of symmetric, skew-symmetric, and orthogonal matrices, all need to be modified in the presence of complex numbers. In every case, the new definition coincides with the old when the vectors and matrices are real.

We have listed these changes at the end of the section, and that list virtually amounts to a dictionary for translating between the real and the complex case. We hope it will be useful to the reader. It also includes, for each class of matrices, the best information known about the location of their eigenvalues. We particularly want to find out about *symmetric matrices*: *Where are their eigenvalues, and what is special about their eigenvectors?* For practical purposes, those are the most important questions in the theory of eigenvalues, and we call attention in advance to the two main results:

1. *A symmetric matrix has real eigenvalues.*
2. *Its eigenvectors can be chosen orthonormal.*

Strangely, to prove that the eigenvalues are real we begin with the opposite possibility—and that takes us to complex numbers and complex vectors and complex matrices. The changes are easy and there is one extra benefit: We find some complex matrices that also have real eigenvalues and orthonormal eigenvectors. Those are the Hermitian matrices, and we see below how they include—and share all the essential properties of—real symmetric matrices.

Complex Numbers and Their Conjugates

Probably the reader is already acquainted with complex numbers; but since only the most basic facts are needed, a brief review is easy to give.† Everyone knows that whatever i is, it satisfies the equation $i^2 = -1$. It is a pure imaginary number, and so are its multiples ib; b is real. The sum of a real and an imaginary number

---

† The important ideas are the complex conjugate and the absolute value.

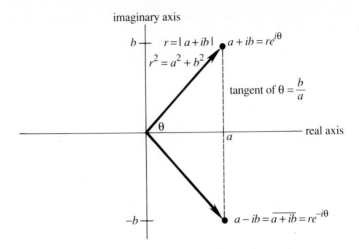

Fig. 5.4. The complex plane, with a complex number and its conjugate.

is a complex number $a + ib$; and it is plotted in a natural way on the complex plane (Fig. 5.4).

The real numbers (for which $b = 0$) and the imaginary numbers ($a = 0$) are included as special cases of complex numbers; they lie on one coordinate axis or the other. Two complex numbers are added by

$$(a + ib) + (c + id) = (a + c) + i(b + d),$$

and multiplied using the rule that $i^2 = -1$:

$$(a + ib)(c + id) = ac + ibc + iad + i^2 bd = (ac - bd) + i(bc + ad).$$

The **complex conjugate** of $a + ib$ is the number $a - ib$, *with the sign of the imaginary part reversed.* Geometrically, it is the mirror image on the other side of the real axis; any real number is its own conjugate. The conjugate is denoted by a bar, $\overline{a + ib} = a - ib$, and it has three important properties:

(1) The conjugate of a product equals the product of the conjugates:

$$\overline{(a + ib)(c + id)} = (ac - bd) - i(bc + ad) = \overline{(a + ib)}\,\overline{(c + id)}. \tag{1}$$

(2) The conjugate of a sum equals the sum of the conjugates:

$$\overline{(a + c) + i(b + d)} = (a + c) - i(b + d) = \overline{(a + ib)} + \overline{(c + id)}.$$

(3) Multiplying any $a + ib$ by its own conjugate $a - ib$ produces a real number, which is the square of the hypotenuse in Fig. 5.4:

$$(a + ib)(a - ib) = a^2 + b^2 = r^2. \tag{2}$$

This distance r is called the **absolute value** of the original $a + ib$ (or the *modulus*), and it is denoted by vertical bars: $|a + ib| = r = \sqrt{a^2 + b^2}$.

Finally, trigonometry connects the sides a and b to the hypotenuse r by

$$a = r \cos \theta, \qquad b = r \sin \theta.$$

By combining these equations, we move into polar coordinates:

$$a + ib = r(\cos \theta + i \sin \theta) = re^{i\theta}. \tag{3}$$

There is an important special case, when the modulus r equals one. Then the complex number is just $e^{i\theta} = \cos \theta + i \sin \theta$, and it falls on the **unit circle** in the complex plane. As θ varies from 0 to 2π, this number $e^{i\theta}$ circles around the origin at the constant radial distance $|e^{i\theta}| = \sqrt{\cos^2\theta + \sin^2\theta} = 1$.

EXAMPLE $x = 3 + 4i$ times its conjugate $\bar{x}$ is the absolute value squared:

$$x\bar{x} = (3 + 4i)(3 - 4i) = 25 = |x|^2 \quad \text{so} \quad r = |x| = 5.$$

To divide by $3 + 4i$, multiply numerator and denominator by its conjugate:

$$\frac{2 + i}{3 + 4i} = \frac{2 + i}{3 + 4i} \frac{3 - 4i}{3 - 4i} = \frac{10 - 5i}{25}.$$

In polar coordinates multiplication and division are easy:

$r_1 e^{i\theta_1}$ times $r_2 e^{i\theta_2}$ has absolute value $r_1 r_2$ and angle $\theta_1 + \theta_2$
$r_1 e^{i\theta_1}$ divided by $r_2 e^{i\theta_2}$ has absolute value r_1/r_2 and angle $\theta_1 - \theta_2$.

Lengths and Transposes in the Complex Case

We return to linear algebra, and make the conversion from real to complex. The first step is to admit complex vectors, and that is no problem: By definition, *the space $\mathbf{C}^n$ contains all vectors x with n complex components*:

$$x = \begin{bmatrix} x_1 \\ x_2 \\ \vdots \\ x_n \end{bmatrix}, \qquad x_j = a_j + ib_j.$$

Vectors x and y are still added component by component, but scalar multiplication is now done with complex numbers. As before, the vectors $v_1, \ldots, v_k$ are linearly dependent if some nontrivial combination $c_1 v_1 + \cdots + c_k v_k$ produces the zero vector; the c_j may now be complex. The unit coordinate vectors are still in $\mathbf{C}^n$; they are still independent; and they still form a basis. Therefore $\mathbf{C}^n$ is also a vector space of dimension n.

We have already emphasized that the definition of length has to be changed. The square of a complex number is not necessarily positive, and $\|x\|^2 = x_1^2 + \cdots + x_n^2$ is of no use. The new definition is completely natural: x_j^2 is replaced by its modulus

$|x_j|^2$, and the length satisfies

$$\|x\|^2 = |x_1|^2 + \cdots + |x_n|^2. \tag{4}$$

EXAMPLES $x = \begin{bmatrix} 1 \\ i \end{bmatrix}$ and $\|x\|^2 = 2;$ $y = \begin{bmatrix} 2 + i \\ 2 - 4i \end{bmatrix}$ and $\|y\|^2 = 25.$

For real vectors there was a close connection between the length and the inner product: $\|x\|^2 = x^{\mathrm{T}}x$. This connection we want to preserve. Therefore the inner product must be modified to match the new definition of length, and the standard modification is *to conjugate the first vector in the inner product*. This means that x is replaced by $\bar{x}$, and **the inner product of x and y becomes**

$$\bar{x}^{\mathrm{T}}y = \bar{x}_1 y_1 + \cdots + \bar{x}_n y_n. \tag{5}$$

A typical inner product in $\mathbf{C}^2$ has $x = (1 + i, 3i)$ and $y = (4, 2 - i)$:

$$\bar{x}^{\mathrm{T}}y = (1 - i)4 + (-3i)(2 - i) = 1 - 10i.$$

And if we take the inner product of x with itself, we are back to the length squared:

$$\bar{x}^{\mathrm{T}}x = \overline{(1 + i)}(1 + i) + \overline{(3i)}(3i) = 2 + 9 = \|x\|^2.$$

Note that $\bar{y}^{\mathrm{T}}x$ is different from $\bar{x}^{\mathrm{T}}y$; from now on we have to watch the order of the vectors in inner products. And there is a further novelty: If x is changed to cx, then the inner product of x and y is multiplied not by c, but by $\bar{c}$.

This leaves only one more change to make. It is a change in notation more than anything else, and it condenses two symbols into one: Instead of a bar for the conjugate and a T for the transpose, these operations are combined into the **conjugate transpose**, and denoted by a superscript H. Thus $\bar{x}^{\mathrm{T}} = x^{\mathrm{H}}$, and the same notation applies to matrices: The conjugate transpose of A is

$$\bar{A}^{\mathrm{T}} = A^{\mathrm{H}}, \quad \text{with entries} \quad (A^{\mathrm{H}})_{ij} = \overline{A_{ji}}. \tag{6}$$

If A is an m by n matrix, then A^{H} is n by m. For example,

$$\begin{bmatrix} 2 + i & 3i \\ 4 - i & 5 \\ 0 & 0 \end{bmatrix}^{\mathrm{H}} = \begin{bmatrix} 2 - i & 4 + i & 0 \\ -3i & 5 & 0 \end{bmatrix}.$$

This symbol A^{H} gives official recognition of the fact that, with complex entries, it is very seldom that we want only the transpose of A. It is the conjugate transpose, or *Hermitian transpose*, which becomes appropriate in virtually every case.† The modifications required by complex numbers are easily summarized:

5N (i) The inner product of x and y is $x^H y$, and they are orthogonal if $x^H y = 0$.
(ii) The length of x is $\|x\| = (x^H x)^{1/2}$.
(iii) $(AB)^T = B^T A^T$, after conjugating every entry, turns into $(AB)^H = B^H A^H$.

Hermitian Matrices

We spoke in earlier chapters about symmetric matrices: $A = A^T$. Now, in the presence of matrices with complex entries, this idea of symmetry has to be extended. The right generalization is not to matrices that equal their transpose, but to *matrices that equal their conjugate transpose*. These are the Hermitian matrices, and a typical example is

$$A = \begin{bmatrix} 2 & 3 - 3i \\ 3 + 3i & 5 \end{bmatrix} = A^H. \tag{7}$$

Notice that *the diagonal entries must be real*; they have to be unchanged by the process of conjugation. Each off-diagonal entry is matched with its mirror image across the main diagonal, and $3 - 3i$ is the conjugate of $3 + 3i$. *In every case $a_{ij} = \overline{a_{ji}}$*. This example will illustrate the three basic properties of Hermitian matrices.

Our main goal is to establish those three properties, and it needs to be emphasized again that they apply equally well to symmetric matrices. *A real symmetric matrix is certainly Hermitian.* For real matrices there is no difference between A^T and A^H; the key question is whether transposing the matrix leaves it unchanged. If so, its eigenvalues are real—as we now prove.

Property 1 If $A = A^H$, then for all complex vectors x, the number $x^H A x$ is real.

There is a contribution to $x^H A x$ from every entry of A:

$$x^H A x = \begin{bmatrix} \bar{u} & \bar{v} \end{bmatrix} \begin{bmatrix} 2 & 3 - 3i \\ 3 + 3i & 5 \end{bmatrix} \begin{bmatrix} u \\ v \end{bmatrix}$$
$$= 2\bar{u}u + 5\bar{v}v + (3 - 3i)\bar{u}v + (3 + 3i)u\bar{v}.$$

The "diagonal terms" are real, because $2\bar{u}u = 2|u|^2$ and $5\bar{v}v = 5|v|^2$. The off-diagonal terms are complex conjugates of one another, so they combine to give twice the real part of $(3 - 3i)\bar{u}v$. Therefore the whole expression $x^H A x$ is real.

---

† The matrix A^H is often referred to as "A Hermitian." You have to listen closely to distinguish that name from the phrase "A is Hermitian," which means that A equals A^H.

For a proof in general, we can compute $(x^H A x)^H$. We should get the conjugate of the 1 by 1 matrix $x^H A x$, but we actually get the same number back again: $(x^H A x)^H = x^H A^H x^{HH} = x^H A x$. So that number must be real.

Property 2   Every eigenvalue of a Hermitian matrix is real.

Proof   Suppose λ is an eigenvalue, and x is a corresponding eigenvector: $Ax = \lambda x$. Then *the trick is to multiply by* x^H: $x^H A x = \lambda x^H x$. The left side is real by Property 1, and the right side $x^H x = \|x\|^2$ is real and positive, because $x \neq 0$. Therefore λ must be real. In our example, $\lambda = 8$ or $\lambda = -1$:

$$|A - \lambda I| = \begin{vmatrix} 2 - \lambda & 3 - 3i \\ 3 + 3i & 5 - \lambda \end{vmatrix} = \lambda^2 - 7\lambda + 10 - |3 - 3i|^2$$

$$= \lambda^2 - 7\lambda - 8 = (\lambda - 8)(\lambda + 1). \tag{8}$$

Note   It seems that the proof could be made easier when A is real:

$$Ax = \lambda x \quad \text{gives} \quad x^T A x = \lambda x^T x, \quad \text{so} \quad \lambda = \frac{x^T A x}{x^T x} \quad \text{is real.}$$

But that looks OK for any real matrix, and there must be a catch: *The eigenvector x might be complex.* It is only when $A = A^T$ that we can be sure λ and x stay real. More than that, *the eigenvectors are perpendicular*: $x^T y = 0$ in the real symmetric case and $x^H y = 0$ in the complex Hermitian case.

Property 3   The eigenvectors of a real symmetric matrix or a Hermitian matrix, if they come from different eigenvalues, are orthogonal to one another.

The proof starts with the information given, $Ax = \lambda_1 x$ and $Ay = \lambda_2 y$ and $A = A^H$:

$$(\lambda_1 x)^H y = (Ax)^H y = x^H A y = x^H (\lambda_2 y). \tag{9}$$

The outside numbers are $\lambda_1 x^H y = \lambda_2 x^H y$, since the λ's are real. Now we use the assumption $\lambda_1 \neq \lambda_2$, *which forces the conclusion that $x^H y = 0$.* In our example,

$$(A - 8I)x = \begin{bmatrix} -6 & 3 - 3i \\ 3 + 3i & -3 \end{bmatrix} \begin{bmatrix} x_1 \\ x_2 \end{bmatrix} = \begin{bmatrix} 0 \\ 0 \end{bmatrix}, \qquad x = \begin{bmatrix} 1 \\ 1 + i \end{bmatrix}$$

$$(A + I)y = \begin{bmatrix} 3 & 3 - 3i \\ 3 + 3i & 6 \end{bmatrix} \begin{bmatrix} y_1 \\ y_2 \end{bmatrix} = \begin{bmatrix} 0 \\ 0 \end{bmatrix}, \qquad y = \begin{bmatrix} 1 - i \\ -1 \end{bmatrix}.$$

These two eigenvectors are orthogonal:

$$x^H y = \begin{bmatrix} 1 & 1 - i \end{bmatrix} \begin{bmatrix} 1 - i \\ -1 \end{bmatrix} = 0.$$

Of course any multiples x/α and y/β would be equally good as eigenvectors. Suppose we pick $\alpha = \|x\|$ and $\beta = \|y\|$, so that x/α and y/β are unit vectors; the eigenvectors have been normalized to have length one. Since they were already orthogonal, they are now *orthonormal*. If they are chosen to be the columns of S, then (as always, when the eigenvectors are the columns) we have $S^{-1}AS = \Lambda$. **The diagonalizing matrix has orthonormal columns.**

In case A is real and symmetric, its eigenvalues are real by Property 2. Its eigenvectors are orthogonal by Property 3. Those eigenvectors are also real (they solve $(A - \lambda I)x = 0$ as in the nullspace of Chapter 2) and their lengths can be normalized to 1. Therefore they go into an orthogonal matrix:

If $A = A^T$, the diagonalizing matrix S can be an orthogonal matrix Q.

Orthonormal columns are equivalent to $Q^TQ = I$ or $Q^T = Q^{-1}$. The usual diagonalization $S^{-1}AS = \Lambda$ becomes special—it is $Q^{-1}AQ = \Lambda$ or $A = Q\Lambda Q^{-1} = Q\Lambda Q^T$. We have reached one of the great theorems of linear algebra:

50 A real symmetric matrix can be factored into $A = Q\Lambda Q^T$—with the orthonormal eigenvectors in Q and the eigenvalues in Λ.

In geometry or mechanics, this is the *principal axis theorem*. It gives the right choice of axes for an ellipse. Those axes are perpendicular, and they point along the eigenvectors of the corresponding matrix. (Section 6.2 connects symmetric matrices to n-dimensional ellipses.) In mechanics the eigenvectors give the principal directions, along which there is pure compression or pure tension—in other directions there is "shear."

In mathematics the formula $A = Q\Lambda Q^T$ is known as the *spectral theorem*. If we multiply columns by rows, the matrix A becomes a combination of one-dimensional projections—which are the special matrices xx^T of rank one:

$$A = Q\Lambda Q^T = \begin{bmatrix} \big| & & \big| \\ x_1 & & x_n \\ \big| & & \big| \end{bmatrix} \begin{bmatrix} \lambda_1 & & \\ & \ddots & \\ & & \lambda_n \end{bmatrix} \begin{bmatrix} -\!\!\!-\!\!\!- x_1^T -\!\!\!-\!\!\!- \\ \\ \\ -\!\!\!-\!\!\!- x_n^T -\!\!\!-\!\!\!- \end{bmatrix}$$

$$= \lambda_1 x_1 x_1^T + \lambda_2 x_2 x_2^T + \cdots + \lambda_n x_n x_n^T. \tag{10}$$

Our 2 by 2 example has eigenvalues 3 and 1:

EXAMPLE $\quad A = \begin{bmatrix} 2 & -1 \\ -1 & 2 \end{bmatrix} = 3 \begin{bmatrix} \frac{1}{2} & -\frac{1}{2} \\ -\frac{1}{2} & \frac{1}{2} \end{bmatrix} + \begin{bmatrix} \frac{1}{2} & \frac{1}{2} \\ \frac{1}{2} & \frac{1}{2} \end{bmatrix}.$

The eigenvectors, with length scaled to one, are

$$x_1 = \frac{1}{\sqrt{2}} \begin{bmatrix} 1 \\ -1 \end{bmatrix} \quad \text{and} \quad x_2 = \frac{1}{\sqrt{2}} \begin{bmatrix} 1 \\ 1 \end{bmatrix}.$$

Then the matrices on the right side are $x_1 x_1^T$ and $x_2 x_2^T$—columns times rows—and they are projections onto the line through x_1 and the line through x_2.

We emphasize that this is the same diagonalization $A = S \Lambda S^{-1}$ as always. It has been specialized to $S = Q$ and $S^{-1} = Q^T$, and then split apart to give a piece from each eigenvector. The result is to build all symmetric matrices out of one-dimensional projections—which are symmetric matrices of rank one.

Remark   If a matrix is real and its eigenvalues happen to be real, then its eigenvectors are also real. They solve $(A - \lambda I)x = 0$ and can be computed by elimination. But they will not be orthogonal unless A is symmetric: $A = Q \Lambda Q^T$ leads to $A^T = A$.

If the matrix is real but some eigenvalues are complex, those eigenvalues come in conjugate pairs. *If $a + ib$ is an eigenvalue of a real matrix, so is $a - ib$.* The determinant of $A - \lambda I$ is a polynomial with real coefficients, and for such polynomials the complex roots are paired. In the 2 by 2 case the quadratic formula contains $\pm(b^2 - 4ac)^{1/2}$.

Note   Strictly speaking, the spectral theorem $A = Q \Lambda Q^T$ has been proved only when the eigenvalues of A are distinct. Then there are certainly n independent eigenvectors, and A can be safely diagonalized. Nevertheless it is true (see Section 5.6) that *even with repeated eigenvalues, a symmetric matrix still has a complete set of orthonormal eigenvectors.* The extreme case is the identity matrix, which has $\lambda = 1$ repeated n times—and no shortage of eigenvectors.

To finish the complex case we need the analogue of a real orthogonal matrix—and you can guess what happens to the requirement $Q^T Q = I$. The transpose will be replaced by the *conjugate transpose.* The condition will become $U^H U = I$. The new letter U reflects the new name: *A complex matrix with orthonormal columns is called a unitary matrix.*

Unitary Matrices

May we propose an analogy? *A Hermitian matrix can be compared to a real number, and a unitary matrix can be compared to a number on the unit circle*—a complex number of absolute value 1. For the eigenvalues this comparison is more than an analogy: The λ's are real if $A^H = A$, and they are on the unit circle if $U^H U = I$. The eigenvectors are orthogonal, and they can be scaled to unit length and made orthonormal.†

---

† Later we compare "skew-Hermitian" matrices with pure imaginary numbers, and "normal" matrices with complex numbers $a + ib$. A matrix without orthogonal eigenvectors belongs to none of these classes, and is outside the whole analogy.

Those statements have been confirmed for Hermitian (including symmetric) matrices. They are not yet proved for unitary (including orthogonal) matrices. Therefore we go directly to the three properties of U that correspond to the earlier Properties 1–3 of A. Remember that U has orthonormal columns:

$$U^H U = I, \quad \text{or} \quad U U^H = I, \quad \text{or} \quad U^H = U^{-1}.$$

This leads directly to Property 1', that multiplication by U has no effect on inner products or angles or lengths. The proof is on one line, just as it was for Q:

Property 1' $(Ux)^H(Uy) = x^H U^H U y = x^H y$ and (by choosing $y = x$) lengths are preserved:

$$\|Ux\|^2 = \|x\|^2. \tag{11}$$

The next property locates the eigenvalues of U; each λ is on the unit circle.

Property 2' Every eigenvalue of U has absolute value $|\lambda| = 1$.

This follows directly from $Ux = \lambda x$, by comparing the lengths of the two sides: $\|Ux\| = \|x\|$ by Property 1', and always $\|\lambda x\| = |\lambda|\,\|x\|$. Therefore $|\lambda| = 1$.

Property 3' Eigenvectors corresponding to different eigenvalues are orthogonal.

The proof assumes $Ux = \lambda_1 x$, $Uy = \lambda_2 y$, and takes inner products by Property 1':

$$x^H y = (Ux)^H(Uy) = (\lambda_1 x)^H(\lambda_2 y) = \bar{\lambda}_1 \lambda_2 x^H y.$$

Comparing the left to the right, either $\bar{\lambda}_1 \lambda_2 = 1$ or $x^H y = 0$. But Property 2' is $\bar{\lambda}_1 \lambda_1 = 1$, so we cannot also have $\bar{\lambda}_1 \lambda_2 = 1$. Thus $x^H y = 0$ and the eigenvectors are orthogonal.

EXAMPLE 1 $U = \begin{bmatrix} \cos t & -\sin t \\ \sin t & \cos t \end{bmatrix}.$

The eigenvalues of these rotations are e^{it} and e^{-it}, of absolute value one. The eigenvectors are $x = (1, -i)$ and $y = (1, i)$ which are orthogonal. (Remember to take conjugates in $x^H y = 1 + i^2 = 0$!) After division by $\sqrt{2}$ they are orthonormal, and the diagonalization of U is

$$\begin{bmatrix} \cos t & -\sin t \\ \sin t & \cos t \end{bmatrix} = \frac{1}{\sqrt{2}}\begin{bmatrix} 1 & 1 \\ -i & i \end{bmatrix}\begin{bmatrix} e^{it} & \\ & e^{-it} \end{bmatrix}\frac{1}{\sqrt{2}}\begin{bmatrix} 1 & i \\ 1 & -i \end{bmatrix}.$$

Notice that the signs on i and $-i$ were reversed when transposing. In fact the right side is a product of three unitary matrices, and it produces the unitary matrix on the left. The next example is, among all complex matrices, the most important one that is unitary.

EXAMPLE 2   The Fourier matrix $U = \dfrac{F}{\sqrt{n}} = \dfrac{1}{\sqrt{n}} \begin{bmatrix} 1 & 1 & \cdot & 1 \\ 1 & w & \cdot & w^{n-1} \\ \cdot & \cdot & \cdot & \cdot \\ 1 & w^{n-1} & \cdot & w^{(n-1)^2} \end{bmatrix}$.

The factor $\sqrt{n}$ shrinks the columns into unit vectors. (Every entry has absolute value 1, so each column of the original F has length $\sqrt{n}$.) The fact that $U^H U = I$ is the fundamental identity of the finite Fourier transform, and we recall the main point from Section 3.5:

row 1 of U^H times column 2 of U is $\dfrac{1}{n}(1 + w + w^2 + \cdots + w^{n-1}) = 0.$

row i of U^H times column j of U is $\dfrac{1}{n}(1 + W + W^2 + \cdots + W^{n-1}) = 0.$

In the first case, the complex number w is an nth root of 1. It is on the unit circle, at the angle $\theta = 2\pi/n$. It equals $e^{2\pi i/n}$, and its powers are spaced evenly around the circle. That spacing assures that the sum of all n powers of w—all the nth roots of 1—is zero. Algebraically the sum is $(w^n - 1)/(w - 1)$ and $w^n - 1$ is zero.

In the second case W is a power of w. It is w^{j-i} and it is again a root of unity. It is not the root at 1, because we are looking *off* the diagonal of $U^H U$ and therefore $j \neq i$. The powers of W again add to $(W^n - 1)/(W - 1) = 0.$

Thus U is a unitary matrix. Earlier we wrote down its inverse—which has exactly the same form except that w is replaced by $w^{-1} = e^{-i\theta} = \bar{w}$. Now we recognize what happened. Since U is unitary, its inverse is found by transposing (which changes nothing) and conjugating (which changes w to $\bar{w}$). The inverse of this U is $\bar{U}$.

By property 1' of unitary matrices, the length of a vector x is the same as the length of Ux. The energy in state space equals the energy in transform space. The energy is the sum of $|x_j|^2$, and it is also the sum of the energies in the separate "harmonics." A vector like $x = (1, 0, \ldots, 0)$ contains equal amounts of every frequency component, and $Ux = (1, 1, \ldots, 1)/\sqrt{n}$ also has length one.

We recall that Ux can be computed quickly by the Fast Fourier Transform.

EXAMPLE 3

$$P = \begin{bmatrix} 0 & 1 & 0 & 0 \\ 0 & 0 & 1 & 0 \\ 0 & 0 & 0 & 1 \\ 1 & 0 & 0 & 0 \end{bmatrix}$$

This is an orthogonal matrix, so by Property 3′ it must have orthogonal eigenvectors. They are the columns of the Fourier matrix! Its eigenvalues must have absolute value 1. They are the numbers $1, w, \ldots, w^{n-1}$ (or $1, i, i^2, i^3$ in this 4 by 4 case). It is a real matrix, but its eigenvalues and eigenvectors are complex.

One final note. Skew-Hermitian matrices satisfy $K^H = -K$ just as skew-symmetric matrices satisfy $K^T = -K$. Their properties follow immediately from their close link to Hermitian matrices:

If A is Hermitian then $K = iA$ is skew-Hermitian.

The eigenvalues of K are purely imaginary instead of purely real; we multiply by i. The eigenvectors are not changed. The Hermitian example on the previous pages would lead to

$$K = iA = \begin{bmatrix} 2i & 3 + 3i \\ -3 + 3i & 5i \end{bmatrix} = -K^H.$$

The diagonal entries are multiples of i (allowing zero). The eigenvalues are $8i$ and $-i$. The eigenvectors are still orthogonal, and we still have $K = U\Lambda U^H$—with a unitary U instead of a real orthogonal Q, and with $8i$ and $-i$ on the diagonal of Λ.

This section is summarized by a table of parallels between real and complex.

Real versus Complex

$\mathbf{R}^n$ = space of vectors with $\leftrightarrow$ $\mathbf{C}^n$ = space of vectors with
n real components $\qquad$ n complex components

length: $\|x\|^2 = x_1^2 + \cdots + x_n^2 \leftrightarrow$ length: $\|x\|^2 = |x_1|^2 + \cdots + |x_n|^2$

transpose: $A_{ij}^T = A_{ji} \leftrightarrow$ Hermitian transpose: $A_{ij}^H = \overline{A_{ji}}$

$(AB)^T = B^T A^T \leftrightarrow (AB)^H = B^H A^H$

inner product: $x^T y = x_1 y_1 + \cdots + x_n y_n \leftrightarrow$ inner product: $x^H y = \bar{x}_1 y_1 + \cdots + \bar{x}_n y_n$

$(Ax)^T y = x^T(A^T y) \leftrightarrow (Ax)^H y = x^H(A^H y)$

orthogonality: $x^T y = 0 \leftrightarrow$ orthogonality: $x^H y = 0$

symmetric matrices: $A^T = A \leftrightarrow$ Hermitian matrices: $A^H = A$

$A = Q\Lambda Q^{-1} = Q\Lambda Q^T$ (real Λ) $\leftrightarrow A = U\Lambda U^{-1} = U\Lambda U^H$ (real Λ)

skew-symmetric matrices: $K^T = -K \leftrightarrow$ skew-Hermitian matrices: $K^H = -K$

orthogonal matrices: $Q^TQ = I$ or $Q^T = Q^{-1} \leftrightarrow$ unitary matrices: $U^HU = I$ or $U^H = U^{-1}$

$$(Qx)^T(Qy) = x^Ty \text{ and } \|Qx\| = \|x\| \leftrightarrow (Ux)^H(Uy) = x^Hy \text{ and } \|Ux\| = \|x\|$$

The columns, rows, and eigenvectors of Q and U are orthonormal, and every $|\lambda| = 1$

EXERCISES

5.5.1   For the complex numbers $3 + 4i$ and $1 - i$
(a)   find their positions in the complex plane;
(b)   find their sum and product;
(c)   find their conjugates and their absolute values.
Do they lie inside or outside the unit circle?

5.5.2   What can you say about

(i)   the sum of a complex number and its conjugate?
(ii)   the conjugate of a number on the unit circle?
(iii) the product of two numbers on the unit circle?
(iv)   the sum of two numbers on the unit circle?

5.5.3   If $x = 2 + i$ and $y = 1 + 3i$ find $\bar{x}$, $x\bar{x}$, xy, $1/x$, and x/y. Check that the absolute value $|xy|$ equals $|x|$ times $|y|$, and the absolute value $|1/x|$ equals 1 divided by $|x|$.

5.5.4   Find a and b for the complex numbers $a + ib$ at the angles $\theta = 30°, 60°, 90°$ on the unit circle. Verify by direct multiplication that the square of the first is the second, and the cube of the first is the third.

5.5.5   (a)   If $x = re^{i\theta}$ what are x^2 and x^{-1} and $\bar{x}$ in polar coordinates? Where are the complex numbers that have $x^{-1} = \bar{x}$?
(b)   At $t = 0$ the complex number $e^{(-1+i)t}$ equals one. Sketch its path in the complex plane as t increases from 0 to 2π.

5.5.6   Find the lengths and the inner product of

$$x = \begin{bmatrix} 2 - 4i \\ 4i \end{bmatrix} \quad \text{and} \quad y = \begin{bmatrix} 2 + 4i \\ 4 \end{bmatrix}.$$

5.5.7   Write out the matrix A^H and compute $C = A^HA$ if

$$A = \begin{bmatrix} 1 & i & 0 \\ i & 0 & 1 \end{bmatrix}.$$

What is the relation between C and C^H? Does it hold whenever C is constructed from some A^HA?

5.5.8   (i)   With the preceding A, use elimination to solve $Ax = 0$.
(ii)   Show that the nullspace you just computed is orthogonal to $\mathcal{R}(A^H)$ *and not to*

the usual row space $\mathscr{R}(A^T)$. The four fundamental spaces in the complex case are $\mathscr{N}(A)$ and $\mathscr{R}(A)$, as before, and then $\mathscr{N}(A^H)$ and $\mathscr{R}(A^H)$.

5.5.9 (a) How is the determinant of A^H related to the determinant of A?

(b) Prove that the determinant of any Hermitian matrix is real.

5.5.10 (a) How many degrees of freedom are there in a real symmetric matrix, a real diagonal matrix, and a real orthogonal matrix? (The first answer is the sum of the other two, because $A = Q\Lambda Q^T$.)

(b) Show that 3 by 3 Hermitian matrices have 9 real degrees of freedom and unitary matrices have 6. (Now columns of U can be multiplied by any $e^{i\theta}$.)

5.5.11 Write the following matrices in the form $\lambda_1 x_1 x_1^H + \lambda_2 x_2 x_2^H$ of the spectral theorem:

$$P = \begin{bmatrix} \frac{1}{2} & \frac{1}{2} \\ \frac{1}{2} & \frac{1}{2} \end{bmatrix}, \quad Q = \begin{bmatrix} 0 & 1 \\ 1 & 0 \end{bmatrix}, \quad R = \begin{bmatrix} 3 & 4 \\ 4 & -3 \end{bmatrix}.$$

5.5.12 Give a reason if true or a counterexample if false:

(1) If A is Hermitian then $A + iI$ is invertible.

(2) If Q is orthogonal then $Q + \frac{1}{2}I$ is invertible.

(3) If A is real then $A + iI$ is invertible.

5.5.13 Suppose A is a symmetric 3 by 3 matrix with eigenvalues 0, 1, 2.

(a) What properties can be guaranteed for the corresponding unit eigenvectors u, v, w?

(b) In terms of u, v, w describe the nullspace, left nullspace, row space, and column space of A.

(c) Find a vector x that satisfies $Ax = v + w$. Is x unique?

(d) Under what conditions on b does $Ax = b$ have a solution?

(e) If u, v, w are the columns of S, what are S^{-1} and $S^{-1}AS$?

5.5.14 In the list below, which classes of matrices contain A and which contain B?

$$A = \begin{bmatrix} 0 & 1 & 0 & 0 \\ 0 & 0 & 1 & 0 \\ 0 & 0 & 0 & 1 \\ 1 & 0 & 0 & 0 \end{bmatrix} \quad \text{and} \quad B = \frac{1}{4}\begin{bmatrix} 1 & 1 & 1 & 1 \\ 1 & 1 & 1 & 1 \\ 1 & 1 & 1 & 1 \\ 1 & 1 & 1 & 1 \end{bmatrix}$$

Orthogonal, invertible, projection, permutation, Hermitian, rank one, diagonalizable, Markov. Find the eigenvalues of A and B.

5.5.15 What is the dimension of the space S of all n by n real symmetric matrices? The spectral theorem says that every symmetric matrix is a combination of n projection matrices. Since the dimension exceeds n, how is this difference explained?

5.5.16 Write down one significant fact about the eigenvalues of

1. A real symmetric matrix

2. A stable matrix: all solutions to $du/dt = Au$ approach zero

3. An orthogonal matrix

4. A Markov matrix
5. A defective matrix (nondiagonalizable)
6. A singular matrix

5.5.17 Show that if U and V are unitary, so is UV. Use the criterion $U^H U = I$.

5.5.18 Show that the determinant of a unitary matrix satisfies $|\det U| = 1$, but the determinant is not necessarily equal to 1. Describe all 2 by 2 matrices that are unitary.

5.5.19 Find a third column so that

$$U = \begin{bmatrix} 1/\sqrt{3} & i/\sqrt{2} \\ 1/\sqrt{3} & 0 \\ i/\sqrt{3} & 1/\sqrt{2} \end{bmatrix}$$

is unitary. How much freedom is there in this choice?

5.5.20 Diagonalize the 2 by 2 skew-Hermitian matrix K, whose entries are all i. Compute $e^{Kt} = Se^{\Lambda t}S^{-1}$, and verify that e^{Kt} is unitary. What is its derivative at $t = 0$?

5.5.21 Describe all 3 by 3 matrices that are simultaneously Hermitian, unitary, and diagonal. How many are there?

5.5.22 Every matrix Z can be split into a Hermitian and a skew-Hermitian part, $Z = A + K$, just as a complex number z is split into $a + ib$. The real part of z is half of $z + \bar{z}$, and the "real part" of Z is half of $Z + Z^H$. Find a similar formula for the "imaginary part" K, and split these matrices into $A + K$:

$$Z = \begin{bmatrix} 3 + i & 4 + 2i \\ 0 & 5 \end{bmatrix} \quad \text{and} \quad Z = \begin{bmatrix} i & i \\ -i & i \end{bmatrix}.$$

5.5.23 Show that the columns of the 4 by 4 Fourier matrix F are eigenvectors of the permutation matrix P in Example 3.

5.5.24 For the same permutation, write out the *circulant matrix* $C = c_0 I + c_1 P + c_2 P^2 + c_3 P^3$. (Its eigenvector matrix is again the Fourier matrix.) Write out also the four components of the matrix-vector product Cx, which is the *convolution* of $c = (c_0, c_1, c_2, c_3)$ and $x = (x_0, x_1, x_2, x_3)$.

5.5.25 For a circulant $C = F\Lambda F^{-1}$, why is it faster to multiply by F^{-1} then Λ then F (the convolution rule) than to multiply directly by C?

5.6 ■ SIMILARITY TRANSFORMATIONS

Virtually every step in this chapter has involved the combination $S^{-1}AS$. The eigenvectors of A went into the columns of S, and the combination $S^{-1}AS$ was diagonal (called Λ). When A was symmetric we wrote Q instead of S, as a reminder that the eigenvectors could be chosen orthonormal. In the complex case, when A was Hermitian, we wrote U—but it was still the matrix of eigenvectors. Now, in this last section, we look at other combinations $M^{-1}AM$—formed in the same way, but with *any invertible M on the right and its inverse on the left*. The eigenvector matrix may fail to exist (the defective case), or we may not know it, or we may not want to use it.

First a new word, to describe the relation between A and $M^{-1}AM$. Those matrices are said to be **similar**. Going from one to the other is a **similarity transformation**. It is the natural step to take, when dealing with differential equations or powers of a matrix or eigenvalues—just as elimination steps were natural when dealing with $Ax = b$. (Elimination multiplied A on the left by L^{-1}, but not on the right by L.) Normally there will be a whole family of matrices $M^{-1}AM$, all similar to A, and there are two key questions:

(1) What do these matrices $M^{-1}AM$ have in common?
(2) With a special choice of M, what special form can be achieved by $M^{-1}AM$?

The final answer is given by the **Jordan form**, with which the chapter ends.

It is worth remembering how these combinations $M^{-1}AM$ arise. Given a differential or difference equation for the unknown u, suppose a "change of variables" $u = Mv$ introduces the new unknown v. Then

$$\frac{du}{dt} = Au \qquad \text{becomes} \qquad M\frac{dv}{dt} = AMv, \qquad \text{or} \qquad \frac{dv}{dt} = M^{-1}AMv$$

$$u_{n+1} = Au_n \qquad \text{becomes} \qquad Mv_{n+1} = AMv_n, \qquad \text{or} \qquad v_{n+1} = M^{-1}AMv_n.$$

The new matrix in the equation is $M^{-1}AM$. In the special case $M = S$ the system is uncoupled because $\Lambda = S^{-1}AS$ is diagonal. The normal modes evolve independently. In the language of linear transformations, which is presented below, the eigenvectors are being chosen as a basis. This is the maximum simplification, but other and less drastic simplifications are also useful. We try to make $M^{-1}AM$ easier to work with than A.

The first question was about the family of matrices $M^{-1}AM$—which includes A itself, by choosing M to be the identity matrix. Any of these matrices can be made to appear in the differential and difference equations, by the change $u = Mv$, so they ought to have something in common and they do: **Similar matrices share the same eigenvalues.**

5P If $B = M^{-1}AM$, then A and B have the same eigenvalues. An eigenvector x of A corresponds to an eigenvector $M^{-1}x$ of B.

The proof is immediate, since $A = MBM^{-1}$:

$$Ax = \lambda x \Rightarrow MBM^{-1}x = \lambda x \Rightarrow B(M^{-1}x) = \lambda(M^{-1}x).$$

The eigenvalue of B is still λ. The eigenvector has been multiplied by M^{-1}.

We can also check that the determinants of $A - \lambda I$ and $B - \lambda I$ are identical, from the product rule for determinants:

$$\det(B - \lambda I) = \det(M^{-1}AM - \lambda I) = \det(M^{-1}(A - \lambda I)M)$$
$$= \det M^{-1} \det(A - \lambda I) \det M = \det(A - \lambda I).$$

The two determinants—the characteristic polynomials of A and B—are equal. Therefore their roots—the eigenvalues of A and B—are the same. The following example finds some matrices that are similar to A.

EXAMPLE $A = \begin{bmatrix} 1 & 0 \\ 0 & 0 \end{bmatrix}$, diagonal with eigenvalues 1 and 0

If $M = \begin{bmatrix} 1 & b \\ 0 & 1 \end{bmatrix}$ then $B = M^{-1}AM = \begin{bmatrix} 1 & b \\ 0 & 0 \end{bmatrix}$: triangular with eigenvalues 1 and 0

If $M = \begin{bmatrix} 1 & 1 \\ -1 & 1 \end{bmatrix}$ then $B = M^{-1}AM = \begin{bmatrix} \frac{1}{2} & \frac{1}{2} \\ \frac{1}{2} & \frac{1}{2} \end{bmatrix}$: projection with eigenvalues 1 and 0

If $M = \begin{bmatrix} a & b \\ c & d \end{bmatrix}$ then $B = M^{-1}AM =$ an arbitrary matrix with eigenvalues 1 and 0.

In this case we can produce any matrix that has the correct eigenvalues. It is an easy case, because the eigenvalues 1 and 0 are distinct. The Jordan form will worry about repeated eigenvalues and a possible shortage of eigenvectors—all we say now is that $M^{-1}AM$ has the same number of independent eigenvectors as A (because the eigenvectors are multiplied by M^{-1}).

The first step is separate and more theoretical—to look at the linear transformations that lie behind the matrices. This takes us back to Section 2.6, in which we thought of a rotation or a reflection or a projection in geometrical terms—as something that happens to n-dimensional space. The transformation can happen without linear algebra, but linear algebra turns it into matrix multiplication.

Change of Basis = Similarity Transformation

The relationship of similar matrices A and $B = M^{-1}AM$ is extremely close, if we go back to linear transformations. Remember the key idea: *Every linear transformation is represented by a matrix*. There was one further point in Section 2.6: *The matrix depends on the choice of basis*. If we change the basis, we change the matrix. Now we are ready to see what the change of basis does to the matrix.

Similar matrices represent the same transformation with respect to different bases.

The algebra is almost straightforward. Suppose we have a transformation T (like rotation) and a basis $v_1, \ldots, v_n$. The jth column of A comes from applying T to v_j:

$$Tv_j = \text{combination of the } v\text{'s} = a_{1j}v_1 + \cdots + a_{nj}v_n.$$

If we have a new basis $V_1, \ldots, V_n$, then the new matrix (call it B) is constructed in the same way: $TV_j = \text{combination of the } V\text{'s} = \sum_{i=1}^{n} b_{ij}V_i$. But also each V must be a combination of the old basis vectors: $V_j = \sum m_{ij}v_i$. That matrix M is really representing *the identity transformation* (!) when the only thing happening is the change of basis. Following the rules of Section 2.6, we just applied the identity transformation (which leaves V_j) and wrote the result as a combination of the v's. The inverse matrix M^{-1} also represents the identity transformation, when the basis is changed from the v's back to the V's. Now the product rule gives the result we want:

5Q The matrices A and B that represent the same linear transformation T with respect to two different bases v and V are **similar**:

$$\begin{array}{cccc} [T]_{V \text{ to } V} = & [I]_{v \text{ to } V} & [T]_{v \text{ to } v} & [I]_{V \text{ to } v} \\ B & = M^{-1} & A & M \end{array}$$

That proof (coming from the product rule) is somewhat mysterious, and an example is the best way to explain it. Suppose T is *projection onto the line L* at angle θ. That is the linear transformation, and it is completely described without the help of a basis. But to represent it by a matrix we do need a basis, and Fig. 5.5 offers two choices. One is the standard basis $v_1 = (1, 0)$, $v_2 = (0, 1)$, and the other is a basis chosen especially for T. In fact $TV_1 = V_1$ (since V_1 is already on the line L) and $TV_2 = 0$ (since V_2 is perpendicular to the line). In that basis, the matrix is diagonal—because V_1 and V_2 are eigenvectors:

$$B = [T]_{V \text{ to } V} = \begin{bmatrix} 1 & 0 \\ 0 & 0 \end{bmatrix}.$$

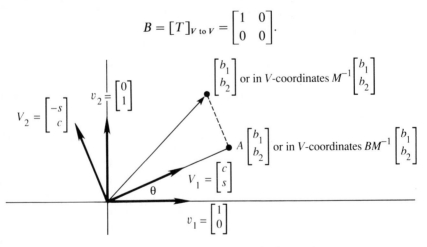

Fig. 5.5. Change of basis to make the projection matrix diagonal.

The other thing is the change of basis matrix M. For that we express V_1 as a combination $v_1 \cos \theta + v_2 \sin \theta$ and put those coefficients into column 1. Similarly V_2 (or IV_2, the transformation is the identity) is $-v_1 \sin \theta + v_2 \cos \theta$, producing column 2:

$$M = [I]_{V \text{ to } v} = \begin{bmatrix} c & -s \\ s & c \end{bmatrix}.$$

The inverse matrix M^{-1} (which is here the transpose) goes from v to V. Combined with B and M it gives the projection matrix that was originally written down in Section 2.6:

$$A = MBM^{-1} = \begin{bmatrix} c^2 & cs \\ cs & s^2 \end{bmatrix}.$$

We can summarize the main point. Suppose we are given a matrix like A. The way to simplify it—in fact to diagonalize it—is to find its eigenvectors. They go into the columns of M (or S) and $M^{-1}AM$ is diagonal. The algebraist says the same thing in the language of linear transformations: The way to have a diagonal matrix representing T is *to choose a basis consisting of eigenvectors*. The standard basis v led to A, which was not simple; the right basis V led to B, which was diagonal.

We emphasize again that combinations like $M^{-1}AM$ do not arise in solving $Ax = b$. There the basic operation was to multiply A (on the left side only!) by a matrix that subtracts a multiple of one row from another. Such a transformation preserved the nullspace and row space of A; it had nothing to do with the eigenvalues. In contrast, similarity transformations leave the eigenvalues unchanged, and in fact those eigenvalues are actually calculated by a sequence of simple similarities. The matrix goes gradually toward a triangular form, and the eigenvalues gradually appear on the main diagonal. (Such a sequence is described in Chapter 7, and one step is illustrated in the seventh exercise below.) This is much better than trying to compute the polynomial $\det(A - \lambda I)$, whose roots should be the eigenvalues. For a large matrix, it is numerically impossible to concentrate all that information into the polynomial and then get it out again.

Triangular Forms with a Unitary M

Our first move beyond the usual case $M = S$ is a little bit crazy: Instead of allowing a more general M, we go the other way and *restrict it to be unitary*. The problem is to find some reasonably simple form that $M^{-1}AM$ can achieve under this restriction. The columns of $M = U$ are required to be orthonormal (in the real case we would write $M = Q$). Unless the eigenvectors are orthogonal, a diagonal Λ is impossible; but the following "Schur's lemma" produces a form which is very useful—at least to the theory.†

5R For any square matrix A, there is a unitary matrix $M = U$ such that $U^{-1}AU = T$ is upper triangular. The eigenvalues of A must be shared by the similar matrix T, and appear along its main diagonal.

Proof Any matrix, say any 4 by 4 matrix, has at least one eigenvalue λ_1; in the worst case, it could be repeated four times. Therefore A has at least one eigenvector x. We normalize x to be a unit vector x_1, and place it in the first column of U. At this stage the other three columns are impossible to determine, so we complete the matrix in any way which leaves it unitary, and call it U_1. (The Gram-Schmidt process guarantees that this can be done.) The product $U_1^{-1}AU_1$ has at least its first column in the right form: $Ax_1 = \lambda_1 x_1$ means that

$$AU_1 = U_1 \begin{bmatrix} \lambda_1 & * & * & * \\ 0 & * & * & * \\ 0 & * & * & * \\ 0 & * & * & * \end{bmatrix}, \quad \text{or} \quad U_1^{-1}AU_1 = \begin{bmatrix} \lambda_1 & * & * & * \\ 0 & * & * & * \\ 0 & * & * & * \\ 0 & * & * & * \end{bmatrix}.$$

At the second step, we work with the 3 by 3 matrix now in the lower right corner. This matrix has an eigenvalue λ_2 and a unit eigenvector x_2, which can be made into the first column of a 3 by 3 unitary matrix M_2. Then

$$U_2 = \begin{bmatrix} 1 & 0 & 0 & 0 \\ 0 & & & \\ 0 & & M_2 & \\ 0 & & & \end{bmatrix} \quad \text{and} \quad U_2^{-1}(U_1^{-1}AU_1)U_2 = \begin{bmatrix} \lambda_1 & * & * & * \\ 0 & \lambda_2 & * & * \\ 0 & 0 & * & * \\ 0 & 0 & * & * \end{bmatrix}.$$

Finally, at the last step, an eigenvector of the 2 by 2 matrix in the lower right corner goes into a unitary M_3, which is put into the corner of U_3, and

$$U_3^{-1}(U_2^{-1}U_1^{-1}AU_1U_2)U_3 = \begin{bmatrix} \lambda_1 & * & * & * \\ 0 & \lambda_2 & * & * \\ 0 & 0 & \lambda_3 & * \\ 0 & 0 & 0 & * \end{bmatrix} = T.$$

The product $U = U_1U_2U_3$ is still a unitary matrix—this was checked in Exercise 5.5.17—so we have the required triangular $U^{-1}AU = T$.

---

† The rest of this chapter is devoted more to theory than to applications. The Jordan form in 5U is independent of the triangular form in 5R.

Because this lemma applies to all matrices, it often allows us to escape the hypothesis that A is diagonalizable. We could use it to prove that ***the powers A^k approach zero when all $|\lambda_i| < 1$, and the exponentials e^{At} approach zero when all*** $\mathrm{Re}\,\lambda_i < 0$—even without the full set of eigenvectors which the theory of stability assumed in Sections 5.3 and 5.4.

EXAMPLE $A = \begin{bmatrix} 2 & -1 \\ 1 & 0 \end{bmatrix}$ has the eigenvalue $\lambda = 1$ (twice).

One line of eigenvectors (in fact the only one) goes through $(1, 1)$. After dividing by $\sqrt{2}$ this is the first column of U, and the other column is orthogonal to it:

$$U^{-1}AU = \begin{bmatrix} 1/\sqrt{2} & 1/\sqrt{2} \\ 1/\sqrt{2} & -1/\sqrt{2} \end{bmatrix} \begin{bmatrix} 2 & -1 \\ 1 & 0 \end{bmatrix} \begin{bmatrix} 1/\sqrt{2} & 1/\sqrt{2} \\ 1/\sqrt{2} & -1/\sqrt{2} \end{bmatrix} = \begin{bmatrix} 1 & 2 \\ 0 & 1 \end{bmatrix}.$$

That is the triangular T, with the eigenvalues on the diagonal. For a larger matrix the first step would have found U_1—to be followed by $U_2, U_3, \ldots$, all multiplied together into U. It is not especially fun to compute T.

Diagonalizing Symmetric and Hermitian Matrices

As an application of this triangular form, we will show that any symmetric or Hermitian matrix—whether its eigenvalues are distinct or not—has a complete set of orthonormal eigenvectors. We need a unitary matrix such that $U^{-1}AU$ is *diagonal*, and Schur's lemma 5R has just found it. There are two steps to get from triangular to diagonal:

1. If A is Hermitian then so is $U^{-1}AU$:

$$(U^{-1}AU)^{\mathrm{H}} = U^{\mathrm{H}}A^{\mathrm{H}}(U^{-1})^{\mathrm{H}} = U^{-1}AU.$$

2. If a symmetric or Hermitian matrix is also triangular, it must be diagonal.

Because $T = U^{-1}AU$ is both Hermitian and triangular, it is automatically diagonal. That completes the proof of a key theorem in linear algebra:

5S (*Spectral theorem*) Every real symmetric matrix can be diagonalized by an orthogonal matrix, and every Hermitian matrix can be diagonalized by a unitary matrix:

$$\text{(real case)} \quad Q^{-1}AQ = \Lambda \qquad \text{(complex case)} \quad U^{-1}AU = \Lambda$$

The columns of Q (or U) contain a complete set of orthonormal eigenvectors.

Remark 1 In the real symmetric case, the eigenvalues and eigenvectors are real at every step of 5R. That produces a *real* unitary U—in other words Q.

Remark 2   It is certainly reasonable, for symmetric matrices, that even with re-peated eigenvalues there is a full set of orthogonal eigenvectors. We can think of A as the limit of symmetric matrices with *distinct* eigenvalues, and as the limit approaches the eigenvectors stay perpendicular. In contrast, the nonsymmetric matrices

$$A(\theta) = \begin{bmatrix} 0 & \cos\theta \\ 0 & \sin\theta \end{bmatrix}$$

have eigenvectors $(1, 0)$ and $(\cos\theta, \sin\theta)$. As $\theta \to 0$, the second eigenvector ap-proaches the first—which is the *only* eigenvector of the nondiagonalizable matrix $\begin{bmatrix} 0 & 1 \\ 0 & 0 \end{bmatrix}$.

EXAMPLE   The spectral theorem is now proved for a symmetric matrix like

$$A = \begin{bmatrix} 0 & 1 & 0 \\ 1 & 0 & 0 \\ 0 & 0 & 1 \end{bmatrix}$$

which has repeated eigenvalues: $\lambda_1 = \lambda_2 = 1$, $\lambda_3 = -1$. One choice of eigenvectors is

$$x_1 = \frac{1}{\sqrt{2}} \begin{bmatrix} 1 \\ 1 \\ 0 \end{bmatrix}, \qquad x_2 = \begin{bmatrix} 0 \\ 0 \\ 1 \end{bmatrix}, \qquad x_3 = \frac{1}{\sqrt{2}} \begin{bmatrix} 1 \\ -1 \\ 0 \end{bmatrix}.$$

These are the columns of an orthogonal matrix Q, and $A = Q\Lambda Q^{-1} = Q\Lambda Q^{\mathrm{T}}$ becomes

$$A = \sum \lambda_i x_i x_i^{\mathrm{T}} = \lambda_1 \begin{bmatrix} \frac{1}{2} & \frac{1}{2} & 0 \\ \frac{1}{2} & \frac{1}{2} & 0 \\ 0 & 0 & 0 \end{bmatrix} + \lambda_2 \begin{bmatrix} 0 & 0 & 0 \\ 0 & 0 & 0 \\ 0 & 0 & 1 \end{bmatrix} + \lambda_3 \begin{bmatrix} \frac{1}{2} & -\frac{1}{2} & 0 \\ -\frac{1}{2} & \frac{1}{2} & 0 \\ 0 & 0 & 0 \end{bmatrix}.$$

But since $\lambda_1 = \lambda_2$, those first two projections (each of rank one) combine to give a projection P_1 of rank two, and A is

$$\begin{bmatrix} 0 & 1 & 0 \\ 1 & 0 & 0 \\ 0 & 0 & 1 \end{bmatrix} = \lambda_1 P_1 + \lambda_3 P_3 = (+1) \begin{bmatrix} \frac{1}{2} & \frac{1}{2} & 0 \\ \frac{1}{2} & \frac{1}{2} & 0 \\ 0 & 0 & 1 \end{bmatrix} + (-1) \begin{bmatrix} \frac{1}{2} & -\frac{1}{2} & 0 \\ -\frac{1}{2} & \frac{1}{2} & 0 \\ 0 & 0 & 0 \end{bmatrix}.$$

There is a whole plane of eigenvectors corresponding to $\lambda = 1$; our x_1 and x_2 were a more or less arbitrary choice. Therefore the separate $x_1 x_1^{\mathrm{T}}$ and $x_2 x_2^{\mathrm{T}}$ were equally arbitrary, and it is only their sum—the projection P_1 onto the whole plane—which is uniquely determined. *Every Hermitian matrix with k different* *eigenvalues has its own "**spectral decomposition**" into $A = \lambda_1 P_1 + \cdots + \lambda_k P_k$, where P_i is the projection onto the eigenspace for λ_i.* Since there is a full set of eigenvectors, the projections add up to the identity. And since the eigenspaces are orthogonal, one projection followed by another must produce zero: $P_j P_i = 0$.

We are very close to answering a natural and important question, and might as well push on the rest of the way: For which matrices is the triangular T the same as the diagonal Λ? Hermitian, skew-Hermitian, and unitary matrices are in this class; they correspond to numbers on the real axis, the pure imaginary axis, and the unit circle. Now we want the whole class, corresponding to all complex numbers. The matrices are called "normal."

5T The matrix N is **normal** if it commutes with N^H: $NN^H = N^H N$. For such matrices, and no others, the triangular $T = U^{-1}NU$ is the diagonal matrix Λ. Normal matrices are exactly those that possess a **complete set of orthonormal eigenvectors**.

Note that Hermitian (or symmetric) matrices are certainly normal: If $A = A^H$, then AA^H and $A^H A$ both equal A^2. Unitary matrices are also normal: UU^H and $U^H U$ both equal the identity. In these special cases we proved that $T = \Lambda$ in two steps, and the same two steps will work for any normal matrix:

(i) If N is normal, then so is $T = U^{-1}NU$:

$$TT^H = U^{-1}NUU^H N^H U = U^{-1}NN^H U = U^{-1}N^H NU = U^H N^H UU^{-1}NU = T^H T.$$

(ii) A triangular T that is normal must be diagonal (Exercises 5.6.19–20).

Thus if N is normal, the triangular $U^{-1}NU$ must be diagonal. Since it has the same eigenvalues as N it must be Λ. The eigenvectors of N are the columns of U, and they are orthonormal. That is the good case, and we turn now to the general case—from the best possible matrices to the worst possible.

The Jordan Form

So far in this section, we have done the best we could with unitary similarities; requiring M to be a unitary matrix U, we got $M^{-1}AM$ into a triangular form T. Now we lift this restriction on M. Any matrix is allowed, and the goal is to make $M^{-1}AM$ as *nearly diagonal as possible*.

The result of this supreme effort at diagonalization is the Jordan form J. If A has a full set of independent eigenvectors, we take $M = S$ and arrive at $J = S^{-1}AS = \Lambda$; the Jordan form coincides with the diagonal Λ. This is impossible for a defective matrix, and for every missing eigenvector the Jordan form will have a 1 just above its main diagonal. The eigenvalues appear on the diagonal itself, because J is triangular. And distinct eigenvalues can always be decoupled. It is only a repeated λ that may (or may not!) require an off-diagonal entry in J.

5U   If A has s independent eigenvectors, it is similar to a matrix with s blocks:

$$J = M^{-1}AM = \begin{bmatrix} J_1 & & & \\ & \cdot & & \\ & & \cdot & \\ & & & J_s \end{bmatrix}.$$

Each Jordan block J_i is a triangular matrix with only a single eigenvalue λ_i and only one eigenvector:

$$J_i = \begin{bmatrix} \lambda_i & 1 & & \\ & \cdot & \cdot & \\ & & \cdot & 1 \\ & & & \lambda_i \end{bmatrix}.$$

When the block has order $m > 1$, the eigenvalue λ_i is repeated m times and there are $m - 1$ 1's above the diagonal. The same eigenvalue λ_i may appear in several blocks, if it corresponds to several independent eigenvectors. Two matrices are similar if they share the same Jordan form J.

Many authors have made this theorem the climax of their linear algebra course. Frankly, I think that is a mistake. It is certainly true that not all matrices are diagonalizable, and that the Jordan form is the most general case; but for that very reason its construction is both technical and extremely unstable. (A slight change in A can put back all the missing eigenvectors, and remove the off-diagonal 1's.) Therefore the right place for the details is in the appendix,† and the best way to start on the Jordan form is to look at some specific and manageable examples.

EXAMPLE 1 $T = \begin{bmatrix} 1 & 2 \\ 0 & 1 \end{bmatrix}$ and $A = \begin{bmatrix} 2 & -1 \\ 1 & 0 \end{bmatrix}$ and $B = \begin{bmatrix} 1 & 0 \\ 1 & 1 \end{bmatrix}$ all lead to

$J = \begin{bmatrix} 1 & 1 \\ 0 & 1 \end{bmatrix}.$

---

† Every author tries to make these details easy to follow, and I believe Filippov's proof is the best. It is almost simple enough to reverse our decision and bring the construction of J back from the appendix.

These four matrices have eigenvalues 1 and 1 with only *one eigenvector*—so J consists of *one block*. We now check that. The determinants all equal 1 and the traces (the sums down the main diagonal) are 2. The eigenvalues satisfy $1 \cdot 1 = 1$ and $1 + 1 = 2$. For A and B and J, which are triangular, the eigenvalues are sitting on the diagonal. We want to show that *these matrices are similar*—they all belong to the same family—and J is the Jordan form for that family.

From T to J, the job is to change 2 to 1, and a diagonal M will do it:

$$M^{-1}TM = \begin{bmatrix} 1 & 0 \\ 0 & 2 \end{bmatrix} \begin{bmatrix} 1 & 2 \\ 0 & 1 \end{bmatrix} \begin{bmatrix} 1 & 0 \\ 0 & \frac{1}{2} \end{bmatrix} = \begin{bmatrix} 1 & 1 \\ 0 & 1 \end{bmatrix} = J.$$

From A to J the job is to make the matrix triangular, and then change 2 to 1:

$$U^{-1}AU = \begin{bmatrix} 1 & 2 \\ 0 & 1 \end{bmatrix} = T \quad \text{and then} \quad M^{-1}TM = \begin{bmatrix} 1 & 1 \\ 0 & 1 \end{bmatrix} = J.$$

$U^{-1}AU$ was the example for Schur's lemma (5R above); U times M reaches J. From B to J the job is to transpose, and a permutation does that:

$$P^{-1}BP = \begin{bmatrix} 0 & 1 \\ 1 & 0 \end{bmatrix} \begin{bmatrix} 1 & 0 \\ 1 & 1 \end{bmatrix} \begin{bmatrix} 0 & 1 \\ 1 & 0 \end{bmatrix} = \begin{bmatrix} 1 & 1 \\ 0 & 1 \end{bmatrix} = J.$$

EXAMPLE 2 $A = \begin{bmatrix} 0 & 1 & 2 \\ 0 & 0 & 1 \\ 0 & 0 & 0 \end{bmatrix}$ and $B = \begin{bmatrix} 0 & 0 & 1 \\ 0 & 0 & 0 \\ 0 & 0 & 0 \end{bmatrix}$.

Since zero is a triple eigenvalue for both matrices, it will appear in every Jordan block; either there is a single 3 by 3 block, or a 2 by 2 and a 1 by 1 block, or three 1 by 1 blocks. Therefore the possible Jordan forms are

$$J_1 = \begin{bmatrix} 0 & 1 & 0 \\ 0 & 0 & 1 \\ 0 & 0 & 0 \end{bmatrix}, \quad J_2 = \begin{bmatrix} 0 & 1 & 0 \\ 0 & 0 & 0 \\ 0 & 0 & 0 \end{bmatrix}, \quad \text{and} \quad J_3 = \begin{bmatrix} 0 & 0 & 0 \\ 0 & 0 & 0 \\ 0 & 0 & 0 \end{bmatrix}.$$

In the case of A, the only eigenvector is $(1, 0, 0)$. Therefore its Jordan form has only one block, and according to the main theorem 5U, A must be similar to J_1. The matrix B has the additional eigenvector $(0, 1, 0)$, and therefore its Jordan form is J_2. There must be two blocks along the diagonal. As for J_3, it is in a class by itself; the only matrix similar to the zero matrix is $M^{-1}0M = 0$.

In these examples, a count of the eigenvectors was enough to determine J—and that is always possible when there is nothing more complicated than a triple eigenvalue. But as a general rule, this counting technique is exploded by the last exercise.

Application to difference and differential equations (*powers and exponentials*). If A can be diagonalized, the powers of $A = S\Lambda S^{-1}$ are easy: $A^k = S\Lambda^k S^{-1}$. If it cannot be diagonalized we still have $A = MJM^{-1}$, with the Jordan form in the middle—and now we need the powers of J:

$$A^k = (MJM^{-1})(MJM^{-1}) \cdots (MJM^{-1}) = MJ^k M^{-1}.$$

J is block-diagonal and the powers of those blocks can be taken separately:

$$J_i^n = \begin{bmatrix} \lambda & 1 & 0 \\ 0 & \lambda & 1 \\ 0 & 0 & \lambda \end{bmatrix}^n = \begin{bmatrix} \lambda^n & n\lambda^{n-1} & n(n-1)\lambda^{n-2} \\ 0 & \lambda^n & n\lambda^{n-1} \\ 0 & 0 & \lambda^n \end{bmatrix}.$$

This will enter the solution to a difference equation, if there is a triple eigenvalue and a single eigenvector. It also leads to the solution to the corresponding differential equation:

$$e^{J_i t} = \begin{bmatrix} e^{\lambda t} & te^{\lambda t} & \frac{1}{2}t^2 e^{\lambda t} \\ 0 & e^{\lambda t} & te^{\lambda t} \\ 0 & 0 & e^{\lambda t} \end{bmatrix}.$$

That comes from summing the series $I + J_i t + (J_i t)^2/2! + \cdots$, which produces $1 + \lambda t + \cdots = e^{\lambda t}$ on the diagonal and $te^{\lambda t}$ just above it.

EXAMPLE   The third column of $e^{J_i t}$ appears in the solution to $du/dt = J_i u$:

$$\begin{bmatrix} du_1/dt \\ du_2/dt \\ du_3/dt \end{bmatrix} = \begin{bmatrix} \lambda & 1 & 0 \\ 0 & \lambda & 1 \\ 0 & 0 & \lambda \end{bmatrix} \begin{bmatrix} u_1 \\ u_2 \\ u_3 \end{bmatrix} \quad \text{starting from} \quad u_0 = \begin{bmatrix} 0 \\ 0 \\ 1 \end{bmatrix}.$$

The system is solved by back-substitution (since the matrix is triangular). The last equation yields $u_3 = e^{\lambda t}$. The equation for u_2 is $du_2/dt = \lambda u_2 + u_3$ and its solution is $te^{\lambda t}$. The top equation is $du_1/dt = \lambda u_1 + u_2$ and its solution is $\frac{1}{2}t^2 e^{\lambda t}$. For a block of size m—from an eigenvalue of multiplicity m with only one eigenvector—the extra factor t appears $m - 1$ times.

These powers and exponentials of J are a part of the solution formula. The other part is the M that connects the original matrix A to the more convenient matrix J:

$$\text{if} \quad u_{k+1} = Au_k \quad \text{then} \quad u_k = A^k u_0 = MJ^k M^{-1} u_0$$
$$\text{if} \quad du/dt = Au \quad \text{then} \quad u = e^{At} u_0 = M e^{Jt} M^{-1} u_0.$$

When M and J are S and Λ (the diagonalizable case) those are the formulas of 5.3 and 5.4. Appendix B returns to the nondiagonalizable case and shows how the Jordan form can be reached.

I hope the following table will be a convenient summary.

Table of Similarity Transformations

1. A is **diagonalizable**: The columns of S are the eigenvectors and $S^{-1}AS = \Lambda$ is *diagonal*.
2. A is **arbitrary**: The columns of M are the eigenvectors and generalized eigenvectors of A, and the Jordan form $M^{-1}AM = J$ is *block diagonal*.
3. A is **arbitrary** and U is unitary: U can be chosen so that $U^{-1}AU = T$ is *triangular*.
4. A is **normal**, $AA^H = A^H A$: U can be chosen so that $U^{-1}AU = \Lambda$.

   Special cases of 4, all with orthonormal eigenvectors:

   a. If A is Hermitian, then Λ is real.
   a'. If A is real symmetric, then Λ is real and $U = Q$ is orthogonal.
   b. If A is skew-Hermitian, then Λ is imaginary.
   c. If A is orthogonal or unitary, then all $|\lambda_i| = 1$.

EXERCISES

5.6.1 If B is similar to A and C is similar to B, show that C is similar to A. (Let $B = M^{-1}AM$ and $C = N^{-1}BN$.) Which matrices are similar to the identity?

5.6.2 Describe in words all matrices that are similar to $\begin{bmatrix} 1 & 0 \\ 0 & -1 \end{bmatrix}$, and find two of them.

5.6.3 Explain why A is never similar to $A + I$.

5.6.4 Find a diagonal M, made up of 1's and -1's, to show that

$$A = \begin{bmatrix} 2 & 1 & & \\ 1 & 2 & 1 & \\ & 1 & 2 & 1 \\ & & 1 & 2 \end{bmatrix} \quad \text{is similar to} \quad B = \begin{bmatrix} 2 & -1 & & \\ -1 & 2 & -1 & \\ & -1 & 2 & -1 \\ & & -1 & 2 \end{bmatrix}.$$

5.6.5 Show (if B is invertible) that BA is similar to AB.

5.6.6 (a) If $CD = -DC$ (and D is invertible) show that C is similar to $-C$.
(b) Deduce that the eigenvalues of C must come in plus-minus pairs.
(c) Show directly that if $Cx = \lambda x$ then $C(Dx) = -\lambda(Dx)$.

5.6.7   Consider any A and a special "plane rotation" M:

$$A = \begin{bmatrix} a & b & c \\ d & e & f \\ g & h & i \end{bmatrix}, \qquad M = \begin{bmatrix} \cos\theta & -\sin\theta & 0 \\ \sin\theta & \cos\theta & 0 \\ 0 & 0 & 1 \end{bmatrix}.$$

Choose the rotation angle θ so as to annihilate the $(3, 1)$ entry of $M^{-1}AM$.

Note   This "annihilation" is not so easy to continue, because the rotations that produce zeros in place of d and h will spoil the new zero in the corner. We have to leave one diagonal below the main one, and finish the eigenvalue calculation in a different way. Otherwise, if we could make A diagonal and see its eigenvalues, we would be finding the roots of the polynomial $\det (A - \lambda I)$ by using only the square roots which determine $\cos\theta$—and that is impossible.

5.6.8   What matrix M changes the basis $V_1 = (1, 1)$, $V_2 = (1, 4)$ to the basis $v_1 = (2, 5)$, $v_2 = (1, 4)$? The columns of M come from expressing V_1 and V_2 as combinations of the v's.

5.6.9   For the same two bases, express the vector $(3, 9)$ as a combination $c_1 V_1 + c_2 V_2$ and also as $d_1 v_1 + d_2 v_2$. Check numerically that M connects c to d: $Mc = d$.

5.6.10   Confirm the last exercise algebraically: If $V_1 = m_{11}v_1 + m_{21}v_2$ and $V_2 = m_{12}v_1 + m_{22}v_2$ and $m_{11}c_1 + m_{12}c_2 = d_1$ and $m_{21}c_1 + m_{22}c_2 = d_2$, then the vectors $c_1 V_1 + c_2 V_2$ and $d_1 v_1 + d_2 v_2$ are the same. This is the "change of basis formula" $Mc = d$.

5.6.11   If the transformation T is a reflection across the $45°$ line in the plane, find its matrix with respect to the standard basis $v_1 = (1, 0)$, $v_2 = (0, 1)$ and also with respect to $V_1 = (1, 1)$, $V_2 = (1, -1)$. Show that those matrices are similar.

5.6.12   The *identity transformation* takes every vector to itself: $Tx = x$. Find the corresponding matrix, if the first basis is $v_1 = (1, 2)$, $v_2 = (3, 4)$ and the second basis is $w_1 = (1, 0)$, $w_2 = (0, 1)$. (It is not the identity matrix!)

5.6.13   The derivative of $a + bx + cx^2$ is $b + 2cx + 0x^2$.
(a)   Write down the 3 by 3 matrix D such that

$$D \begin{bmatrix} a \\ b \\ c \end{bmatrix} = \begin{bmatrix} b \\ 2c \\ 0 \end{bmatrix}.$$

(b)   Compute D^3 and interpret the results in terms of derivatives.
(c)   What are the eigenvalues and eigenvectors of D?

5.6.14   Show that the transformation $Tf(x) = \int_0^x f(t)\,dt$ has no eigenvalues, while for $Tf(x) = df/dx$ every number is an eigenvalue. The functions are to be defined for all x.

5.6.15   On the space of 2 by 2 matrices, let T be the transformation that *transposes every matrix*. Find the eigenvalues and "eigenmatrices" of T (the matrices which satisfy $A^T = \lambda A$).

5.6.16 (a) Find an orthogonal Q so that $Q^{-1}AQ = \Lambda$ if

$$A = \begin{bmatrix} 1 & 1 & 1 \\ 1 & 1 & 1 \\ 1 & 1 & 1 \end{bmatrix} \quad \text{and} \quad \Lambda = \begin{bmatrix} 0 & 0 & 0 \\ 0 & 0 & 0 \\ 0 & 0 & 3 \end{bmatrix}.$$

Then find a second pair of orthonormal eigenvectors x_1, x_2 for $\lambda = 0$.
(b) Verify that $P = x_1 x_1^T + x_2 x_2^T$ is the same for both pairs.

5.6.17 Prove that every *unitary* matrix A is diagonalizable, in two steps:

(i) If A is unitary, and U is too, then so is $T = U^{-1}AU$.
(ii) An upper triangular T that is unitary must be diagonal.

It follows that the triangular T is Λ, and any unitary matrix (distinct eigenvalues or not) has a complete set of orthonormal eigenvectors: $U^{-1}AU = \Lambda$. All eigenvalues satisfy $|\lambda| = 1$.

5.6.18 Find a normal matrix that is not Hermitian, skew-Hermitian, unitary, or diagonal. Show that all permutation matrices are normal.

5.6.19 Suppose T is a 3 by 3 upper triangular matrix, with entries t_{ij}. Compare the entries of TT^H and T^HT, and show that if they are equal then T must be diagonal.

5.6.20 If N is normal, show that $\|Nx\| = \|N^Hx\|$ for every vector x. Deduce that the ith row of N has the same length as the ith column. *Note*: If N is also upper triangular, this leads again to the conclusion that it must be diagonal.

5.6.21 Prove that a matrix with orthonormal eigenvectors has to be normal, as claimed in 5T: If $U^{-1}NU = \Lambda$, or $N = U\Lambda U^H$, then $NN^H = N^HN$.

5.6.22 Find a unitary U and triangular T so that $U^{-1}AU = T$, for

$$A = \begin{bmatrix} 5 & -3 \\ 4 & -2 \end{bmatrix} \quad \text{and} \quad A = \begin{bmatrix} 0 & 1 & 0 \\ 0 & 0 & 0 \\ 1 & 0 & 0 \end{bmatrix}.$$

5.6.23 If A has eigenvalues 0, 1, 2, what are the eigenvalues of $A(A - I)(A - 2I)$?

5.6.24 (a) Show by direct multiplication that a triangular matrix, say 3 by 3, satisfies its own characteristic equation: $(T - \lambda_1 I)(T - \lambda_2 I)(T - \lambda_3 I) = 0$.
(b) Substituting $U^{-1}AU$ for T, deduce the **Cayley-Hamilton theorem: *Any matrix satisfies its own characteristic equation:*** $(A - \lambda_1 I)(A - \lambda_2 I)(A - \lambda_3 I) = 0$.

5.6.25 The characteristic polynomial of $A = \begin{bmatrix} a & b \\ c & d \end{bmatrix}$ is $\lambda^2 - (a + d)\lambda + (ad - bc)$. By direct substitution verify the Cayley-Hamilton theorem: $A^2 - (a + d)A + (ad - bc)I = 0$.

5.6.26 In Example 2 at the end of the chapter, find the M that achieves $M^{-1}BM = J$.

5.6.27 If $a_{ij} = 1$ above the main diagonal and $a_{ij} = 0$ elsewhere, find its Jordan form (say 4 by 4) by finding all its eigenvectors.

5.6.28 Show, by trying for an M and failing, that no two of the Jordan forms in the 3 by 3 example are similar: $J_1 \neq M^{-1}J_2M$, and $J_1 \neq M^{-1}J_3M$, and $J_2 \neq M^{-1}J_3M$.

5.6.29 Solve the first equation by back-substitution and the second by reaching $A = MJM^{-1}$:

$$\frac{du}{dt} = Ju = \begin{bmatrix} 5 & 1 \\ 0 & 5 \end{bmatrix} \begin{bmatrix} u_1 \\ u_2 \end{bmatrix}, \quad u_0 = \begin{bmatrix} 1 \\ 2 \end{bmatrix}, \quad \text{and} \quad \frac{dv}{dt} = Av = \begin{bmatrix} 3 & 1 \\ -4 & 7 \end{bmatrix} \begin{bmatrix} v_1 \\ v_2 \end{bmatrix}, \quad v_0 = \begin{bmatrix} 1 \\ 0 \end{bmatrix}.$$

5.6.30 Compute J^{10} and A^{10} and e^A if $A = MJM^{-1}$:

$$A = \begin{bmatrix} 14 & 9 \\ -16 & -10 \end{bmatrix} = \begin{bmatrix} 3 & -2 \\ -4 & 3 \end{bmatrix} \begin{bmatrix} 2 & 1 \\ 0 & 2 \end{bmatrix} \begin{bmatrix} 3 & 2 \\ 4 & 3 \end{bmatrix}.$$

5.6.31 Write out all possible Jordan forms for a 4 by 4 matrix that has zero as a quadruple eigenvalue. (By convention, the blocks get smaller as we move down the matrix J.) If there are two independent eigenvectors, show that there are two different possibilities for J.

REVIEW EXERCISES: Chapter 5

5.1 Find the eigenvalues and eigenvectors and the diagonalizing matrix S for

$$A = \begin{bmatrix} 1 & 0 \\ 2 & 3 \end{bmatrix} \quad \text{and} \quad B = \begin{bmatrix} 7 & 2 \\ -15 & -4 \end{bmatrix}.$$

5.2 Find the determinants of A and A^{-1} if

$$A = S \begin{bmatrix} \lambda_1 & 0 \\ 0 & \lambda_2 \end{bmatrix} S^{-1}.$$

5.3 If A has eigenvalues 0 and 1, corresponding to the eigenvectors

$$\begin{bmatrix} 1 \\ 2 \end{bmatrix} \quad \text{and} \quad \begin{bmatrix} 2 \\ -1 \end{bmatrix},$$

how can you tell in advance that A is symmetric? What are its trace and determinant? What is A?

5.4 In the previous exercise, what will be the eigenvalues and eigenvectors of A^2? What is the relation of A^2 to A?

5.5 Does there exist a matrix A such that the entire family $A + cI$ is invertible for all complex numbers c? Find a real matrix with $A + rI$ invertible for all real r.

5.6 Solve for both initial values and then find e^{At}:

$$\frac{du}{dt} = \begin{bmatrix} 3 & 1 \\ 1 & 3 \end{bmatrix} u \quad \text{if} \quad u_0 = \begin{bmatrix} 1 \\ 0 \end{bmatrix} \quad \text{and if} \quad u_0 = \begin{bmatrix} 0 \\ 1 \end{bmatrix}.$$

5.7 Would you prefer to have interest compounded quarterly at 40% per year, or annually at 50%?

5.8 True or false (with counterexample if false):
(a) If B is formed from A by exchanging two rows, then B is similar to A.
(b) If a triangular matrix is similar to a diagonal matrix, it is already diagonal.
(c) Any two of these statements imply the third: A is Hermitian, A is unitary, $A^2 = I$.
(d) If A and B are diagonalizable, so is AB.

5.9 What happens to the Fibonacci sequence if we go backward in time, and how is F_{-k} related to F_k? The law $F_{k+2} = F_{k+1} + F_k$ is still in force, so $F_{-1} = 1$.

5.10 Find the general solution to $du/dt = Au$ if

$$A = \begin{bmatrix} 0 & -1 & 0 \\ 1 & 0 & -1 \\ 0 & 1 & 0 \end{bmatrix}.$$

Can you find a time T at which the solution $u(T)$ is guaranteed to return to the initial value u_0?

5.11   If P is the matrix that projects $\mathbf{R}^n$ onto a subspace S, explain why every vector in S is an eigenvector and so is every vector in $S^\perp$. What are the eigenvalues? (Note the connection to $P^2 = P$, which means $\lambda^2 = \lambda$.)

5.12   Show that every matrix of order > 1 is the sum of two singular matrices.

5.13   (a) Show that the matrix differential equation $dX/dt = AX + XB$ has the solution $X(t) = e^{At}X(0)e^{Bt}$. (b) Prove that the solutions of $dX/dt = AX - XA$ keep the same eigenvalues for all time.

5.14   If the eigenvalues of A are 1 and 3 with eigenvectors (5, 2) and (2, 1), find the solutions to $du/dt = Au$ and $u_{k+1} = Au_k$ starting from $u = (9, 4)$.

5.15   Find the eigenvalues and eigenvectors of

$$A = \begin{bmatrix} 0 & -i & 0 \\ i & 1 & i \\ 0 & -i & 0 \end{bmatrix}.$$

What property do you expect for the eigenvectors, and is it true?

5.16   By trying to solve

$$\begin{bmatrix} a & b \\ c & d \end{bmatrix}\begin{bmatrix} a & b \\ c & d \end{bmatrix} = \begin{bmatrix} 0 & 1 \\ 0 & 0 \end{bmatrix} = A$$

show that A has no square root. Change the diagonal entries of A to 4 and find a square root.

5.17   (a)   Find the eigenvalues and eigenvectors of $A = \begin{bmatrix} 0 & 4 \\ \frac{1}{4} & 0 \end{bmatrix}$.
(b)   Solve $du/dt = Au$ starting from $u_0 = (100, 100)$.
(c)   If $v(t) = $ income to stockbrokers and $w(t) = $ income to client, and they help each other by $dv/dt = 4w$ and $dw/dt = \frac{1}{4}v$, what does the ratio v/w approach as $t \to \infty$?

5.18   True or false, with reason if true and counterexample if false:
(a)   For every matrix A there is a solution to $du/dt = Au$ starting from $u_0 = (1, \ldots, 1)$.
(b)   Every invertible matrix can be diagonalized.
(c)   Every diagonalizable matrix can be inverted.
(d)   Exchanging the rows of a 2 by 2 matrix reverses the signs of its eigenvalues.
(e)   If eigenvectors x and y correspond to distinct eigenvalues then $x^H y = 0$.

5.19   If K is a skew-symmetric matrix show that $Q = (I - K)(I + K)^{-1}$ is an orthogonal matrix. Find Q if $K = \begin{bmatrix} 0 & 2 \\ -2 & 0 \end{bmatrix}$.

5.20   If $K^H = -K$ (skew-Hermitian), the eigenvalues are imaginary and the eigenvectors are orthogonal.
(a)   How do you know that $K - I$ is invertible?
(b)   How do you know that $K = U\Lambda U^H$ for a unitary U?
(c)   Why is $e^{\Lambda t}$ unitary?
(d)   Why is e^{Kt} unitary?

5.21 If M is the diagonal matrix with entries d, d^2, d^3, what is $M^{-1}AM$ and what are its eigenvalues if

$$A = \begin{bmatrix} 1 & 1 & 1 \\ 1 & 1 & 1 \\ 1 & 1 & 1 \end{bmatrix}?$$

5.22 If $A^2 = -I$ what are the eigenvalues of A? If A is a real n by n matrix show that n must be even, and give an example.

5.23 If $Ax = \lambda_1 x$ and $A^Ty = \lambda_2 y$ (all real) show that $x^Ty = 0$.

5.24 A variation on the Fourier matrix is the "sine matrix"

$$S = \frac{1}{\sqrt{2}} \begin{bmatrix} \sin \theta & \sin 2\theta & \sin 3\theta \\ \sin 2\theta & \sin 4\theta & \sin 6\theta \\ \sin 3\theta & \sin 6\theta & \sin 9\theta \end{bmatrix} \quad \text{with} \quad \theta = \frac{\pi}{4}.$$

Verify that $S^T = S^{-1}$. (The columns are the eigenvectors of the tridiagonal $-1, 2, -1$ matrix.)

5.25 (a) Find a nonzero matrix N such that $N^3 = 0$.
(b) If $Nx = \lambda x$ show that λ must be zero.
(c) Prove that N (called a "nilpotent" matrix) cannot be symmetric.

5.26 (a) Find the matrix $P = aa^T/a^Ta$ that projects any vector onto the line through $a = (2, 1, 2)$.
(b) What is the only nonzero eigenvalue of P and what is the corresponding eigenvector?
(c) Solve $u_{k+1} = Pu_k$ starting from $u_0 = (9, 9, 0)$.

5.27 Suppose the first row of A is 7, 6 and its eigenvalues are $i, -i$. Find A.

5.28 (a) For which numbers c and d does A have real eigenvalues and orthogonal eigenvectors?

$$A = \begin{bmatrix} 1 & 2 & 0 \\ 2 & d & c \\ 0 & 5 & 3 \end{bmatrix}$$

(b) For which c and d can we find three orthonormal vectors that are combinations of the columns (don't do it!)?

5.29 If the vectors x_1 and x_2 are in the columns of S, what are the eigenvalues and eigenvectors of

$$A = S \begin{bmatrix} 2 & 0 \\ 0 & 1 \end{bmatrix} S^{-1} \quad \text{and} \quad B = S \begin{bmatrix} 2 & 3 \\ 0 & 1 \end{bmatrix} S^{-1}?$$

5.30 What is the limit as $k \to \infty$ (the Markov steady state) of $\begin{bmatrix} .4 & .3 \\ .6 & .7 \end{bmatrix}^k \begin{bmatrix} a \\ b \end{bmatrix}$?

6

POSITIVE DEFINITE MATRICES

6.1 ■ MINIMA, MAXIMA, AND SADDLE POINTS

Up to now, we have had almost no reason to worry about the *signs of the eigenvalues*. In fact, it would have been premature to ask whether λ is positive or negative before it was known to be real. But Chapter 5 established that the most important matrices—symmetric matrices in the real case, and Hermitian matrices in the complex case—do have real eigenvalues. Therefore it makes sense to ask whether they are positive, and one of our goals is the following: to find a test that can be applied directly to the symmetric matrix A, without computing its eigenvalues, which will **guarantee that all those eigenvalues are positive**. The test brings together three of the most basic ideas in the book—pivots, determinants, and eigenvalues.

Before looking for such a test, we want to describe a new situation in which the signs of the eigenvalues are significant. It is completely unlike the question of stability in differential equations, where we needed negative rather than positive eigenvalues. (We should not hurry past that point, but we will: If $-A$ passes the test we are looking for, then $du/dt = Au$ has decaying solutions $e^{\lambda t}x$, with every eigenvalue $\lambda < 0$. And $d^2u/dt^2 = Au$ has pure oscillatory solutions $e^{i\omega t}x$, with $\omega = \sqrt{-\lambda}$.) The new situation is one arising in so many applications to science, to engineering, and to every problem of optimization, that we hope the reader is willing to take the background for granted and start directly with the mathematical problem.

It is the problem of identifying a **minimum**, which we introduce with two examples:

$$F(x, y) = 7 + 2(x + y)^2 - y \sin y - x^3$$

and

$$f(x, y) = 2x^2 + 4xy + y^2.$$

Does either F or f have a minimum at the point $x = y = 0$?

Remark 1   The *zero-order terms* $F(0, 0) = 7$ and $f(0, 0) = 0$ have no effect on the answer. They simply raise or lower the graphs of F and f.

Remark 2   The *linear terms* give a necessary condition: To have any chance of a minimum, we must have a stationary point. The first derivatives must vanish when $x = y = 0$, and they do:

$$\frac{\partial F}{\partial x} = 4(x + y) - 3x^2 = 0 \quad \text{and} \quad \frac{\partial F}{\partial y} = 4(x + y) - y \cos y - \sin y = 0$$

$$\frac{\partial f}{\partial x} = 4x + 4y = 0 \quad \text{and} \quad \frac{\partial f}{\partial y} = 4x + 2y = 0.$$

Thus the origin is a stationary point for both F and f. Geometrically, the surface $z = F(x, y)$ is tangent to the horizontal plane $z = 7$, and the surface $z = f(x, y)$ is tangent to the plane $z = 0$. The question is whether F and f lie above those planes, as we move away from the tangency point $x = y = 0$.

Remark 3   *The quadratic terms, coming from the second derivatives, are decisive:*

$$\frac{\partial^2 F}{\partial x^2} = 4 - 6x = 4 \qquad\qquad \frac{\partial^2 f}{\partial x^2} = 4$$

$$\frac{\partial^2 F}{\partial x\, \partial y} = \frac{\partial^2 F}{\partial y\, \partial x} = 4 \qquad\qquad \frac{\partial^2 f}{\partial x\, \partial y} = \frac{\partial^2 f}{\partial y\, \partial x} = 4$$

$$\frac{\partial^2 F}{\partial y^2} = 4 + y \sin y - 2 \cos y = 2, \qquad \frac{\partial^2 f}{\partial y^2} = 2.$$

These derivatives contain the answer. Since they are the same for F and f, they must contain the same answer. The two functions behave in exactly the same way near the origin, and F **has a minimum if and only if f has a minimum.**

Remark 4   The *higher-degree terms* in F have no effect on the question of a *local* minimum, but they can prevent it from being a *global* minimum. In our example the term $-x^3$ must sooner or later pull F toward $-\infty$, regardless of what happens near $x = y = 0$. Such an eventuality is impossible for f, or for any other "*quadratic form*," which has no higher terms.

   Every quadratic form $f = ax^2 + 2bxy + cy^2$ has a stationary point at the origin, where $\partial f/\partial x = \partial f/\partial y = 0$. If it has a local minimum at $x = y = 0$, then that point is also a global minimum. The surface $z = f(x, y)$ will be shaped like a bowl, resting on the one point at the origin.

To summarize: The question of a local minimum for F is equivalent to the same question for f. If the stationary point of F were at $x = \alpha$, $y = \beta$ instead of $x = y = 0$, the only change would be to use the second derivatives at α, β:

$$f(x, y) = \frac{x^2}{2} \frac{\partial^2 F}{\partial x^2} (\alpha, \beta) + xy \frac{\partial^2 F}{\partial x \partial y} (\alpha, \beta) + \frac{y^2}{2} \frac{\partial^2 F}{\partial y^2} (\alpha, \beta). \tag{1}$$

This quadratic f behaves near $(0, 0)$ in the same way that F behaves near (α, β).

There is one case to be excluded. It corresponds to the possibility $F'' = 0$, which is a tremendous headache even for a function of one variable. The third derivatives are drawn into the problem because the second derivatives fail to give a definite decision. To avoid that difficulty it is usual to require that the quadratic part be nonsingular. For a true minimum, f is allowed to vanish *only at $x = y = 0$*. When f is strictly positive at all other points it is called ***positive definite***.

The problem now comes down to this: For a function of two variables x and y, what is the correct replacement for the condition $F'' > 0$? With only one variable, the sign of the second derivative decides between a minimum or a maximum. Now we have three second derivatives, F_{xx}, $F_{xy} = F_{yx}$, and F_{yy}. These three numbers specify f, and they must determine whether or not F (as well as f) has a minimum. ***What are the conditions on a, b, and c which ensure that $f = ax^2 + 2bxy + cy^2$ is positive definite?***

It is easy to find one necessary condition:

(i)   *If f is positive definite, then necessarily $a > 0$.*

We look at $x = 1$, $y = 0$, where $ax^2 + 2bxy + cy^2$ is equal to a. This must be positive if f is to be positive definite. Translating back to F, this means that $\partial^2 F / \partial x^2 > 0$; we fix $y = 0$, let only x vary, and must have $F'' > 0$ for a minimum. Similarly, if we fix $x = 0$ and look in the y direction, there is a condition on the coefficient c:

(ii)   *If f is positive definite, then necessarily $c > 0$.*

Do these conditions $a > 0$ and $c > 0$ guarantee that f is positive? The answer is no—the cross term $2bxy$ can pull f below zero, if it is large enough.

EXAMPLE $f = x^2 - 10xy + y^2$. In this case $a = 1$ and $c = 1$ are both positive. Suppose, however, that we look at $x = y = 1$; since $f(1, 1) = -8$, this f is not positive definite. The conditions $a > 0$ and $c > 0$ ensure that f is increasing in the x and y directions, but it may still decrease along another line. The function is negative on the line $x = y$, because $b = -10$ overwhelms a and c. It is impossible to test for positive definiteness by looking along any finite number of fixed lines— and this f goes both above and below zero.

Evidently b enters the problem, and in our original f the coefficient b was positive. Does this make f positive, and ensure a minimum? Again the answer is no; the sign of b is of no importance! *Even though all its coefficients were positive, our* *original example $2x^2 + 4xy + y^2$ was not positive definite. **Neither F nor f had a minimum**. On the line $x = -y$, f is negative: $f(1, -1) = 2 - 4 + 1 = -1$.*

It is the size of b, compared to a and c, that must be controlled if f is to be positive definite. We now want to find a precise test, giving a necessary and sufficient condition for positive definiteness. The simplest technique is to "complete the square":

$$f = ax^2 + 2bxy + cy^2 = a\left(x + \frac{b}{a}y\right)^2 + \left(c - \frac{b^2}{a}\right)y^2. \tag{2}$$

The first term on the right is never negative, since the square is multiplied by the positive coefficient a. Necessary condition (i) is still in force. But the square can be zero, and the second term must then be positive. That term is y^2 multiplied by the coefficient $(ac - b^2)/a$. The last requirement for a positive f is that this coefficient must be positive:

(iii)   *If f is positive, then necessarily $ac > b^2$.*

Notice that conditions (i) and (iii), taken together, automatically imply condition (ii). If $a > 0$, and $ac > b^2 \geq 0$, then certainly $c > 0$. The right side of (2) is guaranteed to be positive, and we have answered the question at last:

6A   The quadratic form $f = ax^2 + 2bxy + cy^2$ is positive definite if and only if $a > 0$ and $ac - b^2 > 0$. Correspondingly, F has a (nonsingular) minimum at $x = y = 0$ if and only if its first derivatives are zero and

$$\frac{\partial^2 F}{\partial x^2}(0, 0) > 0, \qquad \left[\frac{\partial^2 F}{\partial x^2}(0, 0)\right]\left[\frac{\partial^2 F}{\partial y^2}(0, 0)\right] > \left[\frac{\partial^2 F}{\partial x\, \partial y}(0, 0)\right]^2.$$

The conditions for a maximum are easy, since f has a maximum whenever $-f$ has a minimum. This means reversing the signs of a, b, and c. It actually leaves the second condition $ac - b^2 > 0$ unchanged: The quadratic form is **negative definite** if and only if $a < 0$ and $ac - b^2 > 0$. The same change applies to F.

The quadratic form f is singular when $ac - b^2 = 0$; this is the case we ruled out. The second term in (2) would disappear, leaving only the first square—which is either **positive semidefinite**, when $a > 0$, or **negative semidefinite**, when $a < 0$. The prefix *semi* allows the possibility that f can equal zero, as it will at the point $x = b$, $y = -a$. Geometrically the surface $z = f(x, y)$ degenerates from a genuine bowl into an infinitely long trough. (Think of the surface $z = x^2$ in three-dimensional space; the trough runs up and down the y axis, and each cross section is the same parabola $z = x^2$.) And a still more singular quadratic form is to have zero everywhere, $a = b = c = 0$, which is both positive semidefinite and negative semidefinite. The bowl has become completely flat.

In one dimension, for a function $F(x)$, the possibilities would now be exhausted: Either there is a minimum, or a maximum, or $F'' = 0$. In two dimensions, however, a very important possibility still remains: *The combination $ac - b^2$ may be*

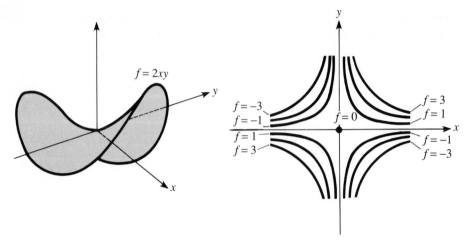

Fig. 6.1. The saddle $f = 2xy$ and its level curves.

negative. This occurred in our example, when the size of b dominated a and c; f was positive in some directions and negative in others. It also occurs, for any b, if a and c are of opposite sign. The x and y directions give opposite results—on one axis f increases, on the other it decreases. It is useful to consider the two special cases

$$f_1 = 2xy \quad \text{and} \quad f_2 = x^2 - y^2.$$

In the first, b is dominating, with $a = c = 0$. In the second, a and c are of opposite sign. Both have $ac - b^2 = -1$.

These quadratic forms are **indefinite**, because they can take either sign; both $f > 0$ and $f < 0$ are possible, depending on x and y. So we have a stationary point that is neither a maximum or a minimum. It is called a **saddle point**. (Presumably because the surface $z = f(x, y)$, say $z = x^2 - y^2$, is shaped like a saddle (Fig. 6.1). It goes down in the direction of the y axis, where the legs fit, and goes up in the direction of the x axis.) You may prefer to think of a road going over a mountain pass; the top of the pass is a minimum as you look along the range of mountains, but it is a maximum as you go along the road.

The saddles $2xy$ and $x^2 - y^2$ are practically the same; if we turn one through $45°$ we get the other. They are also almost impossible to draw.

Calculus would have been enough to find our conditions for a minimum: $F_{xx} > 0$ and $F_{xx}F_{yy} > F_{xy}^2$. But linear algebra is ready to do more, as soon as we recognize how the coefficients of f fit into a symmetric matrix A. The terms ax^2 and cy^2 appear on the diagonal, and the cross derivative $2bxy$ is split between the entry above and the entry below. Then f is identically equal to the matrix product

$$ax^2 + 2bxy + cy^2 = \begin{bmatrix} x & y \end{bmatrix} \begin{bmatrix} a & b \\ b & c \end{bmatrix} \begin{bmatrix} x \\ y \end{bmatrix}. \tag{3}$$

This identity is the key to the whole chapter. It can be rewritten as $f = x^T A x$; it generalizes immediately to n dimensions; and it provides a perfect shorthand for studying the problem of maxima and minima. When the variables are $x_1, \ldots, x_n$, instead of simply x and y, they go into a column vector x. *For any symmetric matrix A, the product $f = x^T A x$ is a pure quadratic form:*

$$x^T A x = \begin{bmatrix} x_1 & x_2 & \cdot & x_n \end{bmatrix} \begin{bmatrix} a_{11} & a_{12} & \cdot & a_{1n} \\ a_{21} & a_{22} & \cdot & a_{2n} \\ \cdot & & \cdot & \cdot \\ a_{n1} & a_{n2} & \cdot & a_{nn} \end{bmatrix} \begin{bmatrix} x_1 \\ x_2 \\ \cdot \\ x_n \end{bmatrix}.$$

$$= a_{11} x_1^2 + a_{12} x_1 x_2 + a_{21} x_2 x_1 + \cdots + a_{nn} x_n^2$$

$$= \sum_{i=1}^{n} \sum_{j=1}^{n} a_{ij} x_i x_j. \tag{4}$$

There are no higher-order terms or lower-order terms—only second-order. The function is zero at $x = 0$, and also its first derivatives are zero. The tangent is flat, and $x = 0$ is a stationary point. We have to decide if it is a minimum or a maximum or a saddle point.

EXAMPLE 1 $f = 2x^2 + 4xy + y^2$ and $A = \begin{bmatrix} 2 & 2 \\ 2 & 1 \end{bmatrix} \rightarrow$ *saddle point*

EXAMPLE 2 $f = 2xy$ and $A = \begin{bmatrix} 0 & 1 \\ 1 & 0 \end{bmatrix} \rightarrow$ *saddle point*

EXAMPLE 3 $f = 2x_1^2 - 2x_1 x_2 + 2x_2^2 - 2x_2 x_3 + 2x_3^2$ corresponds to

$$A = \begin{bmatrix} 2 & -1 & 0 \\ -1 & 2 & -1 \\ 0 & -1 & 2 \end{bmatrix} \rightarrow minimum$$

Those are pure quadratics, but any function $F(x_1, \ldots, x_n)$ is approached in the same way. We look for stationary points, where all first derivatives are zero. At those points, A is the "second-derivative matrix" or "Hessian matrix." Its entries are $a_{ij} = \partial^2 F / \partial x_i \, \partial x_j$. This is automatically symmetric, for the same reason that $\partial^2 F / \partial x \, \partial y = \partial^2 F / \partial y \, \partial x$. Then *$F$ has a minimum when the pure quadratic $f = x^T A x$ is positive.* The second-order terms are isolated in f, and they control F near the stationary point—as the Taylor series shows near $x = 0$:

$$F(x) = F(0) + x^T (\text{grad } F) + \tfrac{1}{2} x^T A x + \text{3rd-order terms.} \tag{5}$$

At a stationary point, grad $F = (\partial F/\partial x_1, \ldots, \partial F/\partial x_n)$ is a vector of zeros. The quadratic $x^T A x$ takes F up or down. If the stationary point is at x_0 instead of 0, the series starts with $F(x_0)$ and x changes to $x - x_0$ on the right side of (5).

The next section contains the tests to decide whether $f = x^T A x$ is positive. Equivalently, they decide whether the matrix A is positive definite—which is the main goal of the chapter.

EXERCISES

6.1.1 Show that the quadratic $f = x^2 + 4xy + 2y^2$ has a saddle point at the origin, despite the fact that its coefficients are positive. Rewrite f as a *difference of two squares*.

6.1.2 Decide for or against the positive definiteness of the following matrices, and write out the corresponding $f = x^T A x$:

(a) $\begin{bmatrix} 1 & 3 \\ 3 & 5 \end{bmatrix}$ (b) $\begin{bmatrix} 1 & -1 \\ -1 & 1 \end{bmatrix}$ (c) $\begin{bmatrix} 2 & 3 \\ 3 & 5 \end{bmatrix}$ (d) $\begin{bmatrix} -1 & 2 \\ 2 & -8 \end{bmatrix}$

The determinant in (b) is zero; along what line is f zero?

6.1.3 If a 2 by 2 symmetric matrix passes the tests $a > 0$, $ac > b^2$, solve the quadratic equation $\det(A - \lambda I) = 0$ and show that the roots are positive.

6.1.4 Decide between a minimum, maximum, or saddle point for the functions
(a) $F = -1 + 4(e^x - x) - 5x \sin y + 6y^2$ at the point $x = y = 0$;
(b) $F = (x^2 - 2x) \cos y$, with stationary point at $x = 1$, $y = \pi$.

6.1.5 (a) For which numbers b is the matrix $A = \begin{bmatrix} 1 & b \\ b & 9 \end{bmatrix}$ positive definite?
(b) Find the factorization $A = LDL^T$ when b is in the range for positive definiteness.
(c) Find the minimum value of $\frac{1}{2}(x^2 + 2bxy + 9y^2) - y$ for b in this range.
(d) What is the minimum if $b = 3$?

6.1.6 Suppose the positive coefficients a and c dominate b in the sense that $a + c > 2b$. Is this enough to guarantee that $ac > b^2$ and the matrix is positive definite? Give a proof or a counterexample.

6.1.7 (a) What 3 by 3 matrices correspond to

$$f_1 = x_1^2 + x_2^2 + x_3^2 - 2x_1x_2 - 2x_1x_3 + 2x_2x_3 \quad \text{and}$$
$$f_2 = x_1^2 + 2x_2^2 + 3x_3^2 - 2x_1x_2 - 2x_1x_3 - 4x_2x_3?$$

(b) Show that f_1 is a *single* perfect square and not positive definite. Where is f_1 equal to 0?
(c) Express f_2 as a sum of three squares and factor its A into LDL^T.

6.1.8 If $A = \begin{bmatrix} a & b \\ b & c \end{bmatrix}$ is positive definite prove that A^{-1} is positive definite.

6.1.9 The quadratic $f = 3(x_1 + 2x_2)^2 + 4x_2^2$ is positive. Find its matrix A, factor it into LDL^T, and connect the entries in D and L to the original f.

6.1.10 If $R = \begin{bmatrix} p & s \\ s & t \end{bmatrix}$ write out R^2 and check that it is positive definite unless R is singular.

6.1.11 (a) If $A = \begin{bmatrix} a & b \\ b & c \end{bmatrix}$ is Hermitian (*with complex b*) find its pivots and its determinant.
   (b) Complete the square for complex $f = x^{\mathrm{H}} A x = a|x_1|^2 + 2\,\mathrm{Re}\,b\bar{x}_1 x_2 + c|x_2|^2 = a|x_1 + (b/a)x_2|^2 +\quad ?$
   (c) What are the tests for $f > 0$, to ensure that A is positive definite?
   (d) Are the matrices $\begin{bmatrix} 1 & 1+i \\ 1-i & 2 \end{bmatrix}$ and $\begin{bmatrix} 3 & 4+i \\ 4-i & 6 \end{bmatrix}$ positive definite?

6.1.12 Decide whether $F = x^2 y^2 - 2x - 2y$ has a minimum at the point $x = y = 1$ (after showing that the first derivatives are zero at that point).

6.1.13 Under what conditions on a, b, c, is $ax^2 + 2bxy + cy^2 \geq x^2 + y^2$ for all x, y?

6.2 ■ TESTS FOR POSITIVE DEFINITENESS

Which symmetric matrices have the property that $x^T A x > 0$ for all nonzero vectors x? There are four or five different ways to answer this question, and we hope to find all of them. The previous section began with some hints about the signs of eigenvalues, but that discussion was left hanging in midair. Instead, the question of eigenvalues gave place to a pair of conditions on a matrix:

$$\text{if} \quad A = \begin{bmatrix} a & b \\ b & c \end{bmatrix}, \quad \text{we need} \quad a > 0, \quad ac - b^2 > 0.$$

Our goal is to generalize those conditions to a matrix of order n, and to find the connection with the signs of the eigenvalues. In the 2 by 2 case, at least, the conditions mean that **both eigenvalues are positive**. Their product is the determinant $ac - b^2 > 0$, so the eigenvalues are either both positive or both negative. They must be positive because their sum is the trace $a + c > 0$.

It is remarkable how closely these two approaches—one direct and computational, the other more concerned with the intrinsic properties of the matrix (its eigenvalues)—reflect the two parts of this book. In fact, looking more closely at the computational test, it is even possible to spot the appearance of the *pivots*. They turned up when we decomposed f into a sum of squares:

$$ax^2 + 2bxy + cy^2 = a\left(x + \frac{b}{a} y\right)^2 + \frac{ac - b^2}{a} y^2. \tag{1}$$

The coefficients a and $(ac - b^2)/a$ are exactly the pivots for a 2 by 2 matrix. If this relationship continues to hold for larger matrices, it will allow a very simple test for positive definiteness: We check the pivots. And at the same time it will have a very natural interpretation: $x^T A x$ is positive definite if and only if it can be written as a sum of n independent squares.

One more preliminary remark. The two parts of this book were linked by the theory of determinants, and therefore we ask what part determinants play. **Certainly it is not enough to require that the determinant of A is positive.** That requirement is satisfied when $a = c = -1$ and $b = 0$, giving $A = -I$, a form that is actually negative definite. The important point is that the determinant test is applied not only to A itself, giving $ac - b^2 > 0$, but also to the 1 by 1 submatrix a in the upper left corner. The natural generalization will involve all n of the *upper left submatrices*

$$A_1 = [a_{11}], \quad A_2 = \begin{bmatrix} a_{11} & a_{12} \\ a_{21} & a_{22} \end{bmatrix}, \quad A_3 = \begin{bmatrix} a_{11} & a_{12} & a_{13} \\ a_{21} & a_{22} & a_{23} \\ a_{31} & a_{32} & a_{33} \end{bmatrix}, \quad \dots, \quad A_n = A.$$

Here is the main theorem, and a detailed proof:

6B   Each of the following tests is a necessary and sufficient condition for the real symmetric matrix A to be *positive definite*:

   (I) $x^T A x > 0$ for all nonzero vectors x.
   (II)   All the eigenvalues of A satisfy $\lambda_i > 0$.
   (III)   All the upper left submatrices A_k have positive determinants.
   (IV)   All the pivots (without row exchanges) satisfy $d_i > 0$.

Proof   Condition I defines a positive definite matrix, and our first step will be to show its equivalence to condition II. Therefore we suppose that I is satisfied, and deduce that each eigenvalue λ_i must be positive. The argument is simple. Suppose x_i is the corresponding unit eigenvector; then

$$Ax_i = \lambda_i x_i, \qquad \text{so} \qquad x_i^T A x_i = x_i^T \lambda_i x_i = \lambda_i,$$

because $x_i^T x_i = 1$. Since condition I holds for all x, it will hold in particular for the eigenvector x_i, and the quantity $x_i^T A x_i = \lambda_i$ must be positive. *A positive definite matrix has positive eigenvalues.*

   Now we go in the other direction, assuming that all $\lambda_i > 0$ and deducing $x^T A x > 0$. (This has to be proved for every vector x, not just the eigenvectors!) Since symmetric matrices have a full set of orthonormal eigenvectors (the spectral theorem), we can write any x as a combination $c_1 x_1 + \cdots + c_n x_n$. Then

$$Ax = c_1 A x_1 + \cdots + c_n A x_n = c_1 \lambda_1 x_1 + \cdots + c_n \lambda_n x_n.$$

Because of the orthogonality, and the normalization $x_i^T x_i = 1$,

$$\begin{aligned} x^T A x &= (c_1 x_1^T + \cdots + c_n x_n^T)(c_1 \lambda_1 x_1 + \cdots + c_n \lambda_n x_n) \\ &= c_1^2 \lambda_1 + \cdots + c_n^2 \lambda_n. \end{aligned} \qquad (2)$$

If every $\lambda_i > 0$, then (2) shows that $x^T A x > 0$. Thus condition II implies condition I.
   We turn to III and IV, whose equivalence to I will be proved in three steps:

If I *holds, so does* III: First, the determinant of any matrix is the product of its eigenvalues. And if I holds, we already know that these eigenvalues are positive:

$$\det A = \lambda_1 \lambda_2 \ldots \lambda_n > 0.$$

To prove the same result for all the submatrices A_k, we check that if A is positive definite, so is every A_k. The trick is to look at the vectors whose last $n - k$ components are zero:

$$x^T A x = \begin{bmatrix} x_k^T & 0 \end{bmatrix} \begin{bmatrix} A_k & * \\ * & * \end{bmatrix} \begin{bmatrix} x_k \\ 0 \end{bmatrix} = x_k^T A_k x_k.$$

If $x^TAx > 0$ for all nonzero x, then in particular $x_k^T A_k x_k > 0$ for all nonzero x_k. Thus condition I holds for A_k, and the submatrix permits the same argument that worked for A itself. Its eigenvalues (which are not the same λ_i!) must be positive, and its determinant is their product, so the upper left determinants are positive.

If III *holds, so does* IV: This is easy to prove because there is a direct relation between the numbers det A_k and the pivots. According to Section 4.4, the kth pivot d_k is exactly the ratio of det A_k to det A_{k-1}. Therefore if the determinants are all positive, so are the pivots—and no row exchanges are needed for positive definite matrices.

If IV *holds, so does* I: We are given that the pivots are positive, and must deduce that $x^TAx > 0$. This is what we did in the 2 by 2 case, by completing the square. The pivots were the numbers outside the squares. To see how that happens for symmetric matrices of any size, we go back to the source of the pivots—Gaussian elimination and $A = LDU$. Here is the essential fact: *In Gaussian elimination of a symmetric matrix, the upper triangular U is the transpose of the lower triangular L. Therefore $A = LDU$ becomes $A = LDL^T$.*

EXAMPLE

$$A = \begin{bmatrix} 2 & -1 & 0 \\ -1 & 2 & -1 \\ 0 & -1 & 2 \end{bmatrix} = \begin{bmatrix} 1 & 0 & 0 \\ -\frac{1}{2} & 1 & 0 \\ 0 & -\frac{2}{3} & 1 \end{bmatrix} \begin{bmatrix} 2 & & \\ & \frac{3}{2} & \\ & & \frac{4}{3} \end{bmatrix} \begin{bmatrix} 1 & -\frac{1}{2} & 0 \\ 0 & 1 & -\frac{2}{3} \\ 0 & 0 & 1 \end{bmatrix} = LDL^T.$$

Multiplying on the left by x^T and on the right by x, we get a sum of squares in which the pivots 2, $\frac{3}{2}$, and $\frac{4}{3}$ are the coefficients:

$$x^TAx = \begin{bmatrix} u & v & w \end{bmatrix} \begin{bmatrix} 1 & 0 & 0 \\ -\frac{1}{2} & 1 & 0 \\ 0 & -\frac{2}{3} & 1 \end{bmatrix} \begin{bmatrix} 2 & & \\ & \frac{3}{2} & \\ & & \frac{4}{3} \end{bmatrix} \begin{bmatrix} 1 & -\frac{1}{2} & 0 \\ 0 & 1 & -\frac{2}{3} \\ 0 & 0 & 1 \end{bmatrix} \begin{bmatrix} u \\ v \\ w \end{bmatrix}$$

$$= 2(u - \tfrac{1}{2}v)^2 + \tfrac{3}{2}(v - \tfrac{2}{3}w)^2 + \tfrac{4}{3}(w)^2.$$

Those positive pivots, multiplying perfect squares, make x^TAx positive. Thus condition IV implies condition I, completing the proof. The theorem would be exactly the same in the complex case, for Hermitian matrices $A = A^H$.

It is beautiful that two basic constructions—elimination and completing the square—are actually the same. The first step of elimination removes x_1 from all later equations. Similarly, the first square accounts for all terms in x^TAx involving x_1. Algebraically, the sum of squares (which gives the proof that $x^TAx > 0$) is

$$x^TAx = (x^TL)(D)(L^Tx) = d_1(L^Tx)_1^2 + d_2(L^Tx)_2^2 + \cdots + d_n(L^Tx)_n^2. \tag{3}$$

The pivots d_i are outside. *The multipliers l_{ij} are inside!* You can see the numbers $-\frac{1}{2}$ and $-\frac{2}{3}$ (and all the columns of L) inside the three squares in the example.

Remark It would be wrong to leave the impression in condition III that the upper submatrices A_k are extremely special. We could equally well test the determinants of the lower right submatrices. Or we could use any chain of principal submatrices, starting with some diagonal entry a_{ii} as the first submatrix, and adding a new row and column pair each time. In particular, a *necessary condition* for positive definiteness is that *every diagonal entry a_{ii} must be positive*. It is just the coefficient of x_i^2. As we know from the examples, however, it is far from sufficient to look only at the diagonal entries.

The pivots d_i are not to be confused with the eigenvalues. For a typical positive definite matrix, they are two completely different sets of positive numbers. In our 3 by 3 example, probably the determinant test is the easiest:

$$\det A_1 = 2, \qquad \det A_2 = 3, \qquad \det A_3 = \det A = 4.$$

For a large matrix, it is simpler to watch the pivots in elimination, which are the ratios $d_1 = 2, d_2 = \frac{3}{2}, d_3 = \frac{4}{3}$. Ordinarily the eigenvalue test is the longest, but for this example we know they are all positive,

$$\lambda_1 = 2 - \sqrt{2}, \qquad \lambda_2 = 2, \qquad \lambda_3 = 2 + \sqrt{2}.$$

Even though it is the hardest to apply to a single matrix A, this is actually the most useful test for theoretical purposes. ***Each test by itself is enough.***

Positive Definite Matrices and Least Squares

I hope you will allow one more test for positive definiteness. It is already very close, and it gives a better understanding of the old tests—and the connections to the rest of the book. We have connected positive definite matrices to pivots (Chapter 1), determinants (Chapter 4), and eigenvalues (Chapter 5). Now we see them in the least squares problems of Chapter 3, coming from the rectangular matrices of Chapter 2.

The rectangular matrix will be R and the least squares problem will be $Rx = b$. It has m equations, with $m \geq n$ (square systems are included). We need the letter A for the symmetric matrix $R^T R$ which comes in the normal equations: *the least squares choice $\bar{x}$ is the solution of $R^T R \bar{x} = R^T b$.* That matrix $R^T R$ is not only symmetric but positive definite, as we now show—provided the n columns of R are linearly independent:

6C A is symmetric positive definite if and only if it satisfies condition

(V) There is a matrix R with independent columns such that $A = R^T R$.

To see that $R^T R$ is positive definite, we recognize $x^T R^T R x$ as the squared length $\|Rx\|^2$. This cannot be negative. Also it cannot be zero (unless $x = 0$), because R has independent columns: If x is nonzero then Rx is nonzero. Thus $x^T A x = \|Rx\|^2$ is positive and $A = R^T R$ is positive definite.

It remains to show that when A satisfies conditions I—IV, it also satisfies V. We have to find an R for which $A = R^{T}R$, and we have almost done it twice already:

(i)   In the last step of the main theorem, Gaussian elimination factored A into LDL^{T}. The pivots are positive, and if $\sqrt{D}$ has their square roots along the diagonal, then $A = L\sqrt{D}\sqrt{D}L^{T}$.† One choice for R is the upper triangular matrix $\sqrt{D}L^{T}$.

(ii)   Another approach, which yields a different R, is to use II instead of IV—positive eigenvalues instead of positive pivots. The eigenvectors go into an orthogonal Q:

$$A = Q\Lambda Q^{T} = (Q\sqrt{\Lambda})(\sqrt{\Lambda}Q^{T}) = R^{T}R. \tag{4}$$

A third possibility is $R = Q\sqrt{\Lambda}Q^{T}$, the **symmetric positive definite square root** of A.

Those three choices of R were all square matrices. We did not need the freedom we were allowed, to let R be rectangular. But there are many other choices, square or rectangular, and we can see why. If you multiply one choice by *any matrix with orthonormal columns* (call it Q), the product satisfies $(QR)^{T}(QR) = R^{T}Q^{T}QR = R^{T}IR = A$. Therefore QR is another choice. The main point is that least squares problems lead to positive definite matrices—and in the applications that is where they come from.††

Remark   My class often asks about *unsymmetric* positive definite matrices. I never use that term. One possible definition is that the symmetric part $\frac{1}{2}(A + A^{T})$ should be positive definite. (The complex case would require the Hermitian part $\frac{1}{2}(A + A^{H})$ to be positive definite.) That is sufficient to guarantee that the real parts of the eigenvalues are positive, but it is not necessary—as the following proof and example show:

Sufficient to make Re $\lambda > 0$: If $Ax = \lambda x$ then $x^{H}Ax = \lambda x^{H}x$ and $x^{H}A^{H}x = \bar{\lambda}x^{H}x$. Adding, the real part is (Re $\lambda)x^{H}x = \frac{1}{2}x^{H}(A + A^{H})x > 0$.

Not necessary: $A = \begin{bmatrix} 1 & 4 \\ 0 & 1 \end{bmatrix}$ has Re $\lambda > 0$ but $\frac{1}{2}(A + A^{T}) = \begin{bmatrix} 1 & 2 \\ 2 & 1 \end{bmatrix}$ is indefinite.

Ellipsoids in *n* Dimensions

Throughout this book, geometrical constructions have been behind the matrix algebra. In Chapter 1, a linear equation represented a plane. The system $Ax = b$ led to an intersection of planes, and to a perpendicular projection (least squares) when the planes did not meet. The determinant was the volume of a parallelepiped.

---

† This is the **Cholesky decomposition**, with the pivots split evenly between the upper and lower triangular pieces. It differs very little from the Gauss $A = LDL^{T}$, and I usually avoid computing those square roots.

†† Applications of positive definite matrices are included in Chapter 1 of my book, *Introduction to Applied Mathematics* (Wellesley-Cambridge Press).

Now, for a positive definite matrix A and for the quadratic function $f = x^T A x$, we finally get a figure that is curved. It is an *ellipse* in two dimensions, and an *ellipsoid* in n dimensions.

The equation to be considered is $x^T A x = 1$. If A is the identity matrix, this simplifies to $x_1^2 + x_2^2 + \cdots + x_n^2 = 1$. It is the equation of the "unit sphere." Notice that if A is multiplied by 4, so that $A = 4I$, the sphere gets smaller. The equation changes to $4x_1^2 + 4x_2^2 + \cdots + 4x_n^2 = 1$, and instead of points like $(1, 0, \ldots, 0)$ it goes through $(\frac{1}{2}, 0, \ldots, 0)$. The center of the ellipse is at the origin, because if a vector x satisfies $x^T A x = 1$ so does the opposite vector $-x$. We are dealing with pure quadratics, and the next step—the important step—is to go from the identity matrix to a *diagonal matrix*:

$$\text{for } A = \begin{bmatrix} 4 & & \\ & 1 & \\ & & \frac{1}{9} \end{bmatrix} \text{ the equation is } x^T A x = 4x_1^2 + x_2^2 + \tfrac{1}{9}x_3^2 = 1.$$

Since the entries are unequal (and positive!) we go from a sphere to an ellipsoid.

You can see the key solutions to the equation. One is $x = (\frac{1}{2}, 0, 0)$, along the first axis. Another is $x = (0, 1, 0)$, on the x_2 axis. In the third direction is $x = (0, 0, 3)$, which is as far from the center as the ellipsoid reaches. The points $(-\frac{1}{2}, 0, 0)$, $(0, -1, 0)$, $(0, 0, -3)$ are at the other ends of the minor axis and middle axis and major axis. It is like a football or a rugby ball, but not quite—those shapes are closer to $x_1^2 + x_2^2 + \frac{1}{2}x_3^2 = 1$. The two equal coefficients make them circular in the x_1-x_2 plane, and the major axis reaches out to $(0, 0, \sqrt{2})$. The ball is upright and more or less ready for kickoff, except that half of it is below ground.

Now comes the final step, to allow nonzeros away from the diagonal of A.

EXAMPLE $A = \begin{bmatrix} 5 & 4 \\ 4 & 5 \end{bmatrix}$ and $x^T A x = 5u^2 + 8uv + 5v^2 = 1$. That is an ellipse in the u-v plane. It is centered at the origin, but the axes are not so clear. The off-diagonal 4's leave the matrix positive definite, but they turn the ellipse—its axes no longer line up with the coordinate axes (Fig. 6.2). Instead we will show that *the axes of the ellipse point toward the eigenvectors of* A. Because the matrix is symmetric, its eigenvectors are orthogonal. Therefore the *major* axis of the ellipse—corresponding to the *smallest* eigenvalue of A—is perpendicular to the minor axis that corresponds to the largest eigenvalue.

To identify the ellipse we compute the eigenvalues $\lambda_1 = 1$ and $\lambda_2 = 9$. The eigenvectors, normalized to unit length, are $(1/\sqrt{2}, -1/\sqrt{2})$ and $(1/\sqrt{2}, 1/\sqrt{2})$. Those are at $45°$ angles with the u-v axes, but they are lined up with the axes of the ellipse. The way to see the ellipse is to *rewrite the equation*:

$$5u^2 + 8uv + v^2 = \left(\frac{u}{\sqrt{2}} - \frac{v}{\sqrt{2}} \right)^2 + 9 \left(\frac{u}{\sqrt{2}} + \frac{v}{\sqrt{2}} \right)^2 = 1. \tag{5}$$

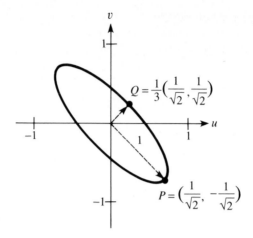

Fig. 6.2. The ellipse $5u^2 + 8uv + 5v^2 = 1$ and its principal axes.

The eigenvalues 1 and 9 are outside the squares. The eigenvectors are inside.† The first square equals one at $(1/\sqrt{2}, -1/\sqrt{2})$, at the end of the major axis. The second square equals one at $(1/\sqrt{2}, 1/\sqrt{2})$, and we need the factor $(\frac{1}{3})^2$ to cancel the 9. The minor axis is one third as long.

Any other ellipsoid $x^{\mathrm{T}}Ax = 1$ can be simplified in the same way, as soon as we understand what happened in equation (5). As always, *the key step was to diagonalize A.* Geometrically, we straightened the picture by rotating the axes. Algebraically, the eigenvectors of A went into an orthogonal matrix Q, and $A = Q\Lambda Q^{\mathrm{T}}$. This was one of the main ideas of Chapter 5, and here the change to $y = Q^{\mathrm{T}}x$ produces a sum of squares:

$$x^{\mathrm{T}}Ax = (x^{\mathrm{T}}Q)\Lambda(Q^{\mathrm{T}}x) = y^{\mathrm{T}}\Lambda y = \lambda_1 y_1^2 + \cdots + \lambda_n y_n^2. \tag{6}$$

The equation is now $\lambda_1 y_1^2 + \cdots + \lambda_n y_n^2 = 1$, and $y_1 = (Q^{\mathrm{T}}x)_1$ is the component of x along the first eigenvector. How large can that component be? At most it is $y_1 = 1/\sqrt{\lambda_1}$, when all the other squares are zero. Therefore *the major axis,* along the eigenvector corresponding to the smallest eigenvalue, *reaches a distance $1/\sqrt{\lambda_1}$ from the center.*

The other principal axes are along the other eigenvectors. Their lengths are $1/\sqrt{\lambda_2}, \ldots, 1/\sqrt{\lambda_n}$. Notice that the λ's must be positive—*the matrix must be positive definite*—or these square roots are in trouble. The equation $y_1^2 - 9y_2^2 = 1$ describes a hyperbola and not an ellipse.

The change from x to $y = Q^{\mathrm{T}}x$ has rotated the axes of the space, to match the axes of the ellipsoid. In the y variables we can see that it *is* an ellipsoid, because the

---

† This is different from completing the square to $5(u + \frac{4}{5}v)^2 + \frac{9}{5}v^2$, with the *pivots* outside.

equation becomes so manageable:

6D Suppose A is positive definite: $A = Q\Lambda Q^T$ with $\lambda_i > 0$. Then the rotation $y = Q^T x$ simplifies $x^T A x = 1$ to

$$x^T Q\Lambda Q^T x = 1 \quad \text{or} \quad y^T \Lambda y = 1 \quad \text{or} \quad \lambda_1 y_1^2 + \cdots + \lambda_n y_n^2 = 1.$$

This is the equation of an ellipsoid. Its axes have lengths $1/\sqrt{\lambda_1}, \ldots, 1/\sqrt{\lambda_n}$ from the center, and in the original x-space they point along the eigenvectors.

EXERCISES

6.2.1 For what range of numbers a and b are the matrices A and B positive definite?

$$A = \begin{bmatrix} a & 1 & 1 \\ 1 & a & 1 \\ 1 & 1 & a \end{bmatrix} \qquad B = \begin{bmatrix} 2 & 2 & 4 \\ 2 & b & 8 \\ 4 & 8 & 7 \end{bmatrix}$$

6.2.2 Decide for or against the positive definiteness of

$$A = \begin{bmatrix} 2 & -1 & -1 \\ -1 & 2 & -1 \\ -1 & -1 & 2 \end{bmatrix}, \qquad B = \begin{bmatrix} 2 & -1 & -1 \\ -1 & 2 & 1 \\ -1 & 1 & 2 \end{bmatrix}, \qquad C = \begin{bmatrix} 0 & 1 & 2 \\ 1 & 0 & 1 \\ 2 & 1 & 0 \end{bmatrix}^2.$$

6.2.3 Construct an *indefinite matrix* with its largest entries on the main diagonal:

$$A = \begin{bmatrix} 1 & b & -b \\ b & 1 & b \\ -b & b & 1 \end{bmatrix} \text{ with } |b| < 1 \text{ can have det } A < 0.$$

6.2.4 Show from the eigenvalues that if A is positive definite, so is A^2 and so is A^{-1}.

6.2.5 Show that if A and B are positive definite, so is $A + B$. Which of the conditions I–IV is useful this time?

6.2.6 From the pivots and eigenvalues and eigenvectors of $A = \begin{bmatrix} 5 & 4 \\ 4 & 5 \end{bmatrix}$, write A as $R^T R$ in all three ways proposed in the proof of 6C: $(L\sqrt{D})(\sqrt{D}L^T)$, $(Q\sqrt{\Lambda})(\sqrt{\Lambda}Q^T)$, and $(Q\sqrt{\Lambda}Q^T)(Q\sqrt{\Lambda}Q^T)$.

6.2.7 If $A = Q\Lambda Q^T$ is symmetric positive definite, then $R = Q\sqrt{\Lambda}Q^T$ is its *symmetric positive definite square root*. Why does R have real eigenvalues? Compute R and verify $R^2 = A$ for

$$A = \begin{bmatrix} 2 & 1 \\ 1 & 2 \end{bmatrix} \quad \text{and} \quad A = \begin{bmatrix} 10 & -6 \\ -6 & 10 \end{bmatrix}.$$

6.2.8   If A is symmetric positive definite and C is nonsingular, prove that $B = C^T A C$ is also symmetric positive definite.

6.2.9   If $A = R^T R$ prove the generalized Schwarz inequality $|x^T A y|^2 \le (x^T A x)(y^T A y)$.

6.2.10 The ellipse $u^2 + 4v^2 = 1$ corresponds to $A = \begin{bmatrix} 1 & 0 \\ 0 & 4 \end{bmatrix}$. Write down the eigenvalues and eigenvectors, and sketch the ellipse.

6.2.11 Reduce the equation $3u^2 - 2\sqrt{2}uv + 2v^2 = 1$ to a sum of squares by finding the eigenvalues of the corresponding A, and sketch the ellipse.

6.2.12 In three dimensions, $\lambda_1 y_1^2 + \lambda_2 y_2^2 + \lambda_3 y_3^2 = 1$ represents an ellipsoid when all $\lambda_i > 0$. Describe all the different kinds of surfaces that appear in the positive semidefinite case when one or more of the eigenvalues is zero.

6.2.13 Write down the five conditions for a 3 by 3 matrix to be *negative definite* ($-A$ is positive definite) with special attention to condition III: How is $\det(-A)$ related to $\det A$?

6.2.14 If a diagonal entry is zero show that A cannot be negative definite.

6.2.15 Decide whether the following matrices are positive definite, negative definite, semidefinite, or indefinite:

$$A = \begin{bmatrix} 1 & 2 & 3 \\ 2 & 5 & 4 \\ 3 & 4 & 9 \end{bmatrix}, \quad B = \begin{bmatrix} 1 & 2 & 0 & 0 \\ 2 & 6 & -2 & 0 \\ 0 & -2 & 5 & -2 \\ 0 & 0 & -2 & 3 \end{bmatrix}, \quad C = -B, \quad D = A^{-1}.$$

Is there a real solution to $-x^2 - 5y^2 - 9z^2 - 4xy - 6xz - 8yz = 1$?

6.2.16 Suppose A is symmetric positive definite and Q is orthogonal:

      T F $Q^T A Q$ is a diagonal matrix

      T F $Q^T A Q$ is symmetric positive definite

      T F $Q^T A Q$ has the same eigenvalues as A

      T F e^{-A} is symmetric positive definite

6.2.17 If A is positive definite and a_{11} is increased, prove from cofactors that the determinant is increased. Show by example that this can fail if A is indefinite.

6.2.18 From $A = R^T R$ show for positive definite matrices that $\det A \le a_{11} a_{22} \cdots a_{nn}$. (*Hint:* The length squared of column j of R is a_{jj}. Use determinant = volume.)

6.2.19 (Lyapunov test for stability of M) Suppose that $AM + M^H A = -I$ with positive definite A. If $Mx = \lambda x$ show that Re $\lambda < 0$. (*Hint:* Multiply the first equation by x^H and x.)

SEMIDEFINITE AND INDEFINITE MATRICES; $Ax = \lambda Mx$ ■ 6.3

The goal of this section, now that the tests for a positive definite matrix are fully established, is to cast our net a little wider. We have three problems in mind:

(1)   tests for a positive semidefinite matrix;
(2)   the connection between eigenvalues and pivots of a symmetric matrix;
(3)   the generalized eigenvalue problem $Ax = \lambda Mx$.

The first will be very quick; all the work has been done. The tests for semidefiniteness will relax the strict inequalities $x^T A x > 0$ and $\lambda > 0$ and $d > 0$ and $\det > 0$, to allow zeros to appear. Then indefinite matrices allow also negative eigenvalues and pivots, and lead to the main result of this section: **The signs of the eigenvalues match the signs of the pivots**. That is the "*law of inertia.*" Finally we go to $Ax = \lambda Mx$, which arises constantly in engineering analysis. As long as M is symmetric positive definite, this generalized problem is parallel to the familiar $Ax = \lambda x$—except that the eigenvectors are orthogonal in a new way. In many applications M is a **mass matrix**. We study a typical example in this section, and present the *finite element method* in Section 6.5.

For semidefinite matrices, the main point is to see the analogies with the positive definite case.

6E   Each of the following tests is a necessary and sufficient condition for A to be *positive semidefinite*:

(I') $x^T A x \geq 0$ for all vectors x (this is the definition).
(II')   All the eigenvalues of A satisfy $\lambda_i \geq 0$.
(III')   All the principal submatrices have nonnegative determinants.
(IV')   No pivots are negative.
(V')   There is a matrix R, possibly with dependent columns, such that $A = R^T R$.

If A has rank r, then $f = x^T A x$ is a sum of r perfect squares.

The connection between $x^T A x \geq 0$ and $\lambda_i \geq 0$, which is the most important, is exactly as before: The diagonalization $A = Q \Lambda Q^T$ leads to

$$x^T A x = x^T Q \Lambda Q^T x = y^T \Lambda y = \lambda_1 y_1^2 + \cdots + \lambda_n y_n^2. \tag{1}$$

This is nonnegative when the λ_i are nonnegative. If A has rank r, there are r non-zero eigenvalues and r perfect squares.

As for the determinant, it is the product of the λ_i and is also nonnegative. And since the principal submatrices are also semidefinite, their eigenvalues and their determinants are also ≥ 0; we have deduced III'. (A *principal* submatrix is formed by throwing away columns and rows together—say column and row 1 and 4. This retains symmetry and also semidefiniteness: If $x^T A x \geq 0$ for all x, then this property still holds when the first and fourth components of x are zero.) The novelty is that III' applies to all the principal submatrices, and not only to those in the upper left corner. Otherwise, we could not distinguish between two matrices whose upper left determinants are all zero:

$$\begin{bmatrix} 0 & 0 \\ 0 & 1 \end{bmatrix} \text{ is positive semidefinite, and } \begin{bmatrix} 0 & 0 \\ 0 & -1 \end{bmatrix} \text{ is negative semidefinite.}$$

Those examples also show that row exchanges may be necessary. In the good case, the first r upper left determinants are nonzero—the r by r corner submatrix is positive definite—and we get r positive pivots without row exchanges. The last $n - r$ rows are zero as usual after elimination. In the not so good case we need row exchanges—and also the corresponding column exchanges to maintain symmetry. The matrix becomes PAP^T, with r positive pivots. In the worst case the main diagonal is full of zeros, and those zeros are still on the diagonal of PAP^T—so no pivots are available. But such a matrix cannot be semidefinite unless it is zero, and in that case we are finished.

The same constructions as before lead to R. Any of the choices $A = (L\sqrt{D})(\sqrt{D}L^T)$ and $A = (Q\sqrt{\Lambda})(\sqrt{\Lambda}Q^T)$ and even $A = (Q\sqrt{\Lambda}Q^T)(Q\sqrt{\Lambda}Q^T)$ produce $A = R^T R$. (In the third choice R is symmetric, the *semidefinite square root* of A.) Finally we get back from V' to I', because every matrix of the form $R^T R$ is positive semidefinite (at least). The quadratic $x^T R^T R x$ is equal to $\|Rx\|^2$ and cannot be negative—which completes the circle.

EXAMPLE

$$A = \begin{bmatrix} 2 & -1 & -1 \\ -1 & 2 & -1 \\ -1 & -1 & 2 \end{bmatrix}$$

is semidefinite, by the following tests:

(I') $x^T A x = (x_1 - x_2)^2 + (x_1 - x_3)^2 + (x_2 - x_3)^2 \geq 0$.
(II')   The eigenvalues are $\lambda_1 = 0$, $\lambda_2 = \lambda_3 = 3$.
(III')   The principal submatrices have determinant 2 if they are 1 by 1; determinant 3 if they are 2 by 2; and det $A = 0$.
(IV')

$$\begin{bmatrix} 2 & -1 & -1 \\ -1 & 2 & -1 \\ -1 & -1 & 2 \end{bmatrix} \rightarrow \begin{bmatrix} 2 & 0 & 0 \\ 0 & \frac{3}{2} & -\frac{3}{2} \\ 0 & -\frac{3}{2} & \frac{3}{2} \end{bmatrix} \rightarrow \begin{bmatrix} 2 & 0 & 0 \\ 0 & \frac{3}{2} & 0 \\ 0 & 0 & 0 \end{bmatrix}.$$

Remark   The conditions for semidefiniteness could also be deduced from the original conditions I–V by the following trick: Add a small multiple of the identity, giving a positive definite matrix $A + \epsilon I$. Then let ϵ approach zero. Since the determinants and eigenvalues depend continuously on ϵ, they will be positive until the last moment; at $\epsilon = 0$ they must still be nonnegative.

Congruence Transformations and the Law of Inertia

Earlier in this book, for elimination and for eigenvalues, we emphasized the elementary operations that make the matrix simpler. In each case, the essential thing was to know which properties of the matrix stayed unchanged. When a multiple of one row was subtracted from another, the list of "invariants" was pretty long: The row space and nullspace and rank and determinant all remain the same. In the case of eigenvalues, the basic operation was a similarity transformation $A \to S^{-1}AS$ (or $A \to M^{-1}AM$). It is the eigenvalues themselves that are unchanged (and also the Jordan form). Now we ask the same question for symmetric matrices: *What are the elementary operations and their invariants for $x^T A x$?*

The basic operation on a quadratic form is to change variables. A new vector y is related to x by some nonsingular matrix, $x = Cy$, and the quadratic form becomes $y^T C^T A C y$. Therefore the matrix operation which is fundamental to the theory of quadratic forms is a *congruence transformation*:

$$A \to C^T A C \qquad \text{for some nonsingular } C. \qquad (2)$$

The symmetry of A is preserved, since $C^T A C$ remains symmetric. The real question is, What other properties of the matrix remain unchanged by a congruence transformation? The answer is given by Sylvester's *law of inertia*.

6F The matrix $C^T A C$ has the same number of positive eigenvalues as A, the same number of negative eigenvalues, and the same number of zero eigenvalues.

In other words, the *signs* of the eigenvalues (and not the eigenvalues themselves) are preserved by a congruence transformation. In the proof, we will suppose for convenience that A is nonsingular. Then $C^T A C$ is also nonsingular, and there are no zero eigenvalues to worry about. (Otherwise we can work with the nonsingular $A + \epsilon I$ and $A - \epsilon I$, and at the end let $\epsilon \to 0$.)

We want to borrow a trick from topology, or more exactly from homotopy theory. Suppose C is linked to an orthogonal matrix Q by a continuous chain of matrices $C(t)$, none of which are singular: At $t = 0$ and $t = 1$, $C(0) = C$ and $C(1) = Q$. Then the eigenvalues of $C(t)^T A C(t)$ will change gradually, as t goes from 0 to 1, from the eigenvalues of $C^T A C$ to the eigenvalues of $Q^T A Q$. Because $C(t)$ is never singular, *none of these eigenvalues can touch zero* (not to mention cross over it!). Therefore the number of eigenvalues to the right of zero, and the number to the left, is the same for $C^T A C$ as for $Q^T A Q$. And this number is also the same for A— which has exactly the same eigenvalues as the similar matrix $Q^{-1} A Q = Q^T A Q$.†
That is the proof.

---

† It is here that we needed Q to be orthogonal. One good choice for Q is to apply Gram-Schmidt to the columns of C. Then $C = QR$, and the chain of matrices is $C(t) = tQ + (1 - t)QR$. The family $C(t)$ goes slowly through Gram-Schmidt, from QR to Q. It is invertible, because Q is invertible and the triangular factor $tI + (1 - t)R$ has positive diagonal.

EXAMPLE 1   If $A = I$ then $C^T A C = C^T C$, and this matrix is positive definite; take $C = R$ in condition V. Both the identity and $C^T C$ have n positive eigenvalues, confirming the law of inertia.

EXAMPLE 2   If $A = \begin{bmatrix} 1 & 0 \\ 0 & -1 \end{bmatrix}$, then $C^T A C$ has a negative determinant:

$$\det C^T A C = (\det C^T)(\det A)(\det C) = -(\det C)^2 < 0.$$

Since this determinant is the product of the eigenvalues, $C^T A C$ must have one positive and one negative eigenvalue, as A has, and again the law is obeyed.

EXAMPLE 3   This application is the important one:

> **6G**   For any symmetric matrix A, *the signs of the pivots agree with the signs of the eigenvalues*. The eigenvalue matrix Λ and the pivot matrix D have the same number of positive entries, negative entries, and zero entries.

We will assume that A allows the usual factorization $A = LDU$ (without row exchanges). Because A is symmetric, U is the transpose of L and $A = LDL^T$. To this we apply the law of inertia: A has the same number of positive eigenvalues as D. But the eigenvalues of D are just its diagonal entries (the pivots). Thus the number of positive pivots matches the number of positive eigenvalues of A.

That is both beautiful and practical. It is beautiful because it brings together (for symmetric matrices) two parts of this book that we have previously kept separate: *pivots* and *eigenvalues*. It is also practical, because the pivots are easy to find—and now they can locate the eigenvalues. Consider the matrices

$$A = \begin{bmatrix} 3 & 3 & 0 \\ 3 & 10 & 7 \\ 0 & 7 & 8 \end{bmatrix} \quad \text{and} \quad A - 2I = \begin{bmatrix} 1 & 3 & 0 \\ 3 & 8 & 7 \\ 0 & 7 & 6 \end{bmatrix}.$$

A has positive pivots; $A - 2I$ has a negative pivot. Therefore A has positive eigenvalues, and *one of them is smaller than 2*—because subtracting 2 dropped it below zero. The next step looks at $A - I$, to see if the eigenvalue is smaller than 1. (It is, because $A - I$ has a negative pivot.) That interval of uncertainty is cut in half at every step, by checking the signs.

This was almost the first practical method of computing eigenvalues. It was dominant about 1960, after one important improvement—to make A tridiagonal first. Then the pivots are computed in $2n$ steps instead of $\frac{1}{6}n^3$. Elimination becomes fast, and the search for eigenvalues becomes simple. It is known as *Givens' method*. The current favorite is the QR method in Chapter 7.

The Generalized Eigenvalue Problem

I am not sure about economics, but physics and engineering and statistics are usually kind enough to produce symmetric matrices in their eigenvalue problems. They may, however, produce *two matrices rather than one*. $Ax = \lambda x$ may be replaced by $Ax = \lambda Mx$, and for an example we look at the motion of two unequal masses.

Newton's law is $F_1 = m_1 a_1$ for the first mass and $F_2 = m_2 a_2$ for the second. We can describe the physics in a sentence: The masses are displaced by v and w, as in Fig. 6.3, and the springs pull them back with strength $F_1 = -2v + w$ and $F_2 = v - 2w$. The new and significant point comes on the left side of the equations, where the two masses appear:

$$m_1 \frac{d^2v}{dt^2} = -2v + w$$
$$m_2 \frac{d^2w}{dt^2} = v - 2w$$
$$\text{or} \qquad \begin{bmatrix} m_1 & 0 \\ 0 & m_2 \end{bmatrix} \frac{d^2u}{dt^2} = \begin{bmatrix} -2 & 1 \\ 1 & -2 \end{bmatrix} u.$$

When the masses were equal, $m_1 = m_2 = 1$, this was the old system $u'' = Au$. Now it is $Mu'' = Au$, with a "*mass matrix*" M. The eigenvalue problem arises when we look for exponential solutions $e^{i\omega t}x$:

$$Mu'' = Au \qquad \text{becomes} \qquad M(i\omega)^2 e^{i\omega t}x = Ae^{i\omega t}x. \qquad (3)$$

Cancelling $e^{i\omega t}$, and writing λ for $(i\omega)^2$, this is

$$Ax = \lambda Mx, \qquad \text{or} \qquad \begin{bmatrix} -2 & 1 \\ 1 & -2 \end{bmatrix} x = \lambda \begin{bmatrix} m_1 & 0 \\ 0 & m_2 \end{bmatrix} x. \qquad (4)$$

There is a solution only if the combination $A - \lambda M$ is singular, so λ must be a root of the polynomial equation $\det(A - \lambda M) = 0$. The special choice $M = I$ brings back the usual equation $\det(A - \lambda I) = 0$. We work out an example with $m_1 = 1$ and $m_2 = 2$:

$$\det(A - \lambda M) = \det \begin{bmatrix} -2 - \lambda & 1 \\ 1 & -2 - 2\lambda \end{bmatrix} = 2\lambda^2 + 6\lambda + 3, \qquad \lambda = \frac{-3 \pm \sqrt{3}}{2}.$$

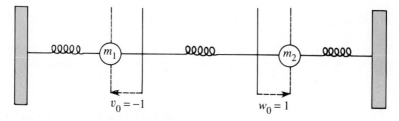

Fig. 6.3. An oscillating system with unequal masses.

Both eigenvalues are negative, and the two natural frequencies are $\omega_i = \sqrt{-\lambda_i}$. The eigenvectors x_i are calculated in the usual way:

$$(A - \lambda_1 M)x_1 = \begin{bmatrix} \dfrac{-1-\sqrt{3}}{2} & 1 \\ 1 & 1-\sqrt{3} \end{bmatrix} x_1 = 0, \qquad x_1 = \begin{bmatrix} \sqrt{3}-1 \\ 1 \end{bmatrix},$$

$$(A - \lambda_2 M)x_2 = \begin{bmatrix} \dfrac{-1+\sqrt{3}}{2} & 1 \\ 1 & 1+\sqrt{3} \end{bmatrix} x_2 = 0, \qquad x_2 = \begin{bmatrix} 1+\sqrt{3} \\ -1 \end{bmatrix}.$$

These eigenvectors give the normal modes of oscillation. At the lower frequency ω_1, the two masses oscillate together—except that the first mass only moves as far as $\sqrt{3} - 1 \approx .73$. In the faster mode, the components of x_2 have opposite signs and the masses move in opposite directions. This time the smaller mass goes much further.

The underlying theory is easier to explain if M is split into $R^T R$. (*M is assumed positive definite*, as in the example.) Then the substitution $y = Rx$ changes

$$Ax = \lambda Mx = \lambda R^T Rx \quad \text{into} \quad AR^{-1}y = \lambda R^T y.$$

Writing C for R^{-1}, and multiplying through by $(R^T)^{-1} = C^T$, this becomes a standard eigenvalue problem for the *single* matrix $C^T AC$:

$$C^T ACy = \lambda y. \tag{5}$$

The eigenvalues λ_j are the same as for the original $Ax = \lambda Mx$, and the eigenvectors are related by $y_j = Rx_j$.† The properties of the symmetric matrix $C^T AC$ lead directly to the corresponding properties of $Ax = \lambda Mx$:

1. The eigenvalues are real.
2. They have the same signs as the ordinary eigenvalues of A, by the law of inertia.
3. The eigenvectors y_j of $C^T AC$ can be chosen orthonormal, so the eigenvectors x_j are "M-orthonormal":

   $$x_i^T M x_j = x_i^T R^T R x_j = y_i^T y_j = 1 \text{ if } i = j, 0 \text{ if } i \neq j. \tag{6}$$

4. Similarly $x_i^T A x_j = \lambda_j x_i^T M x_j$ is either λ_j or zero. The matrices A and M are being *simultaneously diagonalized*. If S has the x_j in its columns, then $S^T AS = \Lambda$ and $S^T MS = I$. Notice that it is a *congruence* transformation, with S^T on the left, and not a similarity transformation with S^{-1}.

---

† A quicker way to produce a single matrix would have been $M^{-1}Ax = \lambda x$, but $M^{-1}A$ is not symmetric; we preferred to put half of M^{-1} on each side of A, giving a symmetric $C^T AC$.

Geometrically, this has a meaning which we do not understand very well. In the positive definite case, the two surfaces $x^T A x = 1$ and $x^T M x = 1$ are ellipsoids. Apparently $x = Sz$ gives a new choice of coordinates—not a pure rotation, because S is not an orthogonal matrix—such that these two ellipsoids become correctly aligned. They are

$$x^T A x = z^T S^T A S z = \lambda_1 z_1^2 + \cdots + \lambda_n z_n^2 = 1$$
$$x^T M x = z^T S^T M S z = z_1^2 + \cdots + z_n^2 = 1. \tag{7}$$

The second ellipsoid is a sphere! And the main point of the theory is easy to summarize: As long as M is positive definite, the generalized eigenvalue problem $Ax = \lambda Mx$ behaves exactly as $Ax = \lambda x$.

EXERCISES

6.3.1 For the semidefinite matrices

$$A = \begin{bmatrix} 2 & -1 & -1 \\ -1 & 2 & -1 \\ -1 & -1 & 2 \end{bmatrix} \text{(rank 2)} \quad \text{and} \quad B = \begin{bmatrix} 1 & 1 & 1 \\ 1 & 1 & 1 \\ 1 & 1 & 1 \end{bmatrix} \text{(rank 1)}$$

write $x^T A x$ as a sum of two squares and $x^T B x$ as one square.

6.3.2 Apply any three tests to each of the matrices

$$A = \begin{bmatrix} 1 & 1 & 1 \\ 1 & 1 & 1 \\ 1 & 1 & 0 \end{bmatrix} \quad \text{and} \quad B = \begin{bmatrix} 2 & 1 & 2 \\ 1 & 1 & 1 \\ 2 & 1 & 2 \end{bmatrix},$$

to decide whether they are positive definite, positive semidefinite, or indefinite.

6.3.3 For $C = \begin{bmatrix} 2 & 0 \\ 0 & -1 \end{bmatrix}$ and $A = \begin{bmatrix} 1 & 1 \\ 1 & 1 \end{bmatrix}$ confirm that $C^T A C$ has eigenvalues of the same signs as A. Construct a chain of nonsingular matrices $C(t)$ linking C to an orthogonal Q. Why is it impossible to construct a nonsingular chain linking C to the identity matrix?

6.3.4 If the pivots are all greater than 1, are the eigenvalues all greater than 1? Test on the tridiagonal $-1, 2, -1$ matrices.

6.3.5 Use the pivots of $A - \frac{1}{2}I$ to decide whether A has an eigenvalue smaller than $\frac{1}{2}$:

$$A - \tfrac{1}{2}I = \begin{bmatrix} 2.5 & 3 & 0 \\ 3 & 9.5 & 7 \\ 0 & 7 & 7.5 \end{bmatrix}.$$

6.3.6 An algebraic proof of the *law of inertia* starts with the orthonormal eigenvectors $x_1, \ldots, x_p$ of A corresponding to eigenvalues $\lambda_i > 0$, and the orthonormal eigenvectors $y_1, \ldots, y_q$ of $C^T A C$ corresponding to eigenvalues $\mu_i < 0$.

(a) To prove that $x_1, \ldots, x_p, Cy_1, \ldots, Cy_q$ are independent, assume that some combination gives zero:

$$a_1 x_1 + \cdots + a_p x_p = b_1 Cy_1 + \cdots + b_q Cy_q \, (= z, \text{ say}).$$

Show that $z^T A z = \lambda_1 a_1^2 + \cdots + \lambda_p a_p^2 \geq 0$ and also $z^T A z = \mu_1 b_1^2 + \cdots + \mu_q b_q^2 \leq 0$.
(b) Deduce that the a's and b's are zero (proving linear independence). From that deduce $p + q \leq n$.
(c) The same argument for the $n - p$ negative λ's and the $n - q$ positive μ's gives $n - p + n - q \leq n$. (We again assume no zero eigenvalues—which are handled separately). Show that $p + q = n$, so the number p of positive λ's equals the number $n - q$ of positive μ's—which is the law of inertia.

6.3.7 If C is nonsingular show that A and $C^T A C$ have the same rank. (You may refer to Section 3.6.) Thus they have the same number of zero eigenvalues.

6.3.8 Find by experiment the number of positive, negative, and zero eigenvalues of

$$A = \begin{bmatrix} I & B \\ B^T & 0 \end{bmatrix}$$

when the block B (of order $\frac{1}{2}n$) is nonsingular.

6.3.9 Do A and $C^T A C$ always satisfy the law of inertia when C is not square?

6.3.10 In the worked example with $m_1 = 1$ and $m_2 = 2$, verify that the normal modes are M-orthogonal: $x_1^T M x_2 = 0$. If both masses are displaced a unit distance and then released, with $v_0 = -1$, $w_0 = 1$, find the coefficients a_i in the resulting motion $u = a_1 \cos \omega_1 t \, x_1 + a_2 \cos \omega_2 t \, x_2$. What is the maximum distance from equilibrium achievable by the first mass (when the two cosines are $+1$ and -1)?

6.3.11 Find the eigenvalues and eigenvectors of

$$\begin{bmatrix} 6 & -3 \\ -3 & 6 \end{bmatrix} x = \frac{\lambda}{18} \begin{bmatrix} 4 & 1 \\ 1 & 4 \end{bmatrix} x.$$

6.3.12 If the 2 by 2 symmetric matrices A and M are indefinite, then $Ax = \lambda M x$ might not have real eigenvalues. Construct an example.

MINIMUM PRINCIPLES AND THE RAYLEIGH QUOTIENT ■ 6.4

In this section, as the end of the book approaches, we shall escape for the first time from linear equations. The unknown x will not be given as the solution to $Ax = b$ or $Ax = \lambda x$. Instead, it will be determined by a minimum principle.

It is astonishing how many natural laws can be expressed as minimum principles. Just the fact that heavy liquids sink to the bottom is a consequence of minimizing their potential energy. And when you sit on a chair or lie on a bed, the springs adjust themselves so that the energy is minimized. Certainly there are more highbrow examples: A building is like a very complicated chair, carrying its own weight, and the fundamental principle of structural engineering is the minimization of total energy. In physics, there are "Lagrangians" and "action integrals"; a straw in a glass of water looks bent because light takes the path that reaches your eye as quickly as possible.†

We have to say immediately that these "energies" are nothing but *positive definite quadratic forms*. And the derivative of a quadratic is linear. Therefore minimization will lead us back to the familiar linear equations, when we set the first derivatives to zero. Our first goal in this section is **to find the minimum principle that is equivalent to** $Ax = b$, **and the one equivalent to** $Ax = \lambda x$. We will be doing in finite dimensions exactly what the calculus of variations does in a continuous problem, where the vanishing of the first derivatives gives a differential equation (Euler's equation). In every case, we are free to search for a solution to the linear equation or for a minimum to the quadratic—and in many problems, as the next section will illustrate, the latter possibility ought not to be ignored.

The first step is very straightforward: We want to find the "parabola" P whose minimum occurs at the point where $Ax = b$. If A is just a scalar, that is easy to do:

$$P(x) = \tfrac{1}{2}Ax^2 - bx, \qquad \text{and} \qquad \frac{dP}{dx} = Ax - b.$$

This will be a minimum only if A is positive. The parabola opens upward, and the vanishing of the first derivative gives $Ax = b$ (Fig. 6.4). In several dimensions this parabola turns into a paraboloid, but there is still the same formula for P; and to assure a minimum there is still a condition of positivity.

6H   If A is symmetric positive definite, then $P(x) = \tfrac{1}{2}x^{\mathrm{T}}Ax - x^{\mathrm{T}}b$ reaches its minimum at the point where $Ax = b$. At that point $P_{\min} = -\tfrac{1}{2}b^{\mathrm{T}}A^{-1}b$.

---

† I am convinced that plants and people also develop in accordance with minimum principles; perhaps civilization itself is based on a law of least action. The discovery of such laws is the fundamental step in passing from observations to explanations, and there must be some still to be found in the social sciences and in biology.

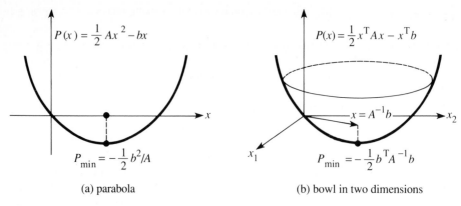

(a) parabola   (b) bowl in two dimensions

Fig. 6.4. The minimum of a quadratic function $P(x)$.

Proof Suppose x is the solution of $Ax = b$. For any vector y, we expand

$$P(y) - P(x) = \tfrac{1}{2}y^{\mathrm{T}}Ay - y^{\mathrm{T}}b - \tfrac{1}{2}x^{\mathrm{T}}Ax + x^{\mathrm{T}}b$$
$$= \tfrac{1}{2}y^{\mathrm{T}}Ay - y^{\mathrm{T}}Ax + \tfrac{1}{2}x^{\mathrm{T}}Ax$$
$$= \tfrac{1}{2}(y - x)^{\mathrm{T}}A(y - x). \tag{1}$$

Since A is positive definite, this quantity can never be negative—and it is zero only if $y - x = 0$. At all other points $P(y)$ is larger than $P(x)$, so the minimum occurs at x.

EXAMPLE   Minimize $P(x) = x_1^2 - x_1 x_2 + x_2^2 - b_1 x_1 - b_2 x_2$. The usual approach, by calculus, is to set the derivatives to zero:

$$\partial P/\partial x_1 = 2x_1 - x_2 - b_1 = 0$$
$$\partial P/\partial x_2 = -x_1 + 2x_2 - b_2 = 0.$$

Linear algebra writes the same result as $Ax = b$, first identifying P as $\tfrac{1}{2}x^{\mathrm{T}}\begin{bmatrix} 2 & -1 \\ -1 & 2 \end{bmatrix}x - x^{\mathrm{T}}b$. The minimum is at $x = A^{-1}b$, where

$$P_{\min} = \tfrac{1}{2}(A^{-1}b)^{\mathrm{T}}A(A^{-1}b) - (A^{-1}b)^{\mathrm{T}}b = -\tfrac{1}{2}b^{\mathrm{T}}A^{-1}b. \tag{2}$$

In applications, $\tfrac{1}{2}x^{\mathrm{T}}Ax$ is the internal energy and $-x^{\mathrm{T}}b$ is the external work. The physical system automatically goes to x, the equilibrium point where the total energy P is a minimum.

Our second goal was to find a minimization problem equivalent to $Ax = \lambda x$. That is not so easy. The function to minimize cannot be simply a quadratic, or its differentiation would lead to a linear equation—and the eigenvalue problem is fundamentally nonlinear. The trick is to look at a *ratio* of quadratics, and the one we need is known as ***Rayleigh's quotient***:

$$R(x) = \frac{x^{\mathrm{T}}Ax}{x^{\mathrm{T}}x}.$$

We go directly to the main theorem:

61 *Rayleigh's principle*: The quotient $R(x)$ is minimized by the first eigenvector $x = x_1$, and its minimum value is the smallest eigenvalue λ_1:

$$R(x_1) = \frac{x_1^{\mathrm{T}} A x_1}{x_1^{\mathrm{T}} x_1} = \frac{x_1^{\mathrm{T}} \lambda_1 x_1}{x_1^{\mathrm{T}} x_1} = \lambda_1.$$

Geometrically, imagine that we fix the numerator at 1, and make the denominator as large as possible. The numerator $x^{\mathrm{T}} A x = 1$ defines an ellipsoid, at least if A is positive definite. The denominator is $x^{\mathrm{T}} x = \|x\|^2$, so we are looking for the point on the ellipsoid farthest from the origin—the vector x of greatest length. From our earlier description of the ellipsoid, its major axis points along the first eigenvector.

Algebraically, this is easy to see (without any requirement of positive definiteness). We diagonalize A by an orthogonal matrix: $Q^{\mathrm{T}} A Q = \Lambda$. With $x = Qy$,

$$R(x) = \frac{(Qy)^{\mathrm{T}} A(Qy)}{(Qy)^{\mathrm{T}}(Qy)} = \frac{y^{\mathrm{T}} \Lambda y}{y^{\mathrm{T}} y} = \frac{\lambda_1 y_1^2 + \cdots + \lambda_n y_n^2}{y_1^2 + \cdots + y_n^2}. \tag{3}$$

The minimum of R certainly occurs at the point where $y_1 = 1$ and $y_2 = \cdots = y_n = 0$. At that point the ratio equals λ_1, and at any other point the ratio is larger:

$$\lambda_1(y_1^2 + y_2^2 + \cdots + y_n^2) \le (\lambda_1 y_1^2 + \lambda_2 y_2^2 + \cdots + \lambda_n y_n^2).$$

The Rayleigh quotient is *never below* λ_1 and *never above* λ_n. Its minimum is at the eigenvector x_1 and its maximum is at x_n. The intermediate eigenvectors are *saddle points*.

One important consequence: Each diagonal entry like a_{11} is between λ_1 and λ_n. The Rayleigh quotient equals a_{11}, when the trial vector is $x = (1, 0, \ldots, 0)$.

Minimax Principles for the Eigenvalues

The difficulty with saddle points is that we have no idea whether $R(x)$ is above or below them. That makes the intermediate eigenvalues $\lambda_2, \ldots, \lambda_{n-1}$ hard to find. In the applications, it is a minimum principle or a maximum principle that is really useful. Therefore we look for such a principle, intending that the minimum or maximum will be reached at the jth eigenvector x_j.[†] The key idea comes from the basic property of symmetric matrices: x_j is perpendicular to the other eigenvectors.

[†] This topic is more special, and it is the minimization of $P(x)$ and $R(x)$ on which the finite element method is based. There is no difficulty in going directly to Section 6.5.

6J   The minimum of $R(x)$, subject to any one constraint $x^T z = 0$, is never above λ_2. If z is the first eigenvector then this constrained minimum of R equals λ_2:

$$\lambda_2 \geq \min_{x^T z = 0} R(x) \quad \text{and} \quad \lambda_2 = \max_z \left[\min_{x^T z = 0} R(x) \right]. \tag{4}$$

This is a "maximin principle" for λ_2. It gives a way to estimate the second eigenvalue without knowing the first one.

EXAMPLE   Throw away the last row and column of any symmetric matrix:

$$A = \begin{bmatrix} 2 & -1 & 0 \\ -1 & 2 & -1 \\ 0 & -1 & 2 \end{bmatrix} \quad \text{becomes} \quad B = \begin{bmatrix} 2 & -1 \\ -1 & 2 \end{bmatrix}.$$

Then *the second eigenvalue of A is above the lowest eigenvalue of B.* (In this example those numbers are 2 and 1.) The lowest eigenvalue of A, which is $2 - \sqrt{2} \approx$ .6, is *below* the lowest eigenvalue of B.

Proof   The Rayleigh quotient for B agrees with the Rayleigh quotient for A whenever $x = (x_1, x_2, 0)$. These are the vectors perpendicular to $z = (0, 0, 1)$, and for them we have $x^T A x = x^T B x$. The minimum over these x's is the smallest eigenvalue $\lambda_1(B)$. It cannot be below the smallest eigenvalue of A, because that is the minimum over *all* x's. But $\lambda_1(B)$ will be below the second eigenvalue of A. Reason: Some combination of the first two eigenvectors of A is perpendicular to z. If we put that combination into the Rayleigh quotient, it gives a number below $\lambda_2(A)$ and above $\lambda_1(B)$. Therefore $\lambda_2(A) \geq \lambda_1(B)$.

The complete picture is an ***intertwining of eigenvalues***:

$$\lambda_1(A) \leq \lambda_1(B) \leq \lambda_2(A) \leq \lambda_2(B) \leq \cdots \leq \lambda_{n-1}(B) \leq \lambda_n(A). \tag{5}$$

Geometrically, this has a natural interpretation. Suppose an ellipsoid is cut by a plane through the origin. Then the cross section is again an ellipsoid, of one lower dimension. The major axis of this cross section cannot be longer than the major axis of the whole ellipsoid: $\lambda_1(B) \geq \lambda_1(A)$. But the major axis of the cross section is *at least as long as the second axis* of the original ellipsoid: $\lambda_1(B) \leq \lambda_2(A)$. Similarly the minor axis of the cross section is smaller than the original second axis, and larger than the original minor axis: $\lambda_2(A) \leq \lambda_2(B) \leq \lambda_3(A)$.

In one special case, when the cutting plane is perpendicular to the original major axis, we have $\lambda_1(B) = \lambda_2(A)$. In general, if we cut by the horizontal plane $x_3 = 0$, we expect inequality. The perpendicular to that cutting plane is $z = (0, 0, 1)$, and we are throwing away the last row and column of A.

We can see the same thing in mechanics. Given an oscillating system of springs and masses, suppose one mass is forced to stay at equilibrium. Then the lowest

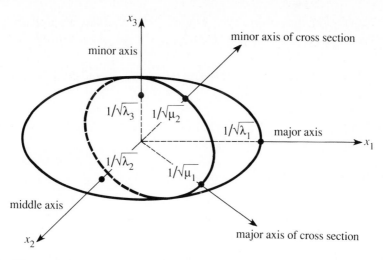

Fig. 6.5. The maximin and minimax principles: $\lambda_1 \leq \mu_1 \leq \lambda_2 \leq \mu_2 \leq \lambda_3$.

frequency is increased, but not above λ_2. The highest frequency is decreased, but not below λ_{n-1}.

It is time to turn these theorems around and arrive at a ***minimax principle***. That means we intend first to *maximize* the Rayleigh quotient, and then to identify the minimum possible value of this maximum.

There are several ways to proceed, depending on which eigenvalue is wanted. To find the largest eigenvalue λ_n, we just maximize $R(x)$ and we have it. But suppose we want to stay with λ_2. Since the Rayleigh quotient equals

$$\frac{\lambda_1 y_1^2 + \lambda_2 y_2^2 + \cdots + \lambda_n y_n^2}{y_1^2 + y_2^2 + \cdots + y_n^2},$$

the way to produce λ_2 as a maximum is to require that $y_3 = \cdots = y_n = 0$. These $n - 2$ constraints leave only a two-dimensional subspace, spanned by the first two eigenvectors. Over this particular two-dimensional subspace, the maximum of $R(x)$ is λ_2—but the eigenvectors x_1 and x_2 are part of the problem, and not known.

When this situation arose for the maximin principle, the key idea was to minimize with an arbitrary constraint $x^T z = 0$. Therefore we copy that idea, and *maximize $R(x)$ over an arbitrary two-dimensional subspace*. We have no way to know whether this subspace contains any eigenvectors; it is very unlikely. Nevertheless, there is still an inequality on the maximum. Previously the minimum with $x^T z = 0$ was below λ_2; now the maximum over the subspace will be above λ_2. If S_2 is any *two-dimensional subspace of* $\mathbf{R}^n$, *then*

$$\max_{x \text{ in } S_2} R(x) \geq \lambda_2. \qquad (6)$$

Proof   Some x in the subspace is orthogonal to the first eigenvector x_1, and for that particular x we know that $R(x) \geq \lambda_2$. The minimax principle follows:

6K   If we maximize $R(x)$ over all possible two-dimensional subspaces S_2, the minimum value for that maximum is λ_2:

$$\lambda_2 = \min_{S_2} \left[\max_{x \text{ in } S_2} R(x) \right]. \tag{7}$$

The quantity in brackets is never below λ_2, by (6), and for the particular subspace spanned by the first two eigenvectors, it equals λ_2.

We close this section with two remarks. I hope that, even without detailed proofs, your intuition says they are correct.

Remark 1   The minimax principle extends from two-dimensional subspaces to j-dimensional subspaces, and produces λ_j:

$$\lambda_j = \min_{S_j} \left[\max_{x \text{ in } S_j} R(x) \right]. \tag{8}$$

Remark 2   For the generalized eigenvalue problem $Ax = \lambda Mx$, all the same principles hold if the denominator in the Rayleigh quotient is changed from $x^T x$ to $x^T Mx$. Substituting $x = Sz$, with the eigenvectors of $M^{-1}A$ in the columns of S, $R(x)$ simplifies to

$$R(x) = \frac{x^T Ax}{x^T Mx} = \frac{\lambda_1 z_1^2 + \cdots + \lambda_n z_n^2}{z_1^2 + \cdots + z_n^2}. \tag{9}$$

This was the point of simultaneously diagonalizing A and M in Section 6.3. Both quadratics become sums of perfect squares, and $\lambda_1 = \min R(x)$ as well as $\lambda_n = \max R(x)$. Even for *unequal* masses in an oscillating system, holding one mass at equilibrium will raise the lowest frequency and lower the highest frequency.

EXERCISES

6.4.1   Consider the system $Ax = b$ given by

$$\begin{bmatrix} 2 & -1 & 0 \\ -1 & 2 & -1 \\ 0 & -1 & 2 \end{bmatrix} \begin{bmatrix} x_1 \\ x_2 \\ x_3 \end{bmatrix} = \begin{bmatrix} 4 \\ 0 \\ 4 \end{bmatrix}.$$

Construct the corresponding quadratic $P(x_1, x_2, x_3)$, compute its partial derivatives $\partial P/\partial x_i$, and verify that they vanish exactly at the desired solution.

6.4.2   Complete the square in $P = \frac{1}{2}x^T Ax - x^T b = \frac{1}{2}(x - A^{-1}b)^T A(x - A^{-1}b) + \text{constant}$. This constant equals $P_{\min}$ because the term before it is never negative. (Why?)

6.4.3   Find the minimum, if there is one, of $P_1 = \frac{1}{2}x^2 + xy + y^2 - 3y$ and $P_2 = \frac{1}{2}x^2 - 3y$. What matrix A is associated with P_2?

6.4.4   Another quadratic that certainly has its minimum at $Ax = b$ is

$$Q(x) = \frac{1}{2}\|Ax - b\|^2 = \frac{1}{2}x^T A^T A x - x^T A^T b + \frac{1}{2}b^T b.$$

Comparing Q with P, and ignoring the constant $\frac{1}{2}b^T b$, what system of equations do we get at the minimum of Q? What are these equations called in the theory of least squares?

6.4.5   For any symmetric matrix A, compute the ratio $R(x)$ for the special choice $x = (1, \ldots, 1)$. What is the relation of the sum of all entries a_{ij} to λ_1 and λ_n?

6.4.6   With $A = \begin{bmatrix} 2 & -1 \\ -1 & 2 \end{bmatrix}$, find a choice of x that gives a smaller $R(x)$ than the bound $\lambda_1 \leq 2$ which comes from the diagonal entries. What is the minimum value of $R(x)$?

6.4.7   If B is positive definite, show from the Rayleigh quotient that the smallest eigenvalue of $A + B$ is larger than the smallest eigenvalue of A.

6.4.8   If λ_1 is the smallest eigenvalue of A and μ_1 is the smallest eigenvalue of B, show that the smallest eigenvalue θ_1 of $A + B$ is at least as large as $\lambda_1 + \mu_1$. (Try the corresponding eigenvector x in the Rayleigh quotients.)

Note   Those two exercises are perhaps the most typical and most important results that come easily from Rayleigh's principle, but only with great difficulty from the eigenvalue equations themselves.

6.4.9   If B is positive definite, show from the minimax principle (7) that the *second* smallest eigenvalue is increased by adding B: $\lambda_2(A + B) > \lambda_2(A)$.

6.4.10   If you throw away *two* columns of A, what inequalities do you expect between the smallest eigenvalue μ of the remaining matrix and the original λ's?

6.4.11   Find the minimum values of

$$R(x) = \frac{x_1^2 - x_1 x_2 + x_2^2}{x_1^2 + x_2^2} \qquad \text{and} \qquad R(x) = \frac{x_1^2 - x_1 x_2 + x_2^2}{2x_1^2 + x_2^2}.$$

6.4.12   Prove from (3) that $R(x)$ is never larger than the largest eigenvalue λ_n.

6.4.13   The minimax principle for the jth eigenvalue involves j-dimensional subspaces S_j:

$$\lambda_j = \min_{S_j} \left[\max_{x \text{ in } S_j} R(x) \right].$$

(a)   If λ_j is positive, infer that every S_j contains a vector x with $R(x) > 0$.
(b)   Deduce that every S_j contains a vector $y = C^{-1}x$ with $\bar{R}(y) = y^T C^T A C y / y^T y > 0$.
(c)   Conclude that the jth eigenvalue of $C^T A C$, from *its* minimax principle, is also positive—proving again the *law of inertia*. (My thanks to the University of Minnesota.)

6.4.14   Show that the smallest eigenvalue λ_1 of $Ax = \lambda M x$ is not larger than the ratio a_{11}/m_{11} of the corner entries.

6.4.15   Which particular subspace in the minimax principle (7) is the one that gives the minimum value λ_2? In other words, over which S_2 is the maximum of $R(x)$ equal to λ_2?

6.4.16   Without computing the eigenvalues, decide how many are positive, negative, and zero for

$$A = \begin{bmatrix} 0 & \cdot & 0 & 1 \\ \cdot & \cdot & 0 & 2 \\ 0 & 0 & 0 & \cdot \\ 1 & 2 & \cdot & n \end{bmatrix}.$$

THE FINITE ELEMENT METHOD ■ 6.5

There were two main ideas in the previous section:

(i) Solving $Ax = b$ is equivalent to minimizing $P(x) = \frac{1}{2}x^{\mathrm{T}}Ax - x^{\mathrm{T}}b$.

(ii) Solving $Ax = \lambda_1 x$ is equivalent to minimizing $R(x) = x^{\mathrm{T}}Ax/x^{\mathrm{T}}x$.

Now we try to explain how these ideas can be applied.

The story is a long one, because these equivalences have been known for more than a century. In some classical problems of engineering and physics, such as the bending of a plate or the ground states (eigenvalues and eigenfunctions) of an atom, these minimizations were used to get rough approximations to the true solution. In a certain sense, the approximations *had* to be rough; the only tools were pencil and paper or a little machine. The mathematical principles were there, but they could not be implemented.

Obviously the digital computer was going to bring a revolution, but the first step was to throw out the minimum principles; they were too old and too slow. It was the method of finite differences that jumped ahead, because it was easy to see how to "discretize" a differential equation. We saw it already, in Section 1.7; every derivative is replaced by a difference quotient. The physical region is covered by a lattice, or a "mesh," and at each mesh point $-u_{j+1} + 2u_j - u_{j-1} = h^2 f_j$. The problem is expressed as $Au = f$—and numerical analysts were completely occupied in the 1950's with the development of quick ways to solve systems that are very large and very sparse.

What we did not fully recognize was that even finite differences become incredibly complicated for real engineering problems, like the stresses on an airplane or the natural frequencies of the human skull. *The real difficulty is not to solve the equations, but to set them up.* For an irregular region we need an irregular mesh, pieced together from triangles or quadrilaterals or tetrahedra, and then we need a systematic way of approximating the underlying physical laws. In other words, the computer has to help not only in the solution of the discrete problem, but in its formulation.

You can guess what has happened. The old methods have come back, but with a new idea and a new name. The new name is the *finite element method*, and the new idea has made it possible to use more of the power of the computer—in constructing a discrete approximation, solving it, and displaying the results—than any other technique in scientific computation.† The key is to keep the basic idea simple; then the applications can be complicated. The emphasis in these applications moved from airplane design to the safety of nuclear reactors, and at this writing there is a violent debate about its extension into fluid dynamics. For problems on this scale, the one undebatable point is their cost—I am afraid a billion dollars would be a conservative estimate of the expense so far. I hope some readers will be interested enough, and vigorous enough, to master the finite element method and put it to good use.

---

† Please forgive this enthusiasm; I know the method may not be immortal.

To explain the method, we start with the classical **Rayleigh-Ritz principle** and then introduce the new idea of finite elements. We can work with the same differential equation $-u'' = f(x)$, and the same boundary conditions $u(0) = u(1) = 0$, which were studied earlier by finite differences. Admittedly this is an "infinite-dimensional" problem, with the vector b replaced by a function f and the matrix A by an operator $-d^2/dx^2$. But we can proceed by analogy, and write down the quadratic whose minimum is required, replacing inner products by integrals:

$$P(v) = \tfrac{1}{2}v^{\mathrm{T}}Av - v^{\mathrm{T}}b = \tfrac{1}{2}\int_0^1 v(x)(-v''(x))\,dx - \int_0^1 v(x)f(x)\,dx. \tag{1}$$

This is to be minimized over all functions v satisfying the boundary conditions, and *the function that gives the minimum will be the solution u*. The differential equation has been converted to a minimum principle, and it only remains to integrate by parts:

$$\int_0^1 v(-v'')\,dx = \int_0^1 (v')^2\,dx - [vv']_{x=0}^{x=1} \quad \text{so} \quad P(v) = \tfrac{1}{2}\int_0^1 (v')^2\,dx - \int_0^1 vf\,dx.$$

The term vv' is zero at both limits, because v is. Now the quadratic term $\int (v')^2$ is symmetric, like $x^{\mathrm{T}}Ax$, and also it is *positive*; we are guaranteed a minimum.

How do we find this minimum? To compute it exactly is equivalent to solving the differential equation exactly, and that problem is infinite-dimensional. *The Rayleigh-Ritz principle produces an n-dimensional problem by choosing only n trial functions $v = V_1, \ldots, v = V_n$*. It admits all combinations $V = y_1 V_1(x) + \cdots + y_n V_n(x)$, and computes the particular combination (call it U) that minimizes $P(V)$. To repeat: The idea is to minimize over a subspace instead of over all possible v, and the function that gives the minimum is U instead of u. We hope the two are close.

Substituting V for v, the quadratic turns into

$$P(V) = \tfrac{1}{2}\int_0^1 (y_1 V_1' + \cdots + y_n V_n')^2\,dx - \int_0^1 (y_1 V_1 + \cdots + y_n V_n)f\,dx. \tag{2}$$

Remember that the V's are chosen in advance; the unknowns are $y_1, \ldots, y_n$. If we put these weights into a vector y, then $P(V) = \tfrac{1}{2}y^{\mathrm{T}}Ay - y^{\mathrm{T}}b$ is recognized as exactly one of the quadratics we are accustomed to. The matrix entries A_{ij} are $\int V_i' V_j'\,dx =$ coefficient of $y_i y_j$, and the components b_j are $\int V_j f\,dx =$ coefficient of y_j. We can certainly find the minimum of $\tfrac{1}{2}y^{\mathrm{T}}Ay - y^{\mathrm{T}}b$; it is equivalent to solving $Ay = b$. Therefore the steps in the Rayleigh-Ritz method are (i) *choose the trial functions* (ii) *compute the coefficients A_{ij} and b_j* (iii) *solve $Ay = b$*, and (iv) *print out the approximate solution $U = y_1 V_1 + \cdots + y_n V_n$*.

Everything depends on step (i). Unless the functions V_j are extremely simple, the other steps will be virtually impossible. And unless some combination of the V_j is close to the true solution u, those steps will be useless. The problem is to combine both computability and accuracy, and *the key idea that has made finite elements successful is the use of piecewise polynomials as the trial functions V*.

The simplest and most widely used element is *piecewise linear*. We begin by placing nodes at the points $x_1 = h$, $x_2 = 2h, \ldots, x_n = nh$, just as for finite differences.

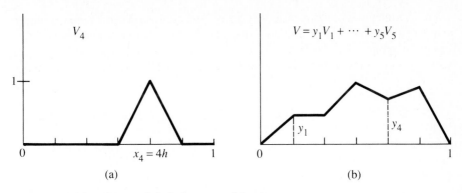

Fig. 6.6. Roof functions and their linear combinations.

At the endpoints $x_0 = 0$ and $x_{n+1} = 1$, the boundary conditions require every V to be zero. Then V_j is the "roof function" which equals 1 at the node x_j, and zero at all the other nodes (Fig. 6.6a). It is concentrated in a small interval around its node, and it is zero everywhere else. Any combination $y_1 V_1 + \cdots + y_n V_n$ must have the value y_j at node j, because the other V's are zero, so its graph is easy to draw (Fig. 6.6b).

That completes step (i). Next we compute the coefficients $A_{ij} = \int V_i' V_j' \, dx$ in the "stiffness matrix" A. The slope V_j' equals $1/h$ in the small interval to the left of x_j, and $-1/h$ in the interval to the right. The same is true of V_i' near its node x_i, and *if these "double intervals" do not overlap then the product $V_i' V_j'$ is identically zero.* The intervals do overlap only when

$$i = j \qquad \text{and} \qquad \int V_i' V_j' \, dx = \int \left(\frac{1}{h} \right)^2 + \int \left(-\frac{1}{h} \right)^2 = \frac{2}{h},$$

or

$$i = j \pm 1 \qquad \text{and} \qquad \int V_i' V_j' \, dx = \int \left(\frac{1}{h} \right) \left(\frac{-1}{h} \right) = \frac{-1}{h}.$$

Therefore the stiffness matrix is actually tridiagonal:

$$A = \frac{1}{h} \begin{bmatrix} 2 & -1 & & & \\ -1 & 2 & -1 & & \\ & -1 & 2 & -1 & \\ & & -1 & 2 & -1 \\ & & & -1 & 2 \end{bmatrix}.$$

This looks just like finite differences! It has led to a thousand discussions about the relation between these two methods. More complicated finite elements—polynomials of higher degree, which are defined on triangles or quadrilaterals for partial differential equations—also produce sparse matrices A. You could think of finite elements as a systematic way to construct accurate difference equations on irregular meshes, so that finite elements fall into the "intersection" of Rayleigh-Ritz methods and finite differences. The essential thing is the simplicity of these piecewise polynomials; on every subinterval their slopes are easy to find and to integrate.

The components b_j on the right side are new. Instead of just the value of f at x_j, as for finite differences, they are now an average of f around that point: $b_j = \int V_j f \, dx$. Then in step (iii) we solve the tridiagonal system $Ay = b$, which gives the coefficients in the trial function $U = y_1 V_1 + \cdots + y_n V_n$ which is minimizing. Finally, connecting all these heights y_j by a broken line, we have a picture of the approximate solution U.

EXAMPLE $-u'' = 2$ with $u(0) = 0$, $u(1) = 0$, and solution $u = x - x^2$. The approximation will use three intervals, with $h = \frac{1}{3}$ and two roof functions. The matrix A is constructed above, and the vector b requires integration of the roof function times f. Since $f = 2$, that produces twice the area under the roof:

$$A = 3 \begin{bmatrix} 2 & -1 \\ -1 & 2 \end{bmatrix} \quad \text{and} \quad b = \begin{bmatrix} \frac{2}{3} \\ \frac{2}{3} \end{bmatrix}.$$

The solution to $Ay = b$ is $y = (\frac{2}{9}, \frac{2}{9})$. Therefore the best U is $\frac{2}{9} V_1 + \frac{2}{9} V_2$. In this case *it agrees with the exact solution* $u = x - x^2 = \frac{1}{3} - \frac{1}{9}$ (or $\frac{2}{3} - \frac{4}{9}$) at the mesh points.

In a more complicated example the approximation will not be exact at the nodes. But it is remarkably close. The underlying theory is explained in the author's book *An Analysis of the Finite Element Method* written jointly with George Fix (Prentice-Hall, 1973). Other books give engineering applications and computer codes. The subject of finite elements has become an important part of engineering education; it is treated more fully in my new book *Introduction to Applied Mathematics* (Wellesley-Cambridge Press, 1986). There we also discuss partial differential equations, where the method really comes into its own.

Eigenvalue Problems

The Rayleigh-Ritz idea—to minimize over a finite-dimensional family of V's in place of all admissible v's—is as useful for eigenvalue problems as for steady-state equations. This time it is the Rayleigh quotient that is minimized. Its true minimum is the fundamental frequency λ_1, and its approximate minimum Λ_1 will be increased when we restrict the class of trial functions from the v's to the V's. Again the discrete problem is manageable, and the principle can actually be applied, but only when the functions V_j are easy to compute. Therefore the step which has been taken in the last 30 years was completely natural and inevitable: to apply the new finite element ideas to this long-established variational form of the eigenvalue problem.

The best example is the simplest one:

$$-u'' = \lambda u, \quad \text{with} \quad u(0) = u(1) = 0.$$

Its first eigenvector is $u = \sin \pi x$, with eigenvalue $\lambda_1 = \pi^2$. The function $v = \sin \pi x$ gives the minimum in the corresponding Rayleigh quotient

$$R(v) = \frac{v^T[-d^2/dx^2]v}{v^T v} = \frac{\int_0^1 v(-v'') \, dx}{\int_0^1 v^2 \, dx} = \frac{\int_0^1 (v')^2 \, dx}{\int_0^1 v^2 \, dx}.$$

Physically this is a ratio of potential to kinetic energy, and they are in balance at the eigenvector. Normally this eigenvector would be unknown, and to approximate it we admit only the candidates $V = y_1 V_1 + \cdots + y_n V_n$:

$$R(V) = \frac{\int_0^1 (y_1 V_1' + \cdots + y_n V_n')^2 \, dx}{\int_0^1 (y_1 V_1 + \cdots + y_n V_n)^2 \, dx} = \frac{y^T A y}{y^T M y}.$$

Now we face the problem of minimizing $y^T A y / y^T M y$. If the matrix M were the identity, this would lead to the standard eigenvalue problem $Ay = \lambda y$. But our matrix M will be tridiagonal, and it is exactly this situation that brings in the *generalized eigenvalue problem*. The minimum value Λ_1 will be the smallest eigenvalue of $Ay = \lambda M y$; it will be close to π^2. The corresponding eigenvector y will give the approximation to the eigenfunction: $U = y_1 V_1 + \cdots + y_n V_n$.

As in the static problem, the method can be summarized in four steps: (i) choose the V_j; (ii) compute A and M; (iii) solve $Ay = \lambda M y$; (iv) print out Λ_1 and U. I don't know why that costs a billion dollars.

EXERCISES

6.5.1 Use three roof functions, with $h = \frac{1}{4}$, to solve $-u'' = 2$ with $u(0) = u(1) = 0$—and verify that the approximation matches $u = x - x^2$ at the nodes.

6.5.2 Solve $-u'' = x$ with $u(0) = u(1) = 0$, and then solve it approximately with two roof functions and $h = \frac{1}{3}$. Where is the largest error?

6.5.3 Suppose $-u'' = 2$ with boundary conditions changed to $u(0) = 0$, $u'(1) = 0$. The boundary condition on u' is "natural" and does not have to be imposed on the trial functions. With $h = \frac{1}{3}$ there is an extra *half-roof* V_3, which goes from 0 to 1 between $x = \frac{2}{3}$ and $x = 1$. Compute $A_{33} = \int (V_3')^2 \, dx$ and $b_3 = \int 2V_3 \, dx$, and solve $Ay = b$ for the finite element solution $y_1 V_1 + y_2 V_2 + y_3 V_3$.

6.5.4 Solve $-u'' = 2$ with a single roof function but put its node at $x = \frac{1}{4}$ instead of $x = \frac{1}{2}$. (Sketch this function V_1.) With boundary conditions $u(0) = u(1) = 0$ compare the finite element approximation with the true $u = x - x^2$.

6.5.5 *Galerkin's method* starts with the differential equation (say $-u'' = f$) instead of minimizing the quadratic P. The trial solution is still $u = y_1 V_1 + y_2 V_2 + \cdots + y_n V_n$, and the y's are chosen to make the difference between $-u''$ and f orthogonal to every V_j:

$$\int(-y_1V_1'' - y_2V_2'' - \cdots - y_nV_n'') V_j \, dx = \int f V_j \, dx.$$

Integrate by parts to reach $Ay = b$, proving that *Galerkin gives the same A and b as Rayleigh-Ritz*.

6.5.6   A basic identity for quadratics (used earlier) is

$$P(y) = \tfrac{1}{2}y^T Ay - y^T b = \tfrac{1}{2}(y - A^{-1}b)^T A(y - A^{-1}b) - \tfrac{1}{2}b^T A^{-1}b.$$

The absolute minimum of P is at $y = A^{-1}b$. The minimum over a subspace of trial functions is at the y *nearest to* $A^{-1}b$— because that makes the first term on the right as small as possible. (This is the key to convergence of U to u.) If $A = I$ and $b = (1, 0, 0)$, which multiple of $V = (1, 1, 1)$ gives the smallest value of P?

6.5.7   For a single roof function V centered at $x = \tfrac{1}{2}$, compute $A = \int (V')^2 \, dx$ and $M = \int V^2 \, dx$. The 1 by 1 eigenvalue problem gives $\lambda = A/M$; is this larger or smaller than the true eigenvalue $\lambda = \pi^2$?

6.5.8   For the roof functions V_1 and V_2 centered at $x = h = \tfrac{1}{3}$ and $x = 2h = \tfrac{2}{3}$, compute the 2 by 2 mass matrix $M_{ij} = \int V_i V_j \, dx$ and solve the eigenvalue problem $Ax = \lambda Mx$ in Exercise 6.3.11.

6.5.9   What is the mass matrix $M_{ij} = \int V_i V_j \, dx$ for n roof functions with $h = 1/(n+1)$?

7

COMPUTATIONS WITH MATRICES

The aim of this book has been to explain some of the applicable parts of matrix theory. In comparison with standard texts in abstract linear algebra, the underlying theory has not been radically changed; one of the best things about the subject is that the theory is really essential for the applications. What is different is the *change in emphasis* which comes with a new point of view. Gaussian elimination becomes more than just a way to find a basis for the row space, and the Gram-Schmidt process is not just a proof that every subspace has an orthonormal basis. Instead, we really *need* these algorithms. And we need a convenient description, $A = LU$ or $A = QR$, of what they do.

This chapter will take a few more steps in the same direction. I suppose these steps are governed by computational necessity, rather than by elegance, and I don't know whether to apologize for that; it makes them sound very superficial, and that is wrong. They deal with the oldest and most fundamental problems of the subject, $Ax = b$ and $Ax = \lambda x$—but virtually every one of them was conceived by the present generation of mathematicians. In numerical analysis there is a kind of survival of the fittest, and we want to describe some ideas that have survived so far. They fall into three groups:

1. **Techniques for solving** $Ax = b$. Gaussian elimination is a perfect algorithm, except perhaps if the particular problem has special properties—as almost every problem has. Our discussion will concentrate on the property of *sparseness*, when most of the entries in A are zero, and on the development

of *iterative rather than direct methods* for solving $Ax = b$. An iterative method is "self-correcting," and it repeats the corrections over and over. It will never reach the exact answer, but the object is to get close more quickly than elimination. In some problems that can be done; in many others, elimination is safer and faster if it takes advantage of the zeros. The competition is far from over, and our first goal is to identify the condition which guarantees convergence to the true solution $A^{-1}b$, and which governs its speed. Then we apply this condition to "overrelaxation" and to other rules of iteration; this is Section 7.4.

2. ***Techniques for solving*** $Ax = \lambda x$. The eigenvalue problem is one of the outstanding successes of numerical analysis. It is clearly defined, its importance is obvious, but until recently no one knew how to solve it. Dozens of algorithms have been suggested, and of course it is impossible to put them in an exact ordering from best to worst; everything depends on the size and the properties of A, and on the number of eigenvalues that are wanted. In other words, it is dangerous just to ask a computation center for an eigenvalue subroutine, without knowing anything about its contents. (I hope you do not need to check every FORTRAN statement.) We have chosen two or three ideas which have superseded almost all of their predecessors: *the QR algorithm*, the family of "*power methods*," and the preprocessing of a symmetric matrix to make it **tridiagonal**.

   The first two methods are iterative, and the last is direct. It does its job in a finite number of steps, but it does not end up with the eigenvalues themselves, only with a much simpler matrix to use in the iterative steps.

3. ***The condition number of a matrix.*** Section 7.2 attempts to measure the sensitivity, or the "vulnerability," of a solution: If A and b are slightly changed, how great is the effect on $x = A^{-1}b$? Before starting on that question, we want to point out one obstacle (which is easy to overcome). There has to be a way to measure the change δA, and to estimate the size of A itself. The length of a vector is already defined, and now we need the *norm of a matrix*. Then the condition number, and the sensitivity of A, will follow directly from the norms of A and A^{-1}. *The matrices in this chapter are square.*

THE NORM AND CONDITION NUMBER OF A MATRIX ■ 7.2

An error and a blunder are very different things. An error is a small mistake, probably unavoidable even by a perfect mathematician or a perfect computer. A blunder is much more serious, and larger by at least an order of magnitude. When the computer rounds off a number to the eighth significant place, that is an error; but when a problem is so excruciatingly sensitive that this roundoff error completely changes the solution, then almost certainly someone has committed a blunder. Our goal in this section is to analyze the effect of errors, so that blunders can be avoided.

We are actually continuing a discussion that began in Chapter 1 with

$$A = \begin{bmatrix} 1 & 1 \\ 1 & 1.0001 \end{bmatrix} \quad \text{and} \quad A' = \begin{bmatrix} .0001 & 1 \\ 1 & 1 \end{bmatrix}.$$

We claimed that A' is well-conditioned, and not particularly sensitive to round-off—except that if Gaussian elimination is applied in a stupid way, the matrix becomes completely vulnerable. It is a blunder to accept .0001 as the first pivot, and we must insist on a larger and safer choice by exchanging the rows of A'. When "partial pivoting" is built into the elimination algorithm, so that the computer automatically looks for the largest pivot, then the natural resistance to roundoff error is no longer compromised.

How do we measure this natural resistance, and decide whether a matrix is well-conditioned or ill-conditioned? If there is a small change in b or in A, how large a change does that produce in the solution x?

We begin with *a change in the right-hand side*, from b to $b + \delta b$. This error might come from experimental data or from roundoff; we may suppose that δb is small, but its direction is outside our control. The solution is changed from x to $x + \delta x$:

$$A(x + \delta x) = b + \delta b, \quad \text{so by subtraction} \quad A(\delta x) = \delta b.$$

This is a particularly simple case; we consider all perturbations δb, and estimate the resulting perturbation $\delta x = A^{-1}\delta b$. There will be a large change in the solution when A^{-1} is large—A is nearly singular—and it is especially large when δb points in the direction that is amplified most by A^{-1}.

To begin with, suppose that A is symmetric and that its eigenvalues are positive: $0 < \lambda_1 \le \lambda_2 \le \cdots \le \lambda_n$. Any vector δb is a combination of the corresponding unit eigenvectors $x_1, \ldots, x_n$, and the worst error is in the direction of the first eigenvector x_1. The factor ϵ is just to suggest that δb and δx are small:

$$\text{if} \quad \delta b = \epsilon x_1, \quad \text{then} \quad \delta x = \frac{\delta b}{\lambda_1}. \tag{1}$$

The error of size $\|\delta b\|$ is amplified by $1/\lambda_1$, which is the largest eigenvalue of A^{-1}. The amplification is greatest when λ_1 is close to zero, so that *nearly singular matrices are the most sensitive.*

There is only one drawback to this measure of sensitivity, and it is serious. Suppose we multiply all the entries of A by 1000; then λ_1 will be multiplied by 1000 and the matrix will look much less singular. This offends our sense of fair play; such a simple rescaling cannot make an ill-conditioned matrix well. It is true that δx will be smaller by a factor of 1000, but so will the solution $x = A^{-1}b$. The relative error $\|\delta x\|/\|x\|$ will be the same. The factor $\|x\|$ in the denominator normalizes the problem against a trivial change of scale. At the same time there is a normalization for δb; our problem is to compare the *relative change* $\|\delta b\|/\|b\|$ with the *relative error* $\|\delta x\|/\|x\|$.

The worst case is when $\|\delta x\|$ is large—the perturbations lie in the direction of the eigenvector x_1—and when the denominator $\|x\|$ is small. The unperturbed solution x should be as small as possible compared to the unperturbed b. This means that *the original problem $Ax = b$ should be at the other extreme*, in the direction of the last eigenvector x_n:

$$\text{if} \quad b = x_n, \quad \text{then} \quad x = A^{-1}b = \frac{b}{\lambda_n}. \tag{2}$$

It is this combination, $b = x_n$ and $\delta b = \epsilon x_1$, that makes the relative error as large as possible. These are the extreme cases in the following inequalities:

7A For a positive definite matrix, the solution $x = A^{-1}b$ and the error $\delta x = A^{-1}\delta b$ always satisfy

$$\|x\| \geq \frac{\|b\|}{\lambda_n} \quad \text{and} \quad \|\delta x\| \leq \frac{\|\delta b\|}{\lambda_1}. \tag{3}$$

Therefore the relative error is bounded by

$$\frac{\|\delta x\|}{\|x\|} \leq \frac{\lambda_n}{\lambda_1} \frac{\|\delta b\|}{\|b\|}. \tag{4}$$

The ratio $c = \lambda_n/\lambda_1 = \lambda_{\max}/\lambda_{\min}$ is called the *condition number* of A.

EXAMPLE 1 The eigenvalues of

$$A = \begin{bmatrix} 1 & 1 \\ 1 & 1.0001 \end{bmatrix}$$

are approximately $\lambda_1 = 10^{-4}/2$ and $\lambda_2 = 2$. Therefore its condition number is about $c = 4 \cdot 10^4$, and we must expect a violent change in the solution from some very ordinary changes in the data. Chapter 1 compared the equations $Ax = b$ and $Ax' = b'$:

$$
\begin{aligned}
u + \quad\quad v &= 2 & u + \quad\quad v &= 2 \\
u + 1.0001v &= 2; & u + 1.0001v &= 2.0001.
\end{aligned}
$$

The right sides are changed only by $\|\delta b\| = 10^{-4}$. At the same time, the solutions went from $u = 2, v = 0$ to $u = v = 1$. This is a relative error of

$$\frac{\|\delta x\|}{\|x\|} = \frac{\|(-1, 1)\|}{\|(2, 0)\|} = \frac{\sqrt{2}}{2}, \quad \text{which equals} \quad 2 \cdot 10^4 \frac{\|\delta b\|}{\|b\|}.$$

Without having made any special choice of the perturbation (our x and δb make $45°$ angles with the worst cases, which accounts for the missing 2 between $2 \cdot 10^4$ and the extreme possibility $c = 4 \cdot 10^4$) there was a tremendous change in the solution.

Notice that the condition number c is not directly affected by the size of the matrix; if $A = I$, or even $A = I/10$, the condition number is $c = \lambda_{max}/\lambda_{min} = 1$. By comparison, *the determinant is a terrible measure of ill-conditioning*. It depends not only on the scaling but also on the order n; if $A = I/10$, then the determinant of A is 10^{-n}. In fact, this "nearly singular" matrix is as well-conditioned as possible.

EXAMPLE 2   Consider the n by n finite difference matrix

$$A = \begin{bmatrix} 2 & -1 & & & \\ -1 & 2 & -1 & & \\ & -1 & 2 & \cdot & \\ & & \cdot & \cdot & -1 \\ & & & -1 & 2 \end{bmatrix}.$$

Its largest eigenvalue is about $\lambda_n = 4$, and its smallest is about $\lambda_1 = \pi^2/n^2$. Therefore the condition number is approximately $c = \frac{1}{2}n^2$, and this time the dependence on the order n is genuine. The better we approximate $-u'' = f$, by increasing the number of unknowns, the harder it is to compute the approximation. It not only takes longer, but it is more affected by roundoff. At a certain crossover point, an increase in n will actually produce a poorer answer.

Fortunately for the engineer, this crossover occurs where the accuracy is already pretty good. Working in single precision, a typical computer might make roundoff errors of order 10^{-9}. If the approximation uses $n = 100$ unknowns, so that $c = 5000$, then such an error is amplified at most to be of order 10^{-5}—which is still more accurate than any ordinary measurements. But there will be trouble with 10,000 unknowns, or with a finite difference approximation to a higher order equation like $d^4u/dx^4 = f(x)$—for which the condition number grows as n^4.†

Our analysis so far has applied exclusively to symmetric matrices with positive eigenvalues. We could easily drop the positivity assumption, and use absolute values; the condition number would become $c = \max|\lambda_i|/\min|\lambda_i|$. But to avoid the assumption of symmetry, as we certainly want to do, there will have to be a major

---

† The usual rule of thumb, experimentally verified, is that the computer can lose $\log c$ decimal places to the roundoff errors in Gaussian elimination.

change. This is easy to see for the matrices

$$A = \begin{bmatrix} 1 & 100 \\ 0 & 1 \end{bmatrix} \quad \text{and} \quad A^{-1} = \begin{bmatrix} 1 & -100 \\ 0 & 1 \end{bmatrix}. \tag{5}$$

The eigenvalues all equal one, but it is certainly not true that the relative change in x is bounded by the relative change in b. The proper condition number is not $\lambda_{max}/\lambda_{min} = 1$. Compare the solutions

$$x = \begin{bmatrix} 0 \\ 1 \end{bmatrix} \quad \text{when} \quad b = \begin{bmatrix} 100 \\ 1 \end{bmatrix}; \quad x' = \begin{bmatrix} 100 \\ 0 \end{bmatrix} \quad \text{when} \quad b' = \begin{bmatrix} 100 \\ 0 \end{bmatrix}.$$

A 1% change in b has produced a hundredfold change in x; the amplification factor is 100^2. Since c represents an upper bound for this amplification, it must be at least 10,000. The difficulty with these matrices is that a large off-diagonal entry in A means an equally large entry in A^{-1}—contradicting the intuitive expectation that A^{-1} should get smaller as A gets bigger.

To find a proper definition of the condition number, we have to look back at equation (3). We were trying to make x small and $b = Ax$ large. (The extreme case occurred at the eigenvector x_n, where the ratio of Ax to x is exactly λ_n.) The change when A is no longer symmetric is that *the maximum of $\|Ax\|/\|x\|$ may be found at a vector x that is not one of the eigenvectors*. This maximum is still an excellent measure of the size of A; it is called the **norm** of the matrix, and denoted by $\|A\|$.

7B   The *norm* of A is the number defined by

$$\|A\| = \max_{x \neq 0} \frac{\|Ax\|}{\|x\|}. \tag{6}$$

In other words, $\|A\|$ bounds the "amplifying power" of the matrix:

$$\|Ax\| \leq \|A\| \|x\| \qquad \text{for all vectors } x. \tag{7}$$

Equality holds for at least one nonzero x.

The matrices A and A^{-1} in equation (5) will have norms somewhere between 100 and 101. In a moment they can be calculated exactly, but first we want to complete the connection between norms and condition numbers. Because $b = Ax$ and $\delta x = A^{-1}\delta b$, the definition (7) of a matrix norm immediately gives

$$\|b\| \leq \|A\| \|x\| \quad \text{and} \quad \|\delta x\| \leq \|A^{-1}\| \|\delta b\|. \tag{8}$$

This is the replacement for (3), when A is not symmetric; in the symmetric case $\|A\|$ is the same as λ_n, and $\|A^{-1}\|$ is the same as $1/\lambda_1$. *The right replacement for λ_n/λ_1 is the product $\|A\| \|A^{-1}\|$*—which is the condition number.

7C   The *condition number* of A is $c = \|A\| \, \|A^{-1}\|$, and the relative error satisfies

$$\frac{\|\delta x\|}{\|x\|} \le c \, \frac{\|\delta b\|}{\|b\|}. \tag{9}$$

If we perturb the matrix A instead of the right side b, then

$$\frac{\|\delta x\|}{\|x + \delta x\|} \le c \, \frac{\|\delta A\|}{\|A\|}. \tag{10}$$

The inequality (9) holds for every b and every δb, and it is just the product of the two inequalities in (8). What is remarkable is that the same condition number appears in (10), when the matrix itself is perturbed: If $Ax = b$ and $(A + \delta A)(x + \delta x) = b$, then by subtraction

$$A \, \delta x + \delta A(x + \delta x) = 0, \qquad \text{or} \qquad \delta x = -A^{-1}(\delta A)(x + \delta x).$$

Multiplying by δA amplifies a vector by no more than the norm $\|\delta A\|$, and then multiplying by A^{-1} amplifies by no more than $\|A^{-1}\|$. Therefore

$$\|\delta x\| \le \|A^{-1}\| \, \|\delta A\| \, \|x + \delta x\|,$$

or

$$\frac{\|\delta x\|}{\|x + \delta x\|} \le \|A^{-1}\| \, \|\delta A\| = c \, \frac{\|\delta A\|}{\|A\|}.$$

These inequalities mean that roundoff error comes from two sources. One is the *natural sensitivity* of the problem, which is measured by c. The other is the actual error δb or δA. This was the basis of Wilkinson's error analysis. Since the elimination algorithm actually produces approximate factors L' and U', it solves the equation with the wrong matrix $A + \delta A = L'U'$ instead of the right matrix $A = LU$. He proved that partial pivoting is adequate to keep δA under control— see his "Rounding Errors in Algebraic Processes"—so *the whole burden of the roundoff error is carried by the condition number c.*

A Formula for the Norm

The norm of A measures the largest amount by which any vector (eigenvector or not) is amplified by matrix multiplication: $\|A\| = \max(\|Ax\|/\|x\|)$. The norm of the identity matrix is 1. To compute this "amplification factor," we square both sides:

$$\|A\|^2 = \max \frac{\|Ax\|^2}{\|x\|^2} = \max \frac{x^T A^T A x}{x^T x}. \tag{11}$$

This brings back a symmetric $A^T A$, and its Rayleigh quotient.

7D   The norm of A is the square root of the largest eigenvalue of $A^T A$: $\|A\|^2 = \lambda_{\max}(A^T A)$. In case A is symmetric, then $A^T A = A^2$ and the norm is the largest eigenvalue: $\|A\| = \max|\lambda_i|$. In every case, the vector that is amplified the most is the corresponding eigenvector of $A^T A$:

$$\frac{x^T A^T A x}{x^T x} = \frac{x^T(\lambda_{\max} x)}{x^T x} = \lambda_{\max} = \|A\|^2.$$

Note 1   The norm and condition number of A are not actually computed in practical problems, but only estimated. There is not time to solve an eigenvalue problem for $\lambda_{\max}(A^T A)$.

Note 2   The condition number applies to the normal equations $A^T A x = A^T b$ in least squares problems. The condition number $c(A^T A)$ is the *square* of $c(A)$. Forming $A^T A$ can turn a healthy problem into a sick one, and it may be necessary to use either Gram-Schmidt or the singular value decomposition $A = Q_1 \Sigma Q_2^T$.

Note 3   The entries σ_i in the diagonal matrix Σ are the **singular values** of A, and *their squares are the eigenvalues of* $A^T A$. Therefore another formula for the norm is $\|A\| = \sigma_{\max}$. The orthogonal Q_1 and Q_2 leave lengths unchanged in $\|Ax\| = \|Q_1 \Sigma Q_2^T x\|$, so the largest amplification factor is the largest σ.

Note 4   Roundoff error enters not only $Ax = b$, but also $Ax = \lambda x$. This raises a new question: What is the "condition number of the eigenvalue problem"? The obvious answer is wrong; it is not the condition number of A itself. Instead *it is the condition number of the diagonalizing S which measures the sensitivity of the eigenvalues.* If μ is an eigenvalue of $A + E$, then its distance from one of the eigenvalues of A is

$$|\mu - \lambda| \le \|S\| \, \|S^{-1}\| \, \|E\| = c(S)\|E\|. \tag{12}$$

In case S is an orthogonal matrix Q, the eigenvalue problem is perfectly conditioned: $c(Q) = 1$, and the change $\mu - \lambda$ in the eigenvalues is no greater than the change E in the matrix A. This happens whenever the eigenvectors are orthonormal—they are the columns of S. Therefore the best case is when A is symmetric, or more generally when $A A^T = A^T A$. Then A is a normal matrix, its diagonalizing S is an orthogonal Q (Section 5.6), and its eigenvalues are perfectly conditioned. You can see the presence of S in the formula for perturbations in each eigenvalue separately: If x_k is the kth column of S and y_k is the kth row of S^{-1}, then

$$\mu_k - \lambda_k = y_k E x_k + \text{terms of order } \|E\|^2. \tag{13}$$

In practice, $y_k E x_k$ is a realistic estimate of the change in the eigenvalue. The idea in every good algorithm is to keep the error matrix E as small as possible—usually by insisting, as QR will do in the next section, on orthogonal matrices at each step of the iteration.

EXERCISES

7.2.1 If A is an orthogonal matrix Q, show that $\|Q\| = 1$ and also $c(Q) = 1$. Orthogonal matrices (and their multiples αQ) are the only perfectly conditioned matrices.

7.2.2 Which "famous" inequality gives $\|(A + B)x\| \le \|Ax\| + \|Bx\|$, and why does it follow from (6) that $\|A + B\| \le \|A\| + \|B\|$?

7.2.3 Explain why $\|ABx\| \le \|A\| \|B\| \|x\|$, and deduce from (6) that $\|AB\| \le \|A\| \|B\|$. Show that this also implies $c(AB) \le c(A)c(B)$.

7.2.4 For the positive definite $A = \left[\begin{smallmatrix} 2 & -1 \\ -1 & 2 \end{smallmatrix}\right]$, compute $\|A^{-1}\| = 1/\lambda_1$, $\|A\| = \lambda_2$, and $c(A) = \lambda_2/\lambda_1$. Find a right side b and a perturbation δb so that the error is worst possible, $\|\delta x\|/\|x\| = c\|\delta b\|/\|b\|$.

7.2.5 Show that if λ is any eigenvalue of A, $Ax = \lambda x$, then $|\lambda| \le \|A\|$.

7.2.6 Find the exact norms of the matrices in (5).

7.2.7 Prove from Exercise 5.6.5, comparing the eigenvalues of $A^{\mathrm{T}}A$ and AA^{T}, that $\|A\| = \|A^{\mathrm{T}}\|$.

7.2.8 For a positive definite A, the Cholesky decomposition is $A = LDL^{\mathrm{T}} = R^{\mathrm{T}}R$, where $R = \sqrt{D}L^{\mathrm{T}}$. Show directly from 7D that the condition number of R is the square root of the condition number of A. It follows that the Gauss algorithm needs no row exchanges for a positive definite matrix; the condition does not deteriorate, since $c(A) = c(R^{\mathrm{T}})c(R)$.

7.2.9 Show that $\lambda_{\max}$, or even $\max|\lambda|$, is not a satisfactory norm, by finding 2 by 2 counterexamples to $\lambda_{\max}(A + B) \le \lambda_{\max}(A) + \lambda_{\max}(B)$ and to $\lambda_{\max}(AB) \le \lambda_{\max}(A)\lambda_{\max}(B)$.

7.2.10 Suppose $\|x\|$ is changed from the Euclidean length $(x_1^2 + \cdots + x_n^2)^{1/2}$ to the "maximum norm" or "L_∞ norm": $\|x\|_\infty = \max|x_i|$. (Example: $\|(1, -2, 1)\| = 2$.) Compute the corresponding matrix norm

$$\|A\|_\infty = \max_{x \ne 0} \frac{\|Ax\|_\infty}{\|x\|_\infty} \qquad \text{if} \qquad A = \begin{bmatrix} 1 & 2 \\ 3 & -4 \end{bmatrix}.$$

7.2.11 Show that the eigenvalues of $B = \left[\begin{smallmatrix} 0 & A^{\mathrm{T}} \\ A & 0 \end{smallmatrix}\right]$ are $\pm\sigma_i$, the singular values of A. *Hint:* Try B^2.

7.2.12 (a) Do A and A^{-1} have the same condition number c?

(b) In parallel with the upper bound (9) on the error, prove a lower bound:

$$\frac{\|\delta x\|}{\|x\|} \ge \frac{1}{c}\frac{\|\delta b\|}{\|b\|}. \quad \text{(Consider } A^{-1}b = x \text{ instead of } Ax = b.)$$

7.3 ■ THE COMPUTATION OF EIGENVALUES

There is no one best way to find the eigenvalues of a matrix. But there are certainly some terrible ways, which should never be tried, and also some ideas that do deserve a permanent place. We begin by describing one very rough and ready approach, the **power method**, whose convergence properties are easy to understand. Then we move steadily toward a much more sophisticated algorithm, which starts by making a symmetric matrix tridiagonal and ends by making it virtually diagonal. Its last step is done by Gram-Schmidt, so the method is known as QR.

    The ordinary power method operates exactly on the principle of a difference equation. It starts with an initial guess u_0 and then successively forms $u_1 = Au_0$, $u_2 = Au_1$, and in general $u_{k+1} = Au_k$. Each step is a matrix-vector multiplication, and after k steps it produces $u_k = A^k u_0$, although the matrix A^k will never appear. In fact the essential thing is that multiplication by A should be easy to do—if the matrix is large, it had better be sparse—because convergence to the eigenvector is often very slow. Assuming A has a full set of eigenvectors $x_1, \ldots, x_n$, the vector u_k will be given by the usual formula for a difference equation:

$$u_k = c_1 \lambda_1^k x_1 + \cdots + c_n \lambda_n^k x_n.$$

Imagine the eigenvalues numbered in increasing order, and suppose that the largest eigenvalue is all by itself; there is no other eigenvalue of the same magnitude, and λ_n is not repeated. Thus $|\lambda_1| \leq \cdots \leq |\lambda_{n-1}| < |\lambda_n|$. Then as long as the initial guess u_0 contained *some* component of the eigenvector x_n, so that $c_n \neq 0$, this component will gradually become dominant:

$$\frac{u_k}{\lambda_n^k} = c_1 \left(\frac{\lambda_1}{\lambda_n}\right)^k x_1 + \cdots + c_{n-1} \left(\frac{\lambda_{n-1}}{\lambda_n}\right)^k x_{n-1} + c_n x_n. \tag{1}$$

The vectors u_k point more and more accurately toward the direction of x_n, and the convergence factor is the ratio $r = |\lambda_{n-1}|/|\lambda_n|$. It is just like convergence to a steady state, which we studied for Markov processes, except that now the largest eigenvalue λ_n may not equal 1. In fact, we do not know the scaling factor λ_n^k in (1), but some scaling factor should be introduced; otherwise u_k can grow very large or very small, in case $|\lambda_n| > 1$ or $|\lambda_n| < 1$. Normally we can just divide each u_k by its first component α_k before taking the next step. With this simple scaling, the power method becomes $u_{k+1} = Au_k/\alpha_k$, and it converges to a multiple of x_n.†

EXAMPLE   (*from California*) with the u_n approaching the eigenvector $\begin{bmatrix} .667 \\ .333 \end{bmatrix}$:

$$A = \begin{bmatrix} .9 & .2 \\ .1 & .8 \end{bmatrix} \text{ was the matrix of population shifts;}$$

$$u_0 = \begin{bmatrix} 1 \\ 0 \end{bmatrix}, \quad u_1 = \begin{bmatrix} .9 \\ .1 \end{bmatrix}, \quad u_2 = \begin{bmatrix} .83 \\ .17 \end{bmatrix}, \quad u_3 = \begin{bmatrix} .781 \\ .219 \end{bmatrix}, \quad u_4 = \begin{bmatrix} .747 \\ .253 \end{bmatrix}.$$

---

† The scaling factors α_k will also converge; they approach λ_n.

The most serious limitation is clear from this example: If r is close to 1, then convergence is very slow. In many applications $r > .9$, which means that more than 20 iterations are needed to reduce $(\lambda_2/\lambda_1)^k$ just by a factor of 10. (The example had $r = .7$, and it was still slow.) Of course if $r = 1$, which means $|\lambda_{n-1}| = |\lambda_n|$, then convergence may not occur at all. There are several ways to get around this limitation, and we shall describe three of them:

(1) The **block power method** works with several vectors at once, in place of a single u_k. If we start with p orthonormal vectors, multiply them all by A, and then apply Gram-Schmidt to orthogonalize them again—that is a single step of the method—then the effect is to reduce the convergence ratio to $r' = |\lambda_{n-p}|/|\lambda_n|$. Furthermore we will simultaneously obtain approximations to p different eigenvalues and their eigenvectors.

(2) The **inverse power method** operates with A^{-1} instead of A. A single step of the difference equation is $v_{k+1} = A^{-1}v_k$, which means that we solve the linear system $Av_{k+1} = v_k$ (and save the factors L and U!). Now the theory guarantees convergence to the *smallest eigenvalue*, provided the convergence factor $r'' = |\lambda_1|/|\lambda_2|$ is less than one. Often it is the smallest eigenvalue that is wanted in the applications, and then inverse iteration is an automatic choice.

(3) The **shifted inverse power method** is the best of all. Suppose that A is replaced by $A - \alpha I$. Then all the eigenvalues λ_i are shifted by the same amount α, and the convergence factor for the inverse method will change to $r''' = |\lambda_1 - \alpha|/|\lambda_2 - \alpha|$. Therefore if α is chosen to be a good approximation to λ_1, r''' will be very small and the convergence is enormously accelerated. Each step of the method solves the system $(A - \alpha I)w_{k+1} = w_k$, and this difference equation is satisfied by

$$w_k = \frac{c_1 x_1}{(\lambda_1 - \alpha)^k} + \frac{c_2 x_2}{(\lambda_2 - \alpha)^k} + \cdots + \frac{c_n x_n}{(\lambda_n - \alpha)^k}.$$

Provided α is close to λ_1, the first of these denominators is so near to zero that only one or two steps are needed to make that first term completely dominant. In particular, if λ_1 has already been computed by another algorithm (such as QR), then α is this computed value. The standard procedure is to factor $A - \alpha I$ into LU† and to solve $Ux_1 = (1, 1, \ldots, 1)^T$ by back-substitution.

If λ_1 is not already approximated by an independent algorithm, then the shifted power method has to generate its own choice of α—or, since we can vary the shift at every step if we want to, it must choose the α_k that enters $(A - \alpha_k I)w_{k+1} = w_k$.

---

† This may look extremely ill-conditioned, since $A - \alpha I$ is as nearly singular as we can make it. Fortunately the error is largely confined to the direction of the eigenvector. Since any multiple of an eigenvector is still an eigenvector, it is only that direction which we are trying to compute.

The simplest possibility is just to work with the scaling factors that bring each w_k back to a reasonable size, but there are other and better ways. In the symmetric case $A = A^T$, the most accurate choice seems to be the **Rayleigh quotient**

$$\alpha_k = R(w_k) = \frac{w_k^T A w_k}{w_k^T w_k}.$$

We already know that this quotient has a minimum at the true eigenvector—the derivatives of R are zero, and its graph is like the bottom of a parabola. Therefore the error $\lambda - \alpha_k$ in the eigenvalue is roughly the square of the error in the eigenvector. The convergence factors $r''' = |\lambda_1 - \alpha_k|/|\lambda_2 - \alpha_k|$ are changing at every step, and in fact r''' itself is converging to zero. The final result, with these Rayleigh quotient shifts, is *cubic convergence*† of α_k to λ_1.

Tridiagonal and Hessenberg Forms

The power method is reasonable only for a matrix that is large and very sparse. When most of the entries are nonzero, the method is a mistake. Therefore we ask whether there is any simple way *to create zeros*. That is the goal of the following paragraphs.

It should be said at the outset that, after computing a similar matrix $U^{-1}AU$ with more zeros than A, we do not intend to go back to the power method. There are much more sophisticated variants, and the best of them seems to be the QR algorithm. (The shifted inverse power method has its place at the very end, in finding the eigenvector.) The first step is to produce as many zeros as possible, and to do it as quickly as possible. Our only restriction on speed will be the use of unitary (or orthogonal) transformations, which preserve symmetry and preserve lengths. If A is symmetric, then so is $U^{-1}AU$, and no entry can become dangerously large.

To go from A to $U^{-1}AU$, there are two main possibilities: Either we can produce one zero at every step (as in Gaussian elimination), or we can work with a whole column at once. For a single zero, it is enough to use a plane rotation as illustrated in equation (7) below; it has $\cos \theta$ and $\sin \theta$ in a 2 by 2 block. Then we could cycle through all the entries below the diagonal, choosing at each step a rotation θ that will produce a zero; this is the principle of **Jacobi's method**. It fails to diagonalize A after a finite number of rotations, since the zeros achieved in the early steps will be destroyed when later zeros are created.

To preserve those zeros, and to stop when they are produced, we have to settle for less than a triangular form. We accept *one nonzero diagonal below the main*

---

† Linear convergence means that every step multiplies the error by a fixed factor $r < 1$. Quadratic convergence means that the error is squared at every step, as in Newton's method $x_{k+1} - x_k = -f(x_k)/f'(x_k)$ for solving $f(x) = 0$. Cubic convergence means that the error is *cubed* at every step, going from 10^{-1} to 10^{-3} to 10^{-9}.

diagonal. This is the **Hessenberg form**. If the matrix is symmetric, then the upper triangular part will copy the lower triangular part, and the matrix will be tridiagonal.

Both these forms can be obtained by a series of rotations in the right planes. That is very effective. Householder has found a new way to accomplish exactly the same thing. His idea gives the "preparation step" for the QR method.† A **Householder transformation**, or an elementary *reflector*, is a matrix of the form

$$H = I - 2\frac{vv^{\mathrm{T}}}{\|v\|^2}.$$

Often v is normalized to become a unit vector $u = v/\|v\|$, and then H becomes $I - 2uu^{\mathrm{T}}$. In either case H is both *symmetric* and *orthogonal*:

$$H^{\mathrm{T}}H = (I - 2uu^{\mathrm{T}})(I - 2uu^{\mathrm{T}}) = I - 4uu^{\mathrm{T}} + 4uu^{\mathrm{T}}uu^{\mathrm{T}} = I.$$

Thus $H = H^{\mathrm{T}} = H^{-1}$. In the complex case, the corresponding matrix $I - 2uu^{\mathrm{H}}$ is both Hermitian and unitary. Householder's plan was to produce zeros with these matrices, and its success depends on the following identity:

7E Suppose z is the column vector $(1, 0, \ldots, 0)$, and $\sigma = \|x\|$, and $v = x + \sigma z$. Then $Hx = -\sigma z = (-\sigma, 0, \ldots, 0)$.

Proof

$$Hx = x - \frac{2vv^{\mathrm{T}}x}{\|v\|^2} = x - (x + \sigma z)\frac{2(x + \sigma z)^{\mathrm{T}}x}{(x + \sigma z)^{\mathrm{T}}(x + \sigma z)}$$

$$= x - (x + \sigma z) \quad (\text{because } x^{\mathrm{T}}x = \sigma^2)$$

$$= -\sigma z. \tag{2}$$

This identity can be used right away. We start with the first column of A, and remember that the final $U^{-1}AU$ is to be tridiagonal in the symmetric case (or Hessenberg in general). Therefore *only the $n - 1$ entries below the diagonal will be involved*:

$$x = \begin{bmatrix} a_{21} \\ a_{31} \\ \vdots \\ a_{n1} \end{bmatrix}, \quad z = \begin{bmatrix} 1 \\ 0 \\ \vdots \\ 0 \end{bmatrix}, \quad Hx = \begin{bmatrix} -\sigma \\ 0 \\ \vdots \\ 0 \end{bmatrix}. \tag{3}$$

† You may want to bypass this preparation, and go directly to the QR algorithm. It is only in actual computations that you need to produce the zeros first.

At this point Householder's matrix H is only of order $n - 1$, so it is imbedded into the lower right corner of a full-size matrix U_1:

$$U_1 = \begin{bmatrix} 1 & 0 & 0 & 0 & 0 \\ 0 & & & & \\ 0 & & H & & \\ 0 & & & & \\ 0 & & & & \end{bmatrix} = U_1^{-1}, \quad \text{and} \quad U_1^{-1}AU_1 = \begin{bmatrix} a_{11} & * & * & * & * \\ -\sigma & * & * & * & * \\ 0 & * & * & * & * \\ 0 & * & * & * & * \\ 0 & * & * & * & * \end{bmatrix}.$$

Because of the 1 in its upper left corner, the matrix U_1 leaves the entry a_{11} completely unchanged—and more important, it does not touch the zeros which appear in (3). Therefore the first stage is complete, and $U_1^{-1}AU_1$ has the required first column.

The second stage is similar: x consists of the last $n - 2$ entries in the second column, z is the unit coordinate vector of matching length, and H_2 is of order $n - 2$. When it is imbedded in U_2, it produces

$$U_2 = \begin{bmatrix} 1 & 0 & 0 & 0 & 0 \\ 0 & 1 & 0 & 0 & 0 \\ 0 & 0 & & & \\ 0 & 0 & & H_2 & \\ 0 & 0 & & & \end{bmatrix} = U_2^{-1}, \quad U_2^{-1}(U_1^{-1}AU_1)U_2 = \begin{bmatrix} * & * & * & * & * \\ * & * & * & * & * \\ 0 & * & * & * & * \\ 0 & 0 & * & * & * \\ 0 & 0 & * & * & * \end{bmatrix}.$$

Finally U_3 will take care of the third column, and for a 5 by 5 matrix the Hessenberg form is achieved. In general U is the product of all the matrices $U_1 U_2 \cdots U_{n-2}$, and the number of operations required to compute it is of order n^3.

EXAMPLE

$$A = \begin{bmatrix} 1 & 0 & 1 \\ 0 & 1 & 1 \\ 1 & 1 & 0 \end{bmatrix}, \quad x = \begin{bmatrix} 0 \\ 1 \end{bmatrix}, \quad v = \begin{bmatrix} 1 \\ 1 \end{bmatrix}, \quad H = \begin{bmatrix} 0 & -1 \\ -1 & 0 \end{bmatrix}.$$

Imbedding H into U, the result is tridiagonal:

$$U = \begin{bmatrix} 1 & 0 & 0 \\ 0 & 0 & -1 \\ 0 & -1 & 0 \end{bmatrix}, \quad U^{-1}AU = \begin{bmatrix} 1 & -1 & 0 \\ -1 & 0 & 1 \\ 0 & 1 & 1 \end{bmatrix}.$$

$U^{-1}AU$ is a matrix that is ready to reveal its eigenvalues—the QR algorithm is ready to begin—but we digress for a moment to mention two other applications of these same Householder transformations.

I.   *The factorization $A = QR$.* This was a shorthand for the Gram-Schmidt process in Chapter 3; now it can be carried out more simply and more stably.

Remember that R is to be upper triangular—we no longer have to accept an extra nonzero diagonal below the main one, since there will be no U's or H's multiplying on the right (as in $U^{-1}AU$) to spoil the zeros already created. Therefore the first step in constructing Q is to work with the whole first column of A:

$$x = \begin{bmatrix} a_{11} \\ a_{21} \\ \vdots \\ a_{n1} \end{bmatrix}, \quad z = \begin{bmatrix} 1 \\ 0 \\ \vdots \\ 0 \end{bmatrix}, \quad v = x + \|x\|z, \quad H_1 = I - 2\frac{vv^{\mathrm{T}}}{\|v\|^2}.$$

The first column of H_1A is exactly as desired; it equals $-\|x\|z$; it is zero below the main diagonal; and it is the first column of R. The second step works with the second column of H_1A, from the pivot on down, and produces an H_2H_1A which is zero below that pivot. (The whole algorithm is very much like Gaussian elimination, and in fact it is a slightly slower alternative.) The result of $n - 1$ steps is again an upper triangular R, but the matrix that records the steps is not a lower triangular L. Instead it is the product $Q = H_1H_2 \cdots H_{n-1}$, which can be stored in this factored form and never computed explicitly. That completes Gram-Schmidt.

II. *The singular value decomposition* $Q_1^{\mathrm{T}}AQ_2 = \Sigma$. The appendix will show how this decomposition gives the optimal solution $\bar{x}$ to any problem in least squares. Σ is a diagonal matrix of the same shape as A, and its entries (the singular values) are the square roots of the eigenvalues of $A^{\mathrm{T}}A$. Since Householder transformations can only *prepare* for the eigenvalue problem, and not solve it, we cannot expect them to produce Σ. Instead, they produce a *bidiagonal matrix*, with zeros everywhere except along the main diagonal and the one above. This preprocessing is numerically stable, because the H's are orthogonal.

The first step is exactly as in QR above: x is the first column of A, and H_1x is zero below the pivot. The next step is to multiply on the right by an $H^{(1)}$ which will produce zeros as indicated along the first row:

$$A \rightarrow H_1A = \begin{bmatrix} * & * & * & * \\ 0 & * & * & * \\ 0 & * & * & * \end{bmatrix} \rightarrow H_1AH^{(1)} = \begin{bmatrix} * & * & 0 & 0 \\ 0 & * & * & * \\ 0 & * & * & * \end{bmatrix}. \quad (4)$$

Then two final Householder transformations achieve

$$H_2H_1AH^{(1)} = \begin{bmatrix} * & * & 0 & 0 \\ 0 & * & * & * \\ 0 & 0 & * & * \end{bmatrix} \quad \text{and} \quad H_2H_1AH^{(1)}H^{(2)} = \begin{bmatrix} * & * & 0 & 0 \\ 0 & * & * & 0 \\ 0 & 0 & * & * \end{bmatrix}.$$

This is the bidiagonal form we wanted, and illustrates again the quick way in which zeros can be produced by Householder transformations.

The QR Algorithm

The algorithm is almost magically simple. It starts with the matrix A_0, factors it by Gram-Schmidt into $Q_0 R_0$, and then reverses the factors: $A_1 = R_0 Q_0$. This new matrix is similar to the original one, $Q_0^{-1} A_0 Q_0 = Q_0^{-1}(Q_0 R_0)Q_0 = A_1$, and the process continues with no change in the eigenvalues:

$$A_k = Q_k R_k \qquad \text{and then} \qquad A_{k+1} = R_k Q_k. \tag{5}$$

This equation describes the *unshifted QR algorithm*, and under fairly general circumstances it converges: A_k approaches a triangular form, and therefore its diagonal entries approach its eigenvalues, which are also the eigenvalues of the original A_0.†

As it stands, the algorithm is good but not very good. To make it special, it needs two refinements: (a) We must allow shifts of origin; and (b) we must ensure that the QR factorization at each step is very quick.

(a)   *The shifted algorithm.* If the number α_k is close to an eigenvalue, step (5) should be shifted immediately to

$$A_k - \alpha_k I = Q_k R_k \qquad \text{and then} \qquad A_{k+1} = R_k Q_k + \alpha_k I. \tag{6}$$

This is justified by the fact that A_{k+1} is similar to A_k:

$$Q_k^{-1} A_k Q_k = Q_k^{-1}(Q_k R_k + \alpha_k I)Q_k = A_{k+1}.$$

What happens in practice is that the (n, n) entry of A_k—the one in the lower right corner—is the first to approach an eigenvalue. Therefore this entry is the simplest and most popular choice for the shift α_k. Normally its effect is to produce quadratic convergence, and in the symmetric case even cubic convergence, to the smallest eigenvalue. After three or four steps of the shifted algorithm, the matrix A_k looks like

$$A_k = \begin{bmatrix} * & * & * & * \\ * & * & * & * \\ 0 & * & * & * \\ 0 & 0 & \epsilon & \lambda_1' \end{bmatrix}, \quad \text{with } \epsilon \ll 1.$$

We accept the computed λ_1' as a very close approximation to the true λ_1. To find the next eigenvalue, the QR algorithm continues with the smaller matrix (3 by 3, in the illustration) in the upper left corner. Its subdiagonal elements will be somewhat reduced by the first QR steps, and another two steps are sufficient to find λ_2. This gives a systematic procedure for finding all the eigenvalues. In fact, *the*

---

† A_0 refers to the matrix with which the QR algorithm begins. If there was already some processing by Householder matrices to obtain a tridiagonal form, then A_0 is connected to the absolutely original A by $U^{-1} A U = A_0$.

QR method is now completely described. It only remains to catch up on the eigenvectors—that is a single inverse power step—and to use the zeros that Householder created.

(b)   The object of the preparatory Householder transformations, which put A_0 into tridiagonal or Hessenberg form, was to make each QR step very fast. Normally the Gram-Schmidt process (which is QR) would take $O(n^3)$ operations, but for a Hessenberg matrix this becomes $O(n^2)$ and for a tridiagonal matrix it is $O(n)$. Without this improvement the algorithm would be impossibly slow, and unless each new A_k is again in Hessenberg or tridiagonal form, the improvement will apply only to the first step.

Fortunately, this does not happen. To show that A_1 has the same form as A_0, look at

$$Q_0 = A_0 R_0^{-1} = \begin{bmatrix} * & * & * & * \\ * & * & * & * \\ 0 & * & * & * \\ 0 & 0 & * & * \end{bmatrix} \begin{bmatrix} * & * & * & * \\ 0 & * & * & * \\ 0 & 0 & * & * \\ 0 & 0 & 0 & * \end{bmatrix}.$$

You can easily check that this multiplication leaves Q_0 with the same three zeros as A_0; Q_0 is itself in Hessenberg form. Then A_1 is constructed by reversing the factors, so

$$A_1 = R_0 Q_0 = \begin{bmatrix} * & * & * & * \\ 0 & * & * & * \\ 0 & 0 & * & * \\ 0 & 0 & 0 & * \end{bmatrix} \begin{bmatrix} * & * & * & * \\ * & * & * & * \\ 0 & * & * & * \\ 0 & 0 & * & * \end{bmatrix}.$$

The same three zeros appear in this product; A_1 *is a Hessenberg matrix whenever* A_0 *is*. The symmetric case is even better, since $A_1 = Q_0^{-1} A_0 Q_0$ remains symmetric:

$$A_1^T = Q_0^T A_0^T (Q_0^{-1})^T = Q_0^{-1} A_0 Q_0 = A_1.$$

By the reasoning just completed, A_1 is also Hessenberg. Then, because it is both symmetric and Hessenberg, A_1 *is tridiagonal*. The same argument applies to each of the matrices $A_2, A_3, \ldots$, so *every QR step begins with a tridiagonal matrix*.

The last point to explain is the factorization itself, producing Q_0 and R_0 from the original A_0 (and Q_k and R_k from each A_k, or really from $A_k - \alpha_k I$). We may use Householder again, but it is simpler to annihilate each subdiagonal element in turn by a "plane rotation." The first one is

$$P_{21} A_0 = \begin{bmatrix} \cos\theta & -\sin\theta & & \\ \sin\theta & \cos\theta & & \\ & & 1 & \\ & & & 1 \end{bmatrix} \begin{bmatrix} a_{11} & * & * & * \\ a_{21} & * & * & * \\ 0 & * & * & * \\ 0 & 0 & * & * \end{bmatrix}. \tag{7}$$

The (2, 1) entry in this product is $a_{11} \sin \theta + a_{21} \cos \theta$, and we just choose the angle θ that makes this combination zero. Then P_{32} is chosen in a similar way, to remove the (3, 2) entry of $P_{32}P_{21}A_0$. After $n - 1$ of these elementary rotations, the final result is the upper triangular factor R_0:

$$R_0 = P_{nn-1} \cdots P_{32}P_{21}A_0. \tag{8}$$

That is as much as we can say—there is a lot more in books on numerical linear algebra—about one of the most remarkable algorithms in scientific computing.

We mention only one more method—the **Lanczos method**—for large sparse matrices. It orthogonalizes the Krylov sequence $x, Ax, A^2x, \ldots$, and is described in the references.

EXERCISES

7.3.1 For the matrix $A = \begin{bmatrix} 2 & -\frac{1}{2} \\ -1 & 2 \end{bmatrix}$, with eigenvalues $\lambda_1 = 1$ and $\lambda_2 = 3$, apply the power method $u_{k+1} = Au_k$ three times to the initial guess $u_0 = \begin{bmatrix} 1 \\ 0 \end{bmatrix}$. What is the limiting vector u_∞?

7.3.2 For the same A and the initial guess $u_0 = \begin{bmatrix} 3 \\ 4 \end{bmatrix}$, compare the results of

(i)   three inverse power steps

$$u_{k+1} = A^{-1}u_k = \frac{1}{3}\begin{bmatrix} 2 & 1 \\ 1 & 2 \end{bmatrix}u_k;$$

(ii)   a single shifted step $u_1 = (A - \alpha I)^{-1}u_0$, with $\alpha = u_0^{\mathsf{T}}Au_0/u_0^{\mathsf{T}}u_0$.

The limiting vector u_∞ is now a multiple of the other eigenvector (1, 1).

7.3.3 Show that for any two different vectors of the same length, $\|x\| = \|y\|$, the choice $v = x - y$ leads to a Householder transformation such that $Hx = y$ and $Hy = x$.

7.3.4 For

$$x = \begin{bmatrix} 3 \\ 4 \end{bmatrix}, \quad \text{and} \quad z = \begin{bmatrix} 1 \\ 0 \end{bmatrix},$$

compute $\sigma = \|x\|$, $v = x + \sigma z$, and the Householder matrix H. Verify that $Hx = -\sigma z$.

7.3.5 Using 7.3.4, find the tridiagonal $U^{-1}AU$ produced from

$$A = \begin{bmatrix} 1 & 3 & 4 \\ 3 & 1 & 0 \\ 4 & 0 & 0 \end{bmatrix}.$$

7.3.6 Show that starting from $A_0 = \begin{bmatrix} 2 & -\frac{1}{2} \\ -1 & 2 \end{bmatrix}$, the unshifted QR algorithm produces only the modest improvement $A_1 = \frac{1}{5}\begin{bmatrix} 14 & -3 \\ -3 & 6 \end{bmatrix}$.

7.3.7 Apply a single QR step to

$$A = \begin{bmatrix} \cos \theta & \sin \theta \\ \sin \theta & 0 \end{bmatrix},$$

with the shift $\alpha = a_{22}$—which in this case means without shift, since $a_{22} = 0$. Show that the off-diagonal entries go from $\sin \theta$ to $-\sin^3 \theta$, an instance of *cubic convergence*.

7.3.8 Show that the tridiagonal $A = \begin{bmatrix} 0 & 1 \\ 1 & 0 \end{bmatrix}$ is left unchanged by every step of the QR algorithm, and is therefore one of the (rare) counterexamples to convergence. It is removed by introducing an arbitrary shift.

7.3.9 Show by induction that without shifts, $(Q_0 Q_1 \cdots Q_k)(R_k \cdots R_1 R_0)$ is exactly the QR factorization of A^{k+1}. This identity connects QR to the power method and leads to an explanation of its convergence; if $|\lambda_1| > |\lambda_2| > \cdots > |\lambda_n|$, then these eigenvalues will gradually appear in descending order on the main diagonal of A_k.

7.4 ■ ITERATIVE METHODS FOR $Ax = b$

In contrast to eigenvalues, where there was no choice, we do not absolutely need an iterative method to solve $Ax = b$. Gaussian elimination will stop at the solution x after a finite number of steps, and as long as that number is reasonable there is no problem. On the other hand, when $n^3/3$ is enormous, we may have to settle for an approximate x that can be obtained more quickly—and it is no use to go part way through elimination and then stop. Our goal is to describe methods that start from any initial guess x_0, that produce an improved approximation x_{k+1} from the previous approximation x_k, and that can be terminated at will.

Such a method is easy to invent, simply by *splitting the matrix* A. If $A = S - T$, then the equation $Ax = b$ is the same as $Sx = Tx + b$. Therefore we can try the iteration

$$\boxed{Sx_{k+1} = Tx_k + b.}$$ (1)

Of course there is no guarantee that this method is any good, and a successful splitting satisfies two different requirements:

1. The new vector x_{k+1} should be *easy to compute*. Therefore S should be a simple (and invertible!) matrix; it may be diagonal or triangular.

2. The sequence x_k should *converge* to the true solution x. If we subtract the iteration (1) from the true equation $Sx = Tx + b$, the result is a formula involving only the errors $e_k = x - x_k$:

$$Se_{k+1} = Te_k.$$ (2)

This is just a difference equation. It starts with the initial error e_0, and after k steps it produces the new error $e_k = (S^{-1}T)^k e_0$. The question of convergence is exactly the same as the question of stability: $x_k \to x$ exactly when $e_k \to 0$.

7F   The iterative method (1) is *convergent* if and only if every eigenvalue λ of $S^{-1}T$ satisfies $|\lambda| < 1$. Its rate of convergence depends on the maximum size of $|\lambda|$, which is known as the *spectral radius* of $S^{-1}T$:

$$\rho(S^{-1}T) = \max_i |\lambda_i|.$$ (3)

Remember that a typical solution to $e_{k+1} = S^{-1}Te_k$ is

$$e_k = c_1 \lambda_1^k x_1 + \cdots + c_n \lambda_n^k x_n.$$ (4)

Obviously, the largest of the $|\lambda_i|$ will eventually be dominant, and will govern the rate at which e_k converges to zero.

The two requirements on the iteration are conflicting. At one extreme, we could achieve immediate convergence with $S = A$ and $T = 0$; the first and only step of the iteration would be $Ax_1 = b$. In that case the error matrix $S^{-1}T$ is zero, its eigenvalues and spectral radius are zero, and the rate of convergence (usually defined as $-\log \rho$) is infinite. But $S = A$ may not be easy to invert; that was the original reason for a splitting. It is remarkable how a simple choice of S can succeed so well, and we discuss three possibilities:

1. $S =$ diagonal part of A (Jacobi's method)
2. $S =$ triangular part of A (Gauss-Seidel method)
3. $S =$ combination of 1 and 2 (successive overrelaxation or SOR).

S is also called a **preconditioner**, and its choice is crucial in numerical analysis.

EXAMPLE 1 (JACOBI)

$$A = \begin{bmatrix} 2 & -1 \\ -1 & 2 \end{bmatrix}, \qquad S = \begin{bmatrix} 2 & \\ & 2 \end{bmatrix}, \qquad T = \begin{bmatrix} 0 & 1 \\ 1 & 0 \end{bmatrix}, \qquad S^{-1}T = \begin{bmatrix} 0 & \frac{1}{2} \\ \frac{1}{2} & 0 \end{bmatrix}.$$

If the components of x are v and w, the Jacobi step $Sx_{k+1} = Tx_k + b$ is

$$\begin{aligned} 2v_{k+1} &= w_k + b_1 \\ 2w_{k+1} &= v_k + b_2, \end{aligned} \qquad \text{or} \qquad \begin{bmatrix} v \\ w \end{bmatrix}_{k+1} = \begin{bmatrix} 0 & \frac{1}{2} \\ \frac{1}{2} & 0 \end{bmatrix} \begin{bmatrix} v \\ w \end{bmatrix}_k + \begin{bmatrix} b_1/2 \\ b_2/2 \end{bmatrix}.$$

The decisive matrix $S^{-1}T$ has eigenvalues $\pm\frac{1}{2}$, which means that the error is cut in half (one more binary digit becomes correct) at every step. In this example, which is much too small to be typical, the convergence is fast.

If we try to imagine a larger matrix A, there is an immediate and very practical difficulty with the Jacobi iteration. *It requires us to keep all the components of x_k until the calculation of x_{k+1} is complete.* A much more natural idea, which requires only half as much storage, is to start using each component of the new vector x_{k+1} as soon as it is computed; x_{k+1} takes the place of x_k a component at a time. Then x_k can be destroyed as fast as x_{k+1} is created. The first component remains as before,

$$a_{11}(x_1)_{k+1} = (-a_{12}x_2 - a_{13}x_3 - \cdots - a_{1n}x_n)_k + b_1.$$

The next step operates immediately with this new value of x_1,

$$a_{22}(x_2)_{k+1} = -a_{21}(x_1)_{k+1} + (-a_{23}x_3 - \cdots - a_{2n}x_n)_k + b_2.$$

And the last equation in the iteration step will use new values exclusively,

$$a_{nn}(x_n)_{k+1} = (-a_{n1}x_1 - a_{n2}x_2 - \cdots - a_{nn-1}x_{n-1})_{k+1} + b_n.$$

This is called the **Gauss-Seidel method**, even though it was apparently unknown to Gauss and not recommended by Seidel. That is a surprising bit of history, because it is not a bad method. Notice that, when all the terms in x_{k+1} are moved to the left side, *the matrix S is now the lower triangular part of A*. On the right side, the other matrix T in the splitting is strictly upper triangular.

EXAMPLE 2 (GAUSS-SEIDEL)

$$A = \begin{bmatrix} 2 & -1 \\ -1 & 2 \end{bmatrix}, \quad S = \begin{bmatrix} 2 & 0 \\ -1 & 2 \end{bmatrix}, \quad T = \begin{bmatrix} 0 & 1 \\ 0 & 0 \end{bmatrix}, \quad S^{-1}T = \begin{bmatrix} 0 & \frac{1}{2} \\ 0 & \frac{1}{4} \end{bmatrix}.$$

A single Gauss-Seidel step takes the components v_k and w_k into

$$\begin{matrix} 2v_{k+1} = w_k + b_1 \\ 2w_{k+1} = v_{k+1} + b_2, \end{matrix} \quad \text{or} \quad \begin{bmatrix} 2 & 0 \\ -1 & 2 \end{bmatrix} x_{k+1} = \begin{bmatrix} 0 & 1 \\ 0 & 0 \end{bmatrix} x_k + b.$$

The eigenvalues of $S^{-1}T$ are again decisive, and they are $\frac{1}{4}$ and 0. The error is divided by 4 every time, so *a single Gauss-Seidel step is worth two Jacobi steps.†* Since both methods require the same number of operations—we just use the new value instead of the old, and actually save on storage—the Gauss-Seidel method is better.

There is a way to make it better still. It was discovered during the years of hand computation (probably by accident) that convergence is faster if we go beyond the Gauss-Seidel correction $x_{k+1} - x_k$. Roughly speaking, the ordinary method converges monotonically; the approximations x_k stay on the same side of the solution x. Therefore it is natural to try introducing an **overrelaxation factor** ω to move closer to the solution. With $\omega = 1$, we recover Gauss-Seidel; with $\omega > 1$, the method is known as **successive overrelaxation** (SOR). The optimal choice of ω depends on the problem, but it never exceeds 2. It is often in the neighborhood of 1.9.

To describe the method more explicitly, let D, L, and U be the diagonal, the strictly lower triangular, and the strictly upper triangular parts of A. (This splitting has nothing to do with the $A = LDU$ of elimination, and in fact we now have $A = L + D + U$.) The Jacobi method has $S = D$ on the left side and $T = -L - U$ on the right side, whereas Gauss-Seidel chose the splitting $S = D + L$ and $T = -U$. Now, to accelerate the convergence, we move to

$$[D + \omega L]x_{k+1} = [(1 - \omega)D - \omega U]x_k + \omega b. \tag{5}$$

Notice that for $\omega = 1$, there is no acceleration and we are back to Gauss-Seidel. But regardless of ω, the matrix on the left is lower triangular and the one on the right is upper triangular. Therefore x_{k+1} can still replace x_k, component by component, as soon as it is computed; a typical step is

$$a_{ii}(x_i)_{k+1} = a_{ii}(x_i)_k + \omega[(-a_{i1}x_1 - \cdots - a_{ii-1}x_{i-1})_{k+1} + (-a_{ii}x_i - \cdots - a_{in}x_n)_k + b_i].$$

If the old guess x_k happened to coincide with the true solution x, then the new guess x_{k+1} would stay the same, and the quantity in brackets would vanish.

---

† This rule holds in many applications, even though it is possible to construct examples in which Jacobi converges and Gauss-Seidel fails (or conversely). The symmetric case is straightforward: If all $a_{ii} > 0$, then Gauss-Seidel converges if and only if A is positive definite.

EXAMPLE 3 (SOR)   For the same matrix $A = \begin{bmatrix} 2 & -1 \\ -1 & 2 \end{bmatrix}$, each SOR step is

$$\begin{bmatrix} 2 & 0 \\ -\omega & 2 \end{bmatrix} x_{k+1} = \begin{bmatrix} 2(1-\omega) & \omega \\ 0 & 2(1-\omega) \end{bmatrix} x_k + \omega b.$$

If we divide by ω, these two matrices are the S and T in the splitting $A = S - T$; the iteration is back to $Sx_{k+1} = Tx_k + b$. Therefore the crucial matrix $S^{-1}T$, whose eigenvalues govern the rate of convergence, is

$$L_\omega = \begin{bmatrix} 2 & 0 \\ -\omega & 2 \end{bmatrix}^{-1} \begin{bmatrix} 2(1-\omega) & \omega \\ 0 & 2(1-\omega) \end{bmatrix} = \begin{bmatrix} 1-\omega & \frac{1}{2}\omega \\ \frac{1}{2}\omega(1-\omega) & 1-\omega+\frac{1}{4}\omega^2 \end{bmatrix}.$$

The optimal choice of ω is the one which makes the largest eigenvalue of L_ω (in other words, its spectral radius) as small as possible. *The whole point of overrelaxation is to discover this optimal ω.* The product of the eigenvalues equals the determinant—and if we look at the two triangular factors that multiply to give L_ω, the first has determinant $\frac{1}{4}$ (after inverting) and the second has determinant $4(1-\omega)^2$. Therefore

$$\lambda_1 \lambda_2 = \det L_\omega = (1-\omega)^2.$$

This is a general rule, that the first matrix $(D + \omega L)^{-1}$ contributes $\det D^{-1}$ because L lies below the diagonal, and the second matrix contributes $\det(1-\omega)D$ because U lies above the diagonal. Their product, in the n by n case, is $\det L_\omega = (1-\omega)^n$. (This already explains why we never go as far as $\omega = 2$; the product of the eigenvalues would be too large to allow all $|\lambda_i| < 1$, and the iteration could not converge.) We also get a clue to the behavior of the eigenvalues, which is this: At $\omega = 1$ the Gauss-Seidel eigenvalues are 0 and $\frac{1}{4}$, and as ω increases these eigenvalues approach one another. *At the optimal ω the two eigenvalues are equal, and at that moment they must both equal $\omega - 1$ in order for their product to match the determinant.†* This value of ω is easy to compute, because the sum of the eigenvalues always agrees with the sum of the diagonal entries (the trace of L_ω). Therefore the best parameter ω_{opt} is determined by

$$\lambda_1 + \lambda_2 = (\omega_{\text{opt}} - 1) + (\omega_{\text{opt}} - 1) = 2 - 2\omega_{\text{opt}} + \tfrac{1}{4}\omega_{\text{opt}}^2. \tag{6}$$

This quadratic equation gives $\omega_{\text{opt}} = 4(2 - \sqrt{3}) \approx 1.07$. Therefore the two equal eigenvalues are approximately $\omega - 1 = .07$, which is a major reduction from the Gauss-Seidel value $\lambda = \frac{1}{4}$ at $\omega = 1$. In this example, the right choice of ω has again doubled the rate of convergence, because $(\frac{1}{4})^2 \approx .07$.

The discovery that such an improvement could be produced so easily, almost as if by magic, was the starting point for 20 years of enormous activity in numerical analysis. The first problem was to develop and extend the theory of overrelaxation,

---

† If ω is further increased, the eigenvalues become a complex conjugate pair—both have $|\lambda| = \omega - 1$, so their product is still $(\omega - 1)^2$ and their modulus is now increasing with ω.

and Young's thesis in 1950 contained the solution—a simple formula for the optimal ω. The key step was to connect the eigenvalues λ of the matrix L_ω to the eigenvalues μ of the original Jacobi matrix $D^{-1}(-L - U)$. That connection is expressed by

$$(\lambda + \omega - 1)^2 = \lambda\omega^2\mu^2. \tag{7}$$

It is valid for a wide class of finite difference matrices, and if we take $\omega = 1$ (Gauss-Seidel) it yields $\lambda^2 = \lambda\mu^2$. Therefore $\lambda = 0$ and $\lambda = \mu^2$. This is confirmed by Examples 1 and 2, in which $\mu = \pm\frac{1}{2}$ and $\lambda = 0$, $\lambda = \frac{1}{4}$. It is completely typical of the relation between Jacobi and Gauss-Seidel: All the matrices in Young's class have eigenvalues μ that occur in plus-minus pairs, and (7) shows that the corresponding λ are 0 and μ^2. By using the latest approximations to x, we double the rate of convergence.

The important problem is to do better still; we want to choose ω so that the largest eigenvalue λ will be minimized. Fortunately, this problem is already solved. Young's equation (7) is nothing but the characteristic equation for the 2 by 2 example L_ω, and the best ω was the one which made the two roots λ both equal to $\omega - 1$. Exactly as in (6), where $\mu^2 = \frac{1}{4}$, this leads to

$$(\omega - 1) + (\omega - 1) = 2 - 2\omega + \mu^2\omega^2, \quad \text{or} \quad \omega = \frac{2(1 - \sqrt{1 - \mu^2})}{\mu^2}.$$

The only difference is that, for a large matrix, this pattern will be repeated for a number of different pairs $\pm\mu_i$—and we can only make a single choice of ω. The largest of these pairs gives the largest Jacobi eigenvalue μ_{max}, and it also gives the largest value of ω and of $\lambda = \omega - 1$. Therefore, since our goal is to make λ_{max} as small as possible, it is that extremal pair which specifies the best choice ω_{opt}:

$$\omega_{opt} = \frac{2(1 - \sqrt{1 - \mu_{max}^2})}{\mu_{max}^2} \quad \text{and} \quad \lambda_{max} = \omega_{opt} - 1. \tag{8}$$

This is Young's formula for the optimal overrelaxation factor.

For our finite difference matrix A, with entries $-1, 2, -1$ down the three main diagonals, we can compute the improvement brought by ω. In the examples this matrix was 2 by 2; now suppose it is n by n, corresponding to the mesh width $h = 1/(n + 1)$. The largest Jacobi eigenvalue, according to Exercise 7.4.3, is $\mu_{max} = \cos \pi h$. Therefore the largest Gauss-Seidel eigenvalue is $\mu_{max}^2 = \cos^2 \pi h$, and the largest SOR eigenvalue is found by substituting in (8):

$$\lambda_{max} = \frac{2(1 - \sin \pi h)}{\cos^2 \pi h} - 1 = \frac{(1 - \sin \pi h)^2}{\cos^2 \pi h} = \frac{1 - \sin \pi h}{1 + \sin \pi h}.$$

This can only be appreciated by an example. Suppose A is of order 21, which is very moderate. Then $h = \frac{1}{22}$, $\cos \pi h = .99$, and the Jacobi method is slow; $\cos^2 \pi h = .98$ means that even Gauss-Seidel will require a great many iterations. But since $\sin \pi h = \sqrt{.02} = .14$, the optimal overrelaxation method will have the convergence factor

$$\lambda_{\max} = \frac{.86}{1.14} = .75, \qquad \text{with} \qquad \omega_{\text{opt}} = 1 + \lambda_{\max} = 1.75.$$

The error is reduced by 25% at every step, and *a single SOR step is the equivalent of* 30 *Jacobi steps*: $(.99)^{30} = .75$.

That is a very striking result from such a simple idea. Its real applications are not in one-dimensional problems (ordinary differential equations); a tridiagonal system $Ax = b$ is already easy to solve. It is in more dimensions, for partial differential equations, that overrelaxation is important. If we replace the unit interval $0 \le x \le 1$ by the unit square $0 \le x, y \le 1$, and change the equation $-u_{xx} = f$ to $-u_{xx} - u_{yy} = f$, then the natural finite difference analogue is the "five-point scheme." The entries $-1, 2, -1$ in the x direction combine with $-1, 2, -1$ in the y direction to give a main diagonal of $+4$ and four off-diagonal entries of -1. But *the matrix A does not have a bandwidth of* 5. There is no way to number the N^2 mesh points in a square so that each point stays close to all four of its neighbors. That is the true *"curse of dimensionality,"* and parallel computers will partly relieve it.

If the ordering goes a row at a time, then every point must wait through a whole row for the neighbor above it to turn up—and the "five-point matrix" has bandwidth N:

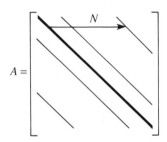

This matrix has had more attention, and been attacked in more different ways, than any other linear equation $Ax = b$. I think that the trend now is back to direct methods, based on an idea of Golub and Hockney; certain special matrices will fall apart when they are dropped the right way. (It is comparable to the Fast Fourier Transform.) Before that came the iterative methods of *alternating direction*, in which the splitting separated the tridiagonal matrix in the x direction from the one in the y direction. And before that came overrelaxation, because the Jacobi eigenvalue $\mu_{\max} = \cos \pi h$ is the same as in one dimension and so is the overrelaxation factor ω_{opt}. In every case the difficulty is to go from model problems to realistic problems, and each of these methods has its own possibilities for coping with equations more general than $-u_{xx} - u_{yy} = f$ and geometries more general than a square.

We cannot close without mentioning the method of *conjugate gradients*, which looked dead but is suddenly very much alive; it is direct rather than iterative, but unlike elimination, it can be stopped part way. And needless to say a completely new idea may still appear and win.† But it seems fair to say that it was the change from .99 to .75 that revolutionized the solution of $Ax = b$.

EXERCISES

7.4.1   For the matrix

$$A = \begin{bmatrix} 2 & -1 & 0 \\ -1 & 2 & -1 \\ 0 & -1 & 2 \end{bmatrix}$$

with eigenvalues $2 - \sqrt{2}, 2, 2 + \sqrt{2}$, find the Jacobi matrix $D^{-1}(-L - U)$ and its eigenvalues, the Gauss-Seidel matrix $(D + L)^{-1}(-U)$ and its eigenvalues, and the numbers ω_{opt} and λ_{max} for SOR. You need not compute the matrix L_{ω}.

7.4.2   For the n by n matrix

$$A = \begin{bmatrix} 2 & -1 & & \\ -1 & \cdot & \cdot & \\ & \cdot & \cdot & -1 \\ & & -1 & 2 \end{bmatrix}$$

describe the Jacobi matrix $J = D^{-1}(-L - U)$. Show that the vector $x_1 = (\sin \pi h, \sin 2\pi h, \ldots, \sin n\pi h)$ is an eigenvector of J with eigenvalue $\lambda_1 = \cos \pi h = \cos \pi/(n + 1)$.

7.4.3   For the same A, show that the vector $x_k = (\sin k\pi h, \sin 2k\pi h, \ldots, \sin nk\pi h)$ is an eigenvector. Multiply x_k by A to find the corresponding eigenvalue α_k. Verify that in the 3 by 3 case these eigenvalues are $2 - \sqrt{2}, 2, 2 + \sqrt{2}$.

Note: The eigenvalues of the Jacobi matrix $J = \frac{1}{2}(-L - U) = I - \frac{1}{2}A$ are $\lambda_k = 1 - \frac{1}{2}\alpha_k = \cos k\pi h$. These λ's do occur in plus-minus pairs and λ_{max} is $\cos \pi h$.

The following exercises require Gershgorin's "circle theorem": *Every eigenvalue of A lies in at least one of the circles $C_1, \ldots, C_n$, where C_i has its center at the diagonal entry a_{ii} and its radius $r_i = \sum_{j \neq i} |a_{ij}|$ equal to the absolute sum along the rest of the row.*

Proof $Ax = \lambda x$ leads to

$$(\lambda - a_{ii})x_i = \sum_{j \neq i} a_{ij}x_j, \quad \text{or} \quad |\lambda - a_{ii}| \leq \sum_{j \neq i} |a_{ij}| \frac{|x_j|}{|x_i|}.$$

---

† A recent choice is $S = L_0 U_0$, where small entries of the true L and U are set to zero while factoring A. It is called "*incomplete LU*" or "*incomplete Cholesky*."

If the largest component of x is x_i, then these last ratios are ≤ 1, and λ lies in the ith circle: $|\lambda - a_{ii}| \leq r_i$.

7.4.4 The matrix

$$A = \begin{bmatrix} 3 & 1 & 1 \\ 0 & 4 & 1 \\ 2 & 2 & 5 \end{bmatrix}$$

is called *diagonally dominant* because every $|a_{ii}| > r_i$. Show that zero cannot lie in any of the circles, and conclude that A is nonsingular.

7.4.5 Write down the Jacobi matrix J for this diagonally dominant A, and find the three Gershgorin circles for J. Show that all the radii satisfy $r_i < 1$ and that the Jacobi iteration converges.

7.4.6 The true solution to $Ax = b$ is slightly different from the elimination solution to $LUx_0 = b$; $A - LU$ misses zero because of roundoff. One possibility is to do everything in double precision, but a better and faster way is *iterative refinement*: Compute only one vector $r = b - Ax_0$ in double precision, solve $LUy = r$, and add the correction y to x_0. Problem: Multiply $x_1 = x_0 + y$ by LU, write the result as a splitting $Sx_1 = Tx_0 + b$, and explain why T is extremely small. This single step brings us almost exactly to x.

7.4.7 For a general 2 by 2 matrix

$$A = \begin{bmatrix} a & b \\ c & d \end{bmatrix},$$

find the Jacobi iteration matrix $S^{-1}T = -D^{-1}(L + U)$ and its eigenvalues μ_i. Find also the Gauss-Seidel matrix $-(D + L)^{-1}U$ and its eigenvalues λ_i, and decide whether $\lambda_{max} = \mu_{max}^2$.

8

LINEAR
PROGRAMMING
AND GAME THEORY

8.1 ■ LINEAR INEQUALITIES

Usually, the difference between algebra and analysis is more or less the difference between equations and inequalities. The line between the two has always seemed clear, but I have finally realized that linear programming is a counterexample: It is about inequalities, but it is unquestionably a part of linear algebra. It is also extremely useful—business decisions are more likely to apply linear programming than determinants or eigenvalues. There are three ways to approach the underlying mathematics: intuitively through the geometry, or computationally through the simplex method, or algebraically through the theory of duality. These approaches are developed in this section and in Sections 8.2 and 8.3. Then Section 8.4 is about problems where the solution is an integer, like marriage. Section 8.5 discusses poker and other matrix games, and at the end we explain the underlying minimax theorem, including its connection with the duality theorem of linear programming.

In Section 8.2, there is something new in this third edition. The simplex method is now in a very lively competition with a completely different way to do the computations, called *Karmarkar's method*. That new idea has received a lot of publicity, partly because the original reports claimed that it was 50 times faster than the simplex method. The claim still stands, but it does not apply to every problem. The new method looks effective on large sparse problems—when the matrix A has a suitable structure—and the simplex method remains on top for many other

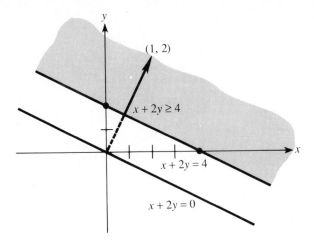

Fig. 8.1. Equations and inequalities.

problems. Nothing is absolutely clear, since the code is under development at AT&T Bell Laboratories and is secret. (Linear programming is such an important and widespread application that the commercial aspects have become serious—this is an unusual situation in scientific computing.) Fortunately the basic ideas were made public, and we can explain the principles of Karmarkar's method in Section 8.2.

This chapter can come any time after Chapter 2. There will be rectangular matrices, but not everything is a linear equation. In fact, one key to this chapter is to see the geometric meaning of *linear inequalities*. An inequality divides n-dimensional space into two *halfspaces*, one where the inequality is satisfied and the other where it is not. A typical example is $x + 2y \geq 4$. The boundary between the two halfspaces is the line $x + 2y = 4$, where the inequality is just barely satisfied. Above this line is the shaded halfspace in Figure 8.1; below it is the opposite halfspace, where the inequality is violated. The picture would be almost the same in three dimensions; the boundary becomes a plane like $x + 2y + z = 4$, and above it is the halfspace $x + 2y + z \geq 4$. In n dimensions we will still call the $n - 1$-dimensional boundary a "plane."

In addition to inequalities of this kind, there is another constraint which is fundamental to linear programming: x and y are required to be *nonnegative*. This requirement is itself a pair of inequalities, $x \geq 0$ and $y \geq 0$. Therefore we have two more halfspaces, with their boundaries on the coordinate axes: $x \geq 0$ admits all points to the right of $x = 0$, and $y \geq 0$ is the halfspace above $y = 0$.

The Feasible Set and the Cost Function

The important step is to impose all three inequalities $x + 2y \geq 4$, $x \geq 0$, and $y \geq 0$ at once. They combine to give the shaded region in Figure 8.2. You can

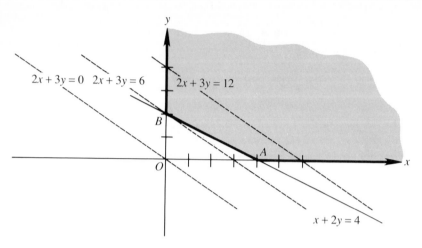

Fig. 8.2. The feasible set and the cost functions $2x + 3y$.

recognize this region as the *intersection* of the three halfspaces. It is no longer a halfspace, but it is typical of what linear programming calls a *feasible set*. To say it more directly, a feasible set is composed of the solutions to a family of linear inequalities.

The system $Ax = b$, with m equations in n unknowns, describes the intersection of m planes—one for each equation. (When $m = n$ and the planes are independent, they intersect in one point, which is $x = A^{-1}b$.) Similarly, a system of m inequalities like $Ax \geq b$ describes the intersection of m halfspaces. If we also require that every component of x is nonnegative (which is written as a vector inequality $x \geq 0$), this adds n more halfspaces. The more constraints we impose, the smaller the feasible set.

It can easily happen that a feasible set is bounded or even empty. If we switch our example to the halfspace $x + 2y \leq 4$, keeping $x \geq 0$ and $y \geq 0$, we get the triangle OAB. By combining both inequalities $x + 2y \geq 4$ and $x + 2y \leq 4$, the set shrinks to a line; the two opposing constraints force $x + 2y = 4$. If we add a contradictory constraint like $x + 2y \leq -2$, the feasible set is empty.

The algebra of linear inequalities (or feasible sets) is one part of our subject. In linear programming, however, there is another essential ingredient: We are interested *not in the set of all feasible points, but rather in the particular point that maximizes or minimizes a certain* "**cost function**." To the example $x + 2y \geq 4$, $x \geq 0$, $y \geq 0$, we add the cost function (or *objective function*) $2x + 3y$. Then the real problem in linear programming is to find the point x, y that *lies in the feasible set and minimizes the cost*.

The problem is illustrated by the geometry of Fig. 8.2. The family of costs $2x + 3y$ gives a family of parallel lines, and we have to find the minimum cost, in other words the first line to intersect the feasible set. That intersection clearly occurs at the point B, where $x^* = 0$ and $y^* = 2$; the minimum cost is $2x^* + 3y^* = 6$. The

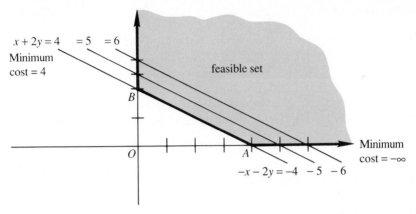

Fig. 8.3. Special cases: All points from A to B minimize $x + 2y$; minimum of $-x - 2y$ is $-\infty$.

vector $(0, 2)$ is called *feasible* because it lies in the feasible set, it is *optimal* because it minimizes the cost function, and the minimum cost 6 is the *value* of the program. We will denote optimal vectors by an asterisk.

You can see that ***the optimal vector occurs at a corner of the feasible set***. This is guaranteed by the geometry, because the lines that give the cost function (or the planes, when we get to more unknowns) are moved steadily up until they intersect the feasible set. The first contact must occur along its boundary! The simplex method will stay on that boundary, going from one corner of the feasible set to the next until it finds the corner with lowest cost. In contrast, Karmarkar's method approaches that optimal solution from ***inside*** the feasible set.

Note   With a different cost function, the intersection might not be just a single point: If the cost happened to be $x + 2y$, then the whole edge between B and A would intersect at the same time, and there would be an infinity of optimal vectors along that edge (Fig. 8.3a). The value is still unique ($x^* + 2y^*$ equals 4 for all these optimal vectors) and therefore the minimum problem still has a definite answer. On the other hand, the maximum problem would have no solution! On our feasible set, the cost can go arbitrarily high and the maximum cost is infinite. Or another way to look at this possibility, staying with minimum problems, is to reverse the cost to $-x - 2y$. Then the minimum is $-\infty$, as in Fig. 8.3b, and again there is no solution. Every linear programming problems falls into one of three possible categories:

(1)   The feasible set is *empty*
(2)   The cost function is *unbounded* on the feasible set
(3)   The cost has a *minimum* (or maximum) on the feasible set.

The first two should be very uncommon for a genuine problem in economics or engineering.

Slack Variables

There is a simple way to change the inequality constraint to an equation, by introducing the *slack variable* $w = x + 2y - 4$. In fact, this is our equation! The old constraint $x + 2y \geq 4$ is converted into $w \geq 0$, which matches perfectly the other constraints $x \geq 0$, $y \geq 0$. The simplex method will begin in exactly this way, using a slack variable for each inequality so as to have only equations and simple nonnegativity constraints. The basic problem in linear programming becomes:

Minimize cx subject to $Ax = b$ *and* $x \geq 0$.

The row vector c contains the costs; in our example $c = \begin{bmatrix} 2 & 3 \end{bmatrix}$. The unknown x contains the original variables and any slack variables. The condition $x \geq 0$ pushes the problem into the nonnegative corner of n-dimensional space. Those inequalities interfere with Gaussian elimination, and a completely new idea is needed.

An Interpretation of the Problem and Its Dual

We want to return to our original example, with cost $2x + 3y$, and put it into words. It is an illustration of the "diet problem" in linear programming, with two sources of protein—say steak and peanut butter. Each pound of peanut butter gives a unit of protein, each pound of steak gives two units, and at least four units are required in the diet. Therefore a diet containing x pounds of peanut butter and y pounds of steak is constrained by $x + 2y \geq 4$, as well as by $x \geq 0$ and $y \geq 0$. (We cannot have negative steak or peanut butter.) This is the feasible set, and the problem is to minimize the cost. If a pound of peanut butter costs \$2 and a pound of steak is \$3, then the cost of the whole diet is $2x + 3y$. Fortunately, the optimal diet is steak: $x^* = 0$ and $y^* = 2$.

Every linear program, including this one, has a **dual**. If the original problem is a minimization, then its dual is a maximization. The solution of one leads directly to the solution of the other; in fact the minimum in the given "primal problem" must equal the maximum in its dual. This is actually the central result in the theory of linear programming, and it will be explained in Section 8.3. Here we stay with the diet problem and try to interpret its dual.

In place of the shopper, who chooses between steak and peanut butter to get protein at minimal cost, the dual problem is faced by a druggist who sells *synthetic protein*. He intends to compete with steak and peanut butter. Immediately we meet the two ingredients of a typical linear program: He wants to maximize the price p, but that price is subject to linear constraints. Synthetic protein must not cost more than the protein in peanut butter (which was \$2 a unit) or the protein in steak (which was \$3 for two units). At the same time, the price must be nonnegative or the druggist will not sell. Since the diet requirement was four units of protein, the revenue to the druggist will be $4p$, and the dual problem is exactly this: *Maximize* $4p$, *subject to* $p \leq 2$, $2p \leq 3$, *and* $p \geq 0$. This is an example in which the dual is easier to solve than the primal; it has only one unknown. The constraint $2p \leq 3$ is the one that is really active, and the maximum price of synthetic protein is $1.50. Therefore the maximum revenue is $4p = \$6$.

Six dollars was also the minimal cost in the original problem, and the shopper ends up paying the same for both natural and synthetic protein. That is the meaning of the duality theorem: ***maximum equals minimum***.

Typical Applications

The next section will concentrate on the solution of linear programs. Therefore, this is the time to describe some practical situations in which the underlying mathematical question can arise: *to minimize or maximize a linear cost function subject to linear constraints*.

1. *Production Planning*. Suppose General Motors makes a profit of $100 on each Chevrolet, $200 on each Buick, and $400 on each Cadillac. They get 20, 17, and 14 miles per gallon, respectively, and Congress insists that the average car produced must get 18. The plant can assemble a Chevrolet in 1 minute, a Buick in 2 minutes, and a Cadillac in 3 minutes. What is the maximum profit in an 8-hour day?

Problem *Maximize* $100x + 200y + 400z$ *subject to*

$$20x + 17y + 14z \geq 18(x + y + z), \qquad x + 2y + 3z \leq 480, \qquad x, y, z \geq 0.$$

2. *Portfolio Selection*. Suppose three types of bonds are available: federal bonds paying 5% and rated A, municipals paying 6% and rated B, and those of a uranium company paying 9% and rated C. We can buy amounts x, y, and z not exceeding a total of $100,000. The problem is to maximize the interest, subject to the conditions that

(i)   no more than $20,000 can be invested in uranium, and
(ii) the portfolio's average quality must be at least B.

Problem *Maximize* $5x + 6y + 9z$ *subject to*

$$x + y + z \leq 100,000, \qquad z \leq 20,000, \qquad z \leq x, \qquad x, y, z \geq 0.$$

EERCISES

8.1.1 Sketch the feasible set with constraints $x + 2y \geq 6$, $2x + y \geq 6$, $x \geq 0$, $y \geq 0$. What points lie at the three "corners" of this set?

8.1.2 (recommended) On the preceding feasible set, what is the minimum value of the cost function $x + y$? Draw the line $x + y =$ constant that first touches the feasible set. What points minimize the cost functions $3x + y$ and $x - y$?

8.1.3   Show that the feasible set constrained by $x \geq 0$, $y \geq 0$, $2x + 5y \leq 3$, $-3x + 8y \leq -5$, is empty.

8.1.4   Show that the following problem is feasible but unbounded, so it has no optimal solution: Maximize $x + y$, subject to $x \geq 0$, $y \geq 0$, $-3x + 2y \leq -1$, $x - y \leq 2$.

8.1.5   Add a single inequality constraint to $x \geq 0$, $y \geq 0$ in such a way that the feasible set contains only one point.

8.1.6   What shape is the feasible set $x \geq 0$, $y \geq 0$, $z \geq 0$, $x + y + z = 1$, and what is the maximum of $x + 2y + 3z$?

8.1.7   Solve the portfolio problem that came before the exercises.

8.1.8   In the feasible set for General Motors, the nonnegativity x, y, $z \geq 0$ leaves an eighth of three-dimensional space (the positive octant). How is this cut by the two planes from the constraints, and what shape is the feasible set? How do its corners show that with only these two constraints, there will be only two kinds of cars in the optimal solution?

8.1.9   *Transportation Problem.*   Suppose Texas, California, and Alaska each produce a million barrels of oil. 800,000 barrels are needed in Chicago at a distance of 1000, 2000, and 3000 miles from the three producers; and 2,200,000 barrels are needed in New England 1500, 3000, and 3700 miles away. If shipments cost one unit for each barrel-mile, what linear program with five equality constraints has to be solved to minimize the shipping cost?

THE SIMPLEX METHOD AND KARMARKAR'S METHOD ■ 8.2

This section is about linear programming with n unknowns and m constraints. In the previous section we had $n = 2$ and $m = 1$; there were two nonnegative variables, and a single constraint $x + 2y \geq 4$. The more general case is not very hard to explain, and not very easy to solve.

The best way is to put the problem directly in matrix form. We are given

(1) an m by n matrix A
(2) a column vector b with m components
(3) a row vector c with n components.

To be "feasible," a vector x has to satisfy $x \geq 0$ and $Ax \geq b$. The "optimal" vector is the feasible vector of least cost—and the cost is $cx = c_1 x_1 + \cdots + c_n x_n$.

Minimum problem: *Minimize cx, subject to $x \geq 0$, $Ax \geq b$.*

The geometric interpretation is straightforward. The first condition $x \geq 0$ restricts the vector to the positive quadrant in n-dimensional space—this is the region common to all the halfspaces $x_j \geq 0$. In two dimensions it is a quarter of the plane, and in three dimensions it is an eighth of the space; a vector has one chance in 2^n of being nonnegative. The other m constraints produce m additional halfspaces, and the feasible vectors meet all of the $m + n$ conditions; they lie in the quadrant $x \geq 0$, and at the same time satisfy $Ax \geq b$. In other words, the feasible set is the intersection of $m + n$ halfspaces. It may be bounded (with flat sides!), or it may be unbounded, or it may be empty.

The cost function cx brings to the problem a complete family of parallel planes. One member of the family, the one that goes through the origin, is the plane $cx = 0$. If x satisfies this equation, it is a vector with "zero cost." The other planes $cx =$ constant give all other possible costs. As the cost varies, these planes sweep out the whole n-dimensional space, and *the optimal vector x^* occurs at the point where they first touch the feasible set.* The vector x^* is feasible, and its cost cx^* is minimal.

Our aim in this section is to compute x^*. We could do it (in principle) by finding all the corners of the feasible set, and computing their costs; the one with the smallest cost would be optimal. In practice such a proposal is impossible. There are billions of corners, and we cannot compute them all. Instead we turn to the *simplex method*, which has become one of the most celebrated ideas in computational mathematics. It was developed by Dantzig as a systematic way to solve linear programs, and either by luck or genius it is an astonishing success. The steps of the simplex method are summarized later, but first we try to explain them. Then we describe the new method proposed by Karmarkar.

The Geometry: Movement Along Edges

I think it is the geometric explanation that gives the method away. The first step simply locates a corner of the feasible set. This is "Phase I," which we suppose

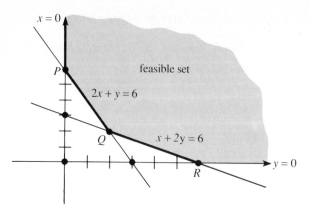

Fig. 8.4. The corners and edges of the feasible set.

to be complete. Then the heart of the method is in its second phase, **which goes from corner to corner along the edges of the feasible set.** At a typical corner there are n edges to choose from, some leading away from the optimal but unknown x^*, and others leading gradually toward it. Dantzig chose to go along an edge that is guaranteed to decrease the cost. That edge leads to a new corner with a lower cost, and there is no possibility of returning to anything more expensive. Eventually a special corner is reached, from which all edges go the wrong way: The cost has been minimized. That corner is the optimal vector x^*, and the method stops.

The real problem is to turn this idea into linear algebra. First we interpret the words *corner* and *edge* in n dimensions. A corner is the meeting point of n different planes, each given by a single equation—just as three planes (or the front wall, the side wall, and the floor) produce a corner in three dimensions. Remember that the feasible set in linear programming is determined by m inequalities $Ax \geq b$ and n inequalities $x \geq 0$. Each corner of the set comes from turning n of these $n + m$ inequalities into equations, and finding the intersection of these n planes. In particular, one possibility is to choose the n equations $x_1 = 0, \ldots, x_n = 0$, and end up with the point at the origin. Like all the other possible choices, *this intersection point will only be a genuine corner if it also satisfies the m remaining constraints.* Otherwise it is not even in the feasible set, and is a complete fake. The example in Exercise 8.1.1 had $n = 2$ variables and $m = 2$ constraints; there are six possible intersections, illustrated in Fig. 8.4. Three of them are actually corners of the feasible set. These three are marked P, Q, R; they are the vectors $(0, 6)$, $(2, 2)$, and $(6, 0)$. One of them must be the optimal vector (unless the minimum cost is $-\infty$). The other three, including the origin, are fakes.

In general there are $(n + m)!/n!m!$ possible intersections, because that counts the number of ways to choose n planes out of $n + m$ candidates.† If the feasible set is empty, then of course none of these intersections will be genuine corners. It is the

---

† The size of that number makes computing all corners totally impractical for large m and n.

task of Phase I either to find one genuine corner or to establish that the set is empty. We continue on the assumption that a corner has been found.

Now, for an edge: Suppose one of the n intersecting planes is removed, leaving only $n - 1$ equations and therefore one degree of freedom. The points that satisfy these $n - 1$ equations form an edge which comes out of the corner. Geometrically, it is the intersection of the $n - 1$ planes. Again linear programming forces us to stay in the feasible set, and we have no choice of direction along the edge; only one direction remains feasible. But we do have a choice of n different edges, and Phase II must make that choice.

To describe this phase, we have to rewrite $Ax \geq b$ in a form completely parallel to the n simple constraints $x_1 \geq 0, \ldots, x_n \geq 0$. This is the role of the **slack variables** $w = Ax - b$. The constraints $Ax \geq b$ are translated into $w_1 \geq 0, \ldots, w_m \geq 0$, with one slack variable for every row of A. Then the equation $w = Ax - b$, or $Ax - w = b$, goes into matrix form as

$$[A \quad -I] \begin{bmatrix} x \\ w \end{bmatrix} = b.$$

The feasible set is governed by these m equations and the $n + m$ simple inequalities $x \geq 0$, $w \geq 0$. We now have **equality constraints** and nonnegativity.

To make the change complete, we want to leave no distinction whatsoever between the original x and the newly arrived w. The simplex method notices no difference, and it would be pointless to go on with the notations

$$[A \quad -I] \qquad \text{and} \qquad \begin{bmatrix} x \\ w \end{bmatrix}.$$

Therefore *we rename that larger matrix A, and we rename that longer vector x*. The equality constraints are then $Ax = b$, and the $n + m$ simple inequalities become just $x \geq 0$.† The original cost vector c needs to be extended by adding m more components, all zero; then the cost cx is the same for the new meaning of x and c as for the old. The only trace left of the slack variable w is in the fact that the new matrix A is m by $n + m$, and the new x has $n + m$ components. We keep this much of the original notation, leaving m and n unchanged as a reminder of what happened. The problem has become: **Minimize** cx, **subject to** $x \geq 0$, $Ax = b$. Notice that A is rectangular.

EXAMPLE In the problem illustrated by Fig. 8.4, with the constraints $x + 2y \geq 6$, $2x + y \geq 6$, and the cost $x + y$, the new system has

$$A = \begin{bmatrix} 1 & 2 & -1 & 0 \\ 2 & 1 & 0 & -1 \end{bmatrix}, \qquad b = \begin{bmatrix} 6 \\ 6 \end{bmatrix}, \qquad \text{and} \qquad c = \begin{bmatrix} 1 & 1 & 0 & 0 \end{bmatrix}.$$

---

† Economics or engineering might have given us equality constraints in the first place, in which case the simplex method starts right out from $Ax = b$—with no need for slack variables.

With this change to equality constraints, the simplex method can begin. Remember that Phase I has already found a corner of the feasible set, where n planes meet: n of the original inequalities $x \geq 0$ and $Ax \geq b$ (alias $w \geq 0$) are turned into equations. In other words, *a corner is a point where n components of the new vector x (the old x and w) are zero.* In $Ax = b$, these n components of x are the *free variables* and the remaining m components are the *basic variables*. Then, setting the n free variables to zero, the m equations $Ax = b$ determine the m basic variables. This solution x is called *basic*, to emphasize its complete dependence on those variables. It is the particular solution of Section 2.2. It will be a genuine corner if its m nonzero components are positive; then it belongs to the feasible set.

> **8A** The *corners of the feasible set* are exactly the *basic feasible solutions* of $Ax = b$: A solution is *basic* when n of its $m + n$ components are zero, and it is *feasible* when it satisfies $x \geq 0$. Phase I of the simplex method finds one basic feasible solution, and Phase II moves step by step to the optimal one.

EXAMPLE The corner point P in Fig. 8.4 is the intersection of $x = 0$ with $2x + y - 6 = 0$, so two components of x are zero and two components equal six; it is basic and feasible.

$$Ax = \begin{bmatrix} 1 & 2 & -1 & 0 \\ 2 & 1 & 0 & -1 \end{bmatrix} \begin{bmatrix} 0 \\ 6 \\ 6 \\ 0 \end{bmatrix} = \begin{bmatrix} 6 \\ 6 \end{bmatrix} = b.$$

The crucial decision is still to be made: Which corner do we go to next? We have found one corner, where $m = 2$ components of x are positive and the rest are zero, and we want to move along an edge to an adjacent corner. The fact that the two corners are neighbors means that of the m basic variables, $m - 1$ *will remain basic and only one will become free* (zero). At the same time one variable which was free will become basic; its value will move up from zero, and the other $m - 1$ basic components will change but stay positive. The real decision is which variable to remove from the basis and which to add. Once we know which variables are basic at the new corner, they are computed by solving $Ax = b$, with the rest of the components of x set to zero.

Example of an entering variable and a leaving variable

$$\text{Minimize} \quad 7x_3 - x_4 - 3x_5 \quad \text{subject to} \quad \begin{aligned} x_1 \quad + x_3 + 6x_4 + 2x_5 &= 8 \\ x_2 + x_3 \quad + 3x_5 &= 9. \end{aligned}$$

We start from the corner where $x_1 = 8$ and $x_2 = 9$ (they are the basic variables) and $x_3 = x_4 = x_5 = 0$. The constraints are satisfied but the cost may not be minimal. In fact, the cost function determines which free variable to bring into the basis. It would be foolish to make x_3 positive, because its coefficient is $+7$ and we are trying to lower the cost. Either x_4 or x_5 is reasonable, and we choose x_5 because its coefficient -3 is the more negative. *The entering variable is x_5.*

With x_5 entering the basis, x_1 or x_2 must leave. In the first equation, we increase x_5 and decrease x_1 while keeping $x_1 + 2x_5 = 8$. When x_5 reaches 4, x_1 will be down to zero. The second equation keeps $x_2 + 3x_5 = 9$; here x_5 can only increase as far as 3, and then $x_2 = 0$. To go further would make x_2 negative. This locates the new corner, where $x_5 = 3$, $x_3 = 0$, and from the first equation $x_1 = 2$. *The leaving variable is x_2,* and the cost is down to -9.

Note   In the constraints, take the ratios of the right sides to the coefficients of the entering variable, $\frac{8}{2}$ and $\frac{9}{3}$. The smallest ratio tells which variable will leave. We consider only positive ratios, because if the coefficient of x_5 were -3 instead of $+3$, then increasing x_5 would actually *increase* x_2 at the same time (at $x_5 = 10$ the second equation would give $x_2 = 39$); now x_2 will never reach zero. If all coefficients of x_5 are negative, this is the *unbounded* case: We can make x_5 arbitrarily large, and bring the cost down toward $-\infty$. But in this example, *the second ratio $\frac{9}{3}$ says that the second variable leaves,* and it also gives $x_5 = 3$.

The current step is ended at the new corner $x_1 = 2$, $x_2 = x_3 = x_4 = 0$, $x_5 = 3$. However, the next step will only be easy if the equations again have a convenient form—with the basic variables standing by themselves as x_1 and x_2 originally did. Therefore, we "pivot" by solving the second equation for x_5, and substituting the result into the cost function and into the first equation: $3x_5 = 9 - x_2 - x_3$. The new problem, starting from the new corner, is:

Minimize $7x_3 - x_4 - (9 - x_2 - x_3) = x_2 + 8x_3 - x_4 - 9$, subject to

$$x_1 - \tfrac{2}{3}x_2 + \tfrac{1}{3}x_3 + 6x_4 \qquad = 2$$
$$\tfrac{1}{3}x_2 + \tfrac{1}{3}x_3 \qquad + x_5 = 3.$$

The next step is easy to take. The only negative coefficient in the cost means that x_4 must be the entering variable. The ratios of $\frac{2}{6}$ and $\frac{3}{0}$, the right sides divided by the x_4 column, mean that x_1 is the departing variable. The new corner is at $x_1 = x_2 = x_3 = 0$, $x_4 = \frac{1}{3}$, $x_5 = 3$, and the new cost is $-9\frac{1}{3}$. This is optimal.

In a large problem, it is possible for a departing variable to reenter the basis later on. But, since the cost keeps going down—except in a degenerate case—it can never happen that all of the m basic variables are the same as before. No corner is ever revisited, and the method must end at the optimal corner (or at $-\infty$ if the cost turns out to be unbounded). What is remarkable is the speed at which the optimum is found.

Summary   The numbers 7, -1, -3 at the first corner and 1, 8, -1 at the second corner decided the entering variables. (These numbers go into r, the crucial vector defined below. When they are positive we stop.) The ratios decided the leaving variables.

Remark on degeneracy   A corner is called *degenerate* if more than the usual n components of x are zero. More than n planes pass through the corner, so some of the basic variables happen to vanish. This is recognized, after pivoting, by the appearance of a zero on the right side of the equations. The ratios which determine the departing variable will include zeros, and the basis might change without actually leaving the corner. In theory, we could cycle forever in the choice of basis.

Fortunately, cycling does not occur. It is so rare that commercial codes ignore it. Unfortunately, degeneracy is extremely common in applications—if you print the cost after each simplex step you see it repeat several times before the simplex method finds a good edge. Then the cost decreases again.

The Tableau

Each simplex step involves decisions and row operations—the entering and leaving variables have to be chosen, and they have to be made to come and go. One way to organize the step is to fit the data into a large matrix, or **tableau.** To explain the operations, we first carry them out in matrix terms, arriving at a formula for the cost (and for the stopping test, which only the optimal corner can pass). We are given the matrix A, the right side b, and the cost vector c, and the starting tableau is just a bigger matrix:

$$T = \left[\begin{array}{c|c} A & b \\ \hline c & 0 \end{array} \right].$$

The tableau is $m + 1$ by $m + n + 1$, with one extra column and row. At the start the basic variables may be mixed with the others, and our first step is to get one basic variable alone on each row; then the decisions are easy to make. Renumbering if necessary, suppose that $x_1, \ldots, x_m$ are the basic variables at the current corner, and the other n x's are free (zero). Then the first m columns of A form a square matrix B (the *basic matrix* for that corner) and the last n columns give an m by n matrix N. Similarly, the cost vector c is split into $[c_B \quad c_N]$, and the unknown x into $[x_B \quad x_N]^T$. At the corner itself, $x_N = 0$. With these free variables out of the way, the equation $Ax = b$ turns into $Bx_B = b$ and determines the basic variables x_B. The cost is $cx = c_B x_B$. To operate with the tableau, we imagine it partitioned with the basic columns first:

$$T = \left[\begin{array}{c|c|c} B & N & b \\ \hline c_B & c_N & 0 \end{array} \right].$$

The basic variables will stand alone if we can get the identity matrix in place of B. That just means ordinary elimination steps (row operations) to reach

$$T' = \left[\begin{array}{c|c|c} I & B^{-1}N & B^{-1}b \\ \hline c_B & c_N & 0 \end{array} \right].$$

The matrix that transforms B into I has to be B^{-1}, even though it will never appear explicitly. Therefore, all the subtractions of one row from another must amount to multiplying through by B^{-1}. (We are really doing a Gauss-Jordan step; the pivots are divided out to give ones on the diagonal, and zeros are produced above them as well as below.) To finish, c_B times the top part is subtracted from the bottom:

$$T'' = \left[\begin{array}{c|c|c} I & B^{-1}N & B^{-1}b \\ \hline 0 & c_N - c_B B^{-1}N & -c_B B^{-1}b \end{array}\right].$$

Now the tableau is ready, if we can interpret it properly. Our problem was to minimize cx subject to $x \geq 0$, $Ax = b$. The equation $Ax = b$ has been multiplied by B^{-1} to give

$$x_B + B^{-1}Nx_N = B^{-1}b, \tag{1}$$

and the cost $cx = c_B x_B + c_N x_N$ has been turned into

$$cx = (c_N - c_B B^{-1}N)x_N + c_B B^{-1}b. \tag{2}$$

The main point is that every important quantity appears in the tableau. On the far right are the basic variables $x_B = B^{-1}b$ (the free variables are just $x_N = 0$). The current cost is $cx = c_B B^{-1}b$, which is in the bottom corner with a minus sign. Most important, we can decide whether the corner is optimal by looking at $r = c_N - c_B B^{-1}N$ in the middle of the bottom row. If any entries in r are negative, the cost can still be reduced. We can make $rx_N < 0$ in equation (2). On the other hand, if $r \geq 0$ the best corner has been found. This is the **stopping test**, or **optimality condition**:

8B If the vector $r = c_N - c_B B^{-1}N$ is nonnegative, then no reduction in cost can be achieved: The corner is already optimal and the minimum cost is $c_B B^{-1}b$. Thus the stopping test is $r \geq 0$. When this test fails, any negative component of r corresponds to an edge along which the cost goes down. The usual strategy is to choose the most negative component of r.

The components of r are the **reduced costs**—what it costs to use a free variable *minus what it saves*. If the direct cost (in c_N) is less than the saving (due to less use of the basic variables), it will pay to try that free variable.[†] Suppose the *most negative* reduced cost is r_i. Then the ith component of x_N will become positive. That is the **entering variable**, which increases from zero to a positive value α.

---

† Computing r is called **pricing out** the variables.

What is α? It is determined by the end of the edge. As the entering component x_i is increased, the other components of x may decrease (in order to maintain $Ax = b$). The first component x_k that decreases to zero becomes the **leaving variable**—it changes from basic to free. *We reach the next corner when a component of x_B reaches zero.*

At that point we have reached a new x which is both feasible and basic: It is feasible because we still have $x \geq 0$, and it is basic because we again have n zero components. The ith component of x_N, which went from zero to α, replaces the kth component of x_B, which dropped to zero. The other components of x_B will have moved around, but remain positive. The one that drops to zero is the one that gives the minimum ratio in (3). We will suppose it is number k.

8C   Suppose u is the ith column of N, where x_i is the entering variable—chosen to go from free to basic. Then its component at the new corner is the minimum ratio

$$\alpha = \min \frac{(B^{-1}b)_j}{(B^{-1}u)_j} = \frac{(B^{-1}b)_k}{(B^{-1}u)_k}. \tag{3}$$

This minimum is taken only over the positive components of $B^{-1}u$. If there are no positive components, the next corner is infinitely far away and the cost can be reduced forever; the minimal cost is $-\infty$. Otherwise, the old kth column of B leaves the basis and the new column u enters.

The columns $B^{-1}b$ and $B^{-1}u$ are in the final tableau ($B^{-1}u$ is a column of $B^{-1}N$, the one above the most negative entry in the bottom row r). This simplex step is summarized after an example, and then the method begins again at the new corner.

EXAMPLE   With the cost function $x + y$ and the constraints $x + 2y - w = 6$ and $2x + y - v = 6$, which we had earlier, the first tableau is

$$\begin{bmatrix} 1 & 2 & -1 & 0 & \vdots & 6 \\ 2 & 1 & 0 & -1 & \vdots & 6 \\ \hline 1 & 1 & 0 & 0 & \vdots & 0 \end{bmatrix}.$$

If we start at the same corner P in Fig. 8.4, where $x = 0$ intersects $2x + y = 6$ (which means $v = 0$), then the basic variables are the other two, y and w. To be organized, we exchange columns 1 and 3 to put basic variables before free variables:

$$T' = \begin{bmatrix} -1 & 2 & \vdots & 1 & 0 & \vdots & 6 \\ 0 & 1 & \vdots & 2 & -1 & \vdots & 6 \\ \hline 0 & 1 & \vdots & 1 & 0 & \vdots & 0 \end{bmatrix}.$$

Then elimination multiplies the first row by -1, to give a unit pivot, and uses the second row to produce zeros in the second column:

$$T'' = \begin{bmatrix} 1 & 0 & \vdots & 3 & -2 & \vdots & 6 \\ 0 & 1 & \vdots & 2 & -1 & \vdots & 6 \\ \hline 0 & 0 & \vdots & -1 & 1 & \vdots & -6 \end{bmatrix}.$$

First we look at $r = \begin{bmatrix} -1 & 1 \end{bmatrix}$, in the bottom row. It has a negative entry, so the current corner $w = y = 6$ and the current cost $+6$ are not optimal. The negative entry is in column 3, so the third variable will enter the basis. The column above that negative entry is $B^{-1}u = \begin{bmatrix} 3 & 2 \end{bmatrix}^T$; its ratios with the last column are $\frac{6}{3}$ and $\frac{6}{2}$. Since the first ratio is the smaller, it is the first unknown w (and the first column of the tableau) that will be pushed out of the basis.

The new tableau exchanges columns 1 and 3, and pivoting by elimination gives

$$\begin{bmatrix} 3 & 0 & \vdots & 1 & -2 & \vdots & 6 \\ 2 & 1 & \vdots & 0 & -1 & \vdots & 6 \\ \hline -1 & 0 & \vdots & 0 & 1 & \vdots & -6 \end{bmatrix} \rightarrow \begin{bmatrix} 1 & 0 & \vdots & \frac{1}{3} & -\frac{2}{3} & \vdots & 2 \\ 0 & 1 & \vdots & -\frac{2}{3} & \frac{1}{3} & \vdots & 2 \\ \hline 0 & 0 & \vdots & \frac{1}{3} & \frac{1}{3} & \vdots & -4 \end{bmatrix}.$$

Now $r = \begin{bmatrix} \frac{1}{3} & \frac{1}{3} \end{bmatrix}$ is positive, and the stopping test is passed. The corner $x = y = 2$ and its cost $+4$ are optimal.

The Organization of a Simplex Step

We have achieved a transition from the geometry of the simplex method to the algebra—from the language of "corners" to "basic feasible solutions." We now know that it is the vector r and the ratio α which are decisive, and we want to look once more at their calculation. This is the heart of the simplex method, and it can be organized in three different ways:

(1) In a tableau.
(2) By computing B^{-1}, and updating it when the ith column u in N replaces the current kth column of B.
(3) By computing $B = LU$, and updating these factors instead of B^{-1}.

This list is really a brief history of the simplex method. In some ways, the most fascinating stage was the first—the *tableau*—which dominated the subject for so many years. For most of us it brought an aura of mystery to linear programming, chiefly because it managed to avoid matrix notation almost completely (by the skillful device of writing out all matrices in full!). For computational purposes, however, the day of the tableau—except for small problems in textbooks—is over.

To see why, remember that after r is computed and its most negative coefficient indicates which column will enter the basis, none of the other columns above r will be used. It was a waste of time to compute them. In a larger problem, hundreds of columns would be computed time and time again, just waiting for their turn to enter the basis. It makes the theory clear, to do the eliminations so completely, but in practice it cannot be justified.

We are ready to leave the tableau. It is quicker, and in the end simpler, to look at the simplex method and see what calculations are really necessary. Each step exchanges a column of N for a column of B, and it has to decide (from r and α) which columns to choose. This step requires the following cycle of operations, beginning with the current basis matrix B and the current solution $x_B = B^{-1}b$:

Table 8.1   A Step of the Simplex Method

(1)   Compute the row vector $\lambda = c_B B^{-1}$ and the reduced costs $r = c_N - \lambda N$.
(2)   If $r \geq 0$ stop: the current solution is optimal. Otherwise, if r_i is the most negative component, choose the ith column of N to enter the basis. Denote it by u.
(3)   Compute $v = B^{-1}u$.
(4)   Calculate the ratios of $B^{-1}b$ to $B^{-1}u$, admitting only the positive components of $B^{-1}u$. If there are no positive components, the minimal cost is $-\infty$; if the smallest ratio occurs at component k, then the kth column of the current B will leave.
(5)   Update B (or B^{-1}) and the solution $x_B = B^{-1}b$. Return to 1.

This is sometimes called the **revised simplex method**, to distinguish it from the operations on a tableau. It is really the simplex method itself, boiled down.

This discussion is finished once we decide how to compute steps 1, 3, and 5:

$$\lambda = c_B B^{-1}, \qquad v = B^{-1}u, \qquad \text{and} \qquad x_B = B^{-1}b. \qquad (4)$$

The most popular way is to work directly with B^{-1}, calculating it explicitly at the first corner. Then, at succeeding corners, the pivoting step is simple. When column k of the identity matrix is replaced by u, column k of B^{-1} is replaced by $v = B^{-1}u$. To recover the identity matrix, elimination will multiply the old B^{-1} by

$$E^{-1} = \begin{bmatrix} 1 & v_1 & \\ & \cdot & \cdot \\ & v_k & \\ & \cdot & \cdot \\ & v_n & 1 \end{bmatrix}^{-1} = \begin{bmatrix} 1 & -v_1/v_k & \\ & \cdot & \cdot \\ & 1/v_k & \\ & \cdot & \cdot \\ & -v_n/v_k & 1 \end{bmatrix} \qquad (5)$$

Many simplex codes use the "**product form of the inverse**," which saves these simple matrices E^{-1} instead of directly updating B^{-1}. When needed, they are applied to b and c_B. At regular intervals (maybe 40 simplex steps) B^{-1} is recomputed and the E^{-1} are erased. Equation (5) is checked in Exercise 8.2.9.

A newer approach is to use the ordinary methods of numerical linear algebra, regarding (4) as three equations sharing the same coefficient matrix B:

$$\lambda B = c_B, \qquad Bv = u, \qquad Bx_B = b. \qquad (6)$$

Then the standard factorization $B = LU$ (or $PB = LU$, with row exchanges for stability) leads to the three solutions. L and U can be updated instead of recomputed.

Just one more thing: *How many simplex steps do we have to take?* This is impossible to answer in advance. Experience shows that the method touches only about m or $3m/2$ different corners, which means an operation count of about $m^2 n$. That is comparable to ordinary elimination for $Ax = b$, and is the reason for the simplex method's success. But mathematics shows that the path length cannot always be bounded by any fixed multiple or power of m. The worst feasible sets (Klee and Minty invented a lopsided cube) can force the simplex method to try every corner. Until recently it was not known if linear programming was in the nice class P—solvable in polynomial time—or in the dreaded class NP, like the traveling salesman problem. For the NP problems it is believed but not proved that all deterministic algorithms must take exponentially long to finish (again in the worst case).

It was **Khachian's method** which showed that linear programming could be solved in polynomial time. It stayed inside the feasible set, and captured the solution in a series of shrinking ellipsoids. Then **Karmarkar's method**, described below, was competitive not only in theory but in practice. All this time, however, the simplex method was doing the job—in an *average* time which is now proved (for a variant of the usual method) to be polynomial. For some reason, hidden in the geometry of many-dimensional polyhedra, bad feasible sets are rare and the simplex method is lucky.

Karmarkar's Method

We come now to the most sensational event in the recent history of linear programming. Karmarkar proposed a method based on two simple ideas, and in his experiments it defeated the simplex method. In other experiments on other problems that has not happened. The choice of problem and the details of the code are both crucial, and the debate is still going on. But the underlying ideas are so natural, and fit so perfectly into the framework of applied linear algebra, that they can be explained in a few paragraphs. I will start with the *rescaling algorithm*, and then add the special devices that allowed Karmarkar to prove convergence in polynomial time.†

The first idea is to start from a point *inside the feasible set*—we will suppose it is $x^0 = (1, 1, \ldots, 1)$—and to move in a *cost-reducing direction*. Since the cost is cx, the best direction is toward $-c$. Normally that takes us off the feasible set; moving in that direction does not maintain $Ax = b$. If $Ax^0 = b$ and $Ax^1 = b$, then $\Delta x = x^1 - x^0$ has to satisfy $A\Delta x = 0$. **The step Δx must lie in the nullspace of A.**

---

† The number of operations is bounded by powers of m and n, as in elimination and Gram-Schmidt. For integer programming and factoring into primes all known algorithms take exponentially long in the worst case—the celebrated problem "$P \neq NP$" is to prove that for such problems no polynomial algorithm can exist.

Therefore we *project* $-c$ onto the nullspace, to find the feasible direction closest to the best direction. This is the natural but expensive step in Karmarkar's method.

Projection onto the nullspace of A was Exercise 3.3.20. The projection matrix is

$$P = I - A^T(AA^T)^{-1}A, \tag{7}$$

which takes every vector into the nullspace because $AP = A - (AA^T)(AA^T)^{-1}A = 0$. If x is already in the nullspace then $Px = x$, because $Ax = 0$. You recognize that $(AA^T)^{-1}$ is not taken literally; the inverse is not computed. Instead we solve a linear equation, and subtract the row space component from c (which is now a column vector):

$$AA^Ty = Ac \qquad \text{and then} \qquad Pc = c - A^Ty. \tag{8}$$

The step Δx is a multiple of the projection $-Pc$. The longer the step, the more the cost is reduced—but we cannot go out of the feasible set. The multiple of $-Pc$ is chosen so that x^1 is close to, but *a little inside*, the boundary at which a component of x reaches zero.

EXAMPLE   Cost $c^Tx = 8x_1 + 2x_2 + 7x_3$; constraint $Ax = 2x_1 + 2x_2 + x_3 = 5 = b$. In this case the feasible set is the triangle PQR in Fig. 8.5. It is a plane in three-dimensional space, cut off by the constraints $x_1 \geq 0$, $x_2 \geq 0$, $x_3 \geq 0$. The optimal

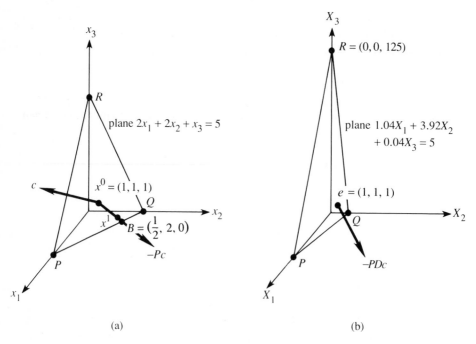

(a)   (b)

Fig. 8.5. A step of the rescaling algorithm: x^0 to x^1 by projection, x^1 to e by scaling.

solution must be one of the corners:

$$P: \quad x = (\tfrac{5}{2}, 0, 0) \quad \text{costs} \quad c^\mathsf{T} x = 20$$
$$Q: \quad x = (0, \tfrac{5}{2}, 0) \quad \text{costs} \quad c^\mathsf{T} x = 5$$
$$R: \quad x = (0, 0, 5) \quad \text{costs} \quad c^\mathsf{T} x = 35$$

Thus $x = (0, \tfrac{5}{2}, 0)$ is optimal, and we hope that the step Δx moves toward Q.

The computation projects $c = (8, 2, 7)$ onto the plane $2x_1 + 2x_2 + x_3 = 5$:

$$A = \begin{bmatrix} 2 & 2 & 1 \end{bmatrix} \quad \text{so} \quad AA^\mathsf{T} y = Ac \quad \text{becomes} \quad 9y = 27$$

$$Pc = c - A^\mathsf{T} y = \begin{bmatrix} 8 \\ 2 \\ 7 \end{bmatrix} - 3 \begin{bmatrix} 2 \\ 2 \\ 1 \end{bmatrix} = \begin{bmatrix} 2 \\ -4 \\ 4 \end{bmatrix}.$$

This vector should be in the nullspace of $A = \begin{bmatrix} 2 & 2 & 1 \end{bmatrix}$, and it is. It indicates the direction of the step Δx. Moving from the starting point $x^0 = (1, 1, 1)$ in this direction gives

$$x^0 - sPc = \begin{bmatrix} 1 - 2s \\ 1 + 4s \\ 1 - 4s \end{bmatrix}.$$

At the steplength $s = \tfrac{1}{4}$, the third component is zero. That choice reaches the boundary point B in the figure, with components $(\tfrac{1}{2}, 2, 0)$. It is closer to the optimal Q, but going all the way to the boundary leaves too little room for the next step. It might happen (although it doesn't happen here) that the zero component of B is not zero at the optimal corner. Therefore we stop short of B, reducing s from .25 to .24. This multiplying factor $\alpha = \tfrac{.24}{.25} = .96$ is typical, and the low value $\alpha = .25$ originally published by Karmarkar was chosen to prove convergence in polynomial time—not to help the reader achieve convergence in real time. When $s = \tfrac{1}{4}(.96) = .24$, the step ends at

$$x^1 = \begin{bmatrix} 0.52 \\ 1.96 \\ 0.04 \end{bmatrix} \quad \text{instead of} \quad B = \begin{bmatrix} 0.5 \\ 2.0 \\ 0.0 \end{bmatrix}.$$

That completes the first idea—the projection which gives the *steepest feasible descent*. The second step needs a new idea, since to continue in the same direction is useless.

Karmarkar's suggestion is to **transform x^1 back to the central position** $e = (1, 1, 1)$. That change of variables will change the problem, and then the second step projects the new c onto the nullspace of the new A. His change of variables (given later) was nonlinear, but the simplest transformation is just a **rescaling** of the axes.

To move $x^1 = (0.52, 1.96, 0.04)$ to the center point e, we divide the components of every vector by those three numbers. In other words, we put those numbers into a diagonal matrix D and **rescale all vectors by** D^{-1}:

$$D = \begin{bmatrix} 0.52 & & \\ & 1.96 & \\ & & 0.04 \end{bmatrix} \quad \text{and} \quad D^{-1}x^1 = \begin{bmatrix} 1 \\ 1 \\ 1 \end{bmatrix} = e.$$

We are back at the center, with room to move. However the feasible set (the triangle PQR) has changed (Fig. 8.5b) and so has the cost vector c. The rescaling from x to $X = D^{-1}x$ has two effects:

the constraint $Ax = b$ becomes $ADX = b$;

the cost c^Tx becomes c^TDX.

Therefore *the matrix AD takes the place of A, and the vector c^TD takes the place of c^T.* In our example that produces

$$AD = \begin{bmatrix} 2 & 2 & 1 \end{bmatrix} D = \begin{bmatrix} 1.04 & 3.92 & 0.04 \end{bmatrix}$$
$$c^TD = \begin{bmatrix} 8 & 2 & 7 \end{bmatrix} D = \begin{bmatrix} 4.16 & 3.92 & 0.28 \end{bmatrix}.$$

The new problem is to minimize c^TDX subject to $ADX = b$ and $X \geq 0$ (which is equivalent to $x \geq 0$). In a larger example the matrix A would have m rows instead of one. The second step projects the new cost vector onto the nullspace of AD. Equation (8) changes to

$$AD^2A^Ty = AD^2c \quad \text{and then} \quad PDc = Dc - DA^Ty. \tag{9}$$

Now you have the whole algorithm, except for starting and stopping. It could start from any point x^0 inside the feasible set. At each step the current guess x^k is rescaled to the point $e = (1, 1, \ldots, 1)$, and then the projection (9) gives the step direction in the rescaled variable. The output from the step is x^{k+1}:

Rescaling Algorithm

(1) Construct a diagonal matrix D from the components of x^k, so that

$$D^{-1}x^k = (1, 1, \ldots, 1) = e.$$

(2) With this D compute the projection PDc in equation (9).
(3) Determine the number s so that $e - sPDc$ has a zero component.
(4) Reduce that s by a factor α (say $\alpha = .96$).
(5) The new vector is $x^{k+1} = x^k - sDPDc$.

The step ΔX in the rescaled variable goes from e to $e - sPDc$. In the original variables it goes from x^k to $x^k - sDPDc$. We do those computations for the example, to go from x^1 to x^2:

$$AD^2A^{\mathrm{T}}y = AD^2c \quad \text{becomes} \quad 16.45y = 19.7 \quad \text{or} \quad y = 1.2$$

$$PDc = \begin{bmatrix} 4.16 \\ 3.92 \\ 0.28 \end{bmatrix} - 1.2 \begin{bmatrix} 1.04 \\ 3.92 \\ 0.04 \end{bmatrix} = \begin{bmatrix} 2.91 \\ -0.78 \\ 0.23 \end{bmatrix}.$$

The step length $s = 1/2.91$ would bring the first component to zero, when we sub-tract $sPDc$ from $e = (1, 1, 1)$. A more conservative step is $s = .96/2.91 = .33$. Then the new X^2 and the new $x^2 = DX^2$ are

$$X^2 = e - (.33)PDc = \begin{bmatrix} 0.040 \\ 1.257 \\ 0.924 \end{bmatrix} \quad \text{and} \quad x^2 = \begin{bmatrix} 0.02 \\ 2.46 \\ 0.04 \end{bmatrix}.$$

We check that x^2 lies in the feasible set: $2x_1 + 2x_2 + x_3 = .04 + 4.92 + .04 = 5$. It is very close to the corner Q, whose components are $(0, 2.5, 0)$. We are certainly close enough for the last step in Karmarkar's method—*to see which corner is best and jump to it.* Everybody knows that the optimal point is one of the vertices P, Q, R, and two steps of the algorithm have identified the right corner Q.

All the work was in step 2 of the algorithm, which is a weighted projection. In the language of Section 3.6, the weighting matrix is $W = D$ and the product $W^{\mathrm{T}}W$ is D^2. *The weighted normal equation yields* y, and its solution is a fundamental problem of numerical linear algebra:

$$(AD^2A^{\mathrm{T}})y = AD^2c. \tag{10}$$

The normal way to compute y is by elimination. That succeeds in a small problem, and also in a large one if all matrices remain sparse. The Gram-Schmidt alternative is to orthogonalize the columns of DA^{T}, which can be expensive to do although it makes the rest of the calculation easy. The favorite for larger problems is the *conjugate gradient method*, which is semi-direct—it gives the exact answer more slowly than elimination, but you can go part way and then stop. (In the middle of elimination you cannot stop.) The conjugate gradient idea, with the "precondi-tioner" that makes it successful, is described in my textbook *Introduction to Applied Mathematics* (Wellesley-Cambridge Press).

Karmarkar's original algorithm went beyond the rescaling $D^{-1}x$, which keeps the cost linear. Instead there was a "projective transformation" to $D^{-1}x/e^{\mathrm{T}}D^{-1}x$. That was needed to prove convergence in polynomial time, but *in practice rescaling is used.* Right now the algorithm is being tested and modified all over the world, and extended to nonlinear optimization. I think Karmarkar's method will survive; it is natural and attractive. Like other new ideas in scientific computing, it will succeed on some problems and lose on others. The simplex method remains tremendously valuable, and so does the whole subject of linear programming—which was discovered centuries after linear equations $Ax = b$, but shares the fundamental ideas of linear algebra. The most far-reaching of those ideas is duality, which comes next.

EXERCISES

8.2.1   Minimize $x_1 + x_2 - x_3$, subject to

$$2x_1 - 4x_2 + x_3 + x_4 \qquad = 4$$
$$3x_1 + 5x_2 + x_3 \qquad + x_5 = 2.$$

Which of x_1, x_2, x_3 should enter the basis, and which of x_4, x_5 should leave? Compute the new pair of basic variables and find the cost at the new corner.

8.2.2   After the preceding simplex step, prepare for and decide on the next step.

8.2.3   In the example after 8C, suppose the cost is $3x + y$. With rearrangement the cost vector is $c = (0, 1, 3, 0)$. Show that $r \geq 0$ and therefore that corner P is optimal.

8.2.4   If the original cost function had been $x - y$, so that after rearrangement $c = (0, -1, 1, 0)$ at the corner P, compute r and decide which column u should enter the basis. Then compute $B^{-1}u$ and show from its sign that you will never meet another corner. We are climbing the y axis in Fig. 8.4, and $x - y$ goes to $-\infty$.

8.2.5   In the same example, change the cost to $x + 3y$. Verify that the simplex method takes you from P to Q to R, and that the corner R is optimal.

8.2.6   Phase I consists in finding a basic feasible solution to $Ax = b$. After changing signs to make $b \geq 0$, consider the auxiliary problem of minimizing $w_1 + w_2 + \cdots + w_m$ subject to $x \geq 0$, $w \geq 0$, $Ax + w = b$. Whenever $Ax = b$ has a nonnegative solution, the minimum cost in this problem will be zero—with $w^* = 0$.
(a)   Show that, for this new problem, the corner $x = 0$, $w = b$ is both basic and feasible. Therefore *its* Phase I is already set, and the simplex method can proceed to find the optimal pair x^*, w^*. If $w^* = 0$, then x^* is the required corner in the original problem.
(b)   With $A = \begin{bmatrix} 1 & -1 \end{bmatrix}$ and $b = \begin{bmatrix} 3 \end{bmatrix}$, write out the auxiliary problem, its Phase I vector $x = 0$, $w = b$, and its optimal vector. Find the corner of the feasible set $x_1 - x_2 = 3$, $x_1 \geq x_2 \geq 0$, and draw a picture of this set.

8.2.7   If we wanted to maximize instead of minimize the cost (with $Ax = b$ and $x \geq 0$), what would be the stopping test on r and what rules would choose the column of N to make basic and the column of B to make free?

8.2.8   Minimize $2x_1 + x_2$ subject to $x_1 + x_2 \geq 4$, $x_1 + 3x_2 \geq 12$, $x_1 - x_2 \geq 0$, $x \geq 0$.

8.2.9   Verify the inverse in (5) and show that BE has $Bv = u$ in its kth column. Then BE is the right basis matrix for the next stop, $E^{-1}B^{-1}$ is its inverse, and E^{-1} updates the basis matrix correctly.

8.2.10 Suppose we want to minimize $x_1 - x_2$, subject to

$$\begin{aligned} 2x_1 - 4x_2 + x_3 \quad\;\;\; &= 6 \\ 3x_1 + 6x_2 \quad\;\;\; + x_4 &= 12 \end{aligned} \quad \text{(all } x_i \geq 0).$$

Starting from $x = (0, 0, 6, 12)$, should x_1 or x_2 be increased from its current value of zero? How far can it be increased until the equations force x_3 or x_4 down to zero? At that point what is x?

8.2.11 For the matrix $P = I - A^T(AA^T)^{-1}A$, show that if x is in the nullspace of A then $Px = x$. The nullspace stays unchanged under the projection.

8.2.12 The change in cost is $c^T \Delta x = -sc^T Pc$ in the first step of Karmarkar's method. Show that this equals $-s\|Pc\|^2$, so the change is negative and the cost is reduced.

8.2.13 (a) Minimize the cost $c^T x = 5x_1 + 4x_2 + 8x_3$ on the plane $x_1 + x_2 + x_3 = 3$, by testing the vertices P, Q, R where the triangle is cut off by the requirement $x \geq 0$.
(b) Project $c = (5, 4, 8)$ onto the nullspace of $A = \begin{bmatrix} 1 & 1 & 1 \end{bmatrix}$ and find the maximum step s that keeps $e - sPc$ nonnegative.

8.2.14 Reduce s in the previous exercise by $\alpha = .98$, to end the step before the boundary, and write down the diagonal matrix D. Sketch the feasible triangle lying on the plane $ADX = 3$ after the first Karmarkar step.

8.2.15 In that example, carry through a second step of Karmarkar's method and compare x^2 with the vector at the optimal corner of the triangle PQR.

8.3 ■ THE THEORY OF DUALITY

Chapter 2 began by saying that although the elimination technique gives one approach to $Ax = b$, a different and deeper understanding is also possible. It is exactly the same for linear programming. The mechanics of the simplex method will solve a linear program, but it is really duality that belongs at the center of the underlying theory. It is an elegant idea, and at the same time fundamental for the applications; we shall explain as much as we understand.

It starts with the standard problem:

PRIMAL *Minimize cx, subject to $x \geq 0$ and $Ax \geq b$.*

But now, instead of creating an equivalent problem with equations in place of inequalities, duality creates an entirely different problem. *The dual problem starts from the same A, b, and c, and reverses everything.* In the primal, c was in the cost function and b was in the constraint; in the dual, these vectors are switched. Furthermore the inequality sign is changed, and the new unknown y is a row vector; the feasible set has $yA \leq c$ instead of $Ax \geq b$. Finally, we maximize rather than minimize. The only thing that stays the same is the requirement of nonnegativity; the unknown y has m components, and it must satisfy $y \geq 0$. In short, the dual of a minimum problem is a maximum problem:

DUAL *Maximize yb, subject to $y \geq 0$ and $yA \leq c$.*

The dual of *this* problem is the original minimum problem.†

Obviously I have to give you some interpretation of all these reversals. They conceal a competition between the minimizer and the maximizer, and the explanation comes from the diet problem; I hope you will follow it through once more. The minimum problem has n unknowns, representing n foods to be eaten in the (nonnegative) amounts $x_1, \ldots, x_n$. The m constraints represent m required vitamins, in place of the one earlier constraint of sufficient protein. The entry a_{ij} is the amount of the ith vitamin in the jth food, and the ith row of $Ax \geq b$ forces the diet to include that vitamin in at least the amount b_i. Finally, if c_j is the cost of the jth food, then $c_1 x_1 + \cdots + c_n x_n = cx$ is the cost of the diet. That cost is to be minimized; this is the primal problem.

In the dual, the druggist is selling vitamin pills rather than food. His prices y_i are adjustable as long as they are nonnegative. The key constraint, however, is

---

† There is complete symmetry between primal and dual. We started with a minimization, but the simplex method applies equally well to a maximization—and anyway both problems get solved at once.

that on each food he cannot charge more than the grocer. Since food j contains vitamins in the amounts a_{ij}, the druggist's price for the equivalent in vitamins cannot exceed the grocer's price c_j. That is the jth constraint in $yA \le c$. Working within this constraint, he can sell an amount b_i of each vitamin for a total income of $y_1b_1 + \cdots + y_mb_m = yb$—which he maximizes.

You must recognize that the feasible sets for the two problems are completely different. The first is a subset of $\mathbf{R}^n$, marked out by the matrix A and the constraint vector b; the second is a subset of $\mathbf{R}^m$, determined by the transpose of A and the other vector c. Nevertheless, when the cost functions are included, the two problems do involve the same input A, b, and c. The whole theory of linear programming hinges on the relation between them, and we come directly to the fundamental result:

8D *Duality Theorem* If either the primal problem or the dual has an optimal vector, then so does the other, and their values are the same: ***The minimum of*** cx ***equals the maximum of*** yb. Otherwise, if optimal vectors do not exist, there are two possibilities: Either both feasible sets are empty, or else one is empty and the other problem is unbounded (the maximum is $+\infty$ or the minimum is $-\infty$).

If both problems have feasible vectors then they have optimal vectors x^* and y^*—and furthermore $cx^* = y^*b$.

Mathematically, this settles the competition between the grocer and the druggist: *The result is always a tie.* We will find a similar "minimax theorem" and a similar equilibrium in game theory. These theorems do not mean that the customer pays nothing for an optimal diet, or that the matrix game is completely fair to both players. They do mean that the customer has no economic reason to prefer vitamins over food, even though the druggist guaranteed to match the grocer on every food—and on expensive foods, like peanut butter, he sells for less. We will show that expensive foods are kept out of the optimal diet, so the outcome can still be (and is) a tie.

This may seem like a total stalemate, but I hope you will not be fooled. The optimal vectors contain the crucial information. In the primal problem, x^* tells the purchaser what to do. In the dual, y^* fixes the natural prices (or "***shadow prices***") at which the economy should run. Insofar as our linear model reflects the true economy, these vectors represent the decisions to be made. They still need to be computed by the simplex method; the duality theorem tells us their most important property.

We want to start on the proof. It may seem obvious that the druggist can raise his prices to meet the grocer's, but it is not. Or rather, only the first part is: Since each food can be replaced by its vitamin equivalent, with no increase in cost, all adequate diets must be at least as expensive as any price the druggist would charge. This is only a one-sided inequality, *druggist's price $\le$ grocer's price*, but it is fundamental. It is called ***weak duality***, and it is easy to prove for any linear program and its dual:

8E   If x and y are any feasible vectors in the minimum and maximum problems, then $yb \leq cx$.

Proof   Since the vectors are feasible, they satisfy

$$Ax \geq b \quad \text{and} \quad yA \leq c.$$

Furthermore, because feasibility also included $x \geq 0$ and $y \geq 0$, we can take inner products without spoiling the inequalities:

$$yAx \geq yb \quad \text{and} \quad yAx \leq cx. \tag{1}$$

Since the left sides are identical, we have weak duality $yb \leq cx$.

The one-sided inequality is easy to use. First of all, it prohibits the possibility that both problems are unbounded. If yb is arbitrarily large, there cannot be a feasible x or we would contradict $yb \leq cx$. Similarly, if the minimization is unbounded—if cx can go down to $-\infty$—then the dual cannot admit a feasible y.

Second, and equally important, we can tell immediately that any vectors which achieve equality, $yb = cx$, must be optimal. At that point the grocer's price equals the druggist's price, and we recognize an optimal diet and optimal vitamin prices by the fact that the consumer has nothing to choose:

8F   If the vectors x and y are feasible and $cx = yb$, then x and y are optimal.

Proof   According to 8E, no feasible y can make yb larger than cx. Since our particular y achieves this value, it is optimal. Similarly no feasible x can bring cx below the number yb, and any x that achieves this minimum must be optimal.

We give an example with two foods and two vitamins. Note how A^T appears when we write out the dual, since $yA \leq c$ for row vectors means $A^T y^T \leq c^T$ for columns.

PRIMAL   Minimize $x_1 + 4x_2$ **DUAL**   Maximize $6y_1 + 7y_2$

subject to $x_1 \geq 0, x_2 \geq 0,$ subject $y_1 \geq 0, y_2 \geq 0,$

$2x_1 + x_2 \geq 6$ $2y_1 + 5y_2 \leq 1$

$5x_1 + 3x_2 \geq 7.$ $y_1 + 3y_2 \leq 4.$

The choice $x_1 = 3$ and $x_2 = 0$ is feasible, with cost $x_1 + 4x_2 = 3$. In the dual problem $y_1 = \frac{1}{2}$ and $y_2 = 0$ give the same value $6y_1 + 7y_2 = 3$. These vectors must be optimal.

That seems almost too simple. Nevertheless it is worth a closer look, to find out what actually happens at the moment when $yb \leq cx$ becomes an equality. It is like calculus, where everybody knows the condition for a maximum or a minimum: *The first derivatives are zero.* On the other hand, everybody forgets that this condition is completely changed by the presence of constraints. The best example is a straight line sloping upward; its derivative is never zero, calculus is almost helpless, and the maximum is certain to occur at the end of the interval. That is exactly the situation that we face in linear programming! There are more variables, and an interval is replaced by a feasible set in several dimensions, but still the maximum is always found at a corner of the feasible set. In the language of the simplex method, there is an optimal x which is *basic*: It has only m nonzero components.

The real problem in linear programming is to decide which corner it is. For this, we have to admit that calculus is not completely helpless. Far from it, because the device of "Lagrange multipliers" will bring back zero derivatives at the maximum and minimum, and in fact *the dual variables y are exactly the Lagrange multipliers* for the problem of minimizing cx. This is also the key to nonlinear programming. The conditions for a constrained minimum and maximum will be stated mathematically in equation (2), but first I want to express them in economic terms: *The diet x and the vitamin prices y are optimal when*

> (i)   The grocer sells zero of any food that is priced above its vitamin equivalent.
> (ii) The druggist charges zero for any vitamin that is oversupplied in the diet.

In the example, $x_2 = 0$ because the second food is too expensive. Its price exceeds the druggist's price, since $y_1 + 3y_2 \leq 4$ is a strict inequality $\frac{1}{2} + 0 < 4$. Similarly, $y_i = 0$ if the ith vitamin is oversupplied; it is a "*free good*," which means it is worthless. The example required seven units of the second vitamin, but the diet actually supplied $5x_1 + 3x_2 = 15$, so we found $y_2 = 0$. You can see how the duality has become complete; it is only when *both* of these conditions are satisfied that we have an optimal pair.

These *optimality conditions* are easy to understand in matrix terms. We are comparing the vector Ax to the vector b (remember that feasibility requires $Ax \geq b$) and we look for any components in which equality fails. This corresponds to a vitamin that is oversupplied, so its price is $y_i = 0$. At the same time we compare yA with c, and expect all strict inequalities (expensive foods) to correspond to $x_j = 0$ (omission from the diet). These are the "*complementary slackness conditions*" of linear programming, and the "*Kuhn-Tucker conditions*" of nonlinear programming:

> **8G**   *Equilibrium Theorem*   Suppose the feasible vectors x and y satisfy the following complementary slackness conditions:
>
> $$\text{if } (Ax)_i > b_i \quad \text{then } y_i = 0, \quad \text{and if } (yA)_j < c_j \quad \text{then } x_j = 0. \tag{2}$$
>
> Then x and y are optimal. Conversely, optimal vectors must satisfy (2).

Proof   The key equations are

$$yb = y(Ax) = (yA)x = cx. \tag{3}$$

Normally only the middle equation is certain. In the first equation, we are sure that $y \geq 0$ and $Ax \geq b$, so we are sure of $yb \leq y(Ax)$. Furthermore, there is only one way in which equality can hold: *Any time there is a discrepancy $b_i < (Ax)_i$, the factor y_i that multiplies these components must be zero.* Then this discrepancy makes no contribution to the inner products, and equality is saved.

The same is true for the remaining equation: Feasibility gives $x \geq 0$ and $yA \leq c$ and therefore $yAx \leq cx$. We get equality only when the second slackness condition is fulfilled: If there is an overpricing $(Ay)_j < c_j$, it must be canceled through multiplication by $x_j = 0$. This leaves us with $yb = cx$ in the key equation (3), and it is this equality that guarantees (and is guaranteed by) the optimality of x and y.

The Proof of Duality

So much for the one-sided inequality $yb \leq cx$. It was easy to prove, it gave a quick test for optimal vectors (they turn it into an equality), and now it has given a set of necessary and sufficient slackness conditions. The only thing it has not done is to show that the equality $yb = cx$ is really possible. Until the optimal vectors are actually produced, which cannot be done by a few simple manipulations, the duality theorem is not complete.

To produce them, we return to the simplex method—which has already computed the optimal x. Our problem is to identify at the same time the optimal y, showing that the method stopped in the right place for the dual problem (even though it was constructed to solve the primal). First we recall how it started. The m inequalities $Ax \geq b$ were changed to equations, by introducing the slack variables $w = Ax - b$ and rewriting feasibility as

$$\begin{bmatrix} A & -I \end{bmatrix} \begin{bmatrix} x \\ w \end{bmatrix} = b, \qquad \begin{bmatrix} x \\ w \end{bmatrix} \geq 0. \tag{4}$$

Then every step of the method picked out m columns of the long matrix $\begin{bmatrix} A & -I \end{bmatrix}$ to be basic columns, and shifted them (at least theoretically, if not physically) to the front. This produced $\begin{bmatrix} B & N \end{bmatrix}$, and the corresponding shift in the long cost vector $\begin{bmatrix} c & 0 \end{bmatrix}$ reordered its components into $\begin{bmatrix} c_B & c_N \end{bmatrix}$. The stopping condition, which brought the simplex method to an end, was $r = c_N - c_B B^{-1} N \geq 0$.

We know that *this condition $r \geq 0$ was finally met*, since the number of corners is finite. At that moment the cost was

$$cx = \begin{bmatrix} c_B & c_N \end{bmatrix} \begin{bmatrix} B^{-1}b \\ 0 \end{bmatrix} = c_B B^{-1}b, \quad \text{the minimum cost.} \tag{5}$$

If we can choose $y = c_B B^{-1}$ ***in the dual, then we certainly have*** $yb = cx$. The minimum and maximum will be equal. Therefore, we must show that this y satisfies the dual constraints $yA \le c$ and $y \ge 0$. We have to show that

$$y[A \quad -I] \le [c \quad 0]. \tag{6}$$

When the simplex method reshuffles the long matrix and vector to put the basic variables first, this rearranges the constraints in (6) into

$$y[B \quad N] \le [c_B \quad c_N]. \tag{7}$$

For our choice $y = c_B B^{-1}$, the first half is an equality and the second half is $c_B B^{-1} N \le c_N$. This is the stopping condition $r \ge 0$ that we know to be satisfied! Therefore our y is feasible, and *the duality theorem is proved.* By locating the critical m by m matrix B, which is nonsingular as long as degeneracy is forbidden, the simplex method has produced the optimal y^* as well as x^*.

Shadow Prices

How does the minimum cost change if we change the right side b or the cost vector c? This is a question in ***sensitivity analysis***, and it allows us to squeeze out of the simplex method a lot of the extra information it contains. For an economist or an executive, these questions about *marginal cost* are the most important.

If we allow large changes in b or c, the solution behaves in a very jumpy way. As the price of eggs increases, there will be a point where they disappear from the diet; in the language of linear programming, the variable x_{egg} will jump from basic to free. To follow it properly, we would have to introduce what is called parametric programming. But if the changes are small, which is much more likely, then ***the corner which was optimal remains optimal***; the choice of basic variables does not change. In other words, B and N stay the same. Geometrically, we have shifted the feasible set a little (by changing b), and we have tilted the family of planes that come up to meet it (by changing c); but if these changes are small, contact occurs first at the same (slightly moved) corner.

At the end of the simplex method, when the right choice of basic variables is known, the corresponding m columns of A make up the basis matrix B. At that corner,

$$\text{minimum cost} = c_B B^{-1} b = y^* b.$$

A shift of size Δb changes the minimum cost by $y^* \Delta b$. The dual solution y^* gives *the rate of change of minimum cost* (its derivative) with respect to changes in b. The components of y^* are the ***shadow prices***, and they make sense; if the requirement for vitamin B_1 goes up by Δ, and the druggist's price is y_1^*, and he is completely competitive with the grocer, then the diet cost (from druggist or grocer) will go up by $y_1^* \Delta$. In case y_1^* is zero, vitamin B_1 is a "free good" and a small change in its requirement has no effect—the diet already contained more than enough.

We now ask a different question. Suppose we want to insist that the diet contain at least some small minimum amount of egg. The nonnegativity condition $x_{egg} \geq 0$ is changed to $x_{egg} \geq \delta$. How does this change the cost?

If eggs were in the optimal diet, there is no change—the new requirement is already satisfied and costs nothing extra. But if they were outside the diet, it will cost something to include the amount δ. The increase will not be the full price $c_{egg}\delta$, since we can cut down on other foods and partially compensate. The increase is actually governed by the "**reduced cost**" of eggs—their own price, *minus* the price we are paying for the equivalent in cheaper foods. To compute it we return to equation (2) of Section 8.2:

$$\text{cost} = (c_N - c_B B^{-1} N)x_N + c_B B^{-1} b = r x_N + c_B B^{-1} b.$$

If egg is the first free variable, then increasing the first component of x_N to δ will increase the cost by $r_1 \delta$. Therefore, the real cost is r_1. Similarly, r gives the actual costs for all other nonbasic foods—the changes in total cost as the zero lower bounds on x (the nonnegativity constraints) are moved upwards. We know that $r \geq 0$, because this was the stopping test; and economics tells us the same thing, that the reduced cost of eggs cannot be negative or they would have entered the diet.

The Theory of Inequalities

There is more than one way to study duality. The approach we followed—to prove $yb \leq cx$, and then use the simplex method to get equality—was convenient because that method had already been established, but overall it was a long proof. Of course it was also a *constructive proof*; x^* and y^* were actually computed. Now we look briefly at a different approach, which leaves behind the mechanics of the simplex algorithm and looks more directly at the geometry. I think the key ideas will be just as clear (in fact, probably clearer) if we omit some of the details.

The best illustration of this approach came in the fundamental theorem of linear algebra. The problem in Chapter 2 was to solve $Ax = b$, in other words to find b in the column space of A. After elimination, and after the four subspaces, this solvability question was answered in a completely different way by Exercise 3.1.11:

8H   Either $Ax = b$ has a solution, or else there is a y such that $yA = 0$, $yb \neq 0$.

This is the **theorem of the alternative**, because to find both x and y is impossible: If $Ax = b$ then $yAx = yb \neq 0$, and this contradicts $yAx = 0x = 0$. In the language of subspaces, either b is in the column space of A or else it has a nonzero component

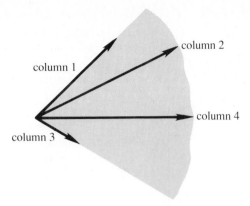

Fig. 8.6. The cone of nonnegative combinations of the columns: $b = Ax$ with $x \geq 0$.

sticking into the perpendicular subspace, which is the left nullspace of A. That component is the required y.[†]

For inequalities, we want to find a theorem of exactly the same kind. The right place to start is with the same system $Ax = b$, but with the added constraint $x \geq 0$. When does there exist not just a solution to $Ax = b$, but a **nonnegative solution**? In other words, when is the feasible set nonempty in the problem with equality constraints?

To answer that question we look again at the combinations of the columns of A. In Chapter 2, when any x was allowed, b was anywhere in the column space. Now we allow only *nonnegative* combinations, and the b's no longer fill out a subspace. Instead, they are represented by the cone-shaped region in Fig. 8.6. For an m by n matrix, there would be n columns in m-dimensional space, and the cone becomes an open-ended pyramid. In the figure, there are four columns in two-dimensional space, and A is 2 by 4. If b lies in this cone, there is a nonnegative solution to $Ax = b$; otherwise there is not.

Our problem is to discover the alternative: *What happens if b lies outside the cone?* That possibility is illustrated in Fig. 8.7, and you can interpret the geometry at a glance. There is a "separating hyperplane," which goes through the origin and has the vector b on one side and the whole cone on the other side. (The prefix *hyper* is only to emphasize that the number of dimensions may be large; the plane consists, as always, of all vectors perpendicular to a fixed vector y.) The inner product between y and b is negative since they make an angle greater than $90°$, whereas the inner product between y and every column of A is positive. In matrix terms this means that $yb < 0$ and $yA \geq 0$, which is the alternative we are looking for.

[†] You see that this proof is not constructive! We only know that a component of b must be in the left nullspace, or b would have been in the column space.

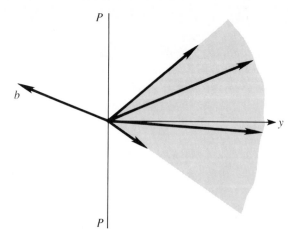

Fig. 8.7. b outside the cone, separated by the hyperplane perpendicular to y.

8I Either $Ax = b$ has a nonnegative solution, or there is a y such that $yA \geq 0$, $yb < 0$.

This is the *theorem of the separating hyperplane*. It is fundamental to mathematical economics, and one reference for its proof is Gale's beautiful book on the theory of linear economic models.

EXAMPLE If A is the identity matrix, its cone is the positive quadrant. Every b in that quadrant is a nonnegative combination of the columns:

$$\text{if} \quad b = \begin{bmatrix} 2 \\ 3 \end{bmatrix}, \quad \text{then} \quad b = 2\begin{bmatrix} 1 \\ 0 \end{bmatrix} + 3\begin{bmatrix} 0 \\ 1 \end{bmatrix}.$$

For every b outside the quadrant, the second alternative must hold:

$$\text{if} \quad b = \begin{bmatrix} 2 \\ -3 \end{bmatrix}, \quad \text{then} \quad y = \begin{bmatrix} 0 & 1 \end{bmatrix} \text{ gives } yA \geq 0 \text{ but } yb = -3.$$

Here the x-axis (perpendicular to y) separates b from the positive quadrant.

This theorem leads to a whole sequence of similar alternatives (see Gale). In fact you almost come to believe that whenever two alternatives are mutually exclusive, one or the other must be true. For example, it is impossible for a subspace S and its orthogonal complement $S^{\perp}$ both to contain positive vectors: Their inner product would be positive, whereas orthogonal vectors have inner product zero. On the other hand, it is not quite certain that either S or $S^\perp$ has to contain a positive vector. S might be the x axis and $S^\perp$ the y axis, in which case they contain only the "semipositive" vectors $[1 \quad 0]$ and $[0 \quad 1]$. What is remarkable is that this slightly weaker alternative does work. Either S contains a positive vector x, or $S^\perp$ contains a semipositive vector y. When S and $S^\perp$ are perpendicular lines in the plane, it is easy to see that one or the other must enter the first quadrant; but I do not see it very clearly in higher dimensions.

For linear programming, the important alternatives come when the constraints are inequalities rather than equations:

8J Either $Ax \geq b$ has a solution with $x \geq 0$, or else there is a vector y such that $yA \geq 0$, $yb < 0$, $y \leq 0$.

8J follows easily from 8I, using the slack variables $w = Ax - b$ to change the inequality into an equation:

$$[A \quad -I] \begin{bmatrix} x \\ w \end{bmatrix} = b.$$

If this has no solution with $x \geq 0$ and $w \geq 0$, then by 8I there must be a y such that

$$y[A \quad -I] \geq [0 \quad 0], \quad yb < 0.$$

This is exactly the other alternative in 8J. It is this result that leads to a "non-constructive proof" of the duality theorem. But we promised to stick to the geometry and omit the algebraic details, so we keep that promise.

EXERCISES

8.3.1 What is the dual of the following problem: Minimize $x_1 + x_2$, subject to $x_1 \geq 0$, $x_2 \geq 0$, $2x_1 \geq 4$, $x_1 + 3x_2 \geq 11$? Find the solution to both this problem and its dual, and verify that minimum equals maximum.

8.3.2 What is the dual of the following problem: Maximize y_2, subject to $y_1 \geq 0$, $y_2 \geq 0$, $y_1 + y_2 \leq 3$? Solve both this problem and its dual.

8.3.3 Suppose A is the identity matrix (so that $m = n$) and the vectors b and c are non-negative. Explain why $x^* = b$ is optimal in the minimum problem, find y^* in the maximum problem, and verify that the two values are the same. If the first component of b is negative, what are x^* and y^*?

8.3.4 Construct a 1 by 1 example in which $Ax \geq b$, $x \geq 0$ is unfeasible, and the dual problem is unbounded.

8.3.5   Starting with the 2 by 2 matrix $A = \begin{bmatrix} 1 & 0 \\ 0 & -1 \end{bmatrix}$, choose b and c so that both of the feasible sets $Ax \geq b$, $x \geq 0$ and $yA \leq c$, $y \geq 0$ are empty.

8.3.6   If all entries of A, b, and c are positive, show that both the primal and the dual are feasible.

8.3.7   Show that $x = (1, 1, 1, 0)$ and $y = (1, 1, 0, 1)$ are feasible in the primal and dual, with

$$
A = \begin{bmatrix} 0 & 0 & 1 & 0 \\ 0 & 1 & 0 & 0 \\ 1 & 1 & 1 & 1 \\ 1 & 0 & 0 & 1 \end{bmatrix}, \qquad
b = \begin{bmatrix} 1 \\ 1 \\ 1 \\ 1 \end{bmatrix}, \qquad
c = \begin{bmatrix} 1 \\ 1 \\ 1 \\ 3 \end{bmatrix}.
$$

Then, after computing cx and yb, explain how you know they are optimal.

8.3.8   Verify that the vectors in the previous exercise satisfy the complementary slackness conditions (2), and find the one slack inequality in both the primal and the dual.

8.3.9   Suppose that $A = \begin{bmatrix} 1 & 0 \\ 0 & 1 \end{bmatrix}$, $b = \begin{bmatrix} -1 \\ 1 \end{bmatrix}$, and $c = \begin{bmatrix} 1 \\ 1 \end{bmatrix}$. Find the optimal x and y, and verify the complementary slackness conditions (as well as $yb = cx$).

8.3.10   If the primal problem is constrained by equations instead of inequalities—*Minimize cx subject to $Ax = b$ and $x \geq 0$*—then the requirement $y \geq 0$ is left out of the dual: *Maximize yb subject to $yA \leq c$.* Show that the one-sided inequality $yb \leq cx$ still holds. Why was $y \geq 0$ needed in (1) but not here? This weak duality can be completed to full duality.

8.3.11   (a)   Without the simplex method, minimize the cost $5x_1 + 3x_2 + 4x_3$ subject to $x_1 + x_2 + x_3 \geq 1$, $x_1 \geq 0$, $x_2 \geq 0$, $x_3 \geq 0$.
(b)   What is the shape of the feasible set?
(c)   What is the dual problem, and what is its solution y?

8.3.12   If the primal has a unique optimal solution x^*, and then c is changed a little, explain why x^* still remains the optimal solution.

8.3.13   If steak costs $c_1 = \$3$ and peanut butter $c_2 = \$2$, and they give two units and one unit of protein (four units are required), find the shadow price of protein and the reduced cost of peanut butter.

8.3.14   If $A = \begin{bmatrix} 1 & 1 \\ 0 & 1 \end{bmatrix}$, describe the cone of nonnegative combinations of the columns. If b lies inside that cone, say $b = (3, 2)$, what is the feasible vector x? If b lies outside, say $b = (0, 1)$, what vector y will satisfy the alternative?

8.3.15   In three dimensions, can you find a set of six vectors whose cone of nonnegative combinations fills the whole space? What about four vectors?

8.3.16   Use 8H to show that there is no solution (because the alternative holds) to

$$
\begin{bmatrix} 2 & 2 \\ 4 & 4 \end{bmatrix} x = \begin{bmatrix} 1 \\ 1 \end{bmatrix}.
$$

8.3.17 Use 8I to show that there is no nonnegative solution (because the alternative holds) to

$$\begin{bmatrix} 1 & 3 & -5 \\ 1 & -4 & -7 \end{bmatrix} x = \begin{bmatrix} 2 \\ 3 \end{bmatrix}.$$

8.3.18 Show that the alternatives in 8J ($Ax \geq b$, $x \geq 0$, $yA \geq 0$, $yb < 0$, $y \leq 0$) cannot both hold. *Hint*: yAx.

8.4 ■ NETWORK MODELS

Some linear programs have a structure that makes their solution very quick. For linear equations, that was true for band matrices. When the nonzeros were close to the main diagonal, $Ax = b$ was easy to solve. In linear programming, where A is rectangular, we are interested in the special class known as ***network programs***. The matrix A is an ***incidence matrix***, its entries are -1 or $+1$ or (mostly) zero, and pivot steps involve only additions and subtractions. Much larger problems than usual can be solved.

Fortunately, networks enter all kinds of applications. The flow of products or people or automobiles satisfies Kirchhoff's current law: cars are not created or destroyed at the nodes. For gas and oil and water, network programming has designed pipeline systems that are millions of dollars cheaper than the intuitive (but not optimized) designs. It even solves the ***marriage problem***—how to maximize the number of marriages, when brides have the power of veto. That may not be the real problem, but it is the one that network programming solves.

One network model is shown in Figure 8.8. There the problem is to *maximize the flow*, going from the source (node 1) to the sink (node 6). The constraints are the *capacities of the edges*. The flows cannot exceed the capacities, and the directions given by the arrows cannot be reversed. This model can be solved without any theory: What is the ***maximal flow*** from left to right?

The unknowns in this problem are the flows x_{ij}, from node i to node j. The capacity constraints are $x_{ij} \le c_{ij}$. The flows are nonnegative (going with the arrows), and *the cost function is* x_{61}—the flow which we can pretend comes back along the dotted line. By maximizing that return flow x_{61}, we maximize the total flow into the sink.

There is another constraint still to be heard from. It is the "conservation law," that *the flow into each node equals the flow out*. That is Kirchhoff's current law:

$$\sum_i x_{ij} - \sum_k x_{jk} = 0 \quad \text{for } j = 1, 2, \ldots, 6. \tag{1}$$

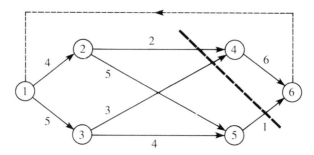

Fig. 8.8. A network with source, sink, capacities, and cut: the maximal flow problem.

The flows into node j are the x_{ij}, coming from all nodes i. The flows out are the x_{jk}, going to all nodes k. The balance in equation (1) can be written as $Ax = 0$, where A is a *node-edge incidence matrix*—with a row for every node and a column for every edge:

$$
A = \begin{bmatrix}
1 & 1 & & & & & & & -1 \\
-1 & & 1 & 1 & & & & & \\
& -1 & & & 1 & 1 & & & \\
& & -1 & & -1 & & 1 & & \\
& & & -1 & & -1 & & 1 & \\
& & & & & & -1 & -1 & 1
\end{bmatrix}
\begin{matrix} 1 \\ 2 \\ 3 \\ 4 \\ 5 \\ 6 \end{matrix}
$$

$$x_{12} \quad x_{13} \quad x_{24} \quad x_{25} \quad x_{34} \quad x_{35} \quad x_{46} \quad x_{56} \quad x_{61}$$

Kirchhoff's law $Ax = 0$ is a 6 by 9 system, and there are certainly solutions. We want nonnegative solutions, not exceeding the capacities.

Looking at the capacities into the sink, the flow cannot exceed $6 + 1 = 7$. Is that achievable? A flow of 2 can go on the path 1-2-4-6-1. A flow of 3 can go along 1-3-4-6-1. An additional flow of 1 can take the lowest path 1-3-5-6-1. The total is 6, and *no more is possible*. How do you prove that the maximum has been reached?

Trial and error is convincing, but mathematics is conclusive: The key is to construct a *cut* in the network, across which all capacities are now full. The cut separates nodes 5 and 6 from the others. The three edges that go forward across the cut have total capacity $2 + 3 + 1 = 6$—and no more can get across. Weak duality says that every cut gives a bound to the total flow, and full duality says that the cut of smallest capacity (*the minimal cut*) can be filled by an achievable flow.

8K **Max flow-min cut theorem.** The maximal flow in a network equals the capacity of the minimal cut.

Strictly speaking, we should say "a minimal cut"—since several cuts might have the same capacity. A cut is a splitting of the nodes into two groups S and T, with the source in S and the sink in T. Its capacity is the sum of the capacities of all edges from S to T.

Proof that max flow = min cut. Certainly the flow cannot be greater than the capacity of any cut, including the minimal cut. The harder problem, here and in all of duality, is to show that equality can be achieved.

Suppose a flow is maximal. Look at all nodes that can be reached from the source by additional flow, without exceeding the capacity on any edge. Those

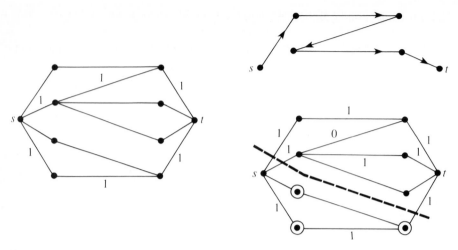

Fig. 8.9. A network for the marriage problem.

nodes go with the source into the set S. The sink must lie in the remaining set T (or it could have received more flow!). Every edge across the cut must be filled, or extra flow could have gone further forward to a node in T. Thus the flow does fill this cut to capacity, and equality has been achieved.

This suggests a way to construct the maximal flow: Check whether any path has unused capacity, and add flow along that "augmenting path." Each step computes the remaining capacities and decides whether the sink is cut off from the source, or additional flow is possible. The processing is organized by the *labeling algorithm*, which labels each node in S by the node that flow can come from—and allows you to backtrack to find the path for extra flow.

Figure 8.9a shows a second network and an achievable flow. All capacities (left to right) equal 1. It might seem that only the top two nodes and the bottom two nodes can be reached by more flow—but that is not so! The extra flow can go *backward* along an edge, and thus cancel an existing flow—provided it eventually reaches the sink (Fig. 8.9b). When this flow is added the total becomes 3, which is maximal. The minimal cut is also sketched in that figure; it crosses 3 solid edges.

The Marriage Problem

Suppose we have four women and four men. Some of those sixteen couples are compatible, others regrettably are not. When is it possible to find a ***complete matching***, with everyone married? If linear algebra can work in 20-dimensional space, it can certainly handle the trivial problem of marriage.

There are two ways to present the problem—in a matrix or on a graph. The matrix contains zeros and ones—$a_{ij} = 0$ if the ith woman and jth man are not compatible, and $a_{ij} = 1$ if they are willing to try. Thus row i gives the choices of the ith woman, and column j corresponds to the jth man. An example is

$$A = \begin{bmatrix} 1 & 0 & 0 & 0 \\ 1 & 1 & 1 & 0 \\ 0 & 0 & 0 & 1 \\ 0 & 0 & 0 & 1 \end{bmatrix}.$$

The corresponding graph was in Figure 8.9. Ignoring the source and sink, there were four nodes at the left and four on the right. *The edges between them correspond to the 1's in the matrix.* There is no edge between the first woman and fourth man, and in the matrix $a_{14} = 0$.

A complete matching (if it is possible) is a set of four 1's in the matrix. They would come from four different rows and four different columns, since bigamy is not allowed. It is like finding a *permutation matrix* within the nonzero entries of A. On the graph, there would be four edges—with no nodes in common. If we put back the source and sink, and send a unit flow along each edge in the matching, the total flow would be 4. The maximal flow is less than 4 exactly when a complete matching is impossible.

In the example the maximal flow was 3. The marriages 1–1, 2–2, 4–4 are allowed (and several other sets of three marriages), but there is no way to reach four. We want to see why not, in a way that applies to any 0–1 matrix.

On the graph, the problem was the minimal cut. It separates the two women at the bottom left from the three men at the top right. The two women have only one man left to choose—not enough. The three men have only two women (the first two). The picture is clearer in the matrix, where

(1)   Rows 3 and 4 are nonzero only in column 4
(2)   Columns 1, 2, 3 are nonzero only in rows 1 and 2.

Whenever there is a subset of k women who among them like fewer than k men, a complete matching is impossible. We want to show that for matrices of any size, that is the decisive test.

The conclusion can be expressed in several different ways:

(1)   (*For chess*)   It is impossible to put four rooks on squares with 1's, in such a way that no rook can take any other rook.
(2)   (*For duality*)   The 1's in the matrix can be covered by fewer than four lines. The minimum number of covering lines (horizontal or vertical) is three, and that equals the maximum number of marriages.
(3)   (*For linear algebra*)   Every matrix with the same zeros as A—any entries can replace the 1's—is singular. Its determinant is zero.

Remember that the determinant is a sum of $4! = 24$ terms. Each term uses all four rows and columns. Because of the zeros in A, all 24 terms are zero and therefore the determinant is zero.

In this example, a block of zeros is preventing a complete matching. The submatrix in rows 3, 4 and columns 1, 2, 3—a 2 by 3 submatrix—is entirely zero. The general rule for an n by n matrix is that *a p by q block of zeros prevents a matching if $p + q > n$*. Exercise 4.3.8 had a 3 by 3 block of zeros with $n = 5$, which forced its determinant to be zero. In the marriage problem, the three women could marry only the two remaining men. If p women can marry only $n - q$ men and $p > n - q$ (which is the same as a zero block with $p + q > n$) then a complete matching is impossible.

The mathematical problem is to prove the converse. If no block of zeros is too large, we will show that the determinant can be nonzero. *If every set of p women likes at least p men, a complete matching is possible*. That is *Hall's condition*, and it applies to sets of every size: each woman must like at least one man, each two women must between them like at least two men, and so on to $p = n$.

8L   A complete matching is possible if and only if Hall's condition holds.

Certainly Hall's condition is necessary: If p women like fewer than p men, a matching is impossible. The harder problem is to find the matching—or to prove that there must be one—when Hall's condition is met. One approach uses mathematical induction on n, without actually constructing the matching. We prefer to look at the network problem and its maximal flow.

The proof is simplest if the capacities are n, instead of 1, on edges across the middle—corresponding to 1's in the matrix. The capacities out of the source at the left, and into the sink at the right, are still 1 (Fig. 8.10). If the maximal flow is n, those edges from the source and into the sink are filled—and the flow must get across, which identifies n marriages. When a complete matching is impossible, and the maximal flow is below n, some cut must be responsible.

That cut will have capacity below n. Suppose the nodes $W_1, \ldots, W_p$ on the left and $M_1, \ldots, M_r$ on the right are in the set S with the source. The capacity across that cut is $n - p$ from the source to the remaining women and r from these men

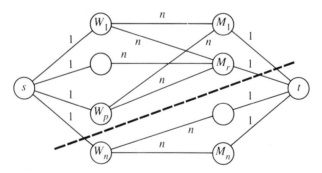

Fig. 8.10. Cut capacity below n; a complete matching is impossible.

to the sink. Since the cut capacity is below n, no edge can cross from $W_1, \ldots, W_p$ to the remaining men. *The p women like only the r men.* But the capacity $n - p + r$ is below n when $p > r$.

When the maximal flow is below n, this shows that Hall's condition fails. When Hall's condition holds, the labeling algorithm which finds a maximal flow will at the same time find—for this special network—a complete matching. In all cases it will find the largest matching possible.

Spanning Trees and the Greedy Algorithm

A fundamental network model is the shortest path problem—when the edges have *lengths* instead of capacities, and we want the shortest path from source to sink. If the edges are telephone lines and the lengths are delay times, we are finding the quickest route for a call. If the nodes are computers, we are looking at a network like ARPANET—which solves the shortest path problem constantly and almost instantaneously.

There is a closely related problem without a source and sink. It finds the **shortest spanning tree**—a set of $n - 1$ edges connecting together *all the nodes* of the network. Instead of getting quickly between two particular nodes, we are now minimizing the cost of connecting all the nodes. There are no loops, because the cost of an edge to close a loop is unnecessary—all we require is a way to get from each node to each other node. *A spanning tree connects the nodes without loops,* and we want the shortest one. Here is one possible algorithm:

(1) *Start from any node s and repeat the following step:*
    Add the shortest edge that connects the current tree to a new node.

In Fig. 8.11, the edge lengths would come in the order 1, 2, 7, 4, 3, 6. The last step skips the edge of length 5, which closes a loop. The total length is 22—but is it minimal? We accepted the edge of length 7 very early, and the second algorithm holds out longer.

(2) *Accept edges in increasing order of length, rejecting any edges that complete a loop.*

Now the edges come in the order 1, 2, 3, 4, 6 (again rejecting 5), and 7. They are the same edges—although that will not always happen. Their total length is the

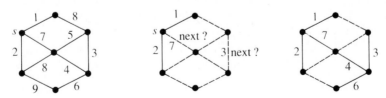

Fig. 8.11. A network and a spanning tree of length 22.

same—and that *does* always happen. The spanning tree problem is exceptional, because it can be solved in one pass.

In the language of linear programming, we are finding the optimal corner first! The maximal flow model was simpler than most linear programs, but there we still started with one flow and gradually improved it. The simplex method was operating, but there was no need to call attention to it—A was an incidence matrix and we dealt with it directly. Now the spanning tree problem is being solved like back-substitution, with no false steps. Almost any method works, as long as it is optimal at each step. This general approach is called the ***greedy algorithm***:

(3)   *Build trees from all n nodes, by repeating the following step: Select **any** tree and add the minimum-length edge going out from that tree.*

The steps depend on the selection order of the trees. To stay with the same tree is algorithm (1). To take the lengths in order is algorithm (2). To sweep through all the trees in turn is a new algorithm (4). It sounds so easy, but for a large problem the data structure becomes critical. With a thousand nodes there might be nearly a million edges, and you don't want to go through that list a thousand times.

That completes our introduction to network models. There are important problems related to matching that are almost as easy:

1.   The ***optimal assignment problem***: Suppose a_{ij} measures the value of applicant i in job j. Assign jobs to maximize the total value—the sum of the a_{ij} on assigned jobs. (If all a_{ij} are 0 or 1, this is the marriage problem.)
2.   The ***transportation problem***: Given costs C_{ij} on the edges and supplies at n points and demands at n markets, choose shipments x_{ij} from suppliers to markets that minimize the total cost $\sum C_{ij}x_{ij}$. (If all supplies and demands are 1, this is the optimal assignment problem—sending one person to each job.)
3.   ***Minimum cost flow***: Now the routes have capacities as well as costs, mixing the maximal flow problem with the transportation problem. What is the cheapest flow, subject to capacity constraints?

A fascinating part of this subject is the development of algorithms. Instead of a theoretical proof of duality, we use *breadth first search* or *depth first search* to find the optimal assignment or the cheapest flow. It is like the simplex method, in starting from a feasible flow (a corner) and adding a new flow (to move to the next corner). The algorithms are special because network problems are so special.

There is also the technique of ***dynamic programming***, which rests on a simple idea: If a path from source to sink is optimal, then each part of the path is also optimal. There cannot be anything better from a node on the path to the sink. The solution is built backwards, with a multistage decision process. At each stage, the distance to the sink is the minimum of a new distance plus an old distance:

$$x\text{-}t \text{ distance} = \text{minimum over } y \text{ of } (x\text{-}y \text{ distance} + y\text{-}t \text{ distance}).$$

That is Bellman's dynamic programming equation, in its simplest form.

I wish there were space for more about networks. They are simple but beautiful.

EXERCISES

8.4.1 In Figure 8.8 add 3 to every capacity. Find by inspection the maximal flow and minimal cut.

8.4.2 Find a maximal flow and minimal cut for the following network:

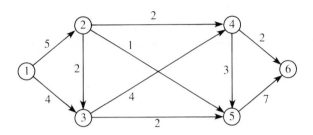

8.4.3 If you could increase the capacity of any one pipe in the network above, which change would produce the largest increase in the maximal flow?

8.4.4 Draw a 5-node network with capacity $|i - j|$ between node i and node j. Find the largest possible flow from node 1 to node 4.

8.4.5 In a graph, the maximum number of paths from s to t with no common edges equals the minimum number of edges whose removal disconnects s from t. Relate this to the max flow-min cut theorem.

8.4.6 Find a maximal set of marriages (a complete matching if possible) for

$$A = \begin{bmatrix} 0 & 0 & 1 & 0 & 0 \\ 1 & 1 & 0 & 1 & 1 \\ 0 & 1 & 1 & 0 & 1 \\ 0 & 0 & 1 & 1 & 0 \\ 0 & 0 & 0 & 1 & 0 \end{bmatrix} \quad \text{and} \quad B = \begin{bmatrix} 1 & 1 & 0 & 0 & 0 \\ 0 & 1 & 0 & 1 & 0 \\ 0 & 0 & 1 & 0 & 1 \\ 1 & 1 & 1 & 0 & 0 \\ 1 & 0 & 0 & 0 & 0 \end{bmatrix}.$$

Sketch the network for B, with heavier lines on the edges in your matching.

8.4.7 For the previous matrix A, which rows violate Hall's condition—by having all their 1's in too few columns? Which p by q submatrix of zeros has $p + q > n$?

8.4.8 How many lines (horizontal and vertical) are needed to cover all the 1's in A above? For any matrix explain why weak duality is true: If k marriages are possible then it takes at least k lines to cover all the 1's.

8.4.9 (a) Suppose every row and every column contain exactly two 1's. Prove that a complete matching is possible. (Show that the 1's cannot be covered by less than n lines).
(b) Find an example with two *or more* 1's in each row and column, for which a complete matching is impossible.

8.4.10 If a 7 by 7 matrix has 15 1's, prove that it allows at least 3 marriages.

8.4.11   For *infinite* sets a complete matching may be impossible even if Hall's condition is passed. If the first row is all 1's and then every $a_{i\,i-1} = 1$, show that any p rows have 1's in at least p columns—and still there is no complete matching.

8.4.12   If Figure 8.8 shows lengths instead of capacities, find the shortest path from s to t and also a minimal spanning tree.

8.4.13   Apply algorithms (1) and (2) to find a shortest spanning tree for the network of Exercise 8.4.2.

8.4.14   (a)   Why does the greedy algorithm work for the spanning tree problem?
(b)   Show by example that it could fail to find the shortest path from s to t, by starting with the shortest edge.

8.4.15   If A is the 5 by 5 matrix with 1's just above and just below the main diagonal, find
(a)   a set of rows with 1's in too few columns
(b)   a set of columns with 1's in too few rows
(c)   a p by q submatrix of zeros with $p + q > 5$
(d)   four lines that cover all the 1's.

8.4.16   The maximal flow problem has slack variables $w_{ij} = c_{ij} - x_{ij}$ for the difference between capacities and flows. State the problem of Fig. 8.8 as a linear program.

8.4.17   Explain how every node below the cut in Fig. 8.9 can be reached by more flow—even the node that already has a flow of 1 directly into it.

GAME THEORY AND THE MINIMAX THEOREM ■ 8.5

The best way to explain a matrix game is to give an example. It has two players, and the rules are the same for every turn:

Player X holds up either one hand or two, and independently, so does player Y. If they make the same decision, Y wins \$10. If they make opposite decisions, then X is the winner—\$10 if he put up one hand, and \$20 if he put up two. The net payoff to X is easy to record in the matrix

$$
A = \begin{bmatrix} -10 & 20 \\ 10 & -10 \end{bmatrix} \qquad \begin{array}{l} \text{one hand by Y} \\ \text{two hands by Y} \end{array}
$$

$$
\begin{array}{cc} \text{one hand} & \text{two hands} \\ \text{by X} & \text{by X} \end{array}
$$

If you think for a moment, you get a rough idea of the best strategy. It is obvious that X will not do the same thing every time, or Y would copy him and win everything. Similarly Y cannot stick to a single strategy, or X will do the opposite. Both players must use a *mixed strategy*, and furthermore the choice at every turn must be absolutely independent of the previous turns. Otherwise, if there is some historical pattern for the opponent to notice, he can take advantage of it. Even a strategy such as "stay with the same choice as long as you win, and switch when you lose" is obviously fatal. After enough plays, your opponent would know exactly what to expect.

This leaves the two players with the following calculation: X can decide that he will put up one hand with frequency x_1 and both hands with frequency $x_2 = 1 - x_1$. At every turn this decision is random. Similarly Y can pick *his* probabilities y_1 and $y_2 = 1 - y_1$. None of these probabilities should be 0 or 1; otherwise the opponent adjusts and wins. At the same time, it is not clear that they should equal $\frac{1}{2}$, since Y would be losing \$20 too often. (He would lose \$20 a quarter of the time, \$10 another quarter of the time, and win \$10 half the time—an average loss of \$2.50—which is more than necessary.) But the more Y moves toward a pure two-hand strategy, the more X will move toward one hand.

The fundamental problem is *to find an equilibrium*. Does there exist a mixed strategy y_1 and y_2 that, if used consistently by Y, offers no special advantage to X? Can X choose probabilities x_1 and x_2 that present Y with no reason to move his own strategy? At such an equilibrium, if it exists, the average payoff to X will have reached a ***saddle point***: It is a maximum as far as X is concerned, and a minimum as far as Y is concerned. To find such a saddle point is to "solve" the game.

One way to look at X's calculations is this: He is combining his two columns with weights x_1 and $1 - x_1$ to produce a new column. Suppose he uses weights $\frac{3}{5}$

and $\frac{2}{5}$; then he produces the column

$$\frac{3}{5}\begin{bmatrix} -10 \\ 10 \end{bmatrix} + \frac{2}{5}\begin{bmatrix} 20 \\ -10 \end{bmatrix} = \begin{bmatrix} 2 \\ 2 \end{bmatrix}.$$

Whatever Y does against this strategy, he will lose $2. On any individual turn, the payoff is still $10 or $20. But if Y consistently holds up one hand, then $\frac{3}{5}$ of the time he wins $10 and $\frac{2}{5}$ of the time he loses $20, an average loss of $2. And the result if Y prefers two hands, or if he mixes strategies in any proportion, remains fixed at $2.

This does not mean that all strategies are optimal for Y! If he is lazy and stays with one hand, X will change and start winning $20. Then Y will change, and then X again. Finally, if as we assume they are both intelligent, Y as well as X will settle down to an optimal mixture. This means that Y will combine his rows with weights y_1 and $1 - y_1$, trying to produce a new row which is as *small* as possible:

$$y_1\begin{bmatrix} -10 & 20 \end{bmatrix} + (1 - y_1)\begin{bmatrix} 10 & -10 \end{bmatrix} = \begin{bmatrix} 10 - 20y_1 & -10 + 30y_1 \end{bmatrix}.$$

The right mixture makes the two components equal: $10 - 20y_1 = -10 + 30y_1$, which means $y_1 = \frac{2}{5}$. With this choice, both components equal 2; the new row becomes $\begin{bmatrix} 2 & 2 \end{bmatrix}$. Therefore, **with this strategy Y cannot lose more than** $2. Y has minimized his maximum loss, and his minimax agrees with the maximin found independently by X. The *value of the game* is this minimax = maximin = $2.

Such a saddle point is remarkable, because it means that X plays his second strategy only $\frac{2}{5}$ of the time, even though it is this strategy that gives him a chance at $20. At the same time Y has been forced to adopt a losing strategy—he would like to match X, but instead he uses the opposite probabilities $\frac{2}{5}$ and $\frac{3}{5}$. You can check that X wins $10 with frequency $\frac{3}{5} \cdot \frac{3}{5} = \frac{9}{25}$, he wins $20 with frequency $\frac{2}{5} \cdot \frac{2}{5} = \frac{4}{25}$, and he loses $10 with the remaining frequency $\frac{12}{25}$. As expected, that gives him an average gain of $2.

We must mention one mistake that is easily made. It is not always true that the optimal mixture of rows is a row with all its entries equal. Suppose X is allowed a third strategy of holding up three hands, and winning $60 when Y puts up one and $80 when Y puts up two. The payoff matrix becomes

$$A = \begin{bmatrix} -10 & 20 & 60 \\ 10 & -10 & 80 \end{bmatrix}.$$

X will choose the new strategy every time; he weights the columns in proportions $x_1 = 0$, $x_2 = 0$, and $x_3 = 1$ (not random at all), and his minimum win is $60. At the same time, Y looks for the mixture of rows which is as small as possible. He always chooses the first row; his maximum loss is $60. Therefore, we still have maximin = minimax, but the saddle point is over in the corner.

The right rule seems to be that in Y's optimal mixture of rows, the value of the game appears (as $2 and $60 did) only in the columns actually used by X. Similarly, in X's optimal mixture of columns, this same value appears in those rows that enter Y's best strategy—the other rows give something higher and Y avoids them. This rule corresponds exactly to the complementary slackness condition of linear programming.

The Minimax Theorem

The most general "matrix game" is exactly like our simple example, with one important difference: X has n possible moves to choose from, and Y has m. The payoff matrix A, which stays the same for every repetition of the game, has m rows and n columns. The entry a_{ij} represents the payment received by X when he chooses his jth strategy and Y chooses his ith; a negative entry simply means a negative payment, which is a win for Y. The result is still a **two-person zero-sum game**; whatever is lost by one player is won by the other. But the existence of a saddle point equilibrium is by no means obvious.

As in the example, player X is free to choose any mixed strategy $x = (x_1, \ldots, x_n)$. This mixture is always a probability vector; the x_i are nonnegative, and they add to 1. These components give the frequencies for the n different pure strategies, and at every repetition of the game X will decide between them by some random device—the device being constructed to produce strategy i with frequency x_i. Y is faced with a similar decision: He chooses a vector $y = (y_1, \ldots, y_m)$, also with $y_i \geq 0$ and $\sum y_i = 1$, which gives the frequencies in his own mixed strategy.

We cannot predict the result of a single play of the game; it is random. On the average, however, the combination of strategy j for X and strategy i for Y will turn up with probability $x_j y_i$—the product of the two separate probabilities. When it does come up, the payoff is a_{ij}. Therefore the expected payoff to X from this particular combination is $a_{ij} x_j y_i$, and *the total expected payoff from each play of the same game is $\sum \sum a_{ij} x_j y_i$*. Again we emphasize that any or all of the entries a_{ij} may be negative; the rules are the same for X and Y, and it is the entries a_{ij} that decide who wins the game.

The expected payoff can be written more easily in matrix notation: The double sum $\sum \sum a_{ij} x_j y_i$ is just yAx, because of the matrix multiplication

$$yAx = \begin{bmatrix} y_1 & \cdots & y_m \end{bmatrix} \begin{bmatrix} a_{11} & a_{12} & \cdots & a_{1n} \\ \vdots & \vdots & & \vdots \\ a_{m1} & a_{m2} & \cdots & a_{mn} \end{bmatrix} \begin{bmatrix} x_1 \\ x_2 \\ \vdots \\ x_n \end{bmatrix} = a_{11}x_1y_1 + \cdots + a_{mn}x_ny_m.$$

It is this payoff yAx that player X wants to maximize and player Y wants to minimize.

EXAMPLE 1   Suppose A is the n by n identity matrix, $A = I$. Then the expected payoff becomes $yIx = x_1y_1 + \cdots + x_ny_n$, and the idea of the game is not hard to explain: X is hoping to hit on the same choice as Y, in which case he receives the payoff $a_{ii} = \$1$. At the same time, Y is trying to evade X, so he will not have to pay. When X picks column i and Y picks a different row j, the payoff is $a_{ij} = 0$.

If X chooses any of his strategies more often than any other, then Y can escape more often; therefore the optimal mixture is $x^* = (1/n, 1/n, \ldots, 1/n)$. Similarly Y cannot overemphasize any strategy or X will discover him, and therefore his optimal choice also has equal probabilities $y^* = (1/n, 1/n, \ldots, 1/n)$. The probability that both will choose strategy i is $(1/n)^2$, and the sum over all such combinations is the expected payoff to X. The total value of the game is n times $(1/n)^2$, or $1/n$, as is confirmed by

$$y^* Ax^* = \begin{bmatrix} 1/n \cdots 1/n \end{bmatrix} \begin{bmatrix} 1 & & \\ & \ddots & \\ & & 1 \end{bmatrix} \begin{bmatrix} 1/n \\ \vdots \\ 1/n \end{bmatrix} = \left(\frac{1}{n}\right)^2 + \cdots + \left(\frac{1}{n}\right)^2 = \frac{1}{n}.$$

As n increases, Y has a better chance to escape.

Notice that the symmetric matrix $A = I$ did not guarantee that the game was fair. In fact, the true situation is exactly the opposite: It is a *skew-symmetric matrix*, $A^T = -A$, which means a completely fair game. Such a matrix faces the two players with identical decisions, since a choice of strategy j by X and i by Y wins a_{ij} for X, and a choice of j by Y and i by X wins the same amount for Y (because $a_{ji} = -a_{ij}$). The optimal strategies x^* and y^* must be the same, and the expected payoff must be $y^* Ax^* = 0$. The value of the game, when $A^T = -A$, is zero. But the strategy is still to be found.

EXAMPLE 2

$$A = \begin{bmatrix} 0 & -1 & -1 \\ 1 & 0 & -1 \\ 1 & 1 & 0 \end{bmatrix}.$$

In words, X and Y both choose a number between 1 and 3, and the one with the smaller number wins \$1. (If X chooses 2 and Y chooses 3, the payoff is $a_{32} = \$1$; if they choose the same number, we are on the main diagonal and nobody wins.) Evidently neither player can choose a strategy involving 2 or 3, or the other can get underneath him. Therefore the pure strategies $x^* = y^* = (1, 0, 0)$ are optimal—both players choose 1 every time—and the value is $y^* Ax^* = a_{11} = 0$.

It is worth remarking that the matrix that leaves all decisions unchanged is not the identity matrix, but the matrix E that has *every* entry e_{ij} equal to 1. Adding a multiple of E to the payoff matrix, $A \to A + \alpha E$, simply means that X wins an additional amount α at every turn. The value of the game is increased by α, but there is no reason to change x^* and y^*.

Now we return to the general theory, putting ourselves first in the place of X. Suppose he chooses the mixed strategy $x = (x_1, \ldots, x_n)$. Then Y will eventually recognize that strategy and choose y to minimize the payment yAx: X will receive $\min_y yAx$. An intelligent player X will select a vector x^* (it may not be unique) that **maximizes this minimum**. By this choice, X guarantees that he will win at least the amount

$$\min_{y} yAx^* = \max_{x} \min_{y} yAx. \tag{1}$$

He cannot expect to win more.

Player Y does the opposite. For any of his own mixed strategies y, he must expect X to discover the vector that will maximize yAx.† Therefore Y will choose the mixture y^* that **minimizes this maximum** and guarantees that he will lose no more than

$$\max_{x} y^*Ax = \min_{y} \max_{x} yAx. \tag{2}$$

Y cannot expect to do better.

I hope you see what the key result will be, if it is true. We want the amount (1) that X is guaranteed to win to coincide with the amount (2) that Y must be satisfied to lose. Then the mixtures x^* and y^* will yield a saddle point equilibrium, and the game will be solved: X can only lose by moving from x^* and Y can only lose by moving from y^*. The existence of this saddle point was proved by von Neumann, and it is known as the **minimax theorem**:

8M   For any m by n matrix A, the minimax over all strategies equals the maximin:

$$\max_{x} \min_{y} yAx = \min_{y} \max_{x} yAx. \tag{3}$$

This quantity is the value of the game. If the maximum on the left is attained at x^*, and the minimum on the right is attained at y^*, then those strategies are optimal and they yield a saddle point from which nobody wants to move:

$$y^*Ax \le y^*Ax^* \le yAx^* \qquad \text{for all } x \text{ and } y. \tag{4}$$

At this saddle point, x^* is at least as good as any other x (since $y^*Ax \le y^*Ax^*$). Similarly, the second player could only pay more by leaving y^*.

Just as in duality theory, we begin with a one-sided inequality: *maximin ≤ minimax*. It is no more than a combination of the definition (1) of x^* and the definition (2) of y^*:

$$\max_{x} \min_{y} yAx = \min_{y} yAx^* \le y^*Ax^* \le \max_{x} y^*Ax = \min_{y} \max_{x} yAx. \tag{5}$$

This only says that if X can guarantee to win at least α, and Y can guarantee to lose no more than β, then necessarily $\alpha \le \beta$. The achievement of von Neumann was to prove that $\alpha = \beta$. That is the minimax theorem. It means that equality must hold throughout (5), and the saddle point property (4) is deduced from it in Exercise 8.5.10.

---

† This may not be x^*. If Y adopts a foolish strategy, then X could get more than he is guaranteed by (1). Game theory has to assume that the players are smart.

For us, the most striking thing about the proof is that *it uses exactly the same mathematics as the theory of linear programming.* Intuitively, that is almost obvious; X and Y are playing "dual" roles, and they are both choosing strategies from the "feasible set" of probability vectors: $x_i \geq 0$, $\sum x_i = 1$, $y_i \geq 0$, $\sum y_i = 1$. What is amazing is that even von Neumann did not immediately recognize the two theories as the same. (He proved the minimax theorem in 1928, linear programming began before 1947, and Gale, Kuhn, and Tucker published the first proof of duality in 1951—based however on von Neumann's notes!) Their proof actually appeared in the same volume where Dantzig demonstrated the equivalence of linear programs and matrix games, so we are reversing history by deducing the minimax theorem from duality.

Briefly, the minimax theorem can be proved as follows. Let b be the column vector of m 1's, and c be the row vector of n 1's. Consider the dual linear programs

(P)   minimize cx **(D)**   maximize yb
       subject to $Ax \geq b$, $x \geq 0$ subject to $yA \leq c$, $y \geq 0$.

To apply duality we have to be sure that both problems are feasible, so if necessary we add the same large number α to all the entries of A. This change cannot affect the optimal strategies in the game, since every payoff goes up by α and so do the minimax and maximin. For the resulting matrix, which we still denote by A, $y = 0$ is feasible in the dual and any large x is feasible in the primal.

Now the duality theorem of linear programming guarantees that there exist feasible x^* and y^* with $cx^* = y^*b$. Because of the ones in b and c, this means that $\sum x_i^* = \sum y_i^*$. If these sums equal θ, then division by θ changes the sums to one— and *the resulting mixed strategies x^*/θ and y^*/θ are optimal.* For any other strategies x and y,

$$Ax^* \geq b \quad \text{implies} \quad yAx^* \geq yb = 1 \quad \text{and} \quad y^*A \leq c \quad \text{implies} \quad y^*Ax \leq cx = 1.$$

The main point is that $y^*Ax \leq 1 \leq yAx^*$. Dividing by θ, this says that player X cannot win more than $1/\theta$ against the strategy y^*/θ, and player Y cannot lose less than $1/\theta$ against x^*/θ. Those are the strategies to be chosen, and maximin = minimax $= 1/\theta$.

This completes the theory, but it leaves unanswered the most natural question of all: Which ordinary games are actually equivalent to "matrix games"? *Do chess and bridge and poker fit into the framework of von Neumann's theory?*

It seems to me that chess does not fit very well, for two reasons. First, a strategy for the white pieces does not consist just of the opening move. It must include a decision on how to respond to the first reply of black, and then how to respond to his second reply, and so on to the end of the game. There are so many alternatives at every step that X has billions of pure strategies, and the same is true for his opponent. Therefore m and n are impossibly large. Furthermore, I do not see much of a role for chance. If white can find a winning strategy or if black can find a drawing strategy—neither has ever been found—that would effectively end the game of chess. Of course it could continue to be played, like tic-tac-toe, but the excitement would tend to go away.

Unlike chess, bridge does contain some deception—for example, deciding what to do in a finesse. It counts as a matrix game, but m and n are again fantastically big. Perhaps separate parts of the game could be analyzed for an optimal strategy. The same is true in baseball, where the pitcher and batter try to outguess each other on the choice of pitch. (Or the catcher tries to guess when the runner will steal, by calling a pitchout; he cannot do it every time without walking the batter, so there must be an optimal frequency—depending on the base runner and on the situation.) Again a small part of the game could be isolated and analyzed.

On the other hand, *blackjack is not a matrix game* (at least in a casino) because the house follows fixed rules. When my friend Ed Thorp found and published a winning strategy—forcing a change in the rules at Las Vegas—his advice depended entirely on keeping track of the cards already seen. There was no element of chance, and therefore no mixed strategy x^*.

There is also the **Prisoner's Dilemma**, in which two accomplices are captured. They are separately offered the same deal: Confess and you are free, provided your accomplice does not confess (the accomplice then gets 10 years). If both confess each gets 6 years. If neither confesses, only a minor crime (2 years each) can be proved. What to do? The temptation to confess is very great, although if they could depend on each other they would hold out. This is a nonzero-sum game; both can lose.

The perfect example of an ordinary game that is also a matrix game is **poker**. Bluffing is essential, and to be effective it has to be unpredictable. The decision to bluff should be made completely at random, if you accept the fundamental assumption of game theory that your opponent is intelligent. (If he finds a pattern, he wins.) The probabilities for and against bluffing will depend on the cards that are seen, and on the bets. In fact, the number of alternatives again makes it impractical to find an absolutely optimal strategy x^*. Nevertheless a good poker player must come pretty close to x^*, and we can compute it exactly if we accept the following enormous simplification of the game:

Player X is dealt a jack or a king, with equal probability, whereas Y always gets a queen. After looking at his hand, X can either fold and concede his ante of $1, or he can bet an additional $2. If X bets, then Y can either fold and concede his own ante of $1, or decide to match the $2 and see whether X is bluffing. In this case the higher card wins the $3 that was bet by the opponent.

Even in this simple game, the "pure strategies" are not all obvious, and it is instructive to write them down. Y has two possibilities, since he only reacts to X:

(1) If X bets, Y folds.
(2) If X bets, Y sees him and tries for $3.

X has four strategies, some reasonable and some foolish:

(1) He bets the extra $2 on a king and folds on a jack.
(2) He bets in either case (bluffing).
(3) He folds in either case, and concedes $1.
(4) He folds on a king and bets on a jack.

The payoff matrix needs a little patience to compute:

$a_{11} = 0$, since X loses \$1 half the time on a jack and wins on a king (Y folds).

$a_{21} = 1$, since X loses \$1 half the time and wins \$3 half the time (Y tries to beat him, even though X has a king).

$a_{12} = 1$, since X bets and Y folds (the bluff succeeds).

$a_{22} = 0$, since X wins \$3 with the king and loses \$3 with the jack (the bluff fails).

The third strategy always loses \$1, and the fourth is also unsuccessful:

$$A = \begin{bmatrix} 0 & 1 & -1 & 0 \\ 1 & 0 & -1 & -2 \end{bmatrix}.$$

The optimal strategy in this game is for X to bluff half the time, $x^* = (\frac{1}{2}, \frac{1}{2}, 0, 0)$, and the underdog Y must choose $y^* = (\frac{1}{2}, \frac{1}{2})$. The value of the game is fifty cents.

That is a strange way to end this book, by teaching you how to play a watered down version of poker, but I guess even poker has its place within linear algebra and its applications. I hope you have enjoyed the book.

EXERCISES

8.5.1   How will the optimal strategies in the original game be affected if the \$20 is increased to \$70, and what is the value (the average win for X) of this new game?

8.5.2   With payoff matrix $A = \begin{bmatrix} 1 & 2 \\ 3 & 4 \end{bmatrix}$, explain the calculation by X of his maximin and by Y of his minimax. What strategies x^* and y^* are optimal?

8.5.3   If a_{ij} is the largest entry in its row and the smallest in its column, why will X always choose column j and Y always choose row i (regardless of the rest of the matrix)? Show that the previous exercise had such an entry, and then construct an A without one.

8.5.4   Find the best strategy for Y by weighting the rows of A with y and $1 - y$ and graphing all three components. X will concentrate on the largest component (dark line) and Y must minimize this maximum. What is y^* and what is the minimax height (dotted line)?

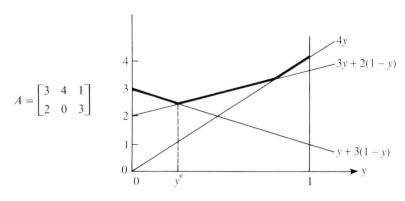

$$A = \begin{bmatrix} 3 & 4 & 1 \\ 2 & 0 & 3 \end{bmatrix}$$

8.5.5   With the same A, find the best strategy for X, and show that he uses only the two columns (the first and third) that met at the minimax point in the graph.

8.5.6   Find both optimal strategies, and the value, if

$$A = \begin{bmatrix} 1 & 0 & -1 \\ -2 & -1 & 2 \end{bmatrix}.$$

8.5.7   Suppose $A = \begin{bmatrix} a & b \\ c & d \end{bmatrix}$. What weights x_1 and $1 - x_1$ will give a column of the form $\begin{bmatrix} u & u \end{bmatrix}^T$, and what weights y_1 and $1 - y_1$ on the two rows will give a new row $\begin{bmatrix} v & v \end{bmatrix}$? Show that $u = v$.

8.5.8   Find x^*, y^*, and the value v for

$$A = \begin{bmatrix} 1 & 0 & 0 \\ 0 & 2 & 0 \\ 0 & 0 & 3 \end{bmatrix}.$$

8.5.9   Compute

$$\min_{\substack{y_i \geq 0 \\ y_1 + y_2 = 1}} \max_{\substack{x_i \geq 0 \\ x_1 + x_2 = 1}} (x_1 y_1 + x_2 y_2).$$

8.5.10   Explain each of the inequalities in (5). Then, once the minimax theorem has turned them into equalities, derive (again in words) the saddle point equations (4).

8.5.11   Show that $x^* = (\frac{1}{2}, \frac{1}{2}, 0, 0)$ and $y^* = (\frac{1}{2}, \frac{1}{2})$ are optimal strategies in poker, by computing yAx^* and y^*Ax and verifying the conditions (4) for a saddle point.

8.5.12   Has it been proved that there cannot exist a chess strategy that always wins for black? I believe this is known only when the players are given two moves at a time; if black had a winning strategy, white could move a knight out and back and then follow that strategy, leading to the impossible conclusion that both would win.

8.5.13   If X chooses a prime number and simultaneously Y guesses whether it is odd or even (with gain or loss of $1), who has the advantage?

8.5.14   If X is a quarterback, with the choice of run or pass, and Y is the defensive captain, who can defend against a run or a pass, suppose the payoff (in yards) is

$$A = \begin{bmatrix} 2 & 8 \\ 6 & -6 \end{bmatrix} \quad \begin{matrix} \text{defense against run} \\ \text{defense against pass} \end{matrix}$$
$$\quad\;\; \text{run} \quad \text{pass}$$

What are the optimal strategies and the average gain on each play?

APPENDIX **A**

THE SINGULAR VALUE DECOMPOSITION AND THE PSEUDOINVERSE

The third great matrix factorization has been saved for this appendix. It joins with LU from elimination and QR from orthogonalization (Gauss and Gram-Schmidt). Nobody's name is attached; it is known as the "SVD" or the *singular value decomposition*. We want to describe it, to prove it, and to discuss its applications—which are many and growing.

The SVD is closely associated with the eigenvalue-eigenvector factorization of a symmetric matrix: $A = Q \Lambda Q^T$. There the eigenvalues are in the diagonal matrix Λ, and the eigenvector matrix Q is *orthogonal*: $Q^T Q = I$ because eigenvectors of a symmetric matrix can be chosen orthonormal. For most matrices that is not true, and for rectangular matrices it is ridiculous. But if we allow the Q on the left and the Q^T on the right to be *any two orthogonal matrices*—not necessarily transposes of each other—the factorization becomes possible again. Furthermore the diagonal (but rectangular) matrix in the middle can be made nonnegative. It will be denoted by Σ, and its positive entries (also called sigma) will be $\sigma_1, \ldots, \sigma_r$. They are the **singular values** of A. They fill the first r places on the main diagonal of Σ—and r is the rank of A.

The key to working with rectangular matrices is, almost always, to consider AA^T and A^TA.

Singular Value Decomposition: Any m by n matrix A can be factored into

$$A = Q_1 \Sigma Q_2^T = \text{(orthogonal)(diagonal)(orthogonal).}$$

The columns of Q_1 (m by m) are eigenvectors of AA^T, and the columns of Q_2 (n by n) are eigenvectors of A^TA. The r singular values on the diagonal of Σ (m by n) are the square roots of the nonzero eigenvalues of both AA^T and A^TA.

Remark 1 For positive definite matrices this factorization is identical to $Q\Lambda Q^T$. For indefinite matrices, any negative eigenvalues in Λ become positive in Σ, and Q_1 is then different from Q_2. For complex matrices Σ remains real but Q_1 and Q_2 become *unitary* (the complex analogue of orthogonal). Then $A = U_1 \Sigma U_2^H$.

Remark 2 The columns of Q_1 and Q_2 give orthonormal bases for *all four fundamental subspaces*:

$$
\begin{array}{lll}
\text{first} & r & \text{columns of } Q_1\text{:} & \text{column space of } A \\
\text{last } m-r & \text{columns of } Q_1\text{:} & \text{left nullspace of } A \\
\text{first} & r & \text{columns of } Q_2\text{:} & \text{row space of } A \\
\text{last } n-r & \text{columns of } Q_2\text{:} & \text{nullspace of } A
\end{array}
$$

Remark 3 The SVD chooses those bases in an extremely special way. They are more than just orthonormal. *If A multiplies a column of Q_2, it produces a multiple of a column of Q_1.* That comes directly from $AQ_2 = Q_1\Sigma$, looked at a column at a time.

Remark 4 The connections with AA^T and A^TA must hold if the formula $Q_1\Sigma Q_2^T$ is correct. That is easy to see:

$$AA^T = (Q_1\Sigma Q_2^T)(Q_2\Sigma^T Q_1^T) = Q_1\Sigma\Sigma^T Q_1^T \quad \text{and similarly} \quad A^TA = Q_2\Sigma^T\Sigma Q_2^T. \quad (1)$$

From the first, Q_1 *must be the eigenvector matrix for AA^T.* The eigenvalue matrix in the middle is $\Sigma\Sigma^T$—which is m by m with $\sigma_1^2, \ldots, \sigma_r^2$ on the diagonal. From the second, Q_2 *must be the eigenvector matrix for A^TA.* The diagonal matrix $\Sigma^T\Sigma$ has the same $\sigma_1^2, \ldots, \sigma_r^2$, but it is n by n.

These checks are easy, but they do not prove the SVD. The proof is not difficult but it can follow the examples and applications.

EXAMPLE 1 (A is diagonal)

$$
\begin{bmatrix} 2 & 0 \\ 0 & -3 \\ 0 & 0 \end{bmatrix} = \begin{bmatrix} 1 & 0 & 0 \\ 0 & -1 & 0 \\ 0 & 0 & 1 \end{bmatrix} \begin{bmatrix} 2 & 0 \\ 0 & 3 \\ 0 & 0 \end{bmatrix} \begin{bmatrix} 1 & 0 \\ 0 & 1 \end{bmatrix}
$$

EXAMPLE 2   (*A* has only one column)

$$
A = \begin{bmatrix} -1 \\ 2 \\ 2 \end{bmatrix} = \begin{bmatrix} -\frac{1}{3} & \frac{2}{3} & \frac{2}{3} \\ \frac{2}{3} & -\frac{1}{3} & \frac{2}{3} \\ \frac{2}{3} & \frac{2}{3} & -\frac{1}{3} \end{bmatrix} \begin{bmatrix} 3 \\ 0 \\ 0 \end{bmatrix} [1]
$$

Here $A^T A$ is 1 by 1 while $A A^T$ is 3 by 3. They both have eigenvalue 9 (whose square root is the 3 in Σ). The two zero eigenvalues of $A A^T$ leave some freedom for the eigenvectors in columns 2 and 3 of Q_1. We made a choice which kept that matrix orthogonal.

EXAMPLE 3   (*A* is already orthogonal)
Either $A = QII$ or $A = IIQ$ or even $A = (QQ_2)IQ_2^T$, but certainly $\Sigma = I$.

EXAMPLE 4   (*A* is an incidence matrix and $A A^T = \begin{bmatrix} 2 & -1 \\ -1 & 2 \end{bmatrix}$ with $\lambda = 3, 1$)

$$
A = \begin{bmatrix} -1 & 1 & 0 \\ 0 & -1 & 1 \end{bmatrix} = \begin{bmatrix} -1 & 1 \\ 1 & 1 \end{bmatrix} \begin{bmatrix} \sqrt{3} & 0 & 0 \\ 0 & 1 & 0 \end{bmatrix} \begin{bmatrix} 1 & -2 & 1 \\ -1 & 0 & 1 \\ 1 & 1 & 1 \end{bmatrix} \begin{matrix} /\sqrt{6} \\ /\sqrt{2} \\ /\sqrt{3} \end{matrix}
$$
$$
\begin{matrix} /\sqrt{2} & /\sqrt{2} \end{matrix}
$$

APPLICATIONS OF THE SVD

We will pick a few of the important applications, after emphasizing one key point. The SVD is terrific for numerically stable computations. The first reason is that Q_1 and Q_2 are orthogonal matrices; they never change the length of a vector. Since $\|Qx\|^2 = x^T Q^T Q x = \|x\|^2$, multiplication by Q cannot destroy the scaling. Such a statement cannot be made for the other factor Σ; we could multiply by a large σ or (more commonly) divide by a small σ, and overflow the computer. But still Σ *is as good as possible.* It reveals exactly what is large and what is small, and the easy availability of that information is the second reason for the popularity of the SVD. We come back to this in the second application.

1. Image processing   Suppose a satellite takes a picture, and wants to send it to earth. The picture may contain 1000 by 1000 "pixels"—little squares each with a definite color. We can code the colors, in a range between black and white, and send back 1,000,000 numbers. It is better to find the essential information in the 1000 by 1000 matrix, and send only that.

Suppose we know the SVD. The key is in the singular values (in Σ). Typically, some are significant and others are extremely small. If we keep 60 and throw away 940, then we send only the corresponding 60 columns of Q_1 and Q_2. The other

940 columns are multiplied in $Q_1 \Sigma Q_2^T$ by the small σ's that are being ignored. In fact, *we can do the matrix multiplication as columns times rows*:

$$Q_1 \Sigma Q_2^T = u_1 \sigma_1 v_1^T + u_2 \sigma_2 v_2^T + \cdots + u_r \sigma_r v_r^T. \tag{2}$$

Here the u's are columns of Q_1 and the v's are columns of Q_2 (v_1^T is the first row of Q_2^T). Any matrix is the sum of r matrices of rank one. If only 60 terms are kept, we send 60 times 2000 numbers instead of a million.

The pictures are really striking, as more and more singular values are included. At first you see nothing, and suddenly you recognize everything.

2. The effective rank   In Chapter 2 we did not hesitate to define the rank of a matrix. It was the number of independent rows, or equivalently the number of independent columns. That can be hard to decide in computations! Our original method was to count the pivots, and in exact arithmetic that is correct. In real arithmetic it can be misleading—but discarding small pivots is not the answer. Consider

$$\begin{bmatrix} \epsilon & 2\epsilon \\ 1 & 2 \end{bmatrix} \quad \text{and} \quad \begin{bmatrix} \epsilon & 1 \\ 0 & 0 \end{bmatrix} \quad \text{and} \quad \begin{bmatrix} \epsilon & 1 \\ \epsilon & 1 + \epsilon \end{bmatrix}.$$

The first has rank 1, although roundoff error will probably produce a second pivot. Both pivots will be small; how many do we ignore? The second has one small pivot, but we cannot pretend that its row is insignificant. The third has two pivots and its rank is 2, but its "effective rank" ought to be 1.

We go to a more stable measure of rank. The first step is to use $A^T A$ or $A A^T$, which are symmetric but share the same rank as A. Their eigenvalues—the singular values squared—are not misleading. Based on the accuracy of the data, we decide on a tolerance like 10^{-6} and count the singular values above it—that is the effective rank. In the examples above, it is always one if ϵ is small. Admittedly it is discontinuous when a singular value passes 10^{-6}, but that is much less troublesome than the ordinary algebraic definition—which is discontinuous at $\sigma = 0$ and allows no control.

3. Polar Decomposition   Every complex number is the product of a nonnegative number r and a number $e^{i\theta}$ on the unit circle: $z = re^{i\theta}$. That is the expression of z in "polar coordinates." If we think of these numbers as 1 by 1 matrices, r corresponds to a *positive semidefinite matrix* and $e^{i\theta}$ corresponds to an *orthogonal matrix*. More exactly, since $e^{i\theta}$ is complex and satisfies $e^{-i\theta} e^{i\theta} = 1$, it forms a 1 by 1 *unitary matrix*: $U^H U = I$. We take the complex conjugate as well as the transpose, for U^H.

The SVD extends this factorization to matrices of any size:

Every real square matrix can be factored into $A = QS$, where Q is **orthogonal** and S is **symmetric positive semidefinite**. If A is invertible then S is positive definite.

For proof we just insert $Q_2^T Q_2 = I$ into the middle of the SVD:

$$A = Q_1 \Sigma Q_2^T = (Q_1 Q_2^T)(Q_2 \Sigma Q_2^T). \tag{3}$$

The factor $S = Q_2 \Sigma Q_2^T$ is symmetric and semidefinite (because Σ is). The factor $Q = Q_1 Q_2^T$ is an orthogonal matrix (because $Q^T Q = Q_2 Q_1^T Q_1 Q_2^T = I$). In the complex case S becomes Hermitian instead of symmetric and Q becomes unitary instead of orthogonal. In the invertible case Σ is definite and so is S.

EXAMPLE OF POLAR DECOMPOSITION $A = QS$

$$\begin{bmatrix} 1 & -2 \\ 3 & -1 \end{bmatrix} = \begin{bmatrix} 0 & -1 \\ 1 & 0 \end{bmatrix} \begin{bmatrix} 3 & -1 \\ -1 & 2 \end{bmatrix}$$

EXAMPLE OF REVERSE POLAR DECOMPOSITION $A = S'Q$

$$\begin{bmatrix} 1 & -2 \\ 3 & -1 \end{bmatrix} = \begin{bmatrix} 2 & 1 \\ 1 & 3 \end{bmatrix} \begin{bmatrix} 0 & -1 \\ 1 & 0 \end{bmatrix}$$

The exercises show how, in the reverse order, S changes but Q remains the same. Both S and S' are symmetric positive definite because this A is invertible.

Note   We could start from $A = QS$ and go backwards to find the SVD. Since S is symmetric it equals $Q_2 \Sigma Q_2^T$—this is the usual eigenvector-eigenvalue factorization $Q \Lambda Q^T$ with only a change of notation. Then

$$A = QS = QQ_2 \Sigma Q_2^T = Q_1 \Sigma Q_2^T$$

is the SVD. Historically, the polar decomposition used to be more prominent—but I believe the singular value decomposition is more fundamental.

Application of $A = QS$: A major use of the polar decomposition is in continuum mechanics (and more recently in robotics). In any deformation it is important to separate stretching from rotation, and that is exactly what QS achieves. The orthogonal matrix Q is a rotation, and possibly a reflection. The material feels no strain. The symmetric matrix S has eigenvalues $\sigma_1, \ldots, \sigma_r$, which are the stretching factors (or compression factors). The diagonalization that displays those eigenvalues is the natural choice of axes—called **principal axes**, as in the ellipses of Section 6.2. It is S that requires work on the material, and stores up elastic energy.

We note that S^2 is $A^T A$, which is symmetric positive definite when A is invertible. S is the symmetric positive definite square root of $A^T A$, and Q is $A S^{-1}$. In fact A *could be rectangular, as long as $A^T A$ is positive definite.* (That is the condition we keep meeting, that A must have independent columns.) In the reverse order $A = S'Q$, the matrix S' is the symmetric positive definite square root of $A A^T$.

4. Least Squares   In Chapter 3 we found the least squares solution to a rectangular system $Ax = b$. It came from the normal equations $A^T A \bar{x} = A^T b$, but there was a firm requirement on A. Its columns had to be independent; the rank had to be n. Otherwise $A^T A$ was not invertible and $\bar{x}$ was not determined—any vector in the nullspace could be added to $\bar{x}$. We can now complete Chapter 3, by choosing a "best" $\bar{x}$ for every linear system $Ax = b$.

There are two possible difficulties with $Ax = b$:

(1)   The rows of A may be dependent.
(2)   The columns of A may be dependent.

In the first case, the equations may have no solution. That happens when b is outside the column space of A, and Chapter 3 gave a remedy: Project b onto the column space. Instead of $Ax = b$, we solve $A\bar{x} = p$. That can be done because p is in the column space. But now case (2) presents a different obstacle. If A has dependent columns, the solution $\bar{x}$ will not be unique. We have to choose a particular solution of $A\bar{x} = p$, and the choice is made according to the following rule:

The optimal solution of $A\bar{x} = p$ is the one that has minimum length.

That optimal solution will be called x^+. It is our preferred choice, as the best solution to $Ax = b$ (which had no solution), and also to $A\bar{x} = p$ (which had too many). The goal is to identify x^+, and we start with an example.

EXAMPLE 1 A is diagonal, with dependent rows and dependent columns:

$$A = \begin{bmatrix} \sigma_1 & 0 & 0 & 0 \\ 0 & \sigma_2 & 0 & 0 \\ 0 & 0 & 0 & 0 \end{bmatrix}.$$

The columns all end with zero. In the column space, the closest vector to $b = (b_1, b_2, b_3)$ is $p = (b_1, b_2, 0)$. That is the projection, and the error $(0, 0, b_3)$ is perpendicular to all the columns. The best we can do with $Ax = b$ is to solve the first two equations, since the third equation is $0 = b_3$. The error in that equation cannot be reduced, but the errors in the first two equations will be zero:

$$A\bar{x} = p \quad \text{is} \quad \begin{bmatrix} \sigma_1 & 0 & 0 & 0 \\ 0 & \sigma_2 & 0 & 0 \\ 0 & 0 & 0 & 0 \end{bmatrix} \begin{bmatrix} \bar{x}_1 \\ \bar{x}_2 \\ \bar{x}_3 \\ \bar{x}_4 \end{bmatrix} = \begin{bmatrix} b_1 \\ b_2 \\ 0 \end{bmatrix}.$$

Now we face the second difficulty. The columns are dependent and $\bar{x}$ is not unique. The first two components are b_1/σ_1 and b_2/σ_2, but the other components $\bar{x}_3$ and $\bar{x}_4$ are totally arbitrary. To make $\bar{x}$ as short as possible, we choose those components to be zero. *The minimum length solution of $A\bar{x} = p$ is x^+:*

$$x^+ = \begin{bmatrix} b_1/\sigma_1 \\ b_2/\sigma_2 \\ 0 \\ 0 \end{bmatrix} = \begin{bmatrix} 1/\sigma_1 & 0 & 0 \\ 0 & 1/\sigma_2 & 0 \\ 0 & 0 & 0 \\ 0 & 0 & 0 \end{bmatrix} \begin{bmatrix} b_1 \\ b_2 \\ b_3 \end{bmatrix}. \tag{4}$$

This equation is important. It finds x^+, and it also displays *the matrix which produces x^+ from b*. That matrix is called the ***pseudoinverse*** of A. It is denoted by A^+, so we have $x^+ = A^+ b$.

Based on this example, we can find the best solution x^+ and the pseudoinverse A^+ for any diagonal matrix.

$$\text{If } A = \begin{bmatrix} \sigma_1 & & \\ & \ddots & \\ & & \sigma_r \end{bmatrix} \quad \text{then } A^+ = \begin{bmatrix} 1/\sigma_1 & & \\ & \ddots & \\ & & 1/\sigma_r \end{bmatrix} \quad \text{and } x^+ = A^+ b = \begin{bmatrix} b_1/\sigma_1 \\ \vdots \\ b_r/\sigma_r \end{bmatrix}.$$

The matrix A is m by n, with r nonzero entries σ_i. Its pseudoinverse A^+ is n by m, with r nonzero entries $1/\sigma_i$. All the blank spaces are zeros.

Notice that $(A^+)^+$ is A again. That is like $(A^{-1})^{-1} = A$, but here A is not invertible. If b has components $(b_1, \ldots, b_m)$, then its projection p has components $(b_1, \ldots, b_r, 0, \ldots, 0)$. The shortest solution to $A\bar{x} = p$ is the x^+ given above. Later we will denote the diagonal matrix by Σ and its pseudoinverse by Σ^+.

To go beyond diagonal matrices, start with the easy case—when A is invertible. The equation $Ax = b$ has one and only one solution: $x = A^{-1}b$. *In that case x is x^+ and A^{-1} is A^+. The pseudoinverse is the same as the inverse, when A is invertible.* We check both conditions: Ax is as close as possible to b (it *equals b*) and x is the shortest solution to $A\bar{x} = b$ (it is the only solution, when A is invertible).

Now we find x^+ in the general case. It is the shortest solution to $A\bar{x} = p$, and we claim that x^+ ***is always in the row space of*** A. Remember that any vector $\bar{x}$ can be split into a row space component and a nullspace component: $\bar{x} = \bar{x}_r + \bar{x}_n$. There are three important points about that splitting:

1. The row space component also solves $A\bar{x}_r = p$, because $A\bar{x}_n = 0$.
2. The components are orthogonal, and they obey Pythagoras' law:

$$\|\bar{x}\|^2 = \|\bar{x}_r\|^2 + \|\bar{x}_n\|^2, \text{ so } \bar{x} \text{ is shortest when } \bar{x}_n = 0.$$

3. All solutions of $A\bar{x} = p$ have the same row space component $\bar{x}_r$. ***That vector is x^+.***

The picture to remember is Fig. 3.4. It displayed the fundamental theorem of linear algebra—by showing how A takes every x into its column space. It also showed how every p in the column space comes from one and only one vector in the row space. (If $A\bar{x}_r = p$ and also $A\bar{x}_r' = p$, then $A(\bar{x}_r - \bar{x}_r') = 0$. The difference lies in the nullspace and also in the row space—so $\bar{x}_r - \bar{x}_r'$ is orthogonal to itself and must be zero.) *All we are doing is to choose that vector $x^+ = \bar{x}_r$ as the best solution to $Ax = b$.* It is the shortest solution of the nearest equation $A\bar{x} = p$.

The pseudoinverse reverses the direction of the figure. It starts with b and comes back to x^+. *It inverts A where A is invertible*—between row space and column space. The pseudoinverse knocks out the left nullspace, by sending it to zero, and it knocks out the nullspace by choosing $\bar{x}_r$ as x^+.

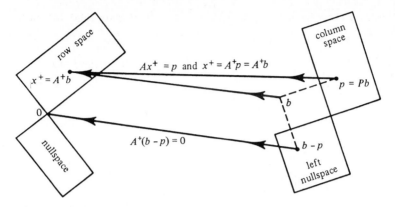

Fig. A.1. The action of the pseudoinverse A^+.

We have not yet shown that there is a matrix A^+ which always gives x^+—but there is. It will be n by m, because it takes b in $\mathbf{R}^m$ back to x^+ in $\mathbf{R}^n$. We look at one more example before finding A^+ in general.

EXAMPLE 2 $Ax = b$ is $-x_1 + 2x_2 + 2x_3 = 18$.

The equation has a whole plane of solutions. According to our theory, the shortest solution should be in the row space of $A = [-1 \ 2 \ 2]$. The multiple of that row which satisfies the equation is $x^+ = (-2, 4, 4)$. There are other solutions like $(-2, 5, 3)$ or $(-2, 7, 1)$ or $(-6, 3, 3)$, but they are all longer than x^+. (They all have nonzero components from the nullspace.) The matrix that produces x^+ from the right side b, which was the very short vector $[18]$, is the pseudoinverse A^+. Where A was 1 by 3, this matrix A^+ is 3 by 1:

$$A^+ = [-1 \ 2 \ 2]^+ = \begin{bmatrix} -\frac{1}{9} \\ \frac{2}{9} \\ \frac{2}{9} \end{bmatrix} \quad \text{and} \quad A^+[18] = \begin{bmatrix} -2 \\ 4 \\ 4 \end{bmatrix}. \tag{5}$$

The column space of A^+ is the row space of A. Now we give a formula for A^+:

Suppose the singular value decomposition of A is $A = Q_1 \Sigma Q_2^T$. Then the pseudo-inverse of A is

$$A^+ = Q_2 \Sigma^+ Q_1^T. \tag{6}$$

The singular values $\sigma_1, \ldots, \sigma_r$ are on the diagonal of Σ (m by n), and the reciprocals $1/\sigma_1, \ldots, 1/\sigma_r$ are on the diagonal of Σ^+ (n by m). The pseudoinverse of A^+ is $A^{++} = A$.

Example 1 contained the diagonal case, when A was Σ and $Q_1 = I$ (m by m) and $Q_2 = I$ (n by n). Example 2 also had $Q_1 = [1]$, and the singular value was 3—the square root of the eigenvalue of AA^T. Please notice Σ and Σ^+ in the following SVD's:

$$[-1\ 2\ 2] = [1][3\ 0\ 0]\begin{bmatrix} -\frac{1}{3} & \frac{2}{3} & \frac{2}{3} \\ \frac{2}{3} & -\frac{1}{3} & \frac{2}{3} \\ \frac{2}{3} & \frac{2}{3} & -\frac{1}{3} \end{bmatrix}$$

$$[-1\ 2\ 2]^+ = \begin{bmatrix} -\frac{1}{3} & \frac{2}{3} & \frac{2}{3} \\ \frac{2}{3} & -\frac{1}{3} & \frac{2}{3} \\ \frac{2}{3} & \frac{2}{3} & -\frac{1}{3} \end{bmatrix}\begin{bmatrix} \frac{1}{3} \\ 0 \\ 0 \end{bmatrix}[1] = \begin{bmatrix} -\frac{1}{9} \\ \frac{2}{9} \\ \frac{2}{9} \end{bmatrix}.$$

We can check directly that this matrix A^+ leads to the optimal solution of $Ax = b$.

The minimum length least squares solution to $Ax = b$ is $x^+ = A^+b = Q_2\Sigma^+Q_1^Tb$.

Proof   Multiplication by the orthogonal matrix Q_1^T leaves lengths unchanged, so

$$\|Ax - b\| = \|Q_1\Sigma Q_2^Tx - b\| = \|\Sigma Q_2^Tx - Q_1^Tb\|.$$

Introduce the new unknown $y = Q_2^Tx = Q_2^{-1}x$, which has the same length as x. Then minimizing $\|Ax - b\|$ is the same as minimizing $\|\Sigma y - Q_1^Tb\|$. This has a diagonal matrix, and we know the best y^+. It is $y^+ = \Sigma^+Q_1^Tb$, so the best x^+ is

$$x^+ = Q_2y^+ = Q_2\Sigma^+Q_1^Tb.$$

We could verify directly that this x^+ is in the row space and that it satisfies $Ax^+ = p$.

Construction of the SVD   With the applications finished, we go back to the singular value decomposition on which they all were based. We have to prove that any matrix can be factored into $A = Q_1\Sigma Q_2^T$, with diagonal Σ and orthogonal Q's.

A symmetric matrix like A^TA has a complete set of orthonormal eigenvectors x_j, which go into the columns of Q_2:

$$A^TAx_j = \lambda_jx_j \text{ with } x_j^Tx_j = 1 \text{ and } x_i^Tx_j = 0 \text{ for } i \neq j. \tag{7}$$

(In the complex case change A^T to A^H and x^T to x^H.) Taking the inner product with x_j, we discover that all $\lambda_j \geq 0$:

$$x_j^TA^TAx_j = \lambda_jx_j^Tx_j \quad \text{or} \quad \|Ax_j\|^2 = \lambda_j. \tag{8}$$

Suppose the eigenvalues $\lambda_1, \ldots, \lambda_r$ are positive, and the remaining $n - r$ of the Ax_j and λ_j are zero. For the positive ones, set $\sigma_j = \sqrt{\lambda_j}$ and $q_j = Ax_j/\sigma_j$. (These numbers $\sigma_1, \ldots, \sigma_r$ will become the singular values on the diagonal of Σ.) Note that the q_j are unit vectors in $\mathbf{R}^m$ from (8), and they are mutually orthogonal from (7):

$$q_i^Tq_j = \frac{x_i^TA^TAx_j}{\sigma_i\sigma_j} = \frac{\lambda_jx_i^Tx_j}{\sigma_i\sigma_j} = 0 \quad \text{for } i \neq j. \tag{9}$$

These r orthonormal q's can be extended, by Gram-Schmidt, to a complete orthonormal basis $q_1, \ldots, q_m$. Those are the columns of Q_1. The i, j entry of the product $Q_1^T A Q_2$ is $q_i^T A x_j$ (row times matrix times column), and we know those entries:

$$q_i^T A x_j = 0 \quad \text{if } j > r \text{ (because } Ax_j = 0)$$
$$q_i^T A x_j = q_i^T \sigma_j q_j \quad \text{if } j \le r \text{ (because } Ax_j = \sigma_j q_j). \tag{10}$$

But $q_i^T q_j = 0$ except when $i = j$. Thus the only nonzeros in the product $\Sigma = Q_1^T A Q_2$ are the first r diagonal entries, which are $\sigma_1, \ldots, \sigma_r$. Since $Q^T = Q^{-1}$ we have $A = Q_1 \Sigma Q_2^T$ as required.

We mention again how these vectors x and q fit the four fundamental subspaces of A. The first r x's are an orthonormal basis for the row space, and the last $n - r$ are an orthonormal basis for the nullspace (where $Ax_j = 0$). The first r q's are an orthonormal basis for the column space, and the last $m - r$ are an orthonormal basis for the left nullspace. Of course it is no surprise that those bases exist. We know that! What is special about the x's and q's is that $Ax_j = \sigma_j q_j$. The matrix takes basis vectors into basis vectors, when the bases are right. By the rules for constructing a matrix, where you apply A to the basis in $\mathbf{R}^n$ and express the result using the basis in $\mathbf{R}^m$, the matrix with respect to these bases is diagonal. It is Σ.

EXERCISES

A.1 Find the singular value decomposition and the pseudoinverse 0^+ of the m by n *zero matrix*.

A.2 Find the singular value decomposition and the pseudoinverse of

$$A = \begin{bmatrix} 1 & 1 & 1 & 1 \end{bmatrix} \quad \text{and} \quad B = \begin{bmatrix} 0 & 1 & 0 \\ 1 & 0 & 0 \end{bmatrix} \quad \text{and} \quad C = \begin{bmatrix} 1 & 1 \\ 0 & 0 \end{bmatrix}.$$

A.3 If an m by n matrix Q has orthonormal columns, what is its pseudoinverse Q^+?

A.4 (a)   Compute $A^T A$ and its eigenvalues and eigenvectors and positive definite square root S if

$$A = \frac{1}{\sqrt{10}} \begin{bmatrix} 10 & 6 \\ 0 & 8 \end{bmatrix}.$$

(b)   Compute the polar decomposition $A = QS$. Knowing Q, find the reverse form $A = S'Q$.

A.5 What is the minimum length least squares solution $x^+ = A^+ b$ to

$$Ax = \begin{bmatrix} 1 & 0 & 0 \\ 1 & 0 & 0 \\ 1 & 1 & 1 \end{bmatrix} \begin{bmatrix} C \\ D \\ E \end{bmatrix} = \begin{bmatrix} 0 \\ 2 \\ 2 \end{bmatrix}?$$

You can compute A^+, or find the general solution to $A^{\mathrm{T}}A\bar{x} = A^{\mathrm{T}}b$ and choose the solution that is in the row space of A. This problem fits the best plane $C + Dt + Ez$ to $b = 0$ at $t = z = 0$, $b = 2$ at the same point, and $b = 2$ at $t = z = 1$.

A.6 (a) If A has independent columns, its left-inverse $(A^{\mathrm{T}}A)^{-1}A^{\mathrm{T}}$ is its pseudoinverse.
(b) If A has independent rows, its right-inverse $A^{\mathrm{T}}(AA^{\mathrm{T}})^{-1}$ is its pseudoinverse.
In both cases verify that $x^+ = A^+b$ is in the row space and $A^{\mathrm{T}}Ax^+ = A^{\mathrm{T}}b$.

A.7 If $A = Q_1 \Sigma Q_2^{-1}$ is split into QS' (reverse polar decomposition) what are Q and S'?

A.8 Is $(AB)^+ = B^+A^+$ always true for pseudoinverses? I believe not.

A.9 Every matrix factors into $A = LU$, where the r columns of L span the column space of A and the r rows of U span the row space (Section 3.6). Verify the explicit formula
$A^+ = \underline{U}^{\mathrm{T}}(\underline{U}\underline{U}^{\mathrm{T}})^{-1}(\underline{L}^{\mathrm{T}}\underline{L})^{-1}\underline{L}^{\mathrm{T}}$.
Hint: Why is A^+b always in the row space, because of $\underline{U}^{\mathrm{T}}$ at the front, and why is $A^{\mathrm{T}}AA^+b = A^{\mathrm{T}}b$, so that A^+b satisfies the normal equation?

A.10 Explain why AA^+ and A^+A are projection matrices (and therefore symmetric). What fundamental subspaces do they project onto?

APPENDIX B

THE JORDAN FORM

Given a square matrix A, we want to choose M so that $M^{-1}AM$ is as nearly diagonal as possible. In the simplest case, A has a complete set of eigenvectors and they become the columns of M—otherwise known as S. The Jordan form is $J = M^{-1}AM = \Lambda$, it is constructed entirely from 1 by 1 blocks $J_i = \lambda_i$, and the goal of a diagonal matrix is completely achieved. In the more general and more difficult case, some eigenvectors are missing and a diagonal form is impossible. That case is now our main concern.

We repeat the theorem that is to be proved:

5S If a matrix has s linearly independent eigenvectors, then it is similar to a matrix which is in *Jordan form* with s square blocks on the diagonal:

$$J = M^{-1}AM = \begin{bmatrix} J_1 & & \\ & \ddots & \\ & & J_s \end{bmatrix}.$$

Each block has one eigenvector, one eigenvalue, and 1's just above the diagonal:

$$J_i = \begin{bmatrix} \lambda_i & 1 & & \\ & \ddots & \ddots & \\ & & \ddots & 1 \\ & & & \lambda_i \end{bmatrix}.$$

An example of such a Jordan matrix is

$$
J = \begin{bmatrix} 8 & 1 & 0 & 0 & 0 \\ 0 & 8 & 0 & 0 & 0 \\ 0 & 0 & 0 & 1 & 0 \\ 0 & 0 & 0 & 0 & 0 \\ 0 & 0 & 0 & 0 & 0 \end{bmatrix} = \begin{bmatrix} \begin{bmatrix} 8 & 1 \\ 0 & 8 \end{bmatrix} & & \\ & \begin{bmatrix} 0 & 1 \\ 0 & 0 \end{bmatrix} & \\ & & [0] \end{bmatrix} = \begin{bmatrix} J_1 & & \\ & J_2 & \\ & & J_3 \end{bmatrix}.
$$

The double eigenvalue $\lambda = 8$ has only a single eigenvector, in the first coordinate direction $e_1 = (1, 0, 0, 0, 0)$; as a result $\lambda = 8$ appears only in a single block J_1. The triple eigenvalue $\lambda = 0$ has two eigenvectors, e_3 and e_5, which correspond to the two Jordan blocks J_2 and J_3. If A had 5 eigenvectors, all blocks would be 1 by 1 and J would be diagonal.

The key question is this: *If A is some other 5 by 5 matrix, under what conditions will its Jordan form be this same matrix J? When will there exist an M such that $M^{-1}AM = J$?* As a first requirement, any similar matrix A must share the same eigenvalues 8, 8, 0, 0, 0. But this is far from sufficient—the diagonal matrix with these eigenvalues is not similar to J—and our question really concerns the eigen-vectors.

To answer it, we rewrite $M^{-1}AM = J$ in the simpler form $AM = MJ$:

$$
A \begin{bmatrix} x_1 & x_2 & x_3 & x_4 & x_5 \end{bmatrix} = \begin{bmatrix} x_1 & x_2 & x_3 & x_4 & x_5 \end{bmatrix} \begin{bmatrix} 8 & 1 & & & \\ 0 & 8 & & & \\ & & 0 & 1 & \\ & & 0 & 0 & \\ & & & & 0 \end{bmatrix}.
$$

Carrying out the multiplications a column at a time,

$$Ax_1 = 8x_1 \quad \text{and} \quad Ax_2 = 8x_2 + x_1 \tag{1}$$

$$Ax_3 = 0x_3 \quad \text{and} \quad Ax_4 = 0x_4 + x_3 \quad \text{and} \quad Ax_5 = 0x_5. \tag{2}$$

Now we can recognize the conditions on A. It must have three genuine eigen-vectors, just as J has. The one with $\lambda = 8$ will go into the first column of M, exactly as it would have gone into the first column of S: $Ax_1 = 8x_1$. The other two, which will be named x_3 and x_5, go into the third and fifth columns of M: $Ax_3 = Ax_5 = 0$. Finally there must be two other special vectors, the "*generalized eigenvectors*" x_2 and x_4. We think of x_2 as belonging to a **string of vectors**, headed by x_1 and described by equation (1). In fact x_2 is the only other vector in the string, and the corresponding block J_1 is of order two. Equation (2) describes *two different strings*, one in which x_4 follows x_3, and another in which x_5 is alone; the blocks J_2 and J_3 are 2 by 2 and 1 by 1.

The search for the Jordan form of A becomes a search for these strings of vectors, each one headed by an eigenvector: For every i,

$$\text{either} \qquad Ax_i = \lambda_i x_i \qquad \text{or} \qquad Ax_i = \lambda_i x_i + x_{i-1}. \tag{3}$$

The vectors x_i go into the columns of M, and each string produces a single block in J. Essentially, we have to show how these strings can be constructed for every matrix A. Then if the strings match the particular equations (1) and (2), our J will be the Jordan form of A.

I think Filippov's idea, published in Volume 26 of the Moscow University Vestnik, makes the construction as clear and simple as possible. It proceeds by mathematical induction, starting from the fact that every 1 by 1 matrix is already in its Jordan form. We may assume that the construction is achieved for all matrices of order less than n—this is the "induction hypothesis"—and then explain the steps for a matrix of order n. There are three steps, and after a general description we apply them to a specific example.

Step 1   If we assume A is singular, then its column space has dimension $r < n$. Looking only within this smaller space, the induction hypothesis guarantees that a Jordan form is possible—there must be r independent vectors w_i in the column space such that

$$\text{either} \qquad Aw_i = \lambda_i w_i \qquad \text{or} \qquad Aw_i = \lambda_i w_i + w_{i-1}. \tag{4}$$

Step 2   Suppose the nullspace and the column space of A have an intersection of dimension p. Of course, every vector in the nullspace is an eigenvector corresponding to $\lambda = 0$. Therefore, there must have been p strings in step 1 which started from this eigenvalue, and we are interested in the vectors w_i that come at the end of these strings. Each of these p vectors is in the column space, so each one is a combination of the columns of A: $w_i = Ay_i$ for some y_i.

Step 3   The nullspace always has dimension $n - r$. Therefore, independent from its p-dimensional intersection with the column space, it must contain $n - r - p$ additional basis vectors z_i lying *outside* that intersection.

Now we put these steps together to give Jordan's theorem:

The r vectors w_i, the p vectors y_i, and the $n - r - p$ vectors z_i form Jordan strings for the matrix A, and these vectors are linearly independent. They go into the columns of M, and $J = M^{-1}AM$ is in Jordan form.

If we want to renumber these vectors as $x_1, \ldots, x_n$, and match them to equation (3), then each y_i should be inserted immediately after the w_i it came from; it completes a string in which $\lambda_i = 0$. The z's come at the very end, each one alone in its own string; again the eigenvalue is zero, since the z's lie in the nullspace. The blocks with nonzero eigenvalues are already finished at step 1, the blocks with zero eigenvalues grow by one row and column at step 2, and step 3 contributes any 1 by 1 blocks $J_i = [0]$.

Now we try an example, and to stay close to the previous pages we take the eigenvalues to be 8, 8, 0, 0, 0:

$$A = \begin{bmatrix} 8 & 0 & 0 & 8 & 8 \\ 0 & 0 & 0 & 8 & 8 \\ 0 & 0 & 0 & 0 & 0 \\ 0 & 0 & 0 & 0 & 0 \\ 0 & 0 & 0 & 0 & 8 \end{bmatrix}.$$

Step 1   The column space has dimension $r = 3$, and its spanned by the coordinate vectors e_1, e_2, e_5. To look within this space we ignore the third and fourth rows and columns of A; what is left has eigenvalues 8, 8, 0, and its Jordan form comes from the vectors

$$w_1 = \begin{bmatrix} 8 \\ 0 \\ 0 \\ 0 \\ 0 \end{bmatrix}, \qquad w_2 = \begin{bmatrix} 0 \\ 1 \\ 0 \\ 0 \\ 1 \end{bmatrix}, \qquad w_3 = \begin{bmatrix} 0 \\ 8 \\ 0 \\ 0 \\ 0 \end{bmatrix}.$$

The w_i are in the column space, they complete the string for $\lambda = 8$, and they start the string for $\lambda = 0$:

$$Aw_1 = 8w_1, \qquad Aw_2 = 8w_2 + w_1, \qquad Aw_3 = 0w_3. \tag{5}$$

Step 2   The nullspace of A contains e_2 and e_3, so its intersection with the column space is spanned by e_2. Therefore $p = 1$ and, as expected, there is one string in equation (5) corresponding to $\lambda = 0$. The vector w_3 comes at the end (as well as the beginning) of that string, and $w_3 = A(e_4 - e_1)$. Therefore $y = e_4 - e_1$.

Step 3   The example has $n - r - p = 5 - 3 - 1 = 1$, and the nullvector $z = e_3$ is outside the column space. It will be this z that produces a 1 by 1 block in J.

If we assemble all five vectors, the full strings are

$$Aw_1 = 8w_1, \qquad Aw_2 = 8w_2 + w_1, \qquad Aw_3 = 0w_3, \qquad Ay = 0y + w_3, \qquad Az = 0z.$$

Comparing with equations (1) and (2), we have a perfect match—the Jordan form of our example will be exactly the J we wrote down earlier. Putting the five vectors into the columns of M must give $AM = MJ$, or $M^{-1}AM = J$:

$$M = \begin{bmatrix} 8 & 0 & 0 & -1 & 0 \\ 0 & 1 & 8 & 0 & 0 \\ 0 & 0 & 0 & 0 & 1 \\ 0 & 0 & 0 & 1 & 0 \\ 0 & 1 & 0 & 0 & 0 \end{bmatrix}.$$

We are sufficiently trustful of mathematics (or sufficiently lazy) not to multiply out $M^{-1}AM$.

In Filippov's construction, the only technical point is to verify the independence of the whole collection w_i, y_i, and z_i. Therefore, we assume that some combination is zero:

$$\sum c_i w_i + \sum d_i y_i + \sum g_i z_i = 0. \tag{6}$$

Multiplying by A, and using equations (4) for the w_i as well as $Az_i = 0$,

$$\sum c_i \begin{bmatrix} \lambda_i w_i \\ \text{or} \\ \lambda_i w_i + w_{i-1} \end{bmatrix} + \sum d_i A y_i = 0. \tag{7}$$

The Ay_i are the special w_i at the end of strings corresponding to $\lambda_i = 0$, so they cannot appear in the first sum. (They are multiplied by zero in $\lambda_i w_i$.) Since (7) is some combination of the w_i, which were independent by the induction hypothesis— they supplied the Jordan form within the column space—we conclude that *each d_i must be zero.* Returning to (6), this leaves $\sum c_i w_i = -\sum g_i z_i$, and the left side is in the column space. Since the z's were independent of that space, each g_i must be zero. Finally $\sum c_i w_i = 0$, and the independence of the w_i produces $c_i = 0$.

If the original A had not been singular, the three steps would have been applied instead to $A' = A - cI$. (The constant c is chosen to make A' singular, and it can be any one of the eigenvalues of A.) The algorithm puts A' into its Jordan form $M^{-1}A'M = J'$, by producing the strings x_i from the w_i, y_i and z_i. Then the Jordan form for A uses the same strings and the same M:

$$M^{-1}AM = M^{-1}A'M + M^{-1}cM = J' + cI = J.$$

This completes the proof that every A is similar to some Jordan matrix J. Except for a reordering of the blocks, *it is similar to only one such J*; there is a unique Jordan form for A. Thus the set of all matrices is split into a number of families, with the following property: *All the matrices in the same family have the same Jordan form, and they are all similar to each other* (and to J), *but no matrices in different families are similar.* In every family J is the most beautiful—if you like matrices to be nearly diagonal. With this classification into families, we stop.

EXAMPLE

$$A = \begin{bmatrix} 0 & 1 & 2 \\ 0 & 0 & 1 \\ 0 & 0 & 0 \end{bmatrix} \quad \text{with } \lambda = 0, 0, 0.$$

This matrix has rank $r = 2$ and only one eigenvector. Within the column space there is a single string w_1, w_2, which happens to coincide with the last two columns:

$$A \begin{bmatrix} 1 \\ 0 \\ 0 \end{bmatrix} = 0 \quad \text{and} \quad A \begin{bmatrix} 2 \\ 1 \\ 0 \end{bmatrix} = \begin{bmatrix} 1 \\ 0 \\ 0 \end{bmatrix},$$

or

$$Aw_1 = 0 \quad \text{and} \quad Aw_2 = 0w_2 + w_1.$$

The nullspace lies entirely within the column space, and it is spanned by w_1. Therefore $p = 1$ in step 2, and the vector y comes from the equation

$$Ay = w_2 = \begin{bmatrix} 2 \\ 1 \\ 0 \end{bmatrix}, \quad \text{whose solution is} \quad y = \begin{bmatrix} 0 \\ 0 \\ 1 \end{bmatrix}.$$

Finally the string w_1, w_2, y goes into the matrix M:

$$M = \begin{bmatrix} 1 & 2 & 0 \\ 0 & 1 & 0 \\ 0 & 0 & 1 \end{bmatrix}, \quad \text{and} \quad M^{-1}AM = \begin{bmatrix} 0 & 1 & 0 \\ 0 & 0 & 1 \\ 0 & 0 & 0 \end{bmatrix} = J.$$

Application to $du/dt = Au$ As always, we simplify the problem by uncoupling the unknowns. This uncoupling is complete only when there is a full set of eigenvectors, and $u = Sv$; the best change of variables in the present case is $u = Mv$. This produces the new equation $M \, dv/dt = AMv$, or $dv/dt = Jv$, which is as simple as the circumstances allow. It is coupled only by the off-diagonal 1's within each Jordan block. In the example just above, which has a single block, $du/dt = Au$ becomes

$$\frac{dv}{dt} = \begin{bmatrix} 0 & 1 & 0 \\ 0 & 0 & 1 \\ 0 & 0 & 0 \end{bmatrix} v \quad \text{or} \quad \begin{aligned} da/dt &= b \\ db/dt &= c \\ dc/dt &= 0 \end{aligned} \quad \text{or} \quad \begin{aligned} a &= a_0 + b_0 t + c_0 t^2/2 \\ b &= b_0 + c_0 t \\ c &= c_0 \end{aligned}$$

The system is solved by working upward from the last equation, and a new power of t enters at every step. (An l by l block has powers as high as t^{l-1}.) The exponentials of J, in this case and in the earlier 5 by 5 example, are

$$e^{Jt} = \begin{bmatrix} 1 & t & t^2/2 \\ 0 & 1 & t \\ 0 & 0 & 1 \end{bmatrix} \quad \text{and} \quad e^{Jt} = \begin{bmatrix} e^{8t} & te^{8t} & 0 & 0 & 0 \\ 0 & e^{8t} & 0 & 0 & 0 \\ 0 & 0 & 1 & t & 0 \\ 0 & 0 & 0 & 1 & 0 \\ 0 & 0 & 0 & 0 & 1 \end{bmatrix}.$$

You can see how the coefficients of a, b, and c appear in the first exponential. And in the second example, you can identify all five of the "special solutions" to $du/dt = Au$. Three of them are the pure exponentials $u_1 = e^{8t}x_1$, $u_3 = e^{0t}x_3$, and $u_5 = e^{0t}x_5$, formed as usual from the three eigenvectors of A. The other two involve the generalized eigenvectors x_2 and x_4:

$$u_2 = e^{8t}(tx_1 + x_2) \quad \text{and} \quad u_4 = e^{0t}(tx_3 + x_4). \tag{8}$$

The most general solution to $du/dt = Au$ is a combination $c_1u_1 + \cdots + c_5u_5$, and the combination which matches u_0 at time $t = 0$ is again

$$u_0 = c_1x_1 + \cdots + c_5x_5, \quad \text{or} \quad u_0 = Mc, \quad \text{or} \quad c = M^{-1}u_0.$$

This only means that $u = Me^{Jt}M^{-1}u_0$, and that the S and Λ in the old formula $Se^{\Lambda t}S^{-1}u_0$ have been replaced by M and J.

EXERCISES

B.1   Find the Jordan forms (in three steps!) of

$$A = \begin{bmatrix} 1 & 1 \\ 1 & 1 \end{bmatrix} \quad \text{and} \quad B = \begin{bmatrix} 0 & 1 & 2 \\ 0 & 0 & 0 \\ 0 & 0 & 0 \end{bmatrix}.$$

B.2   Show that the special solution u_2 in Equation (8) does satisfy $du/dt = Au$, exactly because of the string $Ax_1 = 8x_1$, $Ax_2 = 8x_2 + x_1$.

B.3   For the matrix B above, use $Me^{Jt}M^{-1}$ to compute the exponential e^{Bt}, and compare it with the power series $I + Bt + (Bt)^2/2! + \cdots$.

B.4   Show that each Jordan block J_i is similar to its transpose, $J_i^T = P^{-1}J_iP$, using the permutation matrix P with ones along the cross-diagonal (lower left to upper right). Deduce that every matrix is similar to its transpose.

B.5   Find "by inspection" the Jordan forms of

$$A = \begin{bmatrix} 1 & 2 & 3 \\ 0 & 4 & 5 \\ 0 & 0 & 6 \end{bmatrix} \quad \text{and} \quad B = \begin{bmatrix} 1 & 1 \\ -1 & -1 \end{bmatrix}.$$

B.6 Find the Jordan form J and the matrix M for A and B (B has eigenvalues 1, 1, 1, -1):

$$A = \begin{bmatrix} 0 & 0 & 1 & 0 & 0 \\ 0 & 0 & 0 & 1 & 0 \\ 0 & 0 & 0 & 0 & 1 \\ 0 & 0 & 0 & 0 & 0 \\ 0 & 0 & 0 & 0 & 0 \end{bmatrix} \quad \text{and} \quad B = \begin{bmatrix} 1 & -1 & 0 & -1 \\ 0 & 2 & 0 & 1 \\ -2 & 1 & -1 & 1 \\ 2 & -1 & 2 & 0 \end{bmatrix}$$

What is the solution to $du/dt = Au$ and what is e^{At}?

B.7 Suppose that $A^2 = A$. Show that its Jordan form $J = M^{-1}AM$ satisfies $J^2 = J$. Since the diagonal blocks stay separate, this means $J_i^2 = J_i$ for each block; show by direct computation that J_i can only be a 1 by 1 block, $J_i = [0]$ or $J_i = [1]$. Thus A is similar to a diagonal matrix of zeros and ones.

Note This is typical case of our closing theorem: *The matrix A can be diagonalized if and only if the product $(A - \lambda_1 I)(A - \lambda_2 I) \cdots (A - \lambda_p I)$, without including any repetitions of the λ's, is zero.* One extreme case is a matrix with distinct eigenvalues; the Cayley-Hamilton theorem in Exercise 5.6.23 says that with n factors $A - \lambda I$ we always get zero. The other extreme is the identity matrix, also diagonalizable ($p = 1$ and $A - I = 0$). The nondiagonalizable matrix $A = \begin{bmatrix} 0 & 1 \\ 0 & 0 \end{bmatrix}$ satisfies not $A = 0$ but only $A^2 = 0$—an equation with a repeated root.

APPENDIX C

COMPUTER CODES FOR LINEAR ALGEBRA

Our goal has been to make this book useful for the reader, and we want to take two more steps. One is to indicate sources of good software for numerical linear algebra. The pleasure in this part of mathematics is that it is both practical and beautiful—and it can be learned that way. Part of the practical side is to experience linear algebra in use. The theory has been translated into codes for scientific computing, and we will point to a range of possibilities—from large program libraries to codes freely available by electronic mail to specialized packages that can form part of an applied linear algebra course.

Our second step is to suggest some original experiments that could be run with those codes. Rather than creating standard examples of linear equations or least squares problems or Markov matrices, we try to propose new questions—which the student may be the *first* to answer. It is understood that no solutions are given, or could be given. If you write with news of interesting experiments, I will be very grateful.

One basic set of codes is LINPACK. It has replaced thousands of subroutines, some good and some not, which were created in a multitude of computing environments. LINPACK is designed to be machine-independent, efficient, and simple; it was prepared by Dongarra, Bunch, Moler, and Stewart, and their User's Guide is excellent. The development of these codes was stimulated by the success of EISPACK for the eigenvalue problem; because of EISPACK we know the value of well-written mathematical software.

In LINPACK, all matrices are stored directly in computer memory. On many machines this allows full matrices of order several hundred and band matrices of order several thousand. Almost all problems use two subroutines, one to factor the coefficient matrix and one to operate on a particular right-hand side. As we emphasized in Section 1.5, this saves computer time whenever there is a sequence of right sides associated with the same coefficient matrix. The Guide remarks that "this situation is so common and the savings so important that no provision has been made for solving a single system with just one subroutine."

We give a partial listing of the subroutines. The user can choose between single and double precision, and between real and complex arithmetic. These four options make the package comparatively large (the equivalent of 80,000 cards) but the structure is straightforward. The codes will factor A, or solve $Ax = b$, or compute the determinant or inverse, for the following nine classes of matrices A:

GE general
GB general band
PO positive definite
PB positive definite band
SI symmetric indefinite
HI Hermitian indefinite
TR triangular
GT general tridiagonal
PT positive definite tridiagonal

There are three further codes, which compute the Cholesky decomposition $A = R^{T}R$, the Gram–Schmidt orthogonal–triangular decomposition $A = QR$, and the singular value decomposition $A = Q_1 \Sigma Q_2^{T}$. For Cholesky the matrix is square and positive definite, and R^{T} is just $L\sqrt{D}$. For least squares we note again the choice between QR which is numerically stable, the SVD which is for the very cautious, and the use of $A^{T}A$ and the normal equations which is the fastest of the three. $A^{T}A$ is the usual choice. For symmetric or Hermitian matrices the signs of the eigenvalues match the signs of the pivots by the Law of Inertia (Section 6.3), and this is part of the output from the subroutines for SI and HI. We recall that computing costs are cut nearly in half by symmetry. There is also provision for updating and downdating the Cholesky factors when A is altered by adding or deleting a row.

A basic subroutine is SGEFA, which factors a general matrix in single precision. Partial pivoting is used, so that the kth stage of elimination selects the largest element on or below the diagonal in column k to be the pivot. The integer IPVT(K) records the pivot row. A zero in the pivot does not mean that the factorization fails (we know from Chapter 2 that $PA = LU$ is always possible, and LINPACK mixes P in with L); it does mean that U is singular and there would be division by zero in back-substitution. In practice dividing by zero is due at least as often to a program error, and it is a nearly singular U which indicates numerical instability.

The Guide gives an outline of the factorization algorithm:

for $k = 1, \ldots, n - 1$ *do*
    find l such that $|a_{lk}| = \max_{k \leq i} |a_{ik}|$;
    save l in *ipvt(k)*;
    if $a_{lk} = 0$ *then* skip to end of k loop;

interchange a_{kk} and a_{lk};
for $i = k + 1, \ldots, n$ *do*
    $m_{ik} = -a_{ik}/a_{kk}$ (save in a_{ik});
for $j = k + 1, \ldots, n$ *do*
    interchange a_{kj} and a_{lj};
    for $i = k + 1, \ldots, n$ *do*
        $a_{ij} \leftarrow a_{ij} + m_{ik}*a_{kj}$;
    end j
end k.

The algorithm is generally familiar, but a closer look reveals one surprise. The algorithm does not directly subtract multiples of the pivot row from the rows beneath. In other words, it does not do row operations! Instead the code computes the whole column of multipliers m_{ik}, and then marches column by column to the right using these multipliers. The reason is that ***Fortran stores matrices by columns***, and going down a column gives sequential access to memory; for large matrices stored by a paging scheme there can be significant savings. Therefore the innermost loop (*for* $i = k + 1, \ldots, n$ *do*) is constructed to do a column and not a row operation (compare Exercise 1.4.9 on multiplying Ax by rows or by columns). Similarly the solution codes SGESL, SGBSL, SPOSL, ... carry out back-substitution a column at a time; the last unknown in the triangular system $Ux = c$ is $x_n = c_n/u_{nn}$, and then x_n times the last column of U is subtracted from c. What remains is a system of order $n - 1$, ready for the next step of back-substitution.

This change to column operations is due to Moler, and the programs have been progressively refined to estimate also the *condition number* (defined in Section 7.2). In fact the normal choice in LINPACK will be not SGEFA but SGECO, which includes this estimate and allows the user to decide whether it is safe to proceed with the back-substitution code SGESL. These programs depend on a system of Basic Linear Algebra Subprograms (BLAS) to do operations like the last one in the outline, adding a multiple of one column to another. The final result is a collection of codes which beautifully implements the theory of linear algebra. The User's Guide can be obtained from the

    Society for Industrial and Applied Mathematics (SIAM)
    117 South 17th Street
    Philadelphia PA 19103; (215) 564-2929.

We emphasize also the importance of EISPACK for the eigenvalue problem.

These codes are public. They form part of many subroutine libraries, and we begin with two important sources of software for scientific computing:

    1. International Mathematical and Statistical Libraries (IMSL)
          7500 Bellaire Boulevard
          Houston TX 77036; (800) 222-4675.
    2. Numerical Algorithms Group (NAG)
          1101 31st Street
          Downers Grove IL 60515; (312) 971-2337.

Next we call attention to a new and free source called "***netlib***". It is a quick service to provide specific routines, not libraries. It is entirely by electronic mail sent over ARPANET, CSNET, Telenet, or Phone Net. Since no human looks at the request, it is possible (and probably faster) to get software in the middle of the night. The one-line message "send index" wakes up a server and a listing is sent by return electronic mail. There is no consulting and no guarantee, but in compensation the response is efficient, the cost is minimal, and the most up-to-date version is always available. The addresses are not recognized by the post office:

3. *netlib@anl-mcs* (ARPANET/CSNET) or *research!netlib* (UNIX network)
   EISPACK, ITPACK, LINPACK, MINPACK, QUADPACK,
   TOEPLITZ, TOMS, FFTPACK, BLAS, FISHPACK, . . .

The next source may be the best. It is an interactive program called **MATLAB**, written specifically for matrix computations. It is available for the IBM PC, Macintosh, SUN, APOLLO, and VAX. This code has become a standard instructional tool, and also the basic program for research in numerical linear algebra. We emphasize the speed, the graphics, the Fast Fourier Transform, the "movie" of matrix operations in progress, and the address:

4. Math Works
   20 North Main Street
   Sherborn MA 01770; (617) 653-1415.

Next come six instructional codes for PC's. It is only a partial list, and new programs will continue to appear, but we have heard good reports of these:

5. *Matrix Algebra Calculator* (Herman and Jepson), diskette and booklet with good applications: Brooks-Cole, 555 Abrego Street, Monterey CA 93940.
6. MATRIXPAD (Orzech), convenient matrix entry and rational arithmetic; to be published by D.C. Heath, 125 Spring Street, Lexington MA 02173; information from Mathematics Department, Queen's University, Kingston K7L3N6 Canada.
7. *Lintek* (Fraleigh), associated with text: 0-201-15456-0, Addison-Wesley, Reading MA 01867.
8. *Linear-Kit* (Anton), associated with text; 0-471-83086-0, John Wiley, 605 Third Avenue, New York NY 10158.
9. *Matrix Calculator* (Hogben and Hentzel), basic applications in exact rational arithmetic; Conduit, University of Iowa, Iowa City IA 52242.
10. MINITAB, statistics and basic linear algebra; 3081 Enterprise Drive, State College PA 16801.

Apple and Macintosh codes have developed differently, and more slowly. Some programs are being rewritten, but a full range of linear algebra software seems to be a future certainty rather than a present possibility.

Finally we mention a personal favorite for teaching the simplex method in linear programming. That is an algorithm with decisions at every step; an entering variable replaces a leaving variable. The row operations that follow from the decision are not interesting (this is a major problem in demonstrating linear algebra). By

jumping to the new corner—which may be better, or worse, or optimal—the
"*MacSimplex*" code allows the user to make mistakes and turn back. It is written
by Dr. Andrew Philpott (Engineering School, University of Auckland, New
Zealand) for a Macintosh 512K and Macintosh Plus.

PROPOSED COMPUTER EXPERIMENTS

My goal in these paragraphs is to suggest a direction for computer experiments
which avoids their most common and most serious difficulty. What usually happens
is that a student inputs a matrix, pushes a button, and reads the answer. Not
much learning takes place. It is just an exercise—more of a sit-up than a push-up—
rather than a true experiment. The fun in mathematics is to find something new,
and I believe the computer makes that possible.

One key idea is to use the random number generator to produce random matrices. There the theory is in its infancy, and it is developing fast (partly driven by
fascinating problems in science). The eigenvalues, or the sizes of the pivots and
multipliers, or the determinants, or even singularity vs. invertibility in the 0–1
case—all lead to extremely interesting questions. We indicate some that could be
asked; you will think of others. Random numbers are discussed on page 471.

A second idea is to build matrices in an established pattern, which allows their
size to grow while their input stays easy. One central pattern comes from finite
differences, as approximations to differential equations. By decreasing the stepsize
h, the order of the matrix is increased—but the matrix remains sparse and manageable. Many problems in scientific computing revolve around these matrices: their
properties, their behavior with decreasing h, and ideas for efficient computation.

A number of the packages listed earlier contain particular matrices that can be
called up directly. That avoids the problem of input, but part of mathematics lies
in formulating the problem. The experiments below complement those exercises.

Random 0-1 Matrices (n by n)

1.1 What fraction of them are singular? What are the frequencies of different
determinants? What is the largest possible determinant? What is the largest
pivot (with partial pivoting) and how does its average value increase with n?

1.2 Find the eigenvalues (restricting A to be symmetric if the code requires it
and *plotting them* if the code permits it). What is the largest eigenvalue in
the sample, and what is the average of largest eigenvalues?

1.3 How do the answers change when the probability of 1's moves away from $\frac{1}{2}$?

Random Matrices and Vectors (n by n, m by n, and n by 1)
The entries of A and b are chosen by a random number generator from a range
of values, not only 0-1. Adjust if necessary so that the mean (or expected value,
or average) is zero. There are two possibilities in all experiments:

(i)   to work directly with A and b
(ii) to start with fixed A_0 and b_0, and regard A and b as perturbations—you
control the scaling constant c and work with $A_0 + cA$ and $b_0 + cb$.

There are fascinating questions not yet answered about the statistics of Gaussian elimination (where A and b come from roundoff).

2.1 What is the distribution of det A? Using elimination without pivoting, you can also find the determinants of the upper left submatrices and the pivots and multipliers. All are of major interest.

2.2 What is the average length of $x = A^{-1}b$? Comparing with $x_0 = A_0^{-1}b_0$, what is the average length of $x = (A_0 + cA)^{-1}(b_0 + cb)$ and of the error $x - x_0$?

2.3 Impose symmetry on A, so it has $\frac{1}{2}(n^2 + n)$ independent random entries, and find its *eigenvalues*. Compute also $F(t) =$ fraction of eigenvalues below t. What is the distribution of λ_{max}? How does it change with n? The graph of F should settle down after many experiments. (There is a "semicircle law" which few people have met. F looks like the integral of $(1 - t^2)^{1/2}$, after scaling, and the density of eigenvalues makes a semicircle.) What is the average and the distribution of $\lambda_{min}(A_0 + cA)$?

2.4 Construct random m by n matrices and look at the pivots, determinants, and eigenvalues of $A^T A$—which is guaranteed to be symmetric positive semidefinite.

Random Networks and Incidence Matrices (m by n)

Given n nodes with $\frac{1}{2}(n^2 - n)$ possible edges between them, include each edge with probability p (start with $p = 1$, a complete graph, and then $p = \frac{1}{2}$). A is the *incidence matrix* with a row for each of the m edges (Section 2.5). It contains -1 and $+1$ in the columns corresponding to the two nodes connected by the edge. B is the *reduced incidence matrix*, the same as A but with column n removed. C is the n by n *adjacency matrix*, with $c_{ij} = 1$ if nodes i and j have an edge between them, $c_{ij} = 0$ otherwise.

3.1 What is the probability of a *connected* network? In that case A has rank $n - 1$. How does the probability change with p?

3.2 What power C^k is the first to be entirely positive, with no zeros, and what is the meaning of k in terms of paths between nodes? How does k grow with n?

3.3 The determinant of $B^T B$ counts *the number of spanning trees*. How does this vary with n and p? For smaller graphs ($n = 3, 4, 5$) print out
(a)   a basis for the nullspace of A
(b)   a basis for the nullspace of A^T
(c) m, n, and the rank r
(d) $B^T B$ and its determinant (the number of spanning trees)
Match the results with the graph, especially part (b) with the loops.

3.4 Find the eigenvalues of $B^T B$ and of C—especially the distribution of λ_{max} and the distribution of the *condition number* $c = \lambda_{max}/\lambda_{min}$.

3.5 For the complete graph with all $\frac{1}{2}(n^2 - n)$ edges, what is $B^T B$ and what are its eigenvalues? Print the 3 by 3 and 4 by 4 matrices, with $n = 4$ and $n = 5$

on the diagonal and -1's elsewhere. Find the LU and thus the LDL^T factorization—do you see the pattern of multipliers in L?

Finite Differences: $-u'' + cu' = f$ for $0 \le x \le 1$.

4.1 With $c = 0$, construct the tridiagonal *second-difference matrix* A_n with -1, 2, -1 on the diagonals. Verify that det $A_n = n + 1$ and compute A_n^{-1}. Its i, j entries should have the form $ai(1 - aj)$ above the diagonal; find the number a and look for a pattern below the diagonal.

4.2 Verify that the eigenvalues of A_n are $\lambda_j = 4 \sin^2 j\pi h/2$ and that the eigenvectors look like sines—as in Exercise 7.4.3.

4.3 Construct three possible *first-difference* matrices: F_+ has -1's on the main diagonal and $+1$'s just above, F_- has $+1$'s on the main diagonal and -1's just below, and $F_0 = \frac{1}{2}(F_+ + F_-)$ has $+\frac{1}{2}$ above the diagonal and $-\frac{1}{2}$ below. It is a serious question in engineering problems which F to use. Note that $-u'' + cu'$ is approximated by $M = h^{-2}A + h^{-1}cF$, where h is the stepsize.

Solve $-u'' + cu' = 1$ with $u = 0$ at the endpoints $x = 0$ and $x = 1$. (Try e^{cx} and $ax + b$.) If possible graph the exact solution for $c = 0$, 1, 10, and other values. Then solve $M_+u = e$ and $M_-u = e$ and $M_0u = e$, where $e = (1, 1, \ldots, 1)$ and the M's come from the three F's. With c large, the first-derivative term cu' is important: Which choice of F do you recommend? For small c which is the best? Compare always with the exact u.

4.4 Is there a pattern in det M or in M^{-1} for any of the matrices M_+, M_-, M_0?

Laplace's Equation: $-u_{xx} - u_{yy} = 0$ in a unit square.

There are n^2 unknowns inside the square, when the stepsize is $h = 1/(n + 1)$ in the x and y directions. Number them along the bottom row first, then the next row, and finally the top row. The unknowns in the bottom left, bottom right, and top right corners are numbered 1, n, and n^2.

Construct the -1, 2, -1 second-difference matrices H and V (horizontal and vertical) of order n^2. H takes differences in the x direction, along rows, and will be tridiagonal—but the -1's above and below the diagonal have breaks where the rows end. V takes differences in the y direction, so the -1's are out on the nth diagonals (also with breaks).

5.1 Find the determinant and the inverse of the *five-point difference matrix* $L = H + V$. Can you see any patterns as n increases?

5.2 Find the eigenvalues of L and compare with the sums $\lambda_j + \lambda_k$ of eigenvalues of the usual n by n second-difference matrix A. (That has n eigenvalues, given above, so there are n^2 sums.) Can you connect eigenvectors of A to eigenvectors of L?

5.3 Solve $Lx = e$, where $e = (1, \ldots, 1)$ has length n^2. First use elimination, to get an exact solution. (The number of computations grows like n^4 for this band matrix; how is that predicted by Section 1.7?) Then use the iterative methods of Chapter 7—Jacobi, Gauss-Seidel, and SOR—and compare the decay of the error $\|x - x_k\|$ after k iterations. Experimentally, which ω is best in SOR?

I would like to combine random matrices and difference matrices into $A + cF$, where c is diagonal with random entries. But that is totally new territory and I don't even know the right questions.

CODES FOR A = LU AND BACK-SUBSTITUTION

For the reader's use, we give a FORTRAN listing which is stripped of refinements but will allow the fun of experimenting with $Ax = b$. The translation to BASIC or PL/1 is easy. The input includes N, the order of the matrix, as well as NDIM, the leading dimension of the array, which is an upper bound on N; the computer allocates space in accordance with NDIM, while N may change as different systems are processed. The code overwrites the matrix A, step by step, by U in the upper triangular part and by $I - L$ in the strictly lower triangular part. Thus instead of storing the zeros which arise in elimination we store the multipliers (or rather the *negatives of the multipliers,* since it is $I - L$). It is quite remarkable that elimination requires no extra space in which to maneuver. If there are row exchanges then the columns of $I - L$ undergo permutations which only the most patient reader will want to untangle; what matters is that they are correctly interpreted by the back-substitution code. (L has been described as "psychologically" lower triangular. In case the diagonal of A is sufficiently dominant there will be no pivoting, and $A = LU$.) The last component of IPVT contains ± 1 according as the number of exchanges is even or odd, and the determinant is found from IPVT(N)$*$ A(1, 1)$* \ldots *$A(N, N). If it is zero we stop.

We now list the two subroutines DECOMP and SOLVE, to factor A and find x, and a sample master program which reads the data, calls the subroutines, and prints the results.

```
C
C      SAMPLE PROGRAM FOR N EQUATIONS IN N UNKNOWNS
C
       INTEGER IPVT(10)
       REAL A(10,10),B(10)
       NDIM = 10
C
C      NDIM IS UPPER BOUND ON THE ORDER N
C
       N = 3
       A(1,1)  = 2.0
       A(2,1)  = 4.0
       A(3,1)  = -2.0
       A(1,2)  = 1.0
       A(2,2)  = 1.0
       A(3,2)  = 2.0
       A(1,3)  = 1.0
       A(2,3)  = 0.0
       A(3,3)  = 1.0
```

```
C
      WRITE (6,5) NDIM,N
    5 FORMAT (X,'NDIM = ',I2,' N = ',I2,' --- A:')
      DO 10 I = 1,N
   10 WRITE (6,20) (A(I,J),J=1,N)
   20 FORMAT (X,10F7.3)
      WRITE (6,30)
   30 FORMAT (X)
C
      CALL DECOMP (NDIM,N,A,IPVT,DET)
C
      WRITE (6,35)
   35 FORMAT (X,'AFTER DECOMP:   FACTORS OF A:')
      DO 40 I = 1,N
   40 WRITE (6,20) (A(I,J),J=1,N)
      WRITE (6,45) DET
   45 FORMAT (X,'DET = ',E11.5)
      WRITE (6,30)
      IF (DET.EQ.0.0) STOP
C
      B(1) = 1.0
      B(2) = -2.0
      B(3) = 7.0
      DO 50 I = 1,N
   50 WRITE (6,60) I,B(I)
   60 FORMAT (X,'B(',I2,') =',F7.3)
      CALL SOLVE (NDIM,N,A,IPVT,B)
      DO 70 I = 1,N
   70 WRITE (6,80) I,B(I)
   80 FORMAT (X,'X(',I2,') =',F7.3)
C
      STOP
      END
C
      SUBROUTINE DECOMP (NDIM,N,A,IPVT,DET)
      INTEGER NDIM,N,IPVT(N)
      REAL A(NDIM,N),DET
      REAL P,T
      INTEGER NM1,I,J,K,KP1,M
C
C     INPUT A = COEFFICIENT MATRIX TO BE TRIANGULARIZED
C     OUTPUT A CONTAINS UPPER TRIANGULAR U AND
C        PERMUTED VERSION OF STRICTLY LOWER TRIANGULAR I-L.
C        IN ABSENCE OF ROW EXCHANGES A = LU
C
C     IPVT(K) = INDEX OF KTH PIVOT ROW, EXCEPT
C     IPVT(N) = (-1)**(NUMBER OF INTERCHANGES)
C     DET = IPVT(N)* A(1,1)*...*A(N,N)
C
      DET = 1.0
      IPVT(N) = 1
      IF (N.EQ.1) GO TO 70
      NM1 = N-1
C
      DO 60 K = 1,NM1
      KP1 = K+1
C
C     FIND PIVOT P
```

```
C
      M = K
      DO 10 I = KP1,N
   10 IF (ABS(A(I,K)).GT.ABS(A(M,K))) M = I
      IPVT(K) = M
      IF (M.NE.K) IPVT(N) = -IPVT(N)
      P = A(M,K)
      A(M,K) = A(K,K)
      A(K,K) = P
      DET = DET*P
      IF (P.EQ.0.0) GO TO 60
C
C     COMPUTE MULTIPLIERS
C
   20 DO 30 I = KP1,N
   30 A(I,K) = -A(I,K)/P
C
C     INTERCHANGE AND ELIMINATE BY COLUMNS
C
      DO 50 J = KP1,N
      T = A(M,J)
      A(M,J) = A(K,J)
      A(K,J) = T
      IF (T.EQ.0.0) GO TO 50
      DO 40 I = KP1,N
      A(I,J) = A(I,J)+A(I,K)*T
   40 CONTINUE
   50 CONTINUE
   60 CONTINUE
C
   70 DET = DET*A(N,N)*FLOAT(IPVT(N))
C
      RETURN
      END
C
      SUBROUTINE SOLVE (NDIM,N,A,IPVT,B)
      INTEGER NDIM,N,IPVT(N)
      REAL A(NDIM,N),B(N)
C
C     NDIM IS UPPER BOUND ON THE ORDER N
C     A CONTAINS FACTORS OBTAINED BY DECOMP
C     B IS RIGHT HAND SIDE VECTOR
C     IPVT IS PIVOT VECTOR OBTAINED BY DECOMP
C     ON OUTPUT B CONTAINS SOLUTION X
C
      INTEGER NM1,K,KB,KP1,KM1,M,I
      REAL S
C
C     FORWARD ELIMINATION
C
      IF (N.EQ.1) GO TO 30
      NM1 = N-1
      DO 10 K = 1,NM1
      KP1 = K+1
      M = IPVT(K)
      S = B(M)
      B(M) = B(K)
      B(K) = S
      DO 10 I = KP1,N
   10 B(I) = B(I)+A(I,K)*S
```

```
C
C     BACK SUBSTITUTION
C
      DO 20 KB = 1,NM1
      KM1 = N-KB
      K = KM1+1
      B(K) = B(K)/A(K,K)
      S = -B(K)
      DO 20 I = 1,KM1
   20 B(I) = B(I)+A(I,K).*S
C
   30 B(1) = B(1)/A(1,1)
C
      RETURN
      END
```

Random Number Generator

Since the experiments proposed above require random numbers, we add a few lines about generating them. Many computers will have them directly available (at least if we want a *uniform* distribution on the interval from 0 to 1). Actually they will be *pseudorandom* numbers, which come from a deterministic algorithm but for practical purposes behave randomly. A single line is enough to write down a satisfactory algorithm for single precision computations:

$$m_0 = 0, \quad m_j = (25733m_{j-1} + 13849)(\mathrm{mod}\ 65536), \quad x_j = m_j/65536.$$

The number 65536 is 2^{16}, and all steps are to be in exact arithmetic (which should be automatic on a 32-bit machine). The expression "mod 65536" means that m_j is the remainder after dividing $(25733m_{j-1} + 13849)$ by that number. Then x_j is a fraction between 0 and 1, when m_j is divided by 2^{16}–the binary point is moved 16 places.

There is also a simple way to go from a pair of these uniformly distributed random numbers (say x and y) to a pair of *normally* distributed random numbers u and v (governed by the bell-shaped Gaussian curve). The first step is to shift the uniform distribution to the interval from -1 to 1: set $w = 2x - 1$ and $z = 2y - 1$. Next form $s = w^2 + z^2$. Then u and v are normally distributed with zero mean and unit variance if

$$u = w\left(\frac{-2\log s}{s}\right)^{\frac{1}{2}} \quad \text{and} \quad v = z\left(\frac{-2\log s}{s}\right)^{\frac{1}{2}}.$$

A good reference is volume 2 of Donald Knuth's series *The Art of Computer Programming*.

To choose between 0 and 1 randomly with probabilities p and $1 - p$, compute the uniformly distributed variable x as above and choose zero if x is below p.

REFERENCES

Abstract Linear Algebra

F. R. Gantmacher, "Theory of Matrices." Chelsea, 1959.
P. R. Halmos, "Finite-Dimensional Vector Spaces." Van Nostrand-Reinhold, 1958.
K. Hoffman and R. Kunze, "Linear Algebra." Prentice-Hall, 1971.
T. Muir, "Determinants." Dover, 1960.

Applied Linear Algebra

D. Arganbright, "Mathematical Applications of Electronic Spreadsheets." McGraw-Hill, 1985.
R. Bellman, "Introduction to Matrix Analysis." McGraw-Hill, 1960.
A. Ben-Israel and T. N. E. Greville, "Generalized Inverses: Theory and Applications." Wiley, 1974.
A. Berman and R. J. Plemmons, "Nonnegative Matrices in the Mathematical Sciences." Academic Press, 1979.
V. Chvátal, "Linear Programming." Freeman, 1983.
D. Gale, "The Theory of Linear Economic Models." McGraw-Hill, 1960.
D. G. Luenberger, "Introduction to Linear and Nonlinear Programming." Addison-Wesley, 1973.
B. Noble and J. Daniel, "Applied Linear Algebra." Prentice-Hall, 1977.
G. Strang, "Introduction to Applied Mathematics." Wellesley-Cambridge Press, 1986.

Numerical Linear Algebra

G. Forsythe and C. Moler, "Computer Solution of Linear Algebraic Systems." Prentice-Hall, 1967.
A. George and J. Liu, "Computer Solution of Large Sparse Positive Definite Systems." Prentice-Hall, 1981.
G. Golub and C. Van Loan, "Matrix Computations." Johns Hopkins Press, 1983.
C. L. Lawson and R. J. Hanson, "Solving Least Squares Problems." Prentice-Hall, 1974.
B. N. Parlett, "The Symmetric Eigenvalue Problem." Prentice-Hall, 1981.
G. W. Stewart, "Introduction to Matrix Computations." Academic Press, 1973.
R. S. Varga, "Matrix Iterative Analysis." Prentice-Hall, 1962.
J. M. Wilkinson, "Rounding Errors in Algebraic Processes." Prentice-Hall, 1963.
J. M. Wilkinson, "The Algebraic Eigenvalue Problem." Oxford University Press, 1965.
J. M. Wilkinson and C. Reinsch, Eds., "Handbook for Automatic Computation II, Linear Algebra." Springer, 1971.
D. M. Young, "Iterative Solution of Large Linear Systems." Academic Press, 1971.

SOLUTIONS TO SELECTED EXERCISES

CHAPTER 1

1.2.2 $u = b_1 - b_2,\ v = b_2 - b_3,\ w = b_3$.

1.2.4 No; 3 lines through origin; 3, 0, 1.   **1.2.6** $(-2, 1, 3)$.

1.2.8 $1 \times \text{eq}(1) - 1 \times \text{eq}(2) + 1 \times \text{eq}(3) = 1 \Rightarrow 0 = 1$; -1; $(3, -1, 0)$ is one of the solutions.

1.2.10 $y_1 - y_2 = 1,\ y_1 - y_3 = 2$.   **1.2.12** $(1, 1, 1)$.

1.3.2 $\begin{cases} u + v + w = 2 \\ \quad\ 2v + 2w = -2, \\ \quad\quad\quad 2w = 2 \end{cases}$ $u = 3,\ v = -2,\ w = 1$.

1.3.4 A coefficient $+1$ would make the system singular.

1.3.6 $a = 0$ requires a row exchange, but is nonsingular: $a = 2$ is singular (one pivot, infinity of solutions); $a = -2$ is singular (one pivot, no solution).

1.3.8 $n^3/3 = 72$ million operations; 9000 seconds on a PC, 900 seconds on a VAX, 6 seconds on a CRAY.

1.3.10 The second term $bc + ad$ is $(a + b)(c + d) - ac - bd$ (only 1 additional multiplication).

1.3.12 $.9u + .2v = 200{,}000{,}000$; $.1u + .8v = 30{,}000{,}000$.   **1.4.2** $\begin{bmatrix} 7 \\ 8 \\ 9 \end{bmatrix}, \begin{bmatrix} 2 \\ 5 \\ 8 \end{bmatrix}, \begin{bmatrix} 3 \\ 5 \\ 7 \end{bmatrix}$.

1.4.4 mn and mnp multiplications.   **1.4.6** $A = \begin{bmatrix} 2 & 3 \\ 3 & 4 \\ 4 & 5 \end{bmatrix},\ B = \begin{bmatrix} 1 & -1 \\ -1 & 1 \\ 1 & -1 \end{bmatrix}$.

1.4.9 The first program finds each component of Ax separately, using the rows of A one at a time. The second finds Ax as a combination of the columns of A, computing a column at a time.

1.4.14 $A = \begin{bmatrix} 0 & 1 \\ -1 & 0 \end{bmatrix}$, $B = \begin{bmatrix} 0 & 1 \\ 0 & 0 \end{bmatrix}$, $C = \begin{bmatrix} 0 & 1 \\ 1 & 0 \end{bmatrix}$, $D = A$, $E = F = \begin{bmatrix} 1 & -1 \\ 1 & -1 \end{bmatrix}$.

1.4.18 For this x, Bx is the first column of B and $(AB)x$ is the first column of AB.

1.4.20 $(AB)_{ij} = n$, $(ABC)_{ii} = 2n^2$.

1.5.4 $\begin{bmatrix} 1 & 0 \\ 4 & 1 \end{bmatrix} \begin{bmatrix} 2 & 1 \\ 0 & 3 \end{bmatrix}$; $\begin{bmatrix} 1 & 0 & 0 \\ \frac{1}{3} & 1 & 0 \\ \frac{1}{3} & \frac{1}{4} & 1 \end{bmatrix} \begin{bmatrix} 3 & 1 & 1 \\ 0 & \frac{8}{3} & \frac{2}{3} \\ 0 & 0 & \frac{5}{2} \end{bmatrix}$; $\begin{bmatrix} 1 & 0 & 0 \\ 1 & 1 & 0 \\ 1 & 1 & 1 \end{bmatrix} \begin{bmatrix} 1 & 1 & 1 \\ 0 & 3 & 3 \\ 0 & 0 & 4 \end{bmatrix}$.

1.5.6 $\begin{bmatrix} 1 & 0 \\ 12 & 1 \end{bmatrix}$; $\begin{bmatrix} 1 & 0 \\ 48 & 1 \end{bmatrix}$; $\begin{bmatrix} 1 & 0 \\ -6 & 1 \end{bmatrix}$.

1.5.8 (a) Because by the time a pivot row is used, it is taken from U, not A; (b) Rule (iii).

1.5.10 In $Lc = b$, the unknown c_1 is found in only one operation, c_2 requires two operations, and so on. The total is $1 + 2 + \cdots + n = n(n + 1)/2 \approx n^2/2$.

1.5.12 Gaussian elimination starting from the bottom row (or maybe the first column); no.

1.5.15 $\begin{bmatrix} 0 & 1 & 0 \\ 1 & 0 & 0 \\ 0 & 0 & 1 \end{bmatrix} \begin{bmatrix} 0 & 1 & 1 \\ 1 & 0 & 1 \\ 2 & 3 & 4 \end{bmatrix} = \begin{bmatrix} 1 & 0 & 0 \\ 0 & 1 & 0 \\ 2 & 3 & 1 \end{bmatrix} \begin{bmatrix} 1 & 0 & 0 \\ 0 & 1 & 0 \\ 0 & 0 & -1 \end{bmatrix} \begin{bmatrix} 1 & 0 & 1 \\ 0 & 1 & 1 \\ 0 & 0 & 1 \end{bmatrix}$;

$\begin{bmatrix} 1 & 0 & 0 \\ 0 & 0 & 1 \\ 0 & 1 & 0 \end{bmatrix} \begin{bmatrix} 1 & 2 & 1 \\ 2 & 4 & 2 \\ 1 & 1 & 1 \end{bmatrix} = \begin{bmatrix} 1 & 0 & 0 \\ 1 & 1 & 0 \\ 2 & 0 & 1 \end{bmatrix} \begin{bmatrix} 1 & 0 & 0 \\ 0 & -1 & 0 \\ 0 & 0 & 0 \end{bmatrix} \begin{bmatrix} 1 & 2 & 1 \\ 0 & 1 & 0 \\ 0 & 0 & 0 \end{bmatrix}$.

1.5.17 $\begin{bmatrix} 1 & 0 & 0 \\ 1 & 1 & 0 \\ 2 & 0 & 1 \end{bmatrix}$.

1.5.19 $a = 4$ leads to a row exchange; $3b + 10a = 40$ leads to a singular matrix; $c = 0$ leads to a row exchange; $c = 3$ leads to a singular matrix.

1.6.1 $A_1^{-1} = \begin{bmatrix} 0 & \frac{1}{3} \\ \frac{1}{2} & 0 \end{bmatrix}$; $A_2^{-1} = \begin{bmatrix} \frac{1}{2} & 0 \\ -1 & \frac{1}{2} \end{bmatrix}$; $A_3^{-1} = \begin{bmatrix} \cos\theta & \sin\theta \\ -\sin\theta & \cos\theta \end{bmatrix}$.

1.6.6 $A_1^{-1} = \begin{bmatrix} 1 & 0 & 0 \\ -1 & 1 & -1 \\ 0 & 0 & 1 \end{bmatrix}$; $A_2^{-1} = \begin{bmatrix} \frac{3}{4} & \frac{1}{2} & \frac{1}{4} \\ \frac{1}{2} & 1 & \frac{1}{2} \\ \frac{1}{4} & \frac{1}{2} & \frac{3}{4} \end{bmatrix}$; $A_3^{-1} = \begin{bmatrix} 0 & -1 & 1 \\ -1 & 1 & 0 \\ 1 & 0 & 0 \end{bmatrix}$.

1.6.7 $\begin{bmatrix} \sqrt{3}/2 & 1/2 \\ 1/2 & -\sqrt{3}/2 \end{bmatrix}$; $\begin{bmatrix} -\sqrt{3}/2 & 1/2 \\ 1/2 & \sqrt{3}/2 \end{bmatrix}$; $\begin{bmatrix} 0 & 1 \\ 1 & 0 \end{bmatrix}$.

1.6.9 Suppose the third row of A^{-1} is (a, b, c, d), $A^{-1}A = I \Rightarrow 2a = 0$, $a + 3b = 0$, $4a + 8b = 1 \Rightarrow$ no solution.

1.6.10 $A_1^{-1} = \begin{bmatrix} 0 & 0 & 0 & \frac{1}{4} \\ 0 & 0 & \frac{1}{3} & 0 \\ 0 & \frac{1}{2} & 0 & 0 \\ 1 & 0 & 0 & 0 \end{bmatrix}$; $A_2^{-1} = \begin{bmatrix} 1 & 0 & 0 & 0 \\ \frac{1}{2} & 1 & 0 & 0 \\ \frac{1}{3} & \frac{2}{3} & 1 & 0 \\ \frac{1}{4} & \frac{2}{4} & \frac{3}{4} & 1 \end{bmatrix}$; $A_3^{-1} = \frac{1}{ad - bc} \begin{bmatrix} d & -b & 0 & 0 \\ -c & a & 0 & 0 \\ 0 & 0 & d & -b \\ 0 & 0 & -c & a \end{bmatrix}$.

1.6.12 (1), (2), (5).

1.6.14 $(AA^T)^T = (A^T)^T A^T = AA^T$; $(A^TA)^T = A^T(A^T)^T = A^TA$; take $A = \begin{bmatrix} 1 & 2 \\ 1 & 3 \end{bmatrix}$,

$$AA^T = \begin{bmatrix} 5 & 7 \\ 7 & 10 \end{bmatrix}, A^TA = \begin{bmatrix} 2 & 5 \\ 5 & 10 \end{bmatrix}.$$

1.6.16 (a) $n(n + 1)/2$.   (b) $(n - 1)n/2$.

1.6.18 (a) The inverse of a lower (upper) triangular matrix is still a lower (upper) triangular matrix. The multiplication of two lower (upper) triangular matrices gives a lower (upper) triangular matrix.   (b) The main diagonals of $L_1^{-1}L_2D_2$ and $D_1U_1U_2^{-1}$ are the same as those of D_2 and D_1 respectively. $L_1^{-1}L_2D_2 = D_1U_1U_2^{-1}$, so we have $D_1 = D_2$. By comparing the off-diagonals of $L_1^{-1}L_2D_2 = D_1U_1U_2^{-1}$, they must both be diagonal matrices. $L_1^{-1}L_2D_2 = D_2$, $D_1U_1U_2^{-1} = D_1$, D_1 is invertible $\Rightarrow$ $L_1^{-1}L_2 = I$, $U_1U_2^{-1} = I \Rightarrow L_1 = L_2$, $U_1 = U_2$.

1.6.21 $\begin{bmatrix} 1 & 0 & 0 \\ 3 & 1 & 0 \\ 5 & 1 & 1 \end{bmatrix}\begin{bmatrix} 1 & 0 & 0 \\ 0 & 3 & 0 \\ 0 & 0 & 2 \end{bmatrix}\begin{bmatrix} 1 & 3 & 5 \\ 0 & 1 & 1 \\ 0 & 0 & 1 \end{bmatrix}$; $\begin{bmatrix} 1 & 0 \\ b/a & 1 \end{bmatrix}\begin{bmatrix} a & 0 \\ 0 & d - (b^2/a) \end{bmatrix}\begin{bmatrix} 1 & b/a \\ 0 & 1 \end{bmatrix}$.

1.7.2 $\begin{bmatrix} 33 & -16 & 0 \\ -16 & 33 & -16 \\ 0 & -16 & 33 \end{bmatrix}\begin{bmatrix} u_1 \\ u_2 \\ u_3 \end{bmatrix} = \begin{bmatrix} \frac{1}{4} \\ \frac{1}{2} \\ \frac{3}{4} \end{bmatrix}$.

1.7.4 $(u_1, u_2, u_3) = (\pi^2/8, 0, -\pi^2/8)$ instead of the true values $(1, 0, -1)$.

CHAPTER 2

2.1.2 (a), (d), (e), (f).

2.1.4 The space of all 3 by 3 matrices, since every matrix can be written as a symmetric plus a lower triangular matrix; The space of all diagonal matrices.

2.1.6 $x + 2y + z = 0$; P_0 is a subspace of $\mathbf{R}^3$, P isn't.

2.2.1 Seven patterns, including the matrix with all 0 entries.

2.2.3 $\begin{bmatrix} 1 & 0 & 0 \\ 0 & 1 & 0 \\ 1 & 0 & 1 \end{bmatrix}\begin{bmatrix} 1 & 2 & 0 & 1 \\ 0 & 1 & 1 & 0 \\ 0 & 0 & 0 & 0 \end{bmatrix}$; u and v are basic variables; w and y are free variables.

The general solution is $x = \begin{bmatrix} 2w - y \\ -w \\ w \\ y \end{bmatrix} = w\begin{bmatrix} 2 \\ -1 \\ 1 \\ 0 \end{bmatrix} + y\begin{bmatrix} -1 \\ 0 \\ 0 \\ 1 \end{bmatrix}$; $r = 2$.

2.2.4 $U = \begin{bmatrix} 0 & 1 & 4 & 0 \\ 0 & 0 & 0 & 0 \end{bmatrix}$; u, w, y are basic variables and v is free; the general solution to $Ax = 0$ is $x = (u - 4w, w, y)$; $Ax = b$ is consistent if $b_2 - 2b_1 = 0$; the general solution to $Ax = b$ is $x = \begin{bmatrix} u \\ b_1 - 4w \\ w \\ y \end{bmatrix} = u\begin{bmatrix} 1 \\ 0 \\ 0 \\ 0 \end{bmatrix} + w\begin{bmatrix} 0 \\ -4 \\ 1 \\ 0 \end{bmatrix} + y\begin{bmatrix} 0 \\ 0 \\ 0 \\ 1 \end{bmatrix} + \begin{bmatrix} 0 \\ b_1 \\ 0 \\ 0 \end{bmatrix}$; $r = 1$.

2.2.6 $\begin{bmatrix} u \\ v \\ w \end{bmatrix} = \begin{bmatrix} -2v - 3 \\ v \\ 2 \end{bmatrix} = v \begin{bmatrix} -2 \\ 1 \\ 0 \end{bmatrix} + \begin{bmatrix} -3 \\ 0 \\ 2 \end{bmatrix}.$

2.2.10   (a) $x = \begin{bmatrix} -2x_2 + 2x_4 \\ x_2 \\ -2x_4 \\ x_4 \end{bmatrix} = x_2 \begin{bmatrix} -2 \\ 1 \\ 0 \\ 0 \end{bmatrix} + x_4 \begin{bmatrix} 2 \\ 0 \\ -2 \\ 1 \end{bmatrix}$, where x_2, x_4 can be any numbers.

(b) $x = \begin{bmatrix} a - 3b \\ 0 \\ b \\ 0 \end{bmatrix} + x_2 \begin{bmatrix} -2 \\ 1 \\ 0 \\ 0 \end{bmatrix} + x_4 \begin{bmatrix} 2 \\ 0 \\ -2 \\ 1 \end{bmatrix}$, where x_2, x_4 can be any numbers.

2.2.12 $\begin{bmatrix} 1 & 0 & -1 \\ 0 & 1 & -2 \\ 0 & 0 & 0 \end{bmatrix} \begin{bmatrix} u \\ v \\ w \end{bmatrix} = \begin{bmatrix} 1 \\ 1 \\ 0 \end{bmatrix}.$

2.3.2   (a) Independent   (b) Dependent   (c) Dependent.

2.3.4   Yes; $c_1(v_1 + v_2) + c_2(v_1 + v_3) + c_3(v_2 + v_3) = 0 \Rightarrow (c_1 + c_2)v_1 + (c_1 + c_3)v_2 + (c_2 + c_3)v_3 = 0 \Rightarrow c_1 + c_2 = 0, c_1 + c_3 = 0, c_2 + c_3 = 0 \Rightarrow c_1 = c_2 = c_3 = 0 \Rightarrow w_1, w_2, w_3$ are independent.

2.3.6   (a) The y axis in $\mathbf{R}^3$;   (b) The yz plane in $\mathbf{R}^3$;   (c) The yz plane in $\mathbf{R}^3$;   (d) $\mathbf{R}^3$.

2.3.7   (a) $Ax = b$ has no solution, so b is not in the subspace.   (b) The w's (with or without w_4) span $\mathbf{R}^3$.

2.3.10   One basis is $\begin{bmatrix} 1 & 0 \\ 0 & 0 \end{bmatrix}, \begin{bmatrix} 0 & 1 \\ 0 & 0 \end{bmatrix}, \begin{bmatrix} 0 & 0 \\ 1 & 0 \end{bmatrix}, \begin{bmatrix} 0 & 0 \\ 0 & 1 \end{bmatrix}$; the echelon matrices span the upper triangular matrices.

2.3.12   Let $v_1 = (1, 0, 0, 0), \ldots, v_4 = (0, 0, 0, 1)$ be the coordinate vectors. If W is the line through $(1, 2, 3, 4)$, none of the v's are in W.

2.3.15   (i) If it were not a basis, we could add more independent vectors, which would exceed the given dimension k.   (ii) If it were not a basis, we could delete some vectors, leaving less than the given dimension k.

2.3.17   If v_1, v_2, v_3 is a basis for V, and w_1, w_2, w_3 is a basis for W, then these six vectors cannot be independent and some combination is zero: $\sum c_i v_i + \sum d_i w_i = 0$, or $\sum c_i v_i = -\sum d_i w_i$ is a vector in both subspaces.

2.3.20 $\begin{bmatrix} 1 & 0 \\ 0 & 1 \end{bmatrix}$ and $\begin{bmatrix} 0 & 1 \\ 1 & 0 \end{bmatrix}$ **2.3.22**   (1) True   (2) False.

2.4.2 $\mathscr{R}(A): r = 1, (1, 2); \mathscr{N}(A): n - r = 3, (1, 0, 0, 0), (0, -4, 1, 0), (0, 0, 0, 1); \mathscr{R}(A^\mathsf{T}): r = 1, (0, 1, 4, 0); \mathscr{N}(A^\mathsf{T}): m - r = 1, (-2, 1); \mathscr{R}(U): (1, 0); \mathscr{N}(U): (1, 0, 0, 0), (0, -4, 1, 0), (0, 0, 0, 1); \mathscr{R}(U^\mathsf{T}): (0, 1, 4, 0); \mathscr{N}(U^\mathsf{T}): (0, 1).$

2.4.5 $AB = 0 \Rightarrow A(b_1, \ldots, b_n) = 0 \Rightarrow Ab_1 = 0, \ldots, Ab_n = 0 \Rightarrow b_1 \in \mathscr{N}(A), \ldots, b_n \in \mathscr{N}(A) \Rightarrow \mathscr{R}(B)$ is contained in $\mathscr{N}(A)$.

2.4.6 $Ax = b$ is solvable $\Leftrightarrow b \in \mathscr{R}(A) \Leftrightarrow \mathscr{R}(A) = \mathscr{R}(A') \Leftrightarrow$ rank A = rank A'.

2.4.7   (a) m $= n = r$ (b) $n > m = r$.   **2.4.10** $\begin{bmatrix} 1 & 2 & 4 \end{bmatrix}$; $\begin{bmatrix} 1 & 2 & 4 \\ 2 & 4 & 8 \\ 3 & 6 & 12 \end{bmatrix}$.

2.4.12 $Ax = 0$ has a nonzero solution $\Rightarrow r < n \Rightarrow \mathcal{R}(A^T)$ smaller than $\mathbf{R}^n \Rightarrow A^T y = f$ is not solvable for some f.

2.4.15 A has a right inverse $\begin{bmatrix} \frac{2}{3} & -\frac{1}{3} \\ \frac{1}{3} & \frac{1}{3} \\ -\frac{1}{3} & \frac{2}{3} \end{bmatrix}$; M has a left inverse $\begin{bmatrix} \frac{2}{3} & \frac{1}{3} & -\frac{1}{3} \\ -\frac{1}{3} & \frac{1}{3} & \frac{2}{3} \end{bmatrix}$; If $a \neq 0$,

T has a 2-sided inverse $\begin{bmatrix} 1/a & -b/a^2 \\ 0 & 1/a \end{bmatrix}$.

2.4.17 $A^T A$ is not necessarily invertible.

2.4.20 (a) $\begin{bmatrix} 1 & 1 \\ 0 & 0 \\ 1 & 2 \end{bmatrix}$ (b) No such matrix: $r + (n - r) = 1 + 1 \neq 3$. (c) No such matrix: Column space and row space have the same dimension.

2.5.2 In each column, the sum of the first two entries equals the third. Therefore any combination will have $b_1 + b_2 - b_3 = 0$; $Ax = b \Rightarrow x_1 - x_2 = b_1$, $x_2 - x_3 = b_2$, $x_1 - x_3 = b_3 \Rightarrow b_1 + b_2 - b_3 = 0$. It means that the sum of potential differences around a loop is zero.

2.5.3 The entries in each row add to zero. Therefore any combination will have that same property: $f_1 + f_2 + f_3 = 0$; $A^T y = f \Rightarrow y_1 + y_3 = f_1$, $-y_1 + y_2 = f_2$, $-y_2 - y_3 = f_3 \Rightarrow f_1 + f_2 + f_3 = 0$. It means that the total current entering from outside is zero.

2.5.5 $\begin{bmatrix} c_1 + c_3 & -c_1 & -c_3 \\ -c_1 & c_1 + c_2 & -c_2 \\ -c_3 & -c_2 & c_2 + c_3 \end{bmatrix}$; $\begin{bmatrix} c_1 + c_3 & -c_1 \\ -c_1 & c_1 + c_2 \end{bmatrix}$ has pivots $c_1 + c_3$ and $(c_1 c_3 + c_1 c_2 + c_2 c_3)/(c_1 + c_3)$.

2.5.6 $A = \begin{bmatrix} -1 & 1 & 0 & 0 \\ -1 & 0 & 1 & 0 \\ 0 & -1 & 1 & 0 \\ 0 & -1 & 0 & 1 \\ -1 & 0 & 0 & 1 \\ 0 & 0 & -1 & 1 \end{bmatrix}$; $(1, 0, 0, 1, -1, 0)$, $(0, 0, 1, -1, 0, 1)$, $(0, 1, 0, 0, -1, 1)$.

2.5.7 $b_1 + b_4 - b_5 = 0$, $b_3 - b_4 + b_6 = 0$, $b_2 - b_5 + b_6 = 0$.

2.5.10 No; after removing the last row, the remaining three rows are independent. And they are a basis for the row space.

2.5.11 $\begin{bmatrix} 1 & 0 & 0 & 0 & -1 & 1 & 0 \\ 0 & \frac{1}{2} & 0 & 0 & -1 & 0 & 1 \\ 0 & 0 & \frac{1}{2} & 0 & 0 & 1 & 0 \\ 0 & 0 & 0 & 1 & 0 & 0 & -1 \\ -1 & -1 & 0 & 0 & 0 & 0 & 0 \\ 1 & 0 & 1 & 0 & 0 & 0 & 0 \\ 0 & 1 & 0 & -1 & 0 & 0 & 0 \end{bmatrix} \begin{bmatrix} y_1 \\ y_2 \\ y_3 \\ y_4 \\ x_1 \\ x_2 \\ x_3 \end{bmatrix} = \begin{bmatrix} 0 \\ 0 \\ 0 \\ 0 \\ f_1 \\ f_2 \\ f_3 \end{bmatrix}$; $x = \begin{bmatrix} -4 \\ -\frac{5}{3} \\ -\frac{14}{3} \\ 0 \end{bmatrix}$; $y = \begin{bmatrix} -\frac{7}{3} \\ \frac{4}{3} \\ \frac{10}{3} \\ \frac{14}{3} \end{bmatrix}$.

2.5.12 6; 1; 6; 6.

2.5.16 (a) $-y_1 - y_4 - y_3 = 0$, $y_1 - y_2 = 0$, $y_2 + y_3 - y_5 = 0$ (b) By adding the 3 equations.   (c) 3   (d) They correspond to the two independent loops $y_1 y_2 y_3$ and $y_3 y_5 y_4$.

2.6.4 Ellipse.   **2.6.10** e^t, e^{-t}.   **2.6.11** $\dfrac{x+y}{2} e^t + \dfrac{x-y}{2} e^{-t}$; $\begin{bmatrix} \frac{1}{2} & \frac{1}{2} \\ \frac{1}{2} & -\frac{1}{2} \end{bmatrix}$.

2.6.13 Suppose $u = A^{-1}x$ and $v = A^{-1}y$. Since A is linear, $A(cu + dv) = cAu + dAv = cx + dy$. Therefore $A^{-1}(cx + dy) = cu + dv$ and A^{-1} is linear. Since the transformations satisfy $A^{-1}A = I$, the product rule means that the corresponding matrices satisfy $M^{-1}M = I$.

2.6.16 $\begin{bmatrix} 1 & 0 & 0 & 0 \\ 0 & 0 & 1 & 0 \\ 0 & 1 & 0 & 0 \\ 0 & 0 & 0 & 1 \end{bmatrix}$; the double transpose of a matrix gives the matrix itself.

2.6.19 $p(x), q(x) \in S \Rightarrow \int_0^1 (cp(x) + dq(x))\, dx = c \int_0^1 p(x)\, dx + d \int_0^1 q(x)\, dx = 0 \Rightarrow cp(x) + dq(x) \in S \Rightarrow S$ is a subspace; $-\frac{1}{2} + x$, $-\frac{1}{3} + x^2$, $-\frac{1}{4} + x^3$ is a basis for S.

CHAPTER 3

3.1.2 (1, 1) and (1, 2); (1, 1) and (0, 0).

3.1.3 $(x_2/x_1)(y_2/y_1) = -1 \Rightarrow x_1 y_1 + x_2 y_2 = 0 \Rightarrow x^T y = 0$.

3.1.6 All multiples of $(1, 1, -2)$; $(1/\sqrt{3}, 1/\sqrt{3}, 1/\sqrt{3})$, $(1/\sqrt{2}, -1/\sqrt{2}, 0)$, $(1/\sqrt{6}, 1/\sqrt{6}, -2/\sqrt{6})$.

3.1.8 $x \in V$, $x \in W \Rightarrow x^T x = 0 \Rightarrow x = 0$.   **3.1.10** $x_1 + x_2 - x_3 = 0$.

3.1.11 $A^T y = 0 \Rightarrow y^T b = y^T Ax = (y^T A)x = 0$, which contradicts $y^T b \neq 0$.

3.1.14 $(x - y)^T(x + y) = 0 \Leftrightarrow x^T x + x^T y - y^T x - y^T y = 0 \Leftrightarrow x^T x = y^T y \Leftrightarrow \|x\| = \|y\|$.

3.1.18 $\mathbf{R}^4$; the orthogonal complement is spanned by $(1, 0, 0, 0)$, $(0, 1, 0, 0)$, $(0, 0, 1, 0)$.

3.1.20 It means that every vector which is orthogonal to all vectors orthogonal to S is in S.

3.1.22 One basis is $(1, 1, 1, 1)$.

3.2.1 (a) $(x + y)/2 \geq \sqrt{xy}$.   (b) $\|x + y\|^2 \leq (\|x\| + \|y\|)^2 \Rightarrow (x + y)^T(x + y) \leq \|x\|^2 + 2\|x\|\|y\| + \|y\|^2 \Rightarrow x^T y \leq \|x\|\|y\|$.

3.2.3 $(10/3, 10/3, 10/3)$; $(5/9, 10/9, 10/9)$.   **3.2.5** $\arccos(1/\sqrt{n})$; $\begin{bmatrix} 1 \\ \vdots \\ 1 \end{bmatrix}$ $[1/n \cdots 1/n]$.

3.2.7 $b = (1, \ldots, 1)$; $a_1 = \cdots = a_n$.   **3.2.9** $P^2 = \dfrac{aa^T aa^T}{a^T aa^T a} = \dfrac{a(a^T a)a^T}{a^T aa^T a} = \dfrac{aa^T}{a^T a} = P$.

3.2.11 (a) $\begin{bmatrix} \frac{1}{10} & \frac{3}{10} \\ \frac{3}{10} & \frac{9}{10} \end{bmatrix}$; $\begin{bmatrix} \frac{9}{10} & -\frac{3}{10} \\ -\frac{3}{10} & \frac{1}{10} \end{bmatrix}$.   (b) $\begin{bmatrix} 1 & 0 \\ 0 & 1 \end{bmatrix}$; $\begin{bmatrix} 0 & 0 \\ 0 & 0 \end{bmatrix}$. The sum of the projections onto two perpendicular lines gives the vector itself. The projection onto one line and then another (which is perpendicular to the first one) always gives $\{0\}$.

3.2.13 $\dfrac{a_1 a_1}{a^T a} + \cdots + \dfrac{a_n a_n}{a^T a} = 1$.   **3.2.14** $\begin{bmatrix} \frac{1}{6} & -\frac{1}{3} & \frac{1}{6} \\ -\frac{1}{3} & \frac{2}{3} & -\frac{1}{3} \\ \frac{1}{6} & -\frac{1}{3} & \frac{1}{6} \end{bmatrix}$.

3.2.16 (a) $P^T = P \Rightarrow (Px)^T y = x^T P^T y = x^T(Py)$.   (b) No; $1/\sqrt{15}$, $1/3\sqrt{3}$.
(c) $P^2 = P \Rightarrow (Px)^T Py = x^T P^T Py = x^T P^2 y = x^T(Py)$, $0°$ or $180°$ (on the same line).

3.3.1 2; $(10 - 3x)^2 + (5 - 4x)^2$; $(4, -3)(3, 4)^T = 0$.

3.3.3 $\bar{x} = \begin{bmatrix} \frac{1}{3} \\ \frac{1}{3} \\ \frac{1}{3} \end{bmatrix}$; $p = \begin{bmatrix} \frac{1}{3} \\ \frac{1}{3} \\ \frac{2}{3} \end{bmatrix}$; $b - p = \begin{bmatrix} \frac{2}{3} \\ \frac{2}{3} \\ -\frac{2}{3} \end{bmatrix}$.

3.3.5 $6 + (5/2)t$; $(7/2, 6, 17/2)$.   **3.3.8**   column space S; rank k.

3.3.10 $A^T A = I$, $A^T b = 0$; 0.

3.3.12 (a) $(-1, 1, 0, 0)$, $(-1, 0, 0, 1)$ (b) $\begin{bmatrix} \frac{1}{3} & \frac{1}{3} & 0 & \frac{1}{3} \\ \frac{1}{3} & \frac{1}{3} & 0 & \frac{1}{3} \\ 0 & 0 & 1 & 0 \\ \frac{1}{3} & \frac{1}{3} & 0 & \frac{1}{3} \end{bmatrix}$ (c) $(0, 0, 0, 0)$.

3.3.13 $61/35 - (36/35)t$; $(133/35, 95/35, 61/35, -11/35)$.

3.3.15 $H^2 = (I - 2P)^2 = I - 4P + 4P^2 = I - 4P + 4P = I$.

3.3.18 (1) $C + D + E = 3$, $C + 3E = 6$, $C + 2D + E = 5$, $C = 0$ (No such plane);

(2) $\begin{bmatrix} 4 & 3 & 5 \\ 3 & 5 & 3 \\ 5 & 3 & 11 \end{bmatrix} \begin{bmatrix} C \\ D \\ E \end{bmatrix} = \begin{bmatrix} 14 \\ 13 \\ 26 \end{bmatrix}$

3.3.19 $A^T(AA^T)^{-1}A$.   **3.3.21** $\begin{bmatrix} a_1^T a_1 & -a_1^T a_2 \\ -a_1^T a_2 & a_2^T a_2 \end{bmatrix} \begin{bmatrix} x_1 \\ x_2 \end{bmatrix} = \begin{bmatrix} a_1^T b \\ -a_2^T b \end{bmatrix}$; $x = (2, 1)$.

3.3.24 $-3/10 - (12/5)t$.   **3.3.25** $\begin{bmatrix} 1 & -1 & 1 \\ 1 & 0 & 0 \\ 1 & 1 & 1 \\ 1 & 2 & 4 \end{bmatrix}$; $\begin{bmatrix} C \\ D \\ E \end{bmatrix}$; $\begin{bmatrix} 2 \\ 0 \\ -3 \\ -5 \end{bmatrix}$.

3.4.1 (a) $-4 = C - 2D$, $-3 = C - D$, $-1 = C + D$, $0 = C + 2D$ (b) $-2 + t$; 0
(c) b is in the column space; b itself.

3.4.3 $(-2/3, 1/3, -2/3)$; the sum is b itself; notice that $a_1 a_1^T$, $a_2 a_2^T$, $a_3 a_3^T$ are projections onto three orthogonal directions. Their sum is projection onto the whole space and should be the identity.

3.4.5 $(I - 2uu^T)^T(I - 2uu^T) = I - 4uu^T + 4uu^T uu^T = I$; $Q = \begin{bmatrix} \frac{1}{2} & -\frac{1}{2} & \frac{1}{2} & \frac{1}{2} \\ -\frac{1}{2} & \frac{1}{2} & \frac{1}{2} & \frac{1}{2} \\ \frac{1}{2} & \frac{1}{2} & \frac{1}{2} & -\frac{1}{2} \\ \frac{1}{2} & \frac{1}{2} & -\frac{1}{2} & \frac{1}{2} \end{bmatrix}$.

3.4.7 $(x_1 q_1 + \cdots + x_n q_n)^T(x_1 q_1 + \cdots + x_n q_n) = x_1^2 + \cdots + x_n^2 \Rightarrow \|b\|^2 = b^T b = x_1^2 + \cdots + x_n^2$.

3.4.9 $0q_1 + 0q_2$.   **3.4.12** 2; $\begin{bmatrix} 1/\sqrt{2} & 1/\sqrt{2} \\ 1/\sqrt{2} & -1/\sqrt{2} \end{bmatrix} \begin{bmatrix} \sqrt{2} & 2\sqrt{2} \\ 0 & 2\sqrt{2} \end{bmatrix}$.

3.4.13 $\begin{bmatrix} 0 & 0 & 1 \\ 0 & 1 & 1 \\ 1 & 1 & 1 \end{bmatrix} = \begin{bmatrix} 0 & 0 & 1 \\ 0 & 1 & 0 \\ 1 & 0 & 0 \end{bmatrix} \begin{bmatrix} 1 & 1 & 1 \\ 0 & 1 & 1 \\ 0 & 0 & 1 \end{bmatrix}$.

3.4.16 $\begin{bmatrix} 1 & 1 \\ 2 & 3 \\ 2 & 1 \end{bmatrix} = \begin{bmatrix} 1/3 & 0 \\ 2/3 & 1/\sqrt{2} \\ 2/3 & -1/\sqrt{2} \end{bmatrix} \begin{bmatrix} 3 & 3 \\ 0 & \sqrt{2} \end{bmatrix};\ A_{m \times n} = Q_{m \times n} R_{n \times n}.$

3.4.17 $R\bar{x} = Q^{\mathrm{T}} b \Rightarrow \bar{x} = R^{-1} Q^{\mathrm{T}} b = (5/9, 0).$

3.4.18 Q has the same column space as A, so $P = Q(Q^{\mathrm{T}}Q)^{-1}Q^{\mathrm{T}} = QQ^{\mathrm{T}}.$

3.4.20 $1;\ \sqrt{(e^2 - 1)/2};$ **3.4.22** $\dfrac{\int_0^{2\pi} y(x) \sin x\, dx}{\int_0^{2\pi} \sin^2 x\, dx};\ 0.$

3.4.24 x^3 is already orthogonal to the even functions 1 and $x^2 - \frac{1}{3}$. To subtract off its component in the direction of the odd function x, we compute $(x^3, x) = \int_{-1}^{1} x^3 x\, dx = \frac{2}{5}$ and $(x, x) = \int_{-1}^{1} x^2\, dx = \frac{2}{3}$; the next Legendre polynomial is $x^3 - ((x^3, x)/(x, x))x = x^3 - \frac{3}{5}x.$

3.4.27 $(1/\sqrt{2}, -1/\sqrt{2}, 0, 0), (1/\sqrt{6}, 1/\sqrt{6}, 2/\sqrt{6}, 0), (-1/2\sqrt{3}, -1/2\sqrt{3}, 1/2\sqrt{3}, -1/\sqrt{3}).$

3.5.1 $\begin{bmatrix} 4 & 0 & 0 & 0 \\ 0 & 0 & 0 & 4 \\ 0 & 0 & 4 & 0 \\ 0 & 4 & 0 & 0 \end{bmatrix}, \begin{bmatrix} 16 & 0 & 0 & 0 \\ 0 & 16 & 0 & 0 \\ 0 & 0 & 16 & 0 \\ 0 & 0 & 0 & 16 \end{bmatrix}.$

3.5.2 $1, -1/2 + (\sqrt{3}/2)i, -1/2 - (\sqrt{3}/2)i.$

3.5.5 $x = (2k + 1)\pi, \theta = 2k\pi + \pi/2, k$ is integer.

3.5.7 $(1, 0, 1, 0).$ **3.5.8** $(0, 1, 0, 1).$

3.5.10 $y_0 = y_0' + y_0'', y_1 = y_0' - y_0''; y_0 = y_0' + y_0'', y_1 = y_1' + iy_1'', y_2 = y_0' - y_0'', y_3 = y_1' - iy_1''.$

3.5.11 $\begin{bmatrix} 1 \\ 0 \\ 1 \\ 0 \end{bmatrix} \rightarrow \begin{bmatrix} 1 \\ 1 \\ 0 \\ 0 \end{bmatrix} \rightarrow \begin{bmatrix} 2 \\ 0 \\ 0 \\ 0 \end{bmatrix} \rightarrow \begin{bmatrix} 2 \\ 0 \\ 2 \\ 0 \end{bmatrix}.$

3.5.13 $c_0 = (f_0 + f_1 + f_2 + f_3)/4,\ c_1 = (f_0 - if_1 - f_2 + if_3)/4,\ c_2 = (f_0 - f_1 + f_2 - f_3)/4,$ $c_3 = (f_0 + if_1 - f_2 - if_3)/4; f_0 = 0, f_2 = 0, f_3 = -f_1 \Rightarrow c_0 = 0, c_2 = 0, c_3 = -c_1 \Rightarrow c$ is also odd.

3.6.1 (a) 7 (b) 2 (c) 8 (d) 13.

3.6.3 $V + W$ contains the 4 by 4 matrices $\begin{bmatrix} a_{11} & a_{12} & a_{13} & a_{14} \\ a_{21} & a_{22} & a_{23} & a_{24} \\ 0 & a_{32} & a_{33} & a_{34} \\ 0 & 0 & a_{43} & a_{44} \end{bmatrix}$, $\dim(V + W) = 13,$

$V \cap W$ contains the matrices $\begin{bmatrix} a_{11} & a_{12} & 0 & 0 \\ 0 & a_{22} & a_{23} & 0 \\ 0 & 0 & a_{23} & a_{34} \\ 0 & 0 & 0 & a_{44} \end{bmatrix}$, $\dim(V \cap W) = 7,$

$\dim(V + W) + \dim(V \cap W) = 20 = \dim V + \dim W.$

3.6.6 The line through $(1, 0, 0)$ (it need not be perpendicular to V).

3.6.7 $x = v_1 + w_1 = v_2 + w_2,\ v_1, v_2 \in V,\ w_1, w_2 \in W \Rightarrow v_1 - v_2 = w_2 - w_1 \in V \cap W = \{0\} \Rightarrow v_1 = v_2, w_1 = w_2.$

3.6.9 $A = \begin{bmatrix} 0 & 0 \\ 1 & 0 \end{bmatrix}$, $B = \begin{bmatrix} 0 & 1 \\ 0 & 0 \end{bmatrix}$, nullspace of $AB = \begin{bmatrix} 0 & 0 \\ 0 & 1 \end{bmatrix}$ does not contain $\begin{bmatrix} 0 \\ 1 \end{bmatrix}$, column space of AB is not contained in column space of B.

3.6.11 rank$(AB) \leq$ rank(A), rank$(A) \leq n$, $n < m \Rightarrow$ rank$(AB) < m \Rightarrow AB$ is singular.

3.6.12 rank$(A + B) = \dim(\mathscr{R}(A) + \mathscr{R}(B)) \leq \dim \mathscr{R}(A) + \dim \mathscr{R}(B) = $ rank$(A) + $ rank(B).

3.6.13 $\mathscr{N}(B) = \mathscr{N}(A^{-1}AB) \supseteq \mathscr{N}(AB) \supseteq \mathscr{N}(B) \Rightarrow \mathscr{N}(AB) = = \mathscr{N}(B)$; the row space is the orthogonal complement, and the rank is its dimension—therefore these are the same for B and AB.

3.6.15 $A = \begin{bmatrix} 1 & 0 \\ 0 & 1 \\ 1 & 1 \end{bmatrix} \begin{bmatrix} 1 & -1 & 0 \\ 0 & 1 & -1 \end{bmatrix} = \begin{bmatrix} 1 \\ 0 \\ 1 \end{bmatrix} \begin{bmatrix} 1 & -1 & 0 \end{bmatrix} + \begin{bmatrix} 0 \\ 1 \\ 1 \end{bmatrix} \begin{bmatrix} 0 & 1 & -1 \end{bmatrix}$.

3.6.16 $\dim(U \cap V) = \dim(U) + \dim(V) - \dim(U \cup V) \Rightarrow \dim(U \cap V) \geq 6 + 6 - 8 = 4 > 0$, $\dim(U \cap V \cap W) = \dim(U \cap V) + \dim(W) - \dim((U \cap V) \cup W) \Rightarrow \dim(U \cap V \cap W) \geq 4 + 6 - 8 = 2 > 0$.

3.6.18 $y \in V \cap W \Leftrightarrow y = x_1 v_1 + \cdots + x_k v_k$ and also $y = -x_{k+1} w_1 - \cdots - x_{k+l} w_l$ (*) $\Leftrightarrow x_1 v_1 + \cdots + x_{k+l} w_l = 0 \Leftrightarrow x = (x_1, \ldots, x_{k+l}) \in \mathscr{N}(D)$. Because of the uniqueness of the expressions (*) y yields only one x in $\mathscr{N}(D)$.

3.6.20 $\bar{x}_W = \dfrac{w_1^2 b_1 + \cdots + w_m^2 b_m}{w_1^2 + \cdots + w_m^2}$.   **3.6.21** 11; 5; the line through $(1, -4)$.

3.6.22 $\bar{x}_W = (\frac{1}{21}, \frac{4}{7})$; $A\bar{x}_W = (\frac{1}{21}, \frac{13}{21}, \frac{25}{21})$, $b - A\bar{x}_W = (-\frac{1}{21}, \frac{8}{21}, -\frac{4}{21})$, $(A\bar{x}_W)W^T W(b - A\bar{x}_W) = 0$.

3.6.23 (a) $E(e) = \frac{1}{2}(-2) + \frac{1}{4}(-1) + \frac{1}{4}(5) = 0$, $E(e^2) = \frac{1}{2}(-2)^2 + \frac{1}{4}(-1)^2 + \frac{1}{4}(5^2) = \frac{17}{2}$
(b) $w_1 = \sqrt{\frac{2}{17}}$, $w_2 = 2$.

3.6.24 $A = B + C$, with B containing those p rows and zero elsewhere, $C = A - B$. rank$(A) \leq$ rank$(B) + $ rank$(C) \leq p + q$; 5 by 5.

CHAPTER 4

4.2.1 $2^n \det(A)$; $(-1)^n \det(A)$; $(\det(A))^2$.   **4.2.3** 20; 5.

4.2.6 (a) 0 (b) 16 (c) 16 (d) 1/16 (e) 16.

4.2.7 Multiply the zero row by 2. That multiplies $\det A$ by 2, but A is unchanged $\Rightarrow$ $\det A = 0$.

4.2.8 $(1 - ml)(ad - bc)$.

4.2.10 $\begin{vmatrix} 1 & a & a^2 \\ 1 & b & b^2 \\ 1 & c & c^2 \end{vmatrix} = \begin{vmatrix} 1 & a & a^2 \\ 0 & b - a & b^2 - a^2 \\ 0 & c - a & c^2 - a^2 \end{vmatrix} = \begin{vmatrix} 1 & a & a^2 \\ 0 & b - a & b^2 - a^2 \\ 0 & 0 & (c - a)(c - b) \end{vmatrix} = $
$(b - a)(c - a)(c - b)$.

4.2.12 (a) False; $\det \begin{bmatrix} 4 & 1 \\ 1 & 1 \end{bmatrix} \neq 2 \det \begin{bmatrix} 2 & 1 \\ 1 & 1 \end{bmatrix}$ (b) False; $\det \begin{bmatrix} 0 & 1 \\ 1 & 1 \end{bmatrix} = -1$, its pivots are 1, 1, but there is a row exchange (c) True; $\det(AB) = \det(A) \det(B) = 0$.

4.2.13 Adding every column of A to the first column makes it a zero column, so $\det A = 0$. If every row of A adds to 1, every row of $A - I$ adds to $0 \Rightarrow \det(A - I) = 0$; $A = \begin{bmatrix} \frac{1}{2} & \frac{1}{2} \\ \frac{1}{2} & \frac{1}{2} \end{bmatrix}$, $\det(A - I) = 0$, but $\det A = 0 \neq 1$.

4.2.14 $0; (1 - t^2)^3$.   **4.2.15** $\det(A) = 10$, $\det(A^{-1}) = \frac{1}{10}$, $\lambda = 5$, $\lambda = 2$.

4.2.17 Taking determinants gives $(\det C)(\det D) = (-1)^n(\det D)(\det C)$.

4.3.1 $a_{12}a_{21}a_{34}a_{43}$; it is even, $\det A = 1$.

4.3.2 $\det A = (-1)^{1+2} \det \begin{bmatrix} 1 & 0 & 0 \\ 1 & 0 & 1 \\ 0 & 1 & 0 \end{bmatrix} = (-1) \det \begin{bmatrix} 0 & 1 \\ 1 & 0 \end{bmatrix} = (-1)(-1)^{1+2} \det[1] = 1$.

4.3.3 (1) True   (2) False   (3) False.

4.3.6 (b) 6; $D_{1000} = D_{6 \times 166 + 4} = D_4 = -1$.

4.3.8 In formula (6), $a_{1\alpha} \cdots a_{5\nu}$ is sure to be zero for all possible $(\alpha, \ldots, \nu)$. Or by **3.6.24**, rank $A \le 2 + 2 = 4$.

4.3.9 By row operations on the 4 by 4 matrices, we make A and D into upper triangular matrices:

$$\det \begin{bmatrix} A & B \\ 0 & D \end{bmatrix} = a_{11}a_{22}d_{11}d_{22} = \det(A) \det(D).$$

4.3.10 $\det A_4 = -3$; $\det A_3 = 2$; $\det A_2 = -1$; $\det A_n = (-1)^{n-1}(n - 1)$.

4.3.11 (a) $(n - 1)n!$ (b) $(1 + 1/2! + \cdots + 1/(n - 1)!)n!$ (c) $(n^3 + 2n - 3)/3$.

4.3.13 $\begin{bmatrix} 0 & A \\ -B & I \end{bmatrix}\begin{bmatrix} I & 0 \\ B & I \end{bmatrix} = \begin{bmatrix} AB & A \\ 0 & I \end{bmatrix}$, $\det \begin{bmatrix} I & 0 \\ B & I \end{bmatrix} = 1 \Rightarrow \det \begin{bmatrix} 0 & A \\ -B & I \end{bmatrix} =$
$\det \begin{bmatrix} AB & A \\ 0 & I \end{bmatrix} = \det(AB)$; e.g. $A = [1 \quad 2]$, $B = \begin{bmatrix} 1 \\ 2 \end{bmatrix}$, $\det \begin{bmatrix} 0 & A \\ -B & I \end{bmatrix} = 5 = \det(AB)$;
$A = \begin{bmatrix} 1 \\ 2 \end{bmatrix}$, $B = [1 \quad 2]$, $\det \begin{bmatrix} 0 & A \\ -B & I \end{bmatrix} = 0 = \det(AB)$; because AB is a matrix with
rank$(AB) \le $ rank$(A) \le n < m$.

4.4.2 The (i, j) entry of A_{cof}, which is the cofactor A_{ji}, is zero if $i > j$ and A is upper triangular.

4.4.3 $A^{-1} = \frac{1}{4}\begin{bmatrix} 3 & 2 & 1 \\ 2 & 4 & 2 \\ 1 & 2 & 3 \end{bmatrix}$; $B^{-1} = \begin{bmatrix} 2 & -1 & 0 \\ -1 & 2 & -1 \\ 0 & -1 & 1 \end{bmatrix}$.

4.4.4 A is symmetric $\Rightarrow$ the i, j minor is the transpose of the j, i minor $\Rightarrow A_{ij} = A_{ji}$.

4.4.6 (a) $\det M = x_j$ (b) Look at column j of AM, it is $Ax = b$. All other columns of AM are the same as in A, so $AM = B_j$. (c) $\det A \det M = \det B_j \Rightarrow x_j = \det B_j/\det A$.

4.4.8 (a) The area of that parallelogram is $\det \begin{bmatrix} -2 & 2 \\ -1 & 3 \end{bmatrix}$. (b) The triangle $A'B'C'$ has the same area; it is just moved to the origin.

4.4.10 $AC = CA \Rightarrow ACA^{-1} = C \Rightarrow \det A \det(D - CA^{-1}B) = \det(A(D - CA^{-1}B)) = \det(AD - CB)$.

4.4.11 (a) Columns of $AB = A$ times columns of B. (b) AB is the sum of columns times rows (rank-one matrices).

4.4.15 When multiplied by any odd permutation, a permutation changes its parity. Therefore $\sigma^2 = \sigma\sigma$ is even. If σ^{-1} is even, $\sigma\sigma^{-1}$ would be odd. But $\sigma\sigma^{-1} = I$ is even.

4.4.16 The power of P are all permutation matrices so eventually one must be repeated. If $P^r = P^s$ then $P^{r-s} = I$.

4.4.17 By formula (6), $\det A \le 5! = 120$. Or by $\det A = $ volume, $\det A \le (\sqrt{5})^5$. Or by pivots, $\det A \le 1 \times 2 \times 4 \times 8 \times 16$.

CHAPTER 5

5.1.2 $u = c_1 \begin{bmatrix} 1 \\ -1 \end{bmatrix} e^{2t} + c_2 \begin{bmatrix} 1 \\ -2 \end{bmatrix} e^{3t} = 6 \begin{bmatrix} 1 \\ -1 \end{bmatrix} e^{2t} - 6 \begin{bmatrix} 1 \\ -2 \end{bmatrix} e^{3t}$.

5.1.4 $u = \left[\begin{bmatrix} 1 \\ -1 \end{bmatrix} + 4 \begin{bmatrix} 1 \\ 1 \end{bmatrix} \right] e^t$.

5.1.7 $Ax = \lambda x \Rightarrow (A - 7I)x = (\lambda - 7)x$; $Ax = \lambda x \Rightarrow x = \lambda A^{-1}x \Rightarrow A^{-1}x = (1/\lambda)x$.

5.1.8 Choose $\lambda = 0$.

5.1.9 The coefficient is $\lambda_1 + \cdots + \lambda_n$. In $\det(\lambda - \lambda I)$, a term which includes an off-diagonal a_{ij} excludes both $a_{ii} - \lambda$ and $a_{jj} - \lambda$. Therefore such a term doesn't involve $(-\lambda)^{n-1}$. The coefficient of $(-\lambda)^{n-1}$ must come from the main diagonal and it is $a_{11} + \cdots + a_{nn} = \lambda_1 + \cdots + \lambda_n$.

5.1.12 $\lambda_1 = 5$, $\lambda_2 = -5$, $x_1 = \begin{bmatrix} 2 \\ 1 \end{bmatrix}$, $x_2 = \begin{bmatrix} 1 \\ -2 \end{bmatrix}$; $\lambda_1 = a + b$, $\lambda_2 = a - b$, $x_1 = \begin{bmatrix} 1 \\ 1 \end{bmatrix}$, $x_2 = \begin{bmatrix} 1 \\ -1 \end{bmatrix}$.

5.1.14 $\text{rank}(A) = 1$; $\lambda = 0, 0, 0, 4$; $x_4 = (1, 1, 1, 1)$; $\text{rank}(B) = 2$; $\lambda = 0, 0, 2, -2$; $x_3 = (1, 1, 1, 1)$, $x_4 = (1, -1, 1, -1)$.

5.1.15 $\text{rank}(A) = 1$, $\lambda = 0, \ldots, 0, n$; $\text{rank}(C) = 2$, $\lambda = 0, \ldots, n/2, -n/2$.

5.1.18 The nullspace is spanned by v_0; $x = c_0 v_0 + v_1 + \frac{1}{2}v_2$; v_0, v_1, v_2 are independent (the eigenvalues are distinct), $\Rightarrow v_0$ is not in the column space which is spanned by v_1 and v_2.

5.2.1 $\begin{bmatrix} 1 & 1 \\ 1 & -1 \end{bmatrix} \begin{bmatrix} 2 & 0 \\ 0 & 0 \end{bmatrix} \begin{bmatrix} 1 & 1 \\ 1 & -1 \end{bmatrix}^{-1}$; $\begin{bmatrix} 1 & 1 \\ 0 & -2 \end{bmatrix} \begin{bmatrix} 2 & 0 \\ 0 & 0 \end{bmatrix} \begin{bmatrix} 1 & 1 \\ 0 & -2 \end{bmatrix}^{-1}$.

5.2.4 It has distinct eigenvalues 1, 2, 7; $\Lambda = \text{diag}(1, 2, 7)$.

5.2.5 A_1 and A_3 cannot be diagonalized.

5.2.6 (a) $\lambda = 1$ or -1 (b) trace $= 0$; determinant $= -1$ (c) $(8, -3)$.

5.2.8 (a) $Au = uv^T u = (v^T u)u \Rightarrow \lambda = v^T u$. (b) All other eigenvalues are zero because $\dim \mathcal{N}(A) = n - 1$.

5.2.9 $\text{trace}(AB) = \text{trace}(BA) = aq + bs + cr + dt \Rightarrow \text{trace}(AB - BA) = 0 \Rightarrow AB - BA = I$ is impossible.

5.2.12 F; T; T.   **5.2.13** $A = \begin{bmatrix} 1 & 1 \\ 1 & -1 \end{bmatrix} \begin{bmatrix} 9 & 0 \\ 0 & 1 \end{bmatrix} \begin{bmatrix} 1 & 1 \\ 1 & -1 \end{bmatrix}^{-1}$; $\begin{bmatrix} 2 & 1 \\ 1 & 2 \end{bmatrix}$; 4.

5.2.14 $\det A = \det(S\Lambda S^{-1}) = \det S \det \Lambda \det S^{-1} = \det \Lambda = \lambda_1 \cdots \lambda_n$.

5.3.1 (a) $A^2 = \begin{bmatrix} 2 & 1 \\ 1 & 1 \end{bmatrix}$, $A^3 = \begin{bmatrix} 3 & 2 \\ 2 & 1 \end{bmatrix}$, $A^4 = \begin{bmatrix} 5 & 3 \\ 3 & 2 \end{bmatrix}$, $A^{100} = \begin{bmatrix} F_{101} & F_{100} \\ F_{100} & F_{00} \end{bmatrix}$.

(b) $B = \begin{bmatrix} 2 & 3 \\ -1 & -2 \end{bmatrix} \begin{bmatrix} 1 & 0 \\ 0 & -1 \end{bmatrix} \begin{bmatrix} 2 & 3 \\ -1 & -2 \end{bmatrix}^{-1}$, $1^{-101} = 1, (-1)^{-101} = -1$, so $B^{-101} = B$.

5.3.3 $\begin{bmatrix} G_{k+2} \\ G_{k+1} \end{bmatrix} = \begin{bmatrix} \frac{1}{2} & \frac{1}{2} \\ 1 & 0 \end{bmatrix} \begin{bmatrix} G_{k+1} \\ G_k \end{bmatrix}$, $\lambda_1 = 1, \lambda_2 = -\frac{1}{2}$; $\begin{bmatrix} G_{k+1} \\ G_k \end{bmatrix} = S\Lambda^k S^{-1} \begin{bmatrix} G_1 \\ G_0 \end{bmatrix} = \begin{bmatrix} 1 & 1 \\ 1 & -2 \end{bmatrix} \begin{bmatrix} 1^k & 0 \\ 0 & (-\frac{1}{2})^k \end{bmatrix} \begin{bmatrix} \frac{2}{3} & \frac{1}{3} \\ \frac{1}{3} & -\frac{1}{3} \end{bmatrix} \begin{bmatrix} \frac{1}{2} \\ 0 \end{bmatrix}$, $G_k = [1 - (-\frac{1}{2})^k]/3 \to \frac{1}{3}$.

5.3.5 In the steady state everyone is dead: $d_\infty = 1, s_\infty = 0, w_\infty = 0$.

5.3.6 $\begin{bmatrix} \frac{7}{12} & \frac{1}{6} & 0 \\ \frac{1}{6} & \frac{1}{2} & 0 \\ \frac{1}{4} & \frac{1}{3} & 1 \end{bmatrix}$.   **5.3.7** $y_\infty = 3, z_\infty = 2, y_k = 3(1 - (0.5)^k), z_k = 2 + 3(0.5)^k$.

5.3.10 (a) $0 \le a, b \le 1$ (b) $u_k = \begin{bmatrix} b/(1-a) & 1 \\ 1 & -1 \end{bmatrix} \begin{bmatrix} 1^k & 0 \\ 0 & (a-b)^k \end{bmatrix} \begin{bmatrix} b/(1-a) & 1 \\ 1 & -1 \end{bmatrix}^{-1} \begin{bmatrix} 1 \\ 1 \end{bmatrix}$

$= \begin{bmatrix} \dfrac{2b}{b-a+1} - \dfrac{1-a-b}{b-a+1}(a-b)^k \\ \dfrac{2(1-a)}{b-a+1} - \dfrac{1-a-b}{b-a+1}(a-b)^k \end{bmatrix}$. (c) $u_k \to \begin{bmatrix} \dfrac{2b}{b-a+1} \\ \dfrac{2(1-a)}{b-a+1} \end{bmatrix}$ if $|a-b| < 1$; No.

5.3.11 (a) $A = \begin{bmatrix} \frac{1}{2} & \frac{1}{2} & \frac{1}{2} \\ \frac{1}{4} & \frac{1}{2} & 0 \\ \frac{1}{4} & 0 & \frac{1}{2} \end{bmatrix}$ (b) $\lambda_1 = 1, \lambda_2 = \frac{1}{2}, \lambda_3 = 0, x_1 = (2, 1, 1), x_2 = (0, -1, 1)$,

$x_3 = (2, -1, -1)$ (c) $(2, 1, 1)$ (d) $(2, 1 - (\frac{1}{2})^k, 1 + (\frac{1}{2})^k)$.

5.3.12 The components of Ax add to $x_1 + x_2 + x_3$ (each column adds to 1 and nobody is lost). The components of λx add to $\lambda(x_1 + x_2 + x_3)$. If $\lambda \ne 1$, $x_1 + x_2 + x_3$ must be zero.

5.3.13 The eigenvalues of $I + A$ are $1 \pm i$, so $|\lambda| = \sqrt{2}$, unstable; the eigenvalues of $(I - A)^{-1}$ are $(1 \mp i)^{-1}$, so $|\lambda| = 1/\sqrt{2}$, stable; the third eigenvalues are $(1 \mp \frac{1}{2}i)^{-1}(1 \pm \frac{1}{2}i)$, so $|\lambda| = 1$, neutrally stable; this solution stays on circle.

5.3.15 $a = 1.8, b = 1, c = 0.6$.

5.3.17 $A^0 = I, A^2 = \begin{bmatrix} 0 & 1 \\ 0 & 0.5^2 \end{bmatrix}, A^3 = \begin{bmatrix} 0 & 0.5 \\ 0 & 0.5^3 \end{bmatrix}, \ldots, A^k = \begin{bmatrix} 0 & 0.5^{k-1} \\ 0 & 0.5^k \end{bmatrix}$ $(k \ge 1)$.

5.3.19 $\begin{bmatrix} \frac{1}{4} \\ \frac{3}{4} \end{bmatrix}, \begin{bmatrix} \frac{1}{4} \\ \frac{3}{4} \end{bmatrix}, \begin{bmatrix} \frac{1}{4} & \frac{1}{4} \\ \frac{3}{4} & \frac{3}{4} \end{bmatrix}$.

5.4.1 $\lambda_1 = -2, \lambda_2 = 0; x_1 = (1, -1), x_2 = (1, 1); e^{At} = \frac{1}{2} \begin{bmatrix} e^{-2t} + 1 & -e^{-2t} + 1 \\ -e^{-2t} + 1 & e^{-2t} + 1 \end{bmatrix}$.

5.4.3 $u(t) = \begin{bmatrix} e^{2t} + 2 \\ -e^{2t} + 2 \end{bmatrix}$; as $t \to \infty, e^{2t} \to +\infty$.

5.4.4 $P^2 = P \Rightarrow e^P = I + P + P^2/2! + \cdots = I + (1 + 1/2! + \cdots)P = I + (e - 1)P \approx I + 1.718P$.

5.4.5 (a) $e^{A(t+T)} = Se^{\Lambda(t+T)}S^{-1} = Se^{\Lambda t}e^{\Lambda T}S^{-1} = Se^{\Lambda t}S^{-1}Se^{\Lambda T}S^{-1} = e^{At}e^{AT}$.

5.4.7 $e^{At} = \begin{bmatrix} 1 & t \\ 0 & 1 \end{bmatrix}$; $e^{At}u_0 = \begin{bmatrix} 4t+3 \\ 4 \end{bmatrix}$.

5.4.8 (a) $A = \begin{bmatrix} 4 & -2 \\ 1 & 1 \end{bmatrix}$, $\lambda_1 = 3$, $x_1 = \begin{bmatrix} 2 \\ 1 \end{bmatrix}$, $\lambda_2 = 2$, $x_2 = \begin{bmatrix} 1 \\ 1 \end{bmatrix}$; unstable;

(b) $u = \begin{bmatrix} r \\ w \end{bmatrix} = 100e^{3t}\begin{bmatrix} 2 \\ 1 \end{bmatrix} + 100e^{2t}\begin{bmatrix} 1 \\ 1 \end{bmatrix}$ (c) Ratio approaches 2/1.

5.4.11 A_1 is neutrally stable for $t \le 1$, unstable for $t > 1$. A_2 is unstable for $t < 4$, neutrally stable at $t = 4$, stable with real λ for $4 < t \le 5$, stable with complex λ for $t > 5$. A_3 is stable with real λ for $t < -1$, neutrally stable at $t = -1$, unstable for $t > -1$.

5.4.13 (a) $u_1' = cu_2 - bu_3$, $u_2' = -cu_1 + au_3$, $u_3' = bu_1 - au_2 \Rightarrow u_1'u_1 + u_2'u_2 + u_3'u_3 = 0$.
(b) Because e^{At} is an orthogonal matrix, $\|u(t)\|^2 = \|e^{At}u_0\|^2 = \|u_0\|^2$ is constant.
(c) $\lambda = 0$ or $\pm(\sqrt{a^2 + b^2 + c^2})i$.

5.4.14 $\lambda_1 = -1$, $\lambda_2 = -9$, $\omega_1 = 1$, $\omega_2 = 3$; $u(t) = (a_1 \cos t + b_1 \sin t)x_1 + (a_2 \cos 3t + b_2 \sin 3t)x_2$.

5.4.17 $Ax = \lambda Fx + \lambda^2 x$ or $(A - \lambda F - \lambda^2 I)x = 0$.

5.4.19 Its eigenvalues are real $\Leftrightarrow$ (trace)$^2 - 4$ det $\ge 0 \Leftrightarrow -4(-a^2 - b^2 + c^2) \ge 0 \Leftrightarrow a^2 + b^2 \ge c^2$.

5.4.20 $u = \begin{bmatrix} \frac{8}{3}e^t - 6e^{3t} + \frac{13}{3}e^{4t} \\ -6e^{3t} + 6e^{4t} \\ e^{4t} \end{bmatrix}$.

5.5.2 (i) It is real. (ii) It is on the unit circle. (iii) It is also on the unit circle. (iv) It is on or inside the circle of radius two.

5.5.3 $\bar{x} = 2 - i$, $x\bar{x} = 5$, $xy = -1 + 7i$, $1/x = 2/5 - (1/5)i$, $x/y = 1/2 - (1/2)i$; $|xy| = \sqrt{50} = |x||y|$, $|1/x| = 1/\sqrt{5} = 1/|x|$.

5.5.4 $\sqrt{3}/2 + (1/2)i$, $1/2 + (\sqrt{3}/2)i$, i.

5.5.5 (a) $x^2 = r^2e^{i2\theta}$, $x^{-1} = (1/r)e^{-i\theta}$, $\bar{x} = re^{-i\theta}$; they are on the unit circle.

5.5.8 (i) $\begin{bmatrix} 1 & i & 0 \\ i & 0 & 1 \end{bmatrix} \rightarrow \begin{bmatrix} 1 & i & 0 \\ 0 & 1 & 1 \end{bmatrix} = U$; $Ax = 0$ if x is a multiple of $\begin{bmatrix} i \\ -1 \\ 1 \end{bmatrix}$; this vector is orthogonal not to the columns of A^T (rows of A) but to the columns of A^H.

5.5.10 (a) $n(n+1)/2$, n, $n(n-1)/2$ (b) The 3 entries on the diagonal are real. The 3 entries above the diagonal have 6 real degrees of freedom. The 3 below are the same, so the total real degrees of freedom for Hermitian matrices is 9. For unitary matrices it is also $9 - $ Re u_{11}, Im u_{11}, Re u_{21}, Im u_{21}, Re u_{31}, Re u_{12}, Im u_{12}, Re u_{22}, Re u_{13}. It looks as if $U\Lambda U^H$ has $9 + 3 = 12$ degrees of freedom but 3 are used to multiply the columns (eigenvectors) by arbitrary $e^{i\theta}$ and disappear in the triple product.

5.5.11 P: $\lambda_1 = 0$, $\lambda_2 = 1$, $x_1 = \begin{bmatrix} 1/\sqrt{2} \\ -1/\sqrt{2} \end{bmatrix}$, $x_2 = \begin{bmatrix} 1/\sqrt{2} \\ 1/\sqrt{2} \end{bmatrix}$; Q: $\lambda_1 = 1$, $\lambda_2 = -1$,

$x_1 = \begin{bmatrix} 1/\sqrt{2} \\ 1/\sqrt{2} \end{bmatrix}$, $x_2 = \begin{bmatrix} 1/\sqrt{2} \\ -1/\sqrt{2} \end{bmatrix}$; R: $\lambda_1 = 5$, $\lambda_2 = -5$, $x_1 = \begin{bmatrix} 2/\sqrt{5} \\ 1/\sqrt{5} \end{bmatrix}$, $x_2 = \begin{bmatrix} 1/\sqrt{5} \\ -2/\sqrt{5} \end{bmatrix}$.

5.5.13 (a) They are orthogonal to each other.   (b) The nullspace is spanned by u; the left nullspace is the same as the nullspace; the row space is spanned by v and w; the column space is the same as the row space.   (c) $x = v + \frac{1}{2}w$; no, we can add any multiple of u to x.   (d) $b^T u = 0$.   (e) $S^{-1} = S^T$; $S^{-1}AS = \text{diag}(0, 1, 2)$.

5.5.14 A is orthogonal, invertible, permutation, diagonalizable and Markov. B is projection, Hermitian, rank one, diagonalizable. The eigenvalues of A are ± 1 and $\pm i$. The eigenvalues of B are $0, 0, 0, 4$.

5.5.15 The dimension of S is $n(n + 1)/2$, not n. Every symmetric matrix A is a combination of n projections, but the projections change as A changes. There is no basis of n fixed projection matrices.

5.5.18 The determinant is the product $\lambda_1 \cdots \lambda_n$ with absolute value $|\lambda_1| \cdots |\lambda_n| = 1$ (or $|\det U|^2 = \det U^H U = \det I = 1$). The 1 by 1 $U = [i]$ has det $U = i$; in general $U = \begin{bmatrix} a \cos \theta & b \sin \theta \\ c \sin \theta & d \cos \theta \end{bmatrix}$, where $|a| = |b| = |c| = |d| = 1$ and $\bar{a}b + \bar{c}d = 0$.

5.5.19 The third column can be $(1, -2, i)/\sqrt{6}$, multiplied by any number (real or complex) with absolute value one.

5.5.20 $\Lambda = \begin{bmatrix} 0 & \\ & 2i \end{bmatrix}$, $S = \begin{bmatrix} 1 & 1 \\ -1 & 1 \end{bmatrix}$, $e^{Kt} = Se^{\Lambda t}S^{-1} = \frac{1}{2}\begin{bmatrix} 1 + e^{2it} & -1 + e^{2it} \\ -1 + e^{2it} & 1 + e^{2it} \end{bmatrix}$; at $t = 0$, $\dfrac{de^{Kt}}{dt} = \begin{bmatrix} i & i \\ i & i \end{bmatrix} = K$.

5.5.24 $C = \begin{bmatrix} c_0 & c_1 & c_2 & c_3 \\ c_3 & c_0 & c_1 & c_2 \\ c_2 & c_3 & c_0 & c_1 \\ c_1 & c_2 & c_3 & c_0 \end{bmatrix}$; $Cx = \begin{bmatrix} c_0 x_0 + c_1 x_1 + c_2 x_2 + c_3 x_3 \\ c_3 x_0 + c_0 x_1 + c_1 x_2 + c_2 x_3 \\ c_2 x_0 + c_3 x_1 + c_0 x_2 + c_1 x_3 \\ c_1 x_0 + c_2 x_1 + c_3 x_2 + c_0 x_3 \end{bmatrix}$.

5.6.1 $C = N^{-1}BN = N^{-1}M^{-1}AMN = (MN)^{-1}A(MN)$; the identity matrix.

5.6.2 Those matrices have eigenvalues 1 and -1 with trace 0 and determinant -1, e.g. $\begin{bmatrix} 1 & 0 \\ 2 & -1 \end{bmatrix}$, $\begin{bmatrix} 0 & 1 \\ 1 & 0 \end{bmatrix}$.

5.6.4 $M = \begin{bmatrix} -1 & & & \\ & 1 & & \\ & & -1 & \\ & & & 1 \end{bmatrix}$, $A = MBM^{-1}$.

5.6.6 (a) $CD = -DC \Rightarrow C = D(-C)D^{-1}$ (b) If $\lambda_1, \ldots, \lambda_n$ are eigenvalues of C, then $-\lambda_1, \ldots, -\lambda_n$ are eigenvalues of $-C$. By (a), C and $-C$ have the same eigenvalues. Therefore, $\lambda_1, \ldots, \lambda_n$ must come in plus-minus pairs.   (c) $Cx = \lambda x \Rightarrow DCx = \lambda Dx \Rightarrow -C(Dx) = \lambda(Dx) \Rightarrow C(Dx) = -\lambda(Dx)$.

5.6.7 The $(3, 1)$ entry is $g \cos \theta + h \sin \theta$, which is zero if $\tan \theta = -g/h$.

5.6.8 $M = \begin{bmatrix} 1 & 0 \\ -1 & 1 \end{bmatrix}$.

5.6.10 $c_1 V_1 + c_2 V_2 = (m_{11}c_1 + m_{12}c_2)v_1 + (m_{21}c_1 + m_{22}c_2)v_2 = d_1 v_1 + d_2 v_2$.

5.6.11 Its matrix with respect to v_1 and v_2 is $A = \begin{bmatrix} 0 & 1 \\ 1 & 0 \end{bmatrix}$. Its matrix with respect to V_1 and V_2 is $B = \begin{bmatrix} 1 & 0 \\ 0 & -1 \end{bmatrix}$. Let $M = \begin{bmatrix} 1 & 1 \\ 1 & -1 \end{bmatrix}$, $A = MBM^{-1}$.

5.6.13 (a) $D = \begin{bmatrix} 0 & 1 & 0 \\ 0 & 0 & 2 \\ 0 & 0 & 0 \end{bmatrix}$ (b) $D^3 = \begin{bmatrix} 0 & 0 & 0 \\ 0 & 0 & 0 \\ 0 & 0 & 0 \end{bmatrix}$; D^3 is the third derivative matrix. The third derivatives of 1, x and x^2 are zero, so $D^3 = 0$. (c) $\lambda = 0$ (triple); only one eigenvector $(1, 0, 0)$.

5.6.14 $f = e^{\lambda x}$ is an eigenvector of d/dx with eigenvalue λ. If $\int_0^x f(t)\, dt = \lambda f(x)$ then differentiating both sides forces $f(x) = \lambda f'(x)$ and $f(x) = ce^{x/\lambda}$, but integrating from 0 to x gives $c\lambda(e^{x/\lambda} - 1) \neq \lambda f(x)$.

5.6.15 The eigenvalues are 1 (triple) and -1. The eigenmatrices are $\begin{bmatrix} 1 & 0 \\ 0 & 0 \end{bmatrix}$, $\begin{bmatrix} 0 & 1 \\ 1 & 0 \end{bmatrix}$, $\begin{bmatrix} 0 & 0 \\ 0 & 1 \end{bmatrix}$ and $\begin{bmatrix} 0 & 1 \\ -1 & 0 \end{bmatrix}$.

5.6.17 (i) $TT^H = U^{-1}AUU^HA^H(U^{-1})^H = I$ (ii) If T is triangular and unitary, then its diagonal entries (the eigenvalues) must have absolute value one. Then all off-diagonal entries are zero because the columns are to be unit vectors.

5.6.20 $\|Nx\|^2 = x^HN^HNx = x^HNN^Hx = \|N^Hx\|^2$; with $x = e_i$ this gives $\|$column $i\| = \|$row $i\|$.

5.6.22 $U = \begin{bmatrix} 1/\sqrt{2} & 1/\sqrt{2} \\ 1/\sqrt{2} & -1/\sqrt{2} \end{bmatrix}$, $T = \begin{bmatrix} 2 & 7 \\ 0 & 1 \end{bmatrix}$; $U = \begin{bmatrix} 0 & 1 & 0 \\ 0 & 0 & 1 \\ 1 & 0 & 0 \end{bmatrix}$, $T = \begin{bmatrix} 0 & 1 & 0 \\ 0 & 0 & 1 \\ 0 & 0 & 0 \end{bmatrix}$.

5.6.23 The eigenvalues of $A(A - I)(A - 2I)$ are 0, 0, 0.

5.6.24 From the diagonal zeros in $T - \lambda_1 I$, $T - \lambda_2 I$, $T - \lambda_3 I$, the product $P(T)$ of those matrices is zero. Then $A - \lambda I = U(T - \lambda I)U^{-1}$, so $(A - \lambda_1 I)(A - \lambda_2 I)(A - \lambda_3 I) = U(T - \lambda_1 I)(T - \lambda_2 I)(T - \lambda_3 I)U^{-1} = U0U^{-1} = 0$.

5.6.27 The matrix has only one eigenvector, so its Jordan form has only one block:

$$J = \begin{bmatrix} 0 & 1 & & \\ & 0 & 1 & \\ & & 0 & 1 \\ & & & 0 \end{bmatrix}.$$

5.6.28 $M^{-1}J_3M = 0$, so the last two inequalities are easy; $MJ_1 = J_2M$ forces the first column of M to be zero, so it cannot be invertible.

5.6.30 $J^{10} = \begin{bmatrix} 2^{10} & 10 \cdot 2^9 \\ 0 & 2^{10} \end{bmatrix}$; $A^{10} = 2^{10}\begin{bmatrix} 61 & 45 \\ -80 & -59 \end{bmatrix}$; $e^A = e^2\begin{bmatrix} 13 & 9 \\ -16 & -11 \end{bmatrix}$.

5.6.31 $J_1 = 4$ by 4 block; $J_2 = 3$ by 3 and 1 by 1; $J_3 = 2$ by 2 and 2 by 2; $J_4 = 2$ by 2, 1 by 1, 1 by 1; $J_5 = $ four 1 by 1 blocks $= 0$. Both J_2 and J_3 have two eigenvectors.

CHAPTER 6

6.1.1 $ac - b^2 = 2 - 4 = -2 < 0$; $f = (x + 2y)^2 - 2y^2$.

6.1.2 (a) No (b) No (c) Yes (d) No. In (b), f is zero along the line $x = y$.

6.1.3 $\det(A - \lambda I) = 0 \Rightarrow \lambda^2 - (a + c)\lambda + ac - b^2 = 0 \Rightarrow \lambda_1 = ((a + c) + \sqrt{(a - c)^2 + b^2})/2$, $\lambda_2 = ((a + c) - \sqrt{(a - c)^2 + 4b^2})/2$; $\lambda_1 > 0$ is obvious, $\lambda_2 > 0$ because $(a + c)^2 > (a - c)^2 + 4b^2$ is equivalent to $ac > b^2$.

6.1.4 Second derivative matrices $\begin{bmatrix} 4 & -5 \\ -5 & 12 \end{bmatrix}$ (minimum), $\begin{bmatrix} -2 & 0 \\ 0 & -2 \end{bmatrix}$ (maximum).

6.1.5 (a) $-3 < b < 3$ (b) $\begin{bmatrix} 1 & 0 \\ b & 1 \end{bmatrix}\begin{bmatrix} 1 & 0 \\ 0 & 9-b^2 \end{bmatrix}\begin{bmatrix} 1 & b \\ 0 & 1 \end{bmatrix}$ (c) $-\dfrac{1}{2(9-b^2)}$

(d) No minimum.

6.1.6 No; $\begin{bmatrix} 4 & 2 \\ 2 & \frac{1}{2} \end{bmatrix}$.

6.1.7 (a) $\begin{bmatrix} 1 & -1 & -1 \\ -1 & 1 & 1 \\ -1 & 1 & 1 \end{bmatrix}$ and $\begin{bmatrix} 1 & -1 & -1 \\ -1 & 2 & -2 \\ -1 & -2 & 3 \end{bmatrix}$. (b) $f_1 = (x_1 - x_2 - x_3)^2 = 0$ when

$x_1 - x_2 - x_3 = 0$. (c) $f_2 = (x_1 - x_2 - x_3)^2 + (x_2 - 3x_3)^2 - 7x_3^2$;

$L = \begin{bmatrix} 1 & 0 & 0 \\ -1 & 1 & 0 \\ -1 & -3 & 1 \end{bmatrix}$, $D = \text{diag}(1, 1, -7)$.

6.1.8 A is positive definite $\Leftrightarrow \lambda_1 > 0, \lambda_2 > 0 \Leftrightarrow 1/\lambda_1 > 0, 1/\lambda_2 > 0 \Leftrightarrow A^{-1}$ is positive definite. Direct test: $A_{11}^{-1} = c/(ac - b^2) > 0$ and $\det A^{-1} = 1/\det A > 0$.

6.1.9 $A = \begin{bmatrix} 3 & 6 \\ 6 & 16 \end{bmatrix} = \begin{bmatrix} 1 & 0 \\ 2 & 1 \end{bmatrix}\begin{bmatrix} 3 & 0 \\ 0 & 4 \end{bmatrix}\begin{bmatrix} 1 & 2 \\ 0 & 1 \end{bmatrix}$; the coefficients of the squares are the pivots in D, while the coefficients inside the squares are columns of L.

6.1.10 $R^2 = \begin{bmatrix} p^2 + s^2 & s(p+t) \\ s(p+t) & s^2 + t^2 \end{bmatrix}$ is positive definite $\Leftrightarrow (p^2 + s^2)(s^2 + t^2) > s^2(p+t)^2 \Leftrightarrow$

$(s^2 - pt)^2 > 0 \Leftrightarrow s^2 - pt \neq 0 \Leftrightarrow R$ is nonsingular.

6.1.11 (a) Its pivots are a and $c - |b|^2/a$ and $\det A = ac - |b|^2$. (b) $f = |x_1 + (b/a)x_2|^2 + (c - |b|^2/a)|x_2|^2$.

6.1.12 The second derivative matrix is $\begin{bmatrix} 2 & 4 \\ 4 & 2 \end{bmatrix}$, so F doesn't have a minimum at $(1, 1)$.

6.1.13 $a > 1$ and $(a - 1)(c - 1) > b^2$.

6.2.1 $a > 1$; there is no b that makes B positive definite.

6.2.2 B and C are positive definite, A is not.

6.2.3 $\det A = -2b^3 - 3b^2 + 1 < 0$ at $b = \frac{2}{3}$.

6.2.4 If A has positive eigenvalues λ_i then the eigenvalues of A^2 are λ_i^2 and the eigenvalues of A^{-1} are $1/\lambda_i$, also all positive; by test II, A^2 and A^{-1} are positive definite.

6.2.5 If $x^T Ax > 0$ and $x^T Bx > 0$ for any $x \neq 0$, then $x^T(A + B)x > 0$; condition (I).

6.2.6 $R = \dfrac{1}{\sqrt{5}}\begin{bmatrix} 5 & 4 \\ 0 & 3 \end{bmatrix}$; $R = \dfrac{1}{\sqrt{2}}\begin{bmatrix} 1 & -1 \\ 3 & 3 \end{bmatrix}$; $R = \begin{bmatrix} 2 & 1 \\ 1 & 2 \end{bmatrix}$.

6.2.7 Because $\Lambda > 0$. $R = \dfrac{1}{2}\begin{bmatrix} 1 + \sqrt{3} & -1 + \sqrt{3} \\ -1 + \sqrt{3} & 1 + \sqrt{3} \end{bmatrix}$; $R = \begin{bmatrix} 3 & -1 \\ -1 & 3 \end{bmatrix}$.

6.2.8 If $x^T Ax > 0$ for all $x \neq 0$, then $x^T C^T ACx = (Cx)^T A(Cx) > 0$ (C is nonsingular so $Cx \neq 0$).

6.2.9 $|x^T Ay|^2 = |x^T R^T Ry|^2 = |(Rx)^T Ry|^2 \leq \|Rx\|^2\|Ry\|^2 = (x^T R^T Rx)(y^T R^T Ry) = (x^T Ax)(y^T Ay)$.

6.2.10 $\lambda_1 = 1, \lambda_2 = 4$, semiaxes go to $(\pm 1, 0)$ and $(0, \pm \frac{1}{2})$.

6.2.11 $A = \begin{bmatrix} 3 & -\sqrt{2} \\ -\sqrt{2} & 2 \end{bmatrix}$ with $\lambda = 1$ and 4, $\left(\dfrac{1}{\sqrt{3}}u + \dfrac{\sqrt{2}}{\sqrt{3}}v\right)^2 + 4\left(\dfrac{\sqrt{2}}{\sqrt{3}}u - \dfrac{1}{\sqrt{3}}v\right)^2 = 1$.

6.2.12 One zero eigenvalue pulls the ellipsoid into an infinite cylinder $\lambda_1 y_1^2 + \lambda_2 y_2^2 = 1$ along the third axis; two zero eigenvalues leave only the two planes $y_1 = \pm 1/\sqrt{\lambda_i}$; three zero eigenvalues leave $0 = 1$ (no graph).

6.2.13 (I) $x^T A x < 0$ for all nonzero vectors x. (II) All the eigenvalues of A satisfy $\lambda_i < 0$. (III) $\det A_1 < 0$, $\det A_2 > 0$, $\det A_3 < 0$. (IV) All the pivots (without row exchanges) satisfy $d_i < 0$. (V) There is a matrix R with independent columns such that $A = -R^T R$.

6.2.14 If the ith diagonal entry is zero, let $x = e_i$, then $x^T A x = 0$. So it is impossible that A is negative definite.

6.2.15 B is positive definite, C is negative definite, A and D are indefinite. There is a real solution because the quadratic takes negative values and x can be scaled to give -1.

6.2.16 F; T; T; T.

6.2.17 $\det A = a_{11} A_{11} + \cdots$. If A is positive definite, then $A_{11} > 0$. As a_{11} is increased, $a_{11} A_{11}$ is increased while the others don't change $\Rightarrow \det A$ is increased.

6.2.18 $a_{jj} = $ (row j of R^T)(column j of R) = length squared of column j of R; $\det A = (\det R)^2 = $ (volume of R parallelepiped)$^2 \le$ product of the lengths squared of the columns of $R = a_{11} a_{22} \cdots a_{nn}$.

6.2.19 $x^H A M x + x^H M^H A x = -x^H x \Rightarrow \lambda x^H A x + (\lambda x)^H A x = -x^H x \Rightarrow (\lambda + \bar{\lambda}) x^H A x = -x^H x \Rightarrow \lambda + \bar{\lambda} = 2 \operatorname{Re} \lambda < 0$.

6.3.1 $2(x_1 - \frac{1}{2}x_2 - \frac{1}{2}x_3)^2 + \frac{3}{2}(x_2 - x_3)^2$; $(x_1 + x_2 + x_3)^2$.

6.3.2 A is indefinite, B is positive semidefinite.

6.3.3 A and $C^T A C$ have $\lambda_1 > 0$, $\lambda_2 = 0$. $C(t) = tQ + (1-t)QR$, $Q = \begin{bmatrix} 1 & 0 \\ 0 & -1 \end{bmatrix}$, $R = \begin{bmatrix} 2 & 0 \\ 0 & 1 \end{bmatrix}$; C has one positive and one negative eigenvalue, but the identity matrix has two positive eigenvalues.

6.3.4 No.

6.3.5 The pivots of $A - \frac{1}{2}I$ are 2.5, 5.9, -0.81, so one eigenvalue is negative $\Rightarrow A$ has an eigenvalue smaller than $1/2$.

6.3.6 (b) $p + q \le n$ because more than n independent vectors is impossible.

6.3.7 $\operatorname{rank}(C^T A C) \le \operatorname{rank} A$, $\operatorname{rank}(C^T A C) \ge \operatorname{rank}((C^T)^{-1} C^T A C C^{-1}) = \operatorname{rank} A \Rightarrow \operatorname{rank}(C^T A C) = \operatorname{rank} A$.

6.3.8 A has $\frac{1}{2}n$ positive eigenvalues and $\frac{1}{2}n$ negative eigenvalues.

6.3.9 No.

6.3.10 $\begin{bmatrix} \sqrt{3} - 1 & 1 \end{bmatrix} \begin{bmatrix} 1 & 0 \\ 0 & 2 \end{bmatrix} \begin{bmatrix} 1 + \sqrt{3} \\ -1 \end{bmatrix} = 0$; $u_0 = \begin{bmatrix} -1 \\ 1 \end{bmatrix}$, $a_1 = \frac{1}{2}$, $a_2 = -\frac{1}{2}$; the smaller mass reaches as far as $\sqrt{3}$ although the larger mass never exceeds its initial displacement 1.

6.3.11 $\lambda_1 = 54$, $\lambda_2 = 54/5$, $x_1 = \begin{bmatrix} 1 \\ -1 \end{bmatrix}$, $x_2 = \begin{bmatrix} 1 \\ 1 \end{bmatrix}$.

6.3.12 $A = \begin{bmatrix} 1 & 0 \\ 0 & -1 \end{bmatrix}$, $M = \begin{bmatrix} 0 & 1 \\ 1 & 0 \end{bmatrix}$, $|A - \lambda M| = -\lambda^2 - 1 = 0 \Rightarrow \lambda = \pm i$.

6.4.1 $P = x_1^2 - x_1 x_2 + x_2^2 - x_2 x_3 + x_3^2 - 4x_1 - 4x_3$; $\partial P / \partial x_1 = 2x_1 - x_2 - 4$, $\partial P / \partial x_2 = -x_1 + 2x_2 - x_3$, $\partial P / \partial x_3 = -x_2 + 2x_3 - 4$.

6.4.2 Constant $= -\frac{1}{2}b^T A^{-1} b$. Because A is symmetric positive definite.

6.4.3 $\partial P_1/\partial x = x + y = 0$, $\partial P_1/\partial y = x + 2y - 3 = 0 \Rightarrow x = -3$, $y = 3$. P_2 has no minimum (let $y \to \infty$). It is associated with the semidefinite matrix $\begin{bmatrix} 1 & 0 \\ 0 & 0 \end{bmatrix}$.

6.4.4   Minimizing Q leads to the normal equations $A^T A x = A^T b$.

6.4.5   Put $x = (1, \ldots, 1)$ in Rayleigh's quotient (the denominator becomes n), then $n\lambda_1 \leq$ sum of all $a_{ij} \leq n\lambda_n$.

6.4.6   The minimum value is $R(x) = 1$, with $x = (1, 1)$.

6.4.7   Since $x^T B x > 0$ for all nonzero vectors x, $x^T(A + B)x$ will be larger than $x^T A x$.

6.4.8   If $(A + B)x = 0_1 x$, then $\lambda_1 + \mu_1 \leq \dfrac{x^T A x}{x^T x} + \dfrac{x^T B x}{x^T x} = 0_1$.

6.4.9 $\lambda_2(A + B) = \min_{S_2}\left[\max_{x \text{ in } S_2} \dfrac{x^T(A + B)x}{x^T x}\right] \geq \min_{S_2}\left[\max_{x \text{ in } S_2} \dfrac{x^T A x}{x^T x}\right] = \lambda_2(A)$.

6.4.10 $\lambda_1 \leq \mu \leq \lambda_3$.   **6.4.11** $\frac{1}{2}$; $(3 - \sqrt{3})/4$.

6.4.12 $\lambda_1 y_1^2 + \cdots + \lambda_n y_n^2 \leq \lambda_n(y_1^2 + \cdots + y_n^2) \Rightarrow R(x) \leq \lambda_n$.

6.4.13   (a) $\lambda_j = \min_{S_j}[\max_{x \text{ in } S_j} R(x)] > 0 \Rightarrow$ for all S_j, $\max_{x \text{ in } S_j} R(x) > 0 \Rightarrow S_j$ contains a vector x with $R(x) > 0$.   (b) $y = C^{-1}x \Rightarrow x = Cy \Rightarrow \bar{R}(y) = \dfrac{y^T C^T A C y}{y^T y} = \dfrac{x^T A x}{x^T x} = R(x) > 0$.

6.4.14 $x = e_1$, $\lambda_1 = \min_x \dfrac{x^T A x}{x^T M x} \leq \dfrac{a_{11}}{m_{11}}$.

6.4.15   The one spanned by the eigenvectors x_1 and x_2.

6.4.16   One positive, one negative and $n - 2$ zeros.

6.5.1 $A = 4\begin{bmatrix} 2 & -1 & 0 \\ -1 & 2 & -1 \\ 0 & -1 & 2 \end{bmatrix}$, $b = \begin{bmatrix} \frac{1}{2} \\ \frac{1}{2} \\ \frac{1}{2} \end{bmatrix}$, $Ay = b \Rightarrow y = \begin{bmatrix} \frac{3}{16} \\ \frac{1}{16} \\ \frac{3}{16} \end{bmatrix}$, $U = \frac{3}{16}V_1 + \frac{1}{16}V_2 + \frac{3}{16}V_3 \Rightarrow$ at nodes $x = \frac{1}{4}, \frac{1}{2}, \frac{3}{4}$, $U = u$.

6.5.2 $-u'' = x$, $u(0) = u(1) = 0 \Rightarrow u = -\frac{1}{6}x^3 + \frac{1}{6}x$; $A = 3\begin{bmatrix} 2 & -1 \\ -1 & 2 \end{bmatrix}$, $b = \begin{bmatrix} \frac{1}{9} \\ \frac{2}{9} \end{bmatrix}$, $Ay = b$

$\Rightarrow y = \begin{bmatrix} \frac{4}{81} \\ \frac{5}{81} \end{bmatrix}$, $U = \frac{4}{81}V_1 + \frac{5}{81}V_2 = \begin{cases} \frac{4}{27}x & 0 \leq x \leq \frac{1}{3} \\ \frac{1}{27}x + \frac{1}{27} & \frac{1}{3} \leq x \leq \frac{2}{3} \\ \frac{5}{27} - \frac{5}{27}x & \frac{2}{3} \leq x \leq 1 \end{cases}$; largest error at $x = \dfrac{\sqrt{7}}{3\sqrt{3}}$

6.5.3 $A_{33} = 3$, $b_3 = \frac{1}{3}$, $A = 3\begin{bmatrix} 2 & -1 & 0 \\ -1 & 2 & -1 \\ 0 & -1 & 1 \end{bmatrix}$, $b = \begin{bmatrix} \frac{2}{3} \\ \frac{2}{3} \\ \frac{1}{3} \end{bmatrix}$, $Ay = b \Rightarrow y = \begin{bmatrix} \frac{5}{9} \\ \frac{8}{9} \\ 1 \end{bmatrix}$.

6.5.4 $A = [\frac{16}{3}]$, $b = [2]$, $Ay = b \Rightarrow y = [\frac{3}{8}]$, $U = \frac{3}{8}V_1 = \begin{cases} \frac{3}{2}x & 0 \leq x \leq \frac{1}{4} \\ \frac{1}{2} - \frac{1}{2}x & \frac{1}{4} \leq x \leq 1 \end{cases}$.

6.5.5   Integrate by parts: $\int_0^1 -V_i'' V_j \, dx = \int_0^1 V_i' V_j' \, dx - [V_i' V_j]_{x=0}^{x=1} = \int_0^1 V_i' V_j' \, dx =$ same A_{ij}.

6.5.6 $1/3$ of V gives the smallest value of P.

6.5.7 $A = 4$, $M = \frac{1}{3}$, $\lambda = 12$ is larger than the true eigenvalue.

6.5.8 $M = \begin{bmatrix} \frac{2}{9} & \frac{1}{18} \\ \frac{1}{18} & \frac{2}{9} \end{bmatrix}$, see **6.3.11** for the rest.

6.5.9 $h/6$ times the 1, 4, 1 tridiagonal matrix.

CHAPTER 7

7.2.1   If Q is orthogonal, then $\|Q\| = \max \|Qx\|/\|x\| = 1$ because Q preserves length: $\|Qx\| = \|x\|$ for every x. Also Q^{-1} is orthogonal and has norm one, so $c(Q) = 1$.

7.2.2   The triangle inequality for vectors gives $\|Ax + Bx\| \le \|Ax\| + \|Bx\|$, and when we divide by $\|x\|$ and maximize each term, the result is the triangle inequality for matrix norms.

7.2.3 $\|ABx\| \le \|A\| \|Bx\|$, by the definition of the norm of A, and then $\|Bx\| \le \|B\| \|x\|$. Dividing by $\|x\|$ and maximizing, $\|AB\| \le \|A\| \|B\|$. The same is true for the inverse, $\|B^{-1}A^{-1}\| \le \|B^{-1}\| \|A^{-1}\|$; then $c(AB) \le c(A)c(B)$ by multiplying these two inequalities.

7.2.4 $\|A^{-1}\| = 1$, $\|A\| = 3$, $c(A) = 3$; take $b = x_2 = \begin{bmatrix} 1 \\ -1 \end{bmatrix}$, $\delta b = x_1 = \begin{bmatrix} 1 \\ 1 \end{bmatrix}$.

7.2.5   In the definition $\|A\| = \max \|Ax\|/\|x\|$, choose x to be the particular eigenvector in question; then $\|Ax\| = |\lambda| \|x\|$, and the ratio is $|\lambda| \Rightarrow$ maximum ratio is at least $|\lambda|$.

7.2.6 $A^TA = \begin{bmatrix} 1 & 100 \\ 100 & 10001 \end{bmatrix}$, $\lambda^2 - 10002\lambda + 1 = 0$, $\lambda_{\max} = 5001 + (5001^2 - 1)^{1/2}$. The norm is the square root, and is the same as $\|A^{-1}\|$.

7.2.7 A^TA and AA^T have the same eigenvalues (even if A were singular, which is a limiting case of the exercise). Actually, $A^TAx = \lambda x \Rightarrow AA^T(Ax) = A(A^TAx) = \lambda(Ax)$, and equality of the largest eigenvalues gives $\|A\| = \|A^T\|$.

7.2.8   Since $A = R^TR$ and $A^{-1} = R^{-1}(R^T)^{-1}$, we have $\|A\| = \|R\|^2$ and $\|A^{-1}\| = \|(R^T)^{-1}\|^2 = \|R^{-1}\|^2$. (From the previous exercise, the transpose has the same norm.) Therefore $c(A) = (c(R))^2$.

7.2.9 $A = \begin{bmatrix} 0 & 1 \\ 0 & 0 \end{bmatrix}$, $B = \begin{bmatrix} 0 & 0 \\ 1 & 0 \end{bmatrix}$, $\lambda_{\max}(A + B) > \lambda_{\max}(A) + \lambda_{\max}(B)$, (since $1 > 0 + 0$) and also $\lambda_{\max}(AB) > \lambda_{\max}(A)\lambda_{\max}(B)$.

7.2.10   If $x = \begin{bmatrix} 1 \\ -1 \end{bmatrix}$, then $Ax = \begin{bmatrix} -1 \\ 7 \end{bmatrix}$ and $\|Ax\|_\infty/\|x\|_\infty = 7$. This is the extreme case, and $\|A\|_\infty = 7 =$ largest absolute row sum (extreme x has components ± 1).

7.2.11 $B^2 = \begin{bmatrix} A^TA & 0 \\ 0 & AA^T \end{bmatrix}$ has eigenvalues σ_i^2 (and $|m - n|$ zeros). Therefore the eigenvalues of B are $\pm\sigma_i$ and $|m - n|$ zeros.

7.2.12   (a) Yes. (b) $A^{-1}b = x \Rightarrow \dfrac{\|\delta b\|}{\|b\|} \le c\dfrac{\|\delta x\|}{\|x\|} \Rightarrow \dfrac{\|\delta x\|}{\|x\|} \ge \dfrac{1}{c}\dfrac{\|\delta b\|}{\|b\|}$.

7.3.1 $u_0 = \begin{bmatrix} 1 \\ 0 \end{bmatrix}$, $u_1 = \begin{bmatrix} 2 \\ -1 \end{bmatrix}$, $u_2 = \begin{bmatrix} 5 \\ -4 \end{bmatrix}$, $u_3 = \begin{bmatrix} 14 \\ -13 \end{bmatrix}$; $u_\infty = \begin{bmatrix} -1 \\ -1 \end{bmatrix}$.

7.3.2 $u_0 = \begin{bmatrix} 3 \\ 4 \end{bmatrix}$, $u_1 = \frac{1}{3}\begin{bmatrix} 10 \\ 11 \end{bmatrix}$, $u_2 = \frac{1}{9}\begin{bmatrix} 31 \\ 32 \end{bmatrix}$, $u_3 = \frac{1}{27}\begin{bmatrix} 94 \\ 95 \end{bmatrix}$; with the shift, $\alpha = \frac{26}{25}$ and $u_1 = \frac{25}{49}\begin{bmatrix} 24 & 25 \\ 25 & 24 \end{bmatrix}\begin{bmatrix} 3 \\ 4 \end{bmatrix} \approx \begin{bmatrix} 172 \\ 171 \end{bmatrix}$.

7.3.3 $Hx = x - (x - y)\dfrac{2(x - y)^Tx}{(x - y)^T(x - y)} = x - (x - y) = y$. Then $H(Hx) = Hy$ is $x = Hy$.

7.3.4 $\sigma = 5,\ v = \begin{bmatrix} 8 \\ 4 \end{bmatrix},\ H = \dfrac{1}{5}\begin{bmatrix} -3 & -4 \\ -4 & 3 \end{bmatrix}.$

7.3.5 $U = \begin{bmatrix} 1 & 0 & 0 \\ 0 & -\frac{3}{5} & -\frac{4}{5} \\ 0 & -\frac{4}{5} & \frac{3}{5} \end{bmatrix} = U^{-1},\ U^{-1}AU = \begin{bmatrix} 1 & -5 & 0 \\ -5 & \frac{9}{25} & \frac{12}{25} \\ 0 & \frac{12}{25} & \frac{16}{25} \end{bmatrix}.$

7.3.6 $\begin{bmatrix} 2 & -1 \\ -1 & 2 \end{bmatrix} = Q_0 R_0 = \dfrac{1}{\sqrt{5}}\begin{bmatrix} 2 & 1 \\ -1 & 2 \end{bmatrix}\dfrac{1}{\sqrt{5}}\begin{bmatrix} 5 & -4 \\ 0 & 3 \end{bmatrix},\ A_1 = R_0 Q_0 = \dfrac{1}{5}\begin{bmatrix} 14 & -3 \\ -3 & 6 \end{bmatrix}.$

7.3.7 $\begin{bmatrix} \cos\theta & \sin\theta \\ \sin\theta & 0 \end{bmatrix} = QR = \begin{bmatrix} \cos\theta & -\sin\theta \\ \sin\theta & \cos\theta \end{bmatrix}\begin{bmatrix} 1 & \cos\theta\sin\theta \\ 0 & -\sin^2\theta \end{bmatrix},$

$RQ = \begin{bmatrix} c(1+s^2) & -s^3 \\ -s^3 & -s^2 c \end{bmatrix}.$

7.3.8 A is orthogonal, so $Q = A$, $R = I$, and $RQ = A$ again.

7.3.9   Assume that $(Q_0 \cdots Q_{k-1})(R_{k-1} \cdots R_0)$ is the QR factorization of A^k, which is certainly true if $k = 1$. By construction $A_{k+1} = R_k Q_k$, or $R_k = A_{k+1} Q_k^T = Q_k^T \cdots Q_0^T A Q_0 \cdots Q_k Q_k^T$. Postmultiplying by $(R_{k-1} \cdots R_0)$, and using the assumption, we have $R_k \cdots R_0 = Q_k^T \cdots Q_0^T A^{k+1}$; after moving the Q's to the left side, this is the required result for A^{k+1}.

7.4.1 $D^{-1}(-L-U) = \begin{bmatrix} 0 & \frac{1}{2} & 0 \\ \frac{1}{2} & 0 & \frac{1}{2} \\ 0 & \frac{1}{2} & 0 \end{bmatrix},\ \mu = 0,\ \pm 1/\sqrt{2};\ (D+L)^{-1}(-U) = \begin{bmatrix} 1 & \frac{1}{2} & 0 \\ 0 & \frac{1}{4} & \frac{1}{2} \\ 0 & \frac{1}{8} & \frac{1}{4} \end{bmatrix},$ eigen-

values $0, 0, 1/2$; $\omega_{opt} = 4 - 2\sqrt{2}$, with $\lambda_{max} = 3 - 2\sqrt{2} \approx 0.2$.

7.4.2 J has entries $\frac{1}{2}$ along the diagonals adjacent to the main diagonal, and zeros elsewhere; $Jx_1 = \frac{1}{2}(\sin 2\pi h, \sin 3\pi h + \sin \pi h, \ldots) = (\cos \pi h)x_1$.

7.4.3 $Ax_k = (2 - 2\cos k\pi h)x_k$; $Jx_k = \frac{1}{2}(\sin 2k\pi h, \sin 3k\pi h + \sin k\pi h, \ldots) = (\cos k\pi h)x_k$.

7.4.4   The circle around a_{ii} cannot reach zero if its radius r_i is less than $|a_{ii}|$; therefore zero is not an eigenvalue, and a diagonally dominant matrix cannot be singular.

7.4.5 $J = -\begin{bmatrix} 0 & \frac{1}{3} & \frac{1}{3} \\ 0 & 0 & \frac{1}{4} \\ \frac{2}{5} & \frac{2}{5} & 0 \end{bmatrix}$; the radii are $r_1 = \dfrac{2}{3}$, $r_2 = \dfrac{1}{4}$, $r_3 = \dfrac{4}{5}$, the circles have centers at

zero, so all $|\lambda_i| < 1$.

7.4.7 $-D^{-1}(L+U) = \begin{bmatrix} 0 & -b/a \\ -c/d & 0 \end{bmatrix},\ \mu = \pm\left(\dfrac{bc}{ad}\right)^{1/2};\ -(D+L)^{-1}U = \begin{bmatrix} 0 & -b/a \\ 0 & bc/ad \end{bmatrix},$

$\lambda = 0,\ bc/ad;\ \lambda_{max} = \mu_{max}^2.$

CHAPTER 8

8.1.1   The corners are at $(0, 6)$, $(2, 2)$, $(6, 0)$; see Fig. 8.4.

8.1.2 $x + y$ is minimized at $(2, 2)$, with cost 4; $3x + y$ is minimized at $(0, 6)$, with cost 6; the minimum of $x - y$ is $-\infty$, with $x = 0$, $y \to \infty$.

8.1.3   The constraints give $3(2x + 5y) + 2(-3x + 8y) \le 9 - 10$, or $31y \le -1$, which contradicts $y \ge 0$.

8.1.4   Take x and y to be equal and very large.

8.1.5 $x \geq 0$, $y \geq 0$, $x + y \leq 0$ admits only the point $(0, 0)$.

8.1.6   The feasible set is an equilateral triangle lying on the plane $x + y + z = 1$, with corners at $(x, y, z) = (1, 0, 0)$, $(0, 1, 0)$, $(0, 0, 1)$; the last corner gives the maximum value 3.

8.1.7 $x = z = 20,000$; $y = 60,000$.

8.2.1   At present $x_4 = 4$ and $x_5 = 2$ are in the basis, and the cost is zero. The entering variable should be x_3, to reduce the cost. The leaving variable should be x_5, since $2/1$ is less than $4/1$. With x_3 and x_4 in the basis, the constraints give $x_3 = 2$, $x_4 = 2$, and the cost is now $x_1 + x_2 - x_3 = -2$.

8.2.2   Isolating x_3 in the second constraint gives $x_3 = 2 - 3x_1 - 5x_2 - x_5$. Substituting this into the cost function and the first constraint, the problem at the new corner is: Minimize $-2 + 4x_1 + 6x_2 + x_5$. Since the coefficients 4, 6, and 1 are positive, the minimum occurs where $x_1 = x_2 = x_5 = 0$; we are already at the optimal corner. In other words, the stopping test $r = (4, 6, 1) \geq 0$ is passed, and -2 is the minimum cost.

8.2.3 $r = \begin{bmatrix} 1 & 1 \end{bmatrix}$, so the corner is optimal.

8.2.4 $\begin{bmatrix} B & N & b \\ c_B & c_N & 0 \end{bmatrix} = \begin{bmatrix} -1 & 2 & 1 & 0 & 6 \\ 0 & 1 & 2 & -1 & 6 \\ 0 & -1 & 1 & 0 & 0 \end{bmatrix} \rightarrow \begin{bmatrix} 1 & 0 & 3 & -2 & 6 \\ 0 & 1 & 2 & -1 & 6 \\ 0 & 0 & 3 & -1 & 6 \end{bmatrix}$; $r =$ $\begin{bmatrix} 3 & -1 \end{bmatrix}$, so the second column of N enters the basis; that column is $u = \begin{bmatrix} 0 \\ -1 \end{bmatrix}$, and $B^{-1}u = \begin{bmatrix} -2 \\ -1 \end{bmatrix}$ is negative, so the edge is infinitely long and the minimal cost is $-\infty$.

8.2.5   At P, $r = \begin{bmatrix} -5 & 3 \end{bmatrix}$; at Q, $r = \begin{bmatrix} \frac{5}{3} & -\frac{1}{3} \end{bmatrix}$; at R, $r \geq 0$.

8.2.6   (a) The pair $x = 0$, $w = b$ is nonnegative, it satisfies $Ax + w = b$, and it is basic because $x = 0$ contributes n zero components.   (b) The auxiliary problem minimizes w_1, subject to $x_1 \geq 0$, $x_2 \geq 0$, $w_1 \geq 0$, $x_1 - x_2 + w_1 = 3$. Its Phase I vector is $x_1 = x_2 = 0$, $w_1 = 3$; its optimal vector is $x_1^* = 3$, $x_2^* = w_1^* = 0$. The corner is at $x_1 = 3$, $x_2 = 0$, and the feasible set is a line going up from this point with slope 1.

8.2.7   The stopping test becomes $r \leq 0$; if this fails, and the ith component is the largest, then that column of N enters the basis; the rule **8C** for the vector leaving the basis is the same.

8.2.8   At a corner, two of the constraints become equalities and the other three are satisfied. Therefore the corners are $x_1 = 3$, $x_2 = 3$ and $x_1 = 12$, $x_2 = 0$. The costs are 9 and 24, so the first corner is optimal.

8.2.9 $BE = B[\cdots v \cdots] = [\cdots u \cdots]$, since $Bv = u$.

8.2.10 x_2 should be increased to $x_2 = 2$ at which $x_4 = 0$ and $x = (0, 2, 14, 0)$. We reach the minimum $\frac{3}{2}x_1 + \frac{1}{6}x_4 - 2 = -2$ in one step.

8.2.11 $Ax = 0 \Rightarrow Px = x - A^{\mathrm{T}}(AA^{\mathrm{T}})^{-1}Ax = x$.

8.2.12 $P^2 = P$, $P^{\mathrm{T}} = P \Rightarrow -sc^{\mathrm{T}}Pc = -sc^{\mathrm{T}}P^{\mathrm{T}}Pc = -s(Pc)^{\mathrm{T}}Pc = -s\|Pc\|^2$.

8.2.13   (a) At $P = (3, 0, 0)$, $c^{\mathrm{T}}x = 15$; at $Q = (0, 3, 0)$, $c^{\mathrm{T}}x = 12$; at $R = (0, 0, 3)$, $c^{\mathrm{T}}x = 24$; so $c^{\mathrm{T}}x$ is minimized at $(0, 3, 0)$.   (b) $Pc = (-\frac{2}{3}, -\frac{5}{3}, \frac{7}{3})$, $s = \frac{3}{7}$.

8.2.14   When $s = \frac{3}{7}(0.98) = 0.42$, the step ends at $(1.28, 1.70, 0.02)$. $D = \begin{bmatrix} 1.28 & & \\ & 1.70 & \\ & & 0.02 \end{bmatrix}$.

8.2.15 $AD^2 A^T y = AD^2 c$ becomes $4.53y = 19.76$ or $y = 4.36$. $PDc = (0.82, -0.61, 0.07) \Rightarrow$
$s = 0.98/0.82 = 1.20 \Rightarrow X^2 = e - sPDc = (0.02, 1.73, 0.92) \Rightarrow x^2 = DX^2 =$
$(0.03, 2.94, 0.02)$.

8.3.1 Maximize $4y_1 + 11y_2$, with $y_1 \geq 0$, $y_2 \geq 0$, $2y_1 + y_2 \leq 1$, $3y_2 \leq 1$; $x_1^* = 2$, $x_2^* = 3$,
$y_1^* = \frac{1}{3}$, $y_2^* = \frac{1}{3}$, cost $= 5$.

8.3.2 Minimize $3x_1$, subject to $x_1 \geq 0$, $x_1 \geq 1$; $y_1^* = 0$, $y_2^* = 3$, $x_1^* = 1$, cost $= 3$.

8.3.3 The dual maximizes yb, with $yI \geq c$. Therefore $x = b$ and $y = c$ are feasible, and
give the same value cb for the cost in the primal and dual; by **8F** they must be
optimal. If $b_1 < 0$, then the optimal x^* is changed to $(0, b_2, \ldots, b_n)$ and $y^* =$
$(0, c_2, \ldots, c_n)$.

8.3.4 $A = [-1]$, $b = [1]$, $c = [0]$ is not feasible; the dual maximizes y, with $y \geq 0$, and
$-1y \leq 0$, and is unbounded.

8.3.5 $b = [0 \quad 1]^T$, $c = [-1 \quad 0]$.

8.3.6 If x is very large, then $Ax \geq b$ and $x \geq 0$; if $y = 0$, then $yA \leq c$ and $y \geq 0$. Thus
both are feasible.

8.3.7 Since $cx = 3 = yb$, x and y are optimal by **8F**.

8.3.8 $Ax = [1 \quad 1 \quad 3 \quad 1]^T \geq b = [1 \quad 1 \quad 1 \quad 1]^T$, with strict inequality in the third component; therefore the third component of y is forced to be zero. Similarly $yA =$
$[1 \quad 1 \quad 1 \quad 1] \leq c = [1 \quad 1 \quad 1 \quad 3]$, and the strict inequality forces $x_4 = 0$.

8.3.9 $x^* = [1 \quad 0]^T$, $y^* = [1 \quad 0]$, with $y^*b = 1 = cx^*$. The second inequalities in both
$Ax^* \geq b$ and $y^*A \leq c$ are strict, so the second components of y^* and x^* are zero.

8.3.10 $Ax = b$ gives $yAx = yb$, whether or not $y \geq 0$; $yA \leq c$ gives $yAx \leq cx$ because
$x \geq 0$. Comparing, $yb \leq cx$.

8.3.11 (a) $x_1^* = 0$, $x_2^* = 1$, $x_3^* = 0$, $c^T x = 3$ (b) It is the first quadrant with the tetrahedron
in the corner cut off. (c) Maximize y_1, subject to $y_1 \geq 0$, $y_1 \leq 5$, $y_1 \leq 3$, $y_1 \leq 4$;
$y_1^* = 3$.

8.3.12 The planes $cx =$ constant are slightly tilted, but the first corner of the feasible set
to be touched is the same as before.

8.3.13 As in Section 8.1, the dual maximizes $4p$ subject to $p \leq 2$, $2p \leq 3$, $p \geq 0$. The solution is $p = \$1.50$, the shadow price of protein (this is its price in steak, the optimal
diet). The reduced cost of peanut butter is $\$2 - \$1.50 = 50$ cents; it is positive and
peanut butter is not in the optimal diet.

8.3.14 The columns generate the cone between the positive x axis and the ray $x = y$. In
the first case $x = (1, 2)^T$: $y = (1, -1)$ satisfies the alternative.

8.3.15 The columns of $\begin{bmatrix} 1 & 0 & 0 & -1 & 0 & 0 \\ 0 & 1 & 0 & 0 & -1 & 0 \\ 0 & 0 & 1 & 0 & 0 & -1 \end{bmatrix}$ or $\begin{bmatrix} 1 & 0 & 0 & -1 \\ 0 & 1 & 0 & -1 \\ 0 & 0 & 1 & -1 \end{bmatrix}$.

8.3.16 Take $y = [2 \quad -1]$; then $yA = 0$, $yb \neq 0$.

8.3.17 Take $y = [1 \quad -1]$; then $yA \geq 0$, $yb < 0$.

8.3.18 $yA \geq 0$ gives $yAx \geq 0$; $Ax \geq b$ gives $yAx \leq yb < 0$.

8.4.1 The maximal flow is 13 with the minimal cut separating node 6 from the other
nodes.

8.4.2 The maximal flow is 8 with the minimal cut separating nodes 1–4 from nodes 5–6.

8.4.3   Increasing the capacity of pipes from node 4 to node 6 or node 4 to node 5 will produce the largest increase in the maximal flow. The maximal flow increases from 8 to 9.

8.4.4   The largest possible flow from node 1 to node 4 is 3.

8.4.5   Suppose the capacities of the edges are all 1. The maximum number of disjoint paths from s to t is the maximum flow. The minimum number of edges whose removal disconnects s from t is the minimum cut (because all edges have capacity 1). Then max flow = min cut.

8.4.6   A maximal set of marriages for A is 1–3, 2–1, 3–2, 4–4; a complete matching for B is 1–2, 2–4, 3–5, 4–3, 5–1.

8.4.7   Rows 1, 4 and 5; the submatrix coming from rows 1, 4, 5, and columns 1, 2, 5 has $3 + 3 > 5$.

8.4.8   4 lines are needed. The k marriages correspond to k 1's that are in different rows and columns. It takes k lines to cover those 1's. Therefore it takes at least k lines to cover all the 1's.

8.4.9   (a) The matrix has $2n$ 1's which cannot be covered by less than n lines because each line covers exactly 2 1's.

$$
\text{(b)}\begin{bmatrix}
1 & 1 & 1 & 1 & 1 \\
1 & 0 & 0 & 0 & 1 \\
1 & 0 & 0 & 0 & 1 \\
1 & 0 & 0 & 0 & 1 \\
1 & 1 & 1 & 1 & 1
\end{bmatrix}.
$$

8.4.10   It takes at least 3 lines to cover all the 15 1's.   **8.4.11**   It is obvious.

8.4.12   The shortest path from s to t is 1–2–5–6 or 1–3–5–6. One minimal spanning tree is 1–2–4–3–5–6.

8.4.13   1–3, 3–2, 2–5, 2–4, 4–6 and 2–5, 4–6, 2–4, 3–2, 1–3.

8.4.14   (a) The greedy algorithm starts with n separate trees. At each step, it chooses a tree T and adds a minimum-cost edge e. Suppose that at some step the existing trees (including T) are part of a minimum spanning tree S, but the new edge e (incident to T) is not part of S. Certainly S must contain another edge e' that connects a node inside T to a node outside T; otherwise S is not spanning. And certainly this edge e' is not shorter than e, which was a minimum-cost edge incident to T. Therefore we can remove e' from S and replace it by e without increasing the total length. S is still a minimum spanning tree.   (b) The shortest path does not necessarily contain the shortest edge! See, e.g., **8.4.2**.

8.4.15   (a) Rows 1, 3, 5   (b) Columns 1, 3, 5   (c) The submatrix coming from rows 1, 3, 5 and columns 1, 3, 5   (d) Rows 2, 4 and columns 2, 4.

8.4.16   Maximize x_{61} subject to $\begin{bmatrix} A & 0 \\ I & I \end{bmatrix}\begin{bmatrix} x \\ w \end{bmatrix} = \begin{bmatrix} 0 \\ C \end{bmatrix}, \begin{bmatrix} x \\ w \end{bmatrix} \geq 0.$

8.5.1 $-10x_1 + 70(1 - x_1) = 10x_1 - 10(1 - x_1)$, or $x_1 = \frac{4}{5}, x_2 = \frac{1}{5}; -10y_1 + 10(1 - y_1) = 70y_1 - 10(1 - y_1)$, or $y_1 = \frac{1}{5}, y_2 = \frac{4}{5}$; average payoff 6.

8.5.2 X can guarantee to win 2 by choosing always the second column; Y can guarantee to lose at most 2 by choosing the first row. Therefore $y^* = \begin{bmatrix} 1 & 0 \end{bmatrix}$ and $x^* = \begin{bmatrix} 0 & 1 \end{bmatrix}^{\mathrm{T}}$.

8.5.3 X can guarantee to win a_{ij} by choosing column j, since this is the smallest entry in that column. Y can lose at most a_{ij} by choosing row i, since this is the largest entry in that row. In the previous exercise $a_{12} = 2$ was an equilibrium of this kind, but if we exchange the 2 and 4 below it, no entry has this property and mixed strategies are required.

8.5.4   The intersection y^* occurs when $y + 3(1 - y) = 3y + 2(1 - y)$, so $y^* = \frac{1}{3}$. The height (minimax value) is $\frac{7}{3}$.

8.5.5   The best strategy for X combines the two lines to produce a horizontal line of height $\frac{7}{3}$, guaranteeing this amount. The combination is $\frac{2}{3}(3y + 2(1 - y)) + \frac{1}{3}(y + 3(1 - y)) = \frac{7}{3}$, so X plays the columns with frequencies $\frac{2}{3}, 0, \frac{1}{3}$.

8.5.6 $y^* = [\frac{2}{3} \ \frac{1}{3}]$, $x^* = [\frac{1}{2} \ 0 \ \frac{1}{2}]^T$, value zero.

8.5.8   If $x = (\frac{6}{11}, \frac{3}{11}, \frac{2}{11})$, then X will win $\frac{6}{11}$ against any strategy of Y: if $y = (\frac{6}{11}, \frac{3}{11}, \frac{2}{11})$, then Y loses $\frac{6}{11}$ against any strategy of X; this equilibrium solves the game.

8.5.9   The inner maximum is the larger of y_1 and y_2; we concentrate all of x on that one. Then minimizing the larger subject to $y_1 + y_2 = 1$, gives the answer $\frac{1}{2}$.

8.5.10   In (5), min $yAx^* \leq y^*Ax^*$ because the minimum over all y is not larger than the value for the particular y^*; similarly for $y^*Ax^* \leq$ max y^*Ax. If equality holds in (5), so that min $yAx^* = y^*Ax^*$, then for all y this is less or equal to yAx^*; that is the second half of (4), and the first half follows from max $y^*Ax = y^*Ax^*$.

8.5.11 $Ax^* = [\frac{1}{2} \ \frac{1}{2}]^T$ and $yAx^* = \frac{1}{2}y_1 + \frac{1}{2}y_2$, which equals $\frac{1}{2}$ for all strategies of Y; $y^*A = [\frac{1}{2} \ \frac{1}{2} \ -1 \ -1]$ and $y^*Ax = \frac{1}{2}x_1 + \frac{1}{2}x_2 - x_3 - x_4$, which cannot exceed $\frac{1}{2}$; in between is $y^*Ax^* = \frac{1}{2}$.

8.5.13   It is fair.   **8.5.14** $x^* = (\frac{7}{9}, \frac{2}{9})$, $y^* = (\frac{2}{3}, \frac{1}{3})$; the average gain is $\frac{10}{3}$.

APPENDIX A

A.1 $[0]_{m \times m} = I_{m \times n}[0]_{m \times n} I_{n \times n}$, $[0]^+_{m \times n} = [0]_{n \times m}$.

A.2 $A^+ = \begin{bmatrix} \frac{1}{4} \\ \frac{1}{4} \\ \frac{1}{4} \\ \frac{1}{4} \end{bmatrix}$, $B = \begin{bmatrix} 0 & 1 \\ 1 & 0 \end{bmatrix}\begin{bmatrix} 1 & 0 & 0 \\ 0 & 1 & 0 \end{bmatrix}\begin{bmatrix} 1 & 0 & 0 \\ 0 & 1 & 0 \\ 0 & 0 & 1 \end{bmatrix}$, $B^+ = \begin{bmatrix} 0 & 1 \\ 1 & 0 \\ 0 & 0 \end{bmatrix}$, $C^+ = \begin{bmatrix} \frac{1}{2} & 0 \\ \frac{1}{2} & 0 \end{bmatrix}$.

A.3   Its pseudoinverse is $Q^+ = Q^T$.

A.4   (a) $A^TA = \begin{bmatrix} 10 & 6 \\ 6 & 10 \end{bmatrix} = \frac{1}{2}\begin{bmatrix} 1 & -1 \\ 1 & 1 \end{bmatrix}\begin{bmatrix} 4 & 0 \\ 0 & 16 \end{bmatrix}\begin{bmatrix} 1 & 1 \\ -1 & 1 \end{bmatrix}$,

$S = \frac{1}{2}\begin{bmatrix} 1 & -1 \\ 1 & 1 \end{bmatrix}\begin{bmatrix} 2 & 0 \\ 0 & 4 \end{bmatrix}\begin{bmatrix} 1 & 1 \\ -1 & 1 \end{bmatrix} = \begin{bmatrix} 3 & 1 \\ 1 & 3 \end{bmatrix}$.   (b) $Q = AS^{-1} = \frac{1}{\sqrt{10}}\begin{bmatrix} 3 & 1 \\ -1 & 3 \end{bmatrix}$,

$S' = AQ^{-1} = \frac{1}{5}\begin{bmatrix} 18 & 4 \\ 4 & 12 \end{bmatrix}$.

A.5   General solution $\bar{x} = (1, 1 - E, E)$; Minimum length solution $x^+ = (1, \frac{1}{2}, \frac{1}{2})$.

A.6   (a) With independent columns, the row space is all of R^n; $A^TAx^+ = A^Tb$ is obvious (b) $A^T(AA^T)^{-1}b$ is in the row space because A^T times any vector is in that space; $A^TAx^+ = A^TAA^T(AA^T)^{-1}b = A^Tb$.

A.7 $A = Q_1\Sigma Q_2^{-1} = Q_1 Q_2^{-1}Q_2\Sigma Q_2^{-1} = QS'$, $Q = Q_1 Q_2^{-1}$, $S' = Q_2\Sigma Q_2^{-1}$.

A.9 A^+b is $\underline{U}^\mathrm{T}$ times some vector, and therefore in the row space of $\underline{U}$ = row space of A. Also $A^+AA^\mathrm{T}b = \underline{U}^\mathrm{T}\underline{L}^\mathrm{T}\underline{L}\underline{U}\underline{U}^\mathrm{T}(\underline{U}\underline{U}^\mathrm{T})^{-1}(\underline{L}^\mathrm{T}\underline{L})^{-1}\underline{L}^\mathrm{T}b = A^\mathrm{T}b$.

A.10 $A = Q_1\Sigma Q_2^\mathrm{T} \Rightarrow A^+ = Q_2\Sigma^+Q_1^\mathrm{T} \Rightarrow AA^+ = Q_1\Sigma\Sigma^+Q_1^\mathrm{T} \Rightarrow (AA^+)^2 = Q_1\Sigma\Sigma^+\Sigma\Sigma^+Q_1^\mathrm{T} = Q_1\Sigma\Sigma^+Q_1^\mathrm{T} = AA^+$, the same reason for A^+A; AA^+ and A^+A project onto the column space and row space respectively.

APPENDIX B

B.1 $J = \begin{bmatrix} 2 & 0 \\ 0 & 0 \end{bmatrix}, J = \begin{bmatrix} 0 & 1 & 0 \\ 0 & 0 & 0 \\ 0 & 0 & 0 \end{bmatrix}.$

B.2 $\dfrac{du_2}{dt} = 8e^{8t}(tx_1 + x_2) + e^{8t}x_1,\ Au_2 = e^{8t}(tAx_1 + Ax_2) = e^{8t}(8tx_1 + 8x_2 + x_1).$

B.3 $e^{Bt} = \begin{bmatrix} 1 & t & 2t \\ 0 & 1 & 0 \\ 0 & 0 & 0 \end{bmatrix} = I + Bt$ since $B^2 = 0$.

B.4   Every matrix A is similar to a Jordan matrix $J = M^{-1}AM$, and by part (a), $J = PJ^\mathrm{T}P^{-1}$. (Here P is formed block by block from the cross-diagonal permutations used on each block J_i.) Therefore A is similar to A^T: $M^{-1}AM = J = PJ^\mathrm{T}P^{-1} = PM^\mathrm{T}A^\mathrm{T}(M^\mathrm{T})^{-1}P^{-1}$, or $A = (MPM^\mathrm{T})A^\mathrm{T}(MPM^\mathrm{T})^{-1}$.

B.5 $J = \begin{bmatrix} 1 & 0 & 0 \\ 0 & 4 & 0 \\ 0 & 0 & 6 \end{bmatrix}; J = \begin{bmatrix} 0 & 1 \\ 0 & 0 \end{bmatrix}.$

INDEX

$A = LDL^{\mathrm{T}}$ 48, 195, 332
$A = LDU$ 36, 236
$A = LPU$ 38, 41
$A = LU$ 32, 33, 40, 73, 468
$A = \underline{LU}$ 197, 206, 452
$A = MJM^{-1}$ 312, 453
$A = Q\Lambda Q^{\mathrm{T}}$ 296, 309, 334, 443
$A = QR$ 174, 176, 374
$A = QS$ 445
$A = Q_1 \Sigma Q_2^{\mathrm{T}}$ 375, 443
$A = S\Lambda S^{-1}$ 254
$A = uv^{\mathrm{T}}$ 99
AA^{T} 50, 442
$A^{\mathrm{T}}A$ 50, 156, 210, 442, 446
$A^{\mathrm{T}}CA$ 111
Absolute value 291
Addition of vectors 6
Adjugate matrix 232
Almost periodic 286
Alternative 142, 418, 420
Area 235, 239
Arithmetic mean 147, 150
Associative 25, 29
Associative law 23, 126
Average 405, 465
Axes 132, 296, 335

Back-substitution 12, 15, 468
Band matrix 35, 52, 55, 462
Bandwidth 55
Basic variable 73, 76, 398

Basis 84, 125, 305, 316, 443
Bidiagonal 54
Bit-reversed 193
Block 371
Block elimination 239
Block multiplication 30, 236
Block of zeros 207, 229, 428
Bomb 284
Boolean algebra 215
Boundary condition 52, 59
Breakdown 13
Bridge 205, 248
Buniakowsky 148
Butterfly 193

$\mathbf{C}^n$ 258, 290, 300
California 18, 266, 370
Capacity 424
Cauchy 148
Cayley-Hamilton 317, 460
$CD = -DC$ 29, 221, 242, 315
Chain of matrices 241, 345
Change of basis 125, 305, 316
Change of variables 304, 341
Characteristic equation 246, 317
Characteristic polynomial 246
Checkerboard 129, 253
Chemist 151
Chess 438, 441
Cholesky 195, 334, 462
Circulant 303

Cofactor 226
Cofactor matrix 231
Column at a time 20, 43, 463
Column picture 7, 21
Column space 66, 93, 136, 141
Column vector 5
Columns times rows 29, 206, 445
Combination of columns 21, 25, 66
Combination of rows 22, 25
Commutative 23, 26
Commute 27, 29, 259
Companion matrix 253
Complement 138
Complementary slackness 415
Complete the square 325, 329, 352
Complex conjugate 291, 294, 300
Complex number 183, 257, 290
Complex plane 184, 291
Composition 123
Compound interest 262
Computer 2, 14, 28, 465
Condition number 362, 364, 367, 368,
     463, 466
Cone 419
Congruence 341
Conjugate 291, 301
Conjugate gradient 385, 409
Conjugate transpose 293
Constraint 392
Consumption matrix 269
Convolution 183, 303
Cooley 190
Coordinate 5, 166
Corner 391, 396
Cosine 145, 148, 177
Cost 14
Cost function 390
Covariance matrix 204
Cramer's rule 1, 233, 239
CRAY 17, 28
Cross-product 288
Cross-product matrix 156, 170
Crout 36
CT scanner 209
Current law 104
Curse 385
Cut 425, 428
Cycling 400

Dantzig 395, 438
Defective 249, 255, 277, 310
Degeneracy 399
Degree of freedom 74, 85, 302
Dependent 80, 82
Descartes 5
Determinant 44, 211, 216, 251, 427
     formula 222, 224, 226
Diagonal matrix 36, 254, 442
Diagonalizable 255, 259, 309, 460
Diagonalization 254, 256
Diagonally dominant 387
Diet problem 392, 412
Difference equation 53, 262, 265, 274
Differential equation 52, 244, 277,
     314, 458
Differentiation 118, 120, 316
Diffusion 278, 287
Dimension 74, 85, 86, 95, 200
Direct sum 206
Directed graph 102
Discrete 53
Discrete transform 183
Distance 209
Distinct eigenvalues 255
Distributive 26
Dual 392, 412
Duality 413, 416, 427, 438
Dynamic programming 430

Echelon 72
Echelon matrix 77, 82
Economics 269
Edge 102, 396
Effective rank 445
Eigenfunction 280
Eigenvalue 245, 249, 258, 331, 370
Eigenvalue matrix 254
Eigenvector 245, 247, 295, 315
Eigenvector matrix 254, 443
Einstein 21
EISPACK 461
Elementary matrix 22, 31, 46
Elimination 1, 13, 31, 43, 72, 77, 236
Ellipse 296
Ellipsoid 177, 334, 350
Energy 347
Entering variable 398, 401

Equality constraint 397
Equilibrium 112, 348
Error 153
Error vector 155, 158
Euler's formula 106
Even permutation 237, 468
Existence 96
Experiment 465
Exponential 275, 283, 314, 459

Factorization 2, 32, 192, 468
Fast Fourier Transform 183, 189, 299, 303, 385, 464
Feasible set 390
Fibonacci 228, 263, 272, 319
Filippov 312, 455
Finite difference 53, 280, 355
Finite difference matrix 222, 365, 386
Finite element method 355
First-order 244, 278
Fitting data 159
Five-point matrix 385, 467
Football 106, 335, 441
Formula for pivots 235
FORTRAN 28, 463, 468
Forward elimination 13, 15, 31
Four fundamental subspaces 90, 443
Fourier matrix 183, 188, 299, 303, 321
Fourier series 168, 176
Fredholm 142, 418
Free variable 73, 76, 398
Frequency 286
Frobenius 271
Front wall 136
Function space 64, 177
Fundamental subspace 90, 136, 302, 443
Fundamental theorem 95, 104, 138

Gale 420, 438
Galerkin 360
Galois 251
Game 433
Gauss-Jordan 43, 46, 77
Gauss-Seidel 381, 467
Gaussian elimination 1, 12, 38, 236
Geiger counter 162
General solution 73, 76, 277, 459

Generalized eigenvalue problem 289, 343, 345, 352, 359
Generalized eigenvector 277, 454
Geodesy 145
Gershgorin 386
Givens 342
Golub 385
Gram-Schmidt 166, 172, 176, 181, 196, 341, 375
Graph 102, 106, 466
Greedy algorithm 430
Grounded 110
Group 62

Halfspace 389
Hall's condition 428, 431
Harvard 107
Heat equation 280
Heisenberg 260
Hermitian matrix 294, 300, 309
Hertz 175
Hessenberg 205, 373, 377
Hessian 327
Hilbert matrix 59, 178
Hilbert space 177
Hogben 234
Homogeneous 68, 74
Homotopy 341
Householder 180, 373, 378
Hurwitz 282

IBM 16
Identity matrix 22
Ill-conditioned 56, 363
Image processing 444
Imaginary number 183, 257
Incidence matrix 102, 425, 466
Incomplete LU 386
Inconsistent 8, 71, 153
Indefinite 326, 337
Independence 80, 82, 256
Inequality 389, 418
Infinite-dimensional 64, 177
Infinity of solutions 8
Inhomogeneous 74
Inner product 20, 28, 134, 177, 204, 293
Input-output 269
Integration 118

Interest 262
Intersection 198
Intertwining 350
Invariant 341
Inverse 32, 42
    formula 231
    of product 43
    of transpose 47
Inverse power method 371
Invertible 42, 46
Isomorphism 200
Iterative 361, 381, 387

Jacobi method 372, 381, 467
Jacobian determinant 212, 239
Jordan form 304, 311, 318, 453

Karmarkar 388, 391, 405
Kernel 92
Khachian 405
Kirchhoff 104, 109, 139, 424
Krylov 378
Kuhn-Tucker 415, 438

Labeling algorithm 426
Lagrange multiplier 415
Lanczos 378
Las Vegas 439
Laplace 385, 467
Law of inertia 339, 341, 345, 353
LDL^T factorization 48, 332
LDU factorization 36, 236
Least squares 108, 154, 156, 204, 353,
    446, 450, 462
Leaving variable 398, 401
Left-inverse 42, 90, 96, 100, 121, 170,
    452
Left nullspace 90, 94, 103, 136
Legendre 179
Length 133, 168, 177, 204, 290, 293
Leontief 269
Linear combination 7
Linear independence 80, 82, 256
Linear programming 388, 392, 464
Linear transformation 117, 120, 305
LINPACK 461
Loop 103, 106, 429
Lower triangular 27, 32
LU factorization 32, 40, 73

LU factorization 197, 206
Lyapunov 282, 338

Markov matrix 266, 270, 273, 284
Marriage problem 426
Mass matrix 289, 343, 359
Matching 426, 428
Matrix 19
    0-1 60, 427, 465
    adjacency 466
    band 35, 52, 462
    checkerboard 129, 253
    circulant 303
    coefficient 19
    cofactor 231
    companion 253
    consumption 269
    covariance 204
    cross-product 156, 170
    defective 249, 255, 277, 310
    diagonal 35, 442
    diagonalizable 254, 255, 309, 315,
        460
    difference 365
    echelon 72, 77, 82
    elementary 22, 31, 46
    exponential 276, 459
    finite difference 53, 222, 365, 386,
        467
    Fourier 183, 188, 299, 303, 321
    Hermitian 294, 300, 309
    Hessenberg 205, 373, 377
    Hilbert 59, 178
    identity 22
    ill-conditioned 56, 363
    incidence 102, 425, 466
    indefinite 326, 337
    inverse 32, 42
    invertible 46
    Jordan 304, 311, 318, 453
    lower triangular 32
    Markov 266, 270, 273, 284
    mass 343, 359
    multiplication 20, 23, 24, 25, 29, 123,
        445
    nilpotent 321
    nondiagonalizable 249, 255, 277, 310,
        453
    nonnegative 267, 269

nonsingular 13, 14, 38, 46, 71
normal 297, 311, 315
notation 21
orthogonal 167, 283, 296, 443
payoff 435
permutation 37, 62, 215
positive 271
positive definite 54, 196, 324, 331, 347
projection 124, 148, 158, 170, 250, 406
rank one 98, 261
rectangular 21
reflection 124, 164, 373
rotation 30, 122, 257
semidefinite 339
similar 304, 312
singular 7, 14, 37, 211
skew-Hermitian 297, 300, 320
skew-symmetric 28, 436
square root 261, 334, 337
symmetric 48, 290, 309
transition 267
transpose 47, 150, 217
triangular 27, 216, 308
tridiagonal 53, 377, 462
unitary 297, 301, 308
MacSimplex 465
MATLAB 464
Max flow-min cut 425
Maximal 86
Maximin 350, 434, 437
Mean 165, 171, 465
Middle-aged 165
Millionaire 234
Minimax 351, 434, 437, 440, 441
Minimax theorem 413
Minimum 322, 325
Minimum principle 347
Minor 226
MIT 107
Mixed strategy 433
Modulus 291
M-orthogonal 344, 346
Multiplication 20, 21, 24, 29, 123
Multiplicity 211, 255
Multiplier 12, 32, 196, 332, 468

Natural boundary condition 359

Negative definite 325, 338
Netlib 193, 464
Network 108, 424
Neutrally stable 268, 280
Newton's law 284, 343
Newton's method 372
Nilpotent 321
Node 102
Nondiagonalizable 310, 460
Nonnegative 267, 269, 419
Nonsingular 13, 14, 38, 46, 71
Nontrivial 80
Norm 362, 366, 368, 369
Normal equation 156, 446
Normal matrix 297, 311, 315, 317
Normal mode 249, 280
Nullity 92, 199, 201
Nullspace 68, 74, 92, 136, 246

Objective function 390
Odd permutation 237, 240, 468
Ohm's law 109
Operation 15, 40, 45, 55
Optimality 401, 414, 415, 447
Orange 108
Orthogonal 134, 177, 300
Orthogonal complement 138
Orthogonal eigenvectors 295, 315
Orthogonal matrix 167, 283, 296, 443
Orthogonal subspace 135, 138, 301
Orthogonalization 172, 179, 196
Orthonormal 135, 166, 169, 174, 311
Oscillation 244, 284, 322, 343
Overdetermined 145
Overrelaxation 381, 382

$P \neq NP$ 405
$PA = LU$ 38, 73, 462
Pancake 144
Parallel 4, 16, 76
Parallelepiped 212, 234
Parallelogram 4, 235
Parentheses 23, 26
Partial pivoting 58, 59, 465
Particular solution 76
Payoff 435
PC 15, 17
Peanut butter 392
Permutation 129, 224, 237

Permutation matrix 37, 62, 215
Perpendicular 134, 155
Perron 271
Perturbation 57
Phase I and II 398, 410
Piecewise polynomial 356
Pivot 12, 13, 36, 72, 212, 236, 332, 445
Plane 2, 5
Plane rotation 316, 377
Poker 439
Polar coordinates 212, 292
Polar decomposition 445
Polynomial 98
Polynomial time 407
Positive definite 54, 196, 324, 331, 347, 445
Positive matrix 271
Potential 103
Powers of matrices 258, 314
Power method 370, 379
Preconditioner 381
Premultiplication 26
Pricing out 401
Primal 412
Principal axis 296, 336, 446
Principal submatrix 339
Prisoner 185, 439
Product 123, 201, 217, 240, 259
Product form 404
Product of pivots 222
Professor 207
Projection 117, 123, 144, 164, 250, 306, 406
Projection matrix 148, 158, 170
Pseudoinverse 97, 141, 154, 448, 449
Pure exponential 245, 248
Pure power 264
Pythagoras 133, 169, 180, 448

QR factorization 174, 176, 376
QR method 342, 376
Quadratic 246, 323, 348
Quantum mechanics 260

$\mathbf{R}^n$ 63, 86
Random number 465, 466, 471
Range 92
Rank 76, 80, 100, 129, 201, 202, 445
Rank one 98, 148, 261

Rayleigh quotient 348, 353, 358, 372
Rayleigh-Ritz 356, 360
Reduced cost 401, 418
Reduced factorization 195, 197, 206
Reflection 116, 124, 164, 168, 180, 373
Regression 145
Relativity 5
Repeated eigenvalues 256, 310, 312, 454
Rescaling 59, 407, 408
Revised simplex 404
Right-handed 168, 234
Right-inverse 42, 90, 96, 452
Roof function 357
Roots of unity 184, 186, 299
Rotation 116, 167, 257, 446
Rotation matrix 30, 122
Roundoff error 37, 56, 365, 445
Routh 282
Row at a time 20, 25
Row exchange 24, 36, 37, 215, 225
Row picture 7, 21
Row rank = column rank 93
Row space 90, 91, 136, 141, 302, 448
Row-reduced echelon form 77

$S^{-1}AS$ 125, 254
Saddle point 326, 433, 437
Scalar 6, 134
Schur complement 239
Schur's lemma 307
Schwarz inequality 145, 147, 151, 177, 260, 338
Second-order equation 284
Semicircle law 466
Semidefinite 325, 339, 445
Sensitivity 364, 417
Separating hyperplane 419, 420
Shadow price 413, 417
Shearing 125, 235
Shift 371, 376
Signs of eigenvalues 339, 342
Similar 304, 306, 312, 315, 454, 457
Simplex method 391, 395, 404, 464
Simultaneous 344, 352
Singular 7, 14, 37, 216
Singular value 368, 369
Singular value decomposition 141, 195, 375, 443, 450
Skew-Hermitian 297, 300, 320